Lecture Notes in Computer Science 14286

Founding Editors

Gerhard Goos
Juris Hartmanis

The series Lecture Notes in Computer Science (LNCS), including its subseries Lecture Notes in Artificial Intelligence (LNAI) and Lecture Notes in Bioinformatics (LNBI), has established itself as a medium for the publication of new developments in computer science and information technology research, teaching, and education.

LNCS enjoys close cooperation with the computer science R & D community, the series counts many renowned academics among its volume editors and paper authors, and collaborates with prestigious societies. Its mission is to serve this international community by providing an invaluable service, mainly focused on the publication of conference and workshop proceedings and postproceedings. LNCS commenced publication in 1973.

Meinolf Sellmann · Kevin Tierney
Editors

Learning and Intelligent Optimization

17th International Conference, LION 17
Nice, France, June 4–8, 2023
Revised Selected Papers

 Springer

Editors
Meinolf Sellmann
InsideOpt
Dover, DE, USA

Kevin Tierney 🆔
Bielefeld University
Bielefeld, Germany

ISSN 0302-9743 ISSN 1611-3349 (electronic)
Lecture Notes in Computer Science
ISBN 978-3-031-44504-0 ISBN 978-3-031-44505-7 (eBook)
https://doi.org/10.1007/978-3-031-44505-7

This Springer imprint is published by the registered company Springer Nature Switzerland AG
The registered company address is: Gewerbestrasse 11, 6330 Cham, Switzerland

Paper in this product is recyclable.

Preface

This volume contains the peer-reviewed papers from the 17th Learning and Intelligent Optimization (LION-17) Conference held in Nice, France, from June 4–8, 2023.

LION-17 continued the successful series of internationally recognized LION events (LION-1: Andalo, Italy, 2007; LION-2 and LION-3: Trento, Italy, 2008 and 2009; LION-4: Venice, Italy, 2010; LION-5: Rome, Italy, 2011; LION-6: Paris, France, 2012; LION-7: Catania, Italy, 2013; LION-8: Gainesville, USA, 2014; LION-9: Lille, France, 2015; LION-10: Ischia, Italy, 2016; LION-11: Nizhny Novgorod, Russia, 2017, LION-12: Kalamata, Greece, 2018; LION-13: Chania, Greece, 2019; LION-14 and LION-15: Online, 2020 and 2021; LION-16: Milos Island, Greece, 2022).

The central theme of LION-17 was ML to OR pipelines, which was addressed with four invited talks from leading researchers.

Keynote 1: Wil van der Aalst (RWTH Aachen; Celonis) "Process Mining for Optimization and Optimization for Process Mining"

Keynote 2: Martina Fischetti (European Commission Research Center) "Operations Research + Machine Learning for the design of future offshore wind farms"

Tutorial 1: Elias Khalil (University of Toronto) "Predict-then-Optimize: a Tour of the State-of-the-art using PyEPO"

Tutorial 2: Carlos Ansótegui (University of Lleida) "SAT-based Applications with OptilLog"

There were 83 full papers submitted to LION-17, of which 40 were accepted for presentation after two rounds of double-blind peer review. The editors thank the reviewers for taking the time to evaluate LION-17's submissions.

In total, over 80 participants came to Nice to enjoy exceptional presentations and discussions. Despite a packed schedule, there was still some time left over for participants to appreciate the beautiful city of Nice and relax and reflect on the conference topics in the ocean breeze.

LION-17 would not have been possible without the backing of several generous sponsors. Thank you to **InsideOpt**, **Gurobi Optimization**, **OPTANO**, **NextMv**, and **DecisionBrain** for their support. We also thank a special philanthropist for donating the **Malitsky Registration Assistance Fund**, which reduced the registration fee of nine attendees without sufficient funds to attend.

The editors especially thank our local organizer, Paul Shaw (IBM), for all of his work finding us a memorable venue. Furthermore, we thank Elias Schede (Bielefeld University) and the Bielefeld University staff for all of their support in ensuring the conference was a success.

August 2023

Meinolf Sellmann
Kevin Tierney

Organization

Technical Program Committee

Chairs

Meinolf Sellmann	InsideOpt, USA
Kevin Tierney	Bielefeld University, Germany

Members

Imene Ait Abderrahim	Université Djilali Bounaâma, Algeria
Carlos Ansòtegui	University of Lleida, Spain
Francesco Archetti	Consorzio Milano Ricerche, Italy
Annabella Astorino	ICAR-CNR, Italy
Hendrik Baier	Eindhoven University of Technology, The Netherlands
Roberto Battiti	University of Trento, Italy
Laurens Bliek	Eindhoven University of Technology, The Netherlands
Christian Blum	Spanish National Research Council, Spain
Mauro Brunato	University of Trento, Italy
Zaharah Bukhsh	Eindhoven University of Technology, The Netherlands
Sonia Cafieri	École Nationale de l'Aviation Civile, France
Antonio Candelieri	University of Milano-Bicocca, Italy
John Chinneck	Carleton University, Canada
Konstantinos Chatzilygeroudis	University of Patras, Greece
Philippe Codognet	JFLI/Sorbonne Universitè, Japan/France
Patrick De Causmaecker	Katholieke Universiteit Leuven, Belgium
Renato De Leone	University of Camerino, Italy
Clarisse Dhaenens	Université Lille 1 & Polytech Lille, Inria, France
Luca Di Gaspero	University of Udine, Italy
Bistra Dilkina	University of Southern California, USA
Theresa Elbracht	Bielefeld University, Germany
Adil Erzin	Sobolev Institute of Mathematics, Russia
Paola Festa	University of Napoli Federico II, Italy
Adriana Gabor	Khalifa University, UAE
Jerome Geyer-Klingeberg	Celonis, Germany

Isel Grau	Eindhoven University of Technology, The Netherlands
Vladimir Grishagin	Nizhni Novgorod State University, Russia
Mario Guarracino	ICAR-CNR, Italy
Ioannis Hatzilygeroudis	University of Patras, Greece
Youssef Hamadi	Tempero, France
Andre Hottung	Bielefeld University, Germany
Laetitia Jourdan	Inria/LIFL/CNRS, France
Serdar Kadioglu	Fidelity & Brown University, USA
Marie-Eleonore Kessaci	Université de Lille, France
Michael Khachay	Krasovsky Institute of Mathematics and Mechanics, Russia
Elias B. Khalil	University of Toronto, Canada
Zeynep Kiziltan	University of Bologna, Italy
Yury Kochetov	Sobolev Institute of Mathematics, Russia
Ilias Kotsireas	Wilfrid Laurier University, Waterloo, Canada
Dmitri Kvasov	University of Calabria, Italy
Dario Landa-Silva	University of Nottingham, UK
Hoai An Le Thi	University of Lorraine, France
Daniela Lera	University of Cagliari, Italy
Michele Lombardi	University of Bologna, Italy
Yuri Malitsky	FactSet, USA
Vittorio Maniezzo	University of Bologna, Italy
Silvano Martello	University of Bologna, Italy
Yannis Marinakis	Technical University of Crete, Greece
Nikolaos Matsatsinis	Technical University of Crete, Greece
Laurent Moalic	University of Haute-Alsace, France
Hossein Moosaei	Jan Evangelista Purkyně University, Czech Republic
Tatsushi Nishi	Okayama University, Japan
Panos Pardalos	University of Florida, USA
Axel Parmentier	École Nationale des Ponts et Chaussées, France
Konstantinos Parsopoulos	University of Ioannina, Greece
Vincenzo Piuri	Università degli Studi di Milano, Italy
Till Porrmann	Bielefeld University, Germany
Oleg Prokopyev	University of Pittsburgh, USA
Helena Ramalhinho	Universitat Pompeu Fabra, Spain
Michael Römer	Bielefeld University, Germany
Massimo Roma	Sapienza University of Rome, Italy
Valeria Ruggiero	University of Ferrara, Italy
Frédéric Saubion	University of Angers, France
Andrea Schaerf	University of Udine, Italy

Elias Schede	Bielefeld University, Germany
Marc Schoenauer	Inria Saclay Île-de-France, France
Marc Sevaux	Lab-STICC, Université de Bretagne-Sud, France
Paul Shaw	IBM, France
Dimitris Simos	SBA Research, Austria
Thomas Stützle	Université Libre de Bruxelles, Belgium
Tatiana Tchemisova	University of Aveiro, Portugal
Gerardo Toraldo	Università della Campania "Luigi Vanvitelli", Italy
Paolo Turrini	University of Warwick, UK
Michael Vrahatis	University of Patras, Greece
Om Prakash Vyas	Indian Institute of Information Technology Allahabad, India
Ranjana Vyas	Indian Institute of Information Technology Allahabad, India
Dimitri Weiß	Bielefeld University, Germany
Daniel Wetzel	Bielefeld University, Germany
David Winkelmann	Bielefeld University, Germany
Dachuan Xu	Beijing University of Technology, China
Qingfu Zhang	City University of Hong Kong, China
Anatoly Zhigljavsky	Cardiff University, UK
Antanas Zilinskas	Vilnius University, Lithuania

Local Organization

Paul Shaw   IBM, France

Sponsors

Contents

Anomaly Classification to Enable Self-healing in Cyber Physical Systems Using Process Mining

Uphar Singh[1]([✉]), Deepak Gajjala[1], Rahamatullah Khondoker[2],
Harshit Gupta[1], Ayush Sinha[3], and O. P. Vyas[1]

[1] Indian Institute of Information Technology, Allahabad, Prayagraj, India
{pse2017003,iib2019024,rsi2020501,opvyas}@iiita.ac.in
[2] THM University of Applied Sciences, Friedberg, Friedberg, Germany
rahamatullah.khondoker@mnd.thm.de
[3] Indian Institute of Technology, Kanpur, Kanpur, India
ayush@c3ihub.iitk.ac.in

Abstract. Industrial Cyber Physical Systems (CPS) are large-scale critical infrastructures that are vulnerable to cyberattacks with wide-ranging consequences. Being a combination of heterogeneous devices and protocols, the large-scale CPS anomaly is also exposed to critical vulnerabilities. These vulnerabilities are treated in terms of anomalies and cyberattacks, and their detection and corresponding self-healing mechanisms on large-scale critical infrastructures can be challenging because of their massive size and interconnections. With the objective of process optimization through anomaly detection and conformance-checking approach, the present work addresses different issues, such as event log generation with tools such as PLG 2.0 for data-driven approach. Self-healing is enabled through machine learning models based on anomaly classification ensemble learning-based machine learning models. The work uses process mining to analyze event log files, and then combinations of conformance-checking methods with ensemble classification models were used to best classify anomalies. Finally, the proposed work establishes that, in comparison to techniques like KNN and C-SVC, the proposed ensemble models perform better, with an accuracy of 84.7% using trace alignment as a conformance technique with gradient boosting to classify anomalies, with the end objectives of process improvement.

Keywords: Cybersecurity · Process Mining · Cyber Physical Sytems · Anomaly Detection · Anomaly Classification · Self Healing

1 Introduction

Industrial cyber-physical systems are large-scale infrastructures that manage critical industries, including electric power generation and transmission, communication, transportation and freight systems. A cyber attack on these industries can cause widespread disruption to the daily life of a significant chunk of

© The Author(s), under exclusive license to Springer Nature Switzerland AG 2023
M. Sellmann and K. Tierney (Eds.): LION 2023, LNCS 14286, pp. 1–15, 2023.
https://doi.org/10.1007/978-3-031-44505-7_1

the population, for example, recent attacks like Ransomware attack on Colonial Pipeline in May 2021, which is a major U.S. petroleum pipeline operator, halted the flow of petrol and jet fuel throughout the East Coast of the country. Significant logistical and financial effects resulted from the attack, which led to major gasoline shortages and forced the corporation to temporarily cease operations. [16] And the attack on the SolarWinds supply chain in late 2020, it was revealed that a sophisticated cyberattack had compromised the SolarWinds software, which is extensively used by corporations and government organisations. Unauthorised access to sensitive data from several organisations, particularly those in the industrial and labor-intensive industries, was caused by the attack and lasted for several months. [17] Hence, performing detection and remedial actions, i.e., enabling self-healing, is essential for a cyber attack on these systems. These industries have interconnection between many devices, which means there is a very large surface area for an attack and the device data logs to analyse processes are huge. To overcome this, process mining techniques have recently been used to understand the underlying process model for an ideal process and detect whenever a running process is out of line with the ideal. This detection can be facilitated through various techniques, including machine learning and deep learning, of which machine learning is an interesting area of research. Since there is a lack of a large amount of labelled data, machine learning is a more suited approach to deep learning, and also not much work has been done in investigating the usefulness of ensemble approaches using process mining techniques. We explore the effectiveness of ensemble models in detecting and classifying anomalous event logs. We also further investigate the different conformance checking approaches to compare which conformance checking approach works best in combination with ensemble classification. Now, in the domain of process mining, some related terminology will help to understand the work done on anomaly detection in logs generated by the business processes.

1.1 Process Mining

Process mining is the topic of business process management study that deals with analysing event data that various information systems produce while processes are being carried out. Process mining is a methodology that identifies trends, patterns, and irregularities in a process, enhancing efficiency and effectiveness. It uses data-driven techniques to detect constraints and improve efficiency in various domains, including business processes and cyber security. It involves three separate sets of tasks, such as Process Discovery, Conformance Checking, and Enhancement. [1]

1. **Process Discovery:** The concept of process discovery involves the extraction and visualization of process models, which is obtained from the application of process discovery algorithms to event logs, as depicted in Fig. 1. The Alpha algorithm, Heuristic Miner, Fuzzy Miner, Genetic Miner, and Inductive Miner are widely recognized process discovery algorithms.

Fig. 1. Process Discovery

2. **Conformance Checking:** Conformance checking, also known as delta analysis, is a process to check conformance (Fig. 2) with the expected process model. The inputs are the Process model and Event logs, and we obtain quantitative results by applying conformance algorithms. We check if event logs conform to the expected process model based on the results. Popular Conformance checking algorithms are Token-based replay algorithms, Trace Alignment Algorithms and Casual Foot-prints.

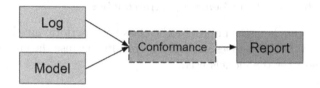

Fig. 2. Conformance Checking

3. **Enhancement:** Enhancement helps (Fig. 3) us to improvise the process model and to align more event logs. The inputs are the Process model and Event logs, and the new model is obtained after this analysis. A new process model can now accommodate more event logs. Enhancing the model can help us conduct a more sophisticated analysis and improve its performance. They were focused on improving the model using additional attributes such as location, costs, timing etc. Popular Enhancement techniques are performance analysis, resource interaction analysis and root-cause analysis.

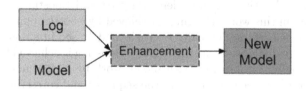

Fig. 3. Enhancement

1.2   Anomalies in Event Logs

For this work, we will be considering two types of anomalies in event logs, i.e., control-flow based anomalies and XOR-based anomalies.

1. There are three types of control-flow anomalies:
   (a) Missed activities: when activities are skipped in the trace flow.
   (b) Duplicated activities: when activities are multiplied in the trace flow.
   (c) Exchanged activities: when activities are interchanged in the trace flow.
2. Based on the conditions at a point in the model, one of two activities is selected (not both, hence an exclusive OR: XOR element are present in the petri-net model). Furthermore, there are two types of XOR anomalies:
   (a) XOR split anomaly: XOR split elements are changed into XOR join
   (b) XOR join anomaly: XOR join elements are changed into XOR split

1.3   Ensemble Machine Learning Approaches

Ensemble Learning helps us improvise the classifier's performance by combining multiple algorithms. This gives us improved results because the ensemble model has advantages over the models used.

Algorithms:

1. **Support Vector Machines:** Support Vector Machine is one of the popular algorithms in classification. Support Vector Machine works on the idea of maximising the distance between the support vectors and separation hyperplane. Support Vector Machine is computationally expensive.

2. **Decision Trees:** These are famous in classification due to their simpleness, high detection accuracy, and fast adaptation. Decision Tree is a tree-structured classifier and classifies based on the features of the data set.

1.4   Models Used

Ensemble learning is a technique for obtaining better results by training multiple models to solve the same problem. Bagging and Boosting are two major techniques used in this work. Bagging is achieved by bootstrapping the data set and using these subsets individually for each model used in ensemble learning. Later all the individual results are aggregated. Boosting is achieved by combining models sequentially. In boosting, each model work by giving more prominence to the errors of the preceding model in the sequence and correcting them.

1. **Bagging meta-estimator.** The bagging meta-estimator adopted the classic bagging technique, Fig. 4. Randomized subsets are created, and each subset is used by each one of the models. All the results are aggregated to get the best results. We can run these models in parallel and independently.

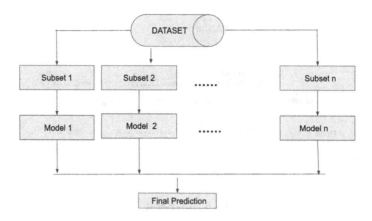

Fig. 4. Bagging Meta-estimator

2. **Random Forest.** Random Forest (Fig. 5) is an extension of the Bagging meta-estimator with its base estimator as a Decision tree. It gives better results with less training time as compared to other algorithms. This solves overfitting data sets. Sampling over features can also be done to reduce the co-relation between the outputs.

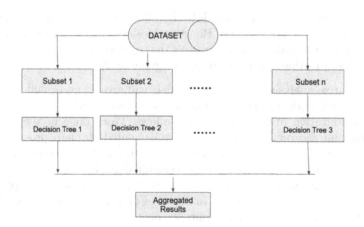

Fig. 5. Random Forest

3. **AdaBoost.** AdaBoost, also known as Adaptive Boost, adopted the classic boosting technique. The idea of Boosting is to exploit the strength of weak learnability (Fig. 6). AdaBoost works sequentially and corrects the errors from the preceding model by assigning weights to the inaccurate predicted data.
4. **Gradient Boosting Model.** Gradient Boosting Model can handle large and complex data sets to obtain great prediction speed and accuracy. GBM tries to minimize the loss function obtained from the preceding model, which is

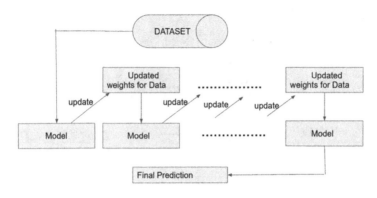

Fig. 6. AdaBoost

iterated until we get the loss function remains the same or till the number of iterations are done.

5. **Extreme Gradient Boosting.** Extreme Gradient Boosting is an efficient implementation of the Gradient Boosting technique. The XGBoost is endorsed because of its fast implementation and great accuracy. It reduces the over-fitting of data and hence produces great results. Extreme Gradient Boosting uses the more regularised model, which helps us to reduce overfitting.

6. **Light Gradient Boosting Model.** Light GBM is a light and faster gradient Boosting Model. Light Gradient Boosting Model splits the tree leaf-wise while other gradient boosting algorithms split the tree level-wise. Leaf-wise split can reduce loss function and hence produces great results.

Comparative analysis of the above models will give us an insight into which ensemble model works better with process conformance output.

The organization of this paper is as follows: Work done in the detection, prevention and recovery of components implemented through process mining techniques are in Sect. 2, Employable problem statement and dataset description is stated in Sect. 3, Steps to be encountered during our experiment and expected outcomes have been mentioned in Sect. 4, Discussions on implementation steps and results are described in Sect. 5. Finally, Sect. 6 states conclusions and mentions the scope for further improvements.

2   Literature Survey

Wil Van der Aalst et al. focused on supporting the ways to inspect audit trails with the operation of process mining approaches for finding security violations. The authors also showed the possible usage of the Alpha algorithm at numerous levels varying from high-level fraud prevention to low-level intrusion recognition. The researchers further highlighted the importance of implementing regulations that would encourage the enforcement of security checks through the development of tools within the framework of business processes [4]. Another work by

Fábio Bezerra et. al. investigated how ProM tools can support anomaly detection in logs from Process Aware Information Systems (PAIS) to identify abnormal cyber-attack behavior. It uses data logs from industrial control systems and applies process mining techniques for anomaly detection and conformance checking analysis vut it is limited to the control-flow perspective only. The model relies on selecting an appropriate model, and further automation can be done by genetic algorithms [3]. Authors Fani Sani et al. proposed a method that uses sequential relations in the process mining for outlier detection in both the ProM and the RapidProM frameworks. They performed some experiments that produced an output that the procedure can identify and eliminate outlier behaviour from the event data having an abundance of irregular, long-term or parallel-reliant behaviour. Fani Sani et al. also exhibit that the sequence filter technique performs better than other modern filtering techniques and the embedded filtering mechanism of the Inductive Miner algorithm for a few actual event logs [6].

Sylvio Barbon et al. [7] examined six varieties of anomalies on top of thirty-eight actual and fabricated event logs, differentiating the prognostic performance of Local Outlier Factor, SVM and One-Class SVM. Proposed the usage of a conventional natural language processing method, namely word2vec, so that business process behaviours can be encoded as the factors of activities in an event log. Besides, this technique examines the usage of One-Class Classification algorithms, showing their superiority in the context of scarcity of labels. Deep learning methods were considered out of the scope as these methods need a considerable quantity of computational resources and actual labels. In addition to this work, Prasannjeet singh et al. [2] discussed the integration of Process Mining, log file analytics, and machine learning techniques is employed to facilitate the timely identification, prediction, and subsequent autonomous recovery of Internet of Things (IoT) devices within Cyber-Physical Systems (CPS). Consistent event logs were produced using processes and a second iteration of the log generator for a sample model, with the smart home serving as a reference. They also explored various anomalies that can be inserted in process traces as control flow disruptions or Xor-based anomalies as separate experiments. It was used to check conformance with the token replay algorithm, which supplied fitness values along with the data of the traces out of the batches when replayed, producing activities missing their corresponding input tokens. By classifying anomalies they have, for the K-Nearest Neighbours ML algorithm, utilizing full feature space on an average produces an accuracy of 99.96%, compared to only fitness data, which on an average produced 59. 56% and without XOR data produced 81.34%. In the case of the C-Support Vector Classifier, average accuracies with full feature space data, only fitness, and no XOR data are 100%, 62.92% and 82.23% individually. They also mentioned a technique for productive conformance checking and process discovery methods to support self-healing of IoT networks. Another domain where the application of process mining was explored by Motahareh Hosseini et al. where they analyzed an Engineering, Procurement, and Construction (EPC) company's performance with the supplier's selection procedure event logs. The authors explored components affecting process implementation like duplicate activities, repetitive loops, and upgrading the company's upcoming production, and examined connections of individuals who participated in the project. They have shown that by the construction of a social network graph using the Degree Centrality metric poses challenges in the context of purchasing disciplines, primarily due to the presence of loops and frequent repetition in activity execution. This study places particular emphasis on the process of preparing technical and engineering documents, such as product requests, technical evaluations, and order request purchasing. Additionally, it addresses the issues of poor communication and the lack of direct communication between consultants and experts, as well as the absence of effective communication between procurement and engineering deputies and experts in these respective fields. These issues have ultimately resulted in a rise in time consumption and subsequently increased expenses, a decline in company credit, a decrease in employee morale and motivation, and similar outcomes. Consequently, it has been suggested that a reevaluation of the procedure and enhancement of effective and pragmatic communication among stakeholders be undertaken in subsequent endeavors [5]. In 2014, an intrusion detection system was developed by Silva and Schukat [12] for the Modbus/TCP protocol, employing a K-Nearest Neighbors (KNN) classifier. However, it displayed a notable rate of false positives. Later, in 2016, Pajouh et al. [14] introduced the implementation of a two-tier classification module within an intrusion detection model that utilizes the Naive Bayes algorithm and the Certainty Factor variant of the K-Nearest Neighbors (KNN) algorithm. It has the capability to detect and identify malicious activities such as User to Root attacks and Remote to Local intrusions. In 2018, a proposal was made by Anthi et al. [13] for the implementation of an intrusion detection system specifically designed for Internet of Things (IoT) devices. This system utilizes a variety of machine learning classifiers to effectively detect and identify network scanning activities, as well as straightforward denial of service attacks. The adaptive intrusion detection system developed by Stewart et al. (2015) was specifically designed to effectively handle the dynamic topologies of SCADA systems. The researchers utilized multiple One-Class Support Vector Machine (OCSVM) models for this purpose in order to select the best one for accurately identifying different attacks. Nevertheless, the proposed model required a lot of computational resources despite having a high rate of false alarms for detection.

3   Problem Statement and Dataset Description

We aim to detect and classify anomalous processes in cyber-physical systems using an ensemble learning approach. To the best of our knowledge, little work has been done in investigating the usefulness of ensemble approaches using process mining techniques. We explore the effectiveness of ensemble models in detecting and classifying anomalous event logs. We also further investigate the different conformance-checking approaches to compare which conformance-checking approach works best in combination with ensemble classification. The dataset taken as a reference for this article is a power grid testbed called Electric

Power, and Intelligent Control (EPIC) consists of 4 stages: transmission, generation, smart homes, and microgrids [9], Fig. 7. In a fibre optic ring network, each of EPIC's 4 stages comprises its own set of power supply unit, protection, communication systems, switches and PLCs. Three generators and a power supply from the SUTD grid make up the generation stage. The maximum power output from the three generators, each with a 10KW rating, is 30KW. During the transmission stage, an autotransformer increases or decreases the voltage supplied to the microgrid or smart home. Two 15 and 30-kVA load banks with programmable changeable resistive and electrical capacity loads make up a smart house. Photovoltaic cells and batteries make up the Microgrid.

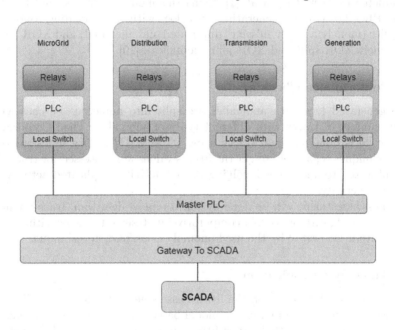

Fig. 7. EPIC Architecture [9]

The five categories of communications in EPIC are generation, transmission, micro-grid, smart homes, and controls. On the roof, 110 PV cells with inverters are placed to change solar energy into electrical energy that can be fed into the setup. The highest power output from the cells is 34 KW in total. During the blackout or low energy conversion due to cloud cover, the integration of a battery bank with inverters serves to supplement power supply to the EPIC system. In order to maintain network communications in the case of a complete blackout, backup power is implemented and made accessible from a separate battery bank to the SCADA workstation. EPIC provides power to run the SWaT and WADI testbeds simultaneously as needed. This relationship is helpful for studies into how cyberattacks on a power station can affect downstream infrastructure. The generators and transformers are supported by EPIC's experimental research into the cyber security elements of those components.

4   Methodology

In this section, our proposed data-driven approach is illustrated by optimizing the business process with the help of anomaly classification and conformance checking.

4.1   Event Logs and Process Discovery

We will be using a reference Petri-net model of microgrid event logs in the Electric Power and Intelligent Control(EPIC) dataset [9] on which process discovery is conducted by Wei Deng et al. [1]. Using this model in Process and Log Generator (PLG 2.0) [15], for generating event logs with corresponding control flow anomalies (namely activity missing, duplicated and alienated) and XOR anomalies (namely XOR split and XOR join anomalies). [10]

4.2   Conformance Checking

These generated logs with anomalies are split into numerous batches with a certain number of traces with different types of anomalies. Then Conformance checking [8] is done on these traces using PM4Py [11] library functions by all three techniques separately to get the fitness values were assessed for traces that were subjected to anomalies involving missed activity, duplicated activity, and exchanged activity.

The fitness assessment will be done by token replay algorithm, trace alignment and footprint algorithms to do a comparative analysis of these algorithms based on the output generated by the used anomaly classification technique.

4.3   Anomaly Classification

To classify anomalies based on the fitness assessment of traces, we will explore ensemble learning with an ensemble model based on Decision Trees. Finally, we will analyze the results with different fitness information and event log anomalies. The work aims to experiment and test the effectiveness of the ensemble learning approach in process conformance checking to detect and classify anomalous traces. These results will include the percentage accuracy and ratio comparison of nonconformance traces.

Additionally, the second outcome is to compare the difference between the various conformance-checking approaches and to find the best combination of conformance algorithm with ensemble learning.

5   Results and Discussions

5.1   Model Preparation

Traditional modelling techniques require lots of expertise, and a large-scale system, such as a power grid, requires much time. As the dataset description mentioned earlier, we have taken the petri-net model generated through process discovery by [1]. In the article, the authors have generated the petri-net model of the microgrid part of EPIC dataset through ProM software using an Inductive miner (Fig. 8). They also analysed the modelling results and finally checked the consistency of the model through a validation set (which is also average data) to prove that the generated model can effectively model the grid business.

We converted the selected Petri-net model to the BPMN model by hand to feed into PLG2.0 to generate process logs. To check the conversion's fidelity, we used the Inductive mining technique of process discovery with generated logs to get the corresponding Petri-net model, which in our case, exactly matched the original model.

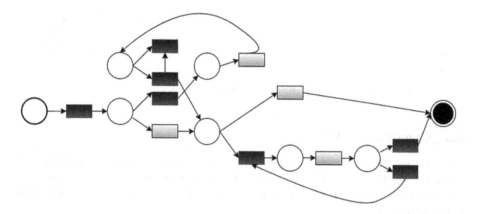

Fig. 8. Petri-net model of microgrid using the EPIC dataset [1]

5.2   Dataset Generation

The objective of our experiment is to evaluate whether traces generated with consideration of the reference process model (as mentioned in Fig. 8 on Process and Log Generator 2.0) injected with various control-flow anomalies, giving an outcome mentioning consequential fitness differences. We also aim to detect with good accuracy to analyze traces and to classify control-flow anomalies using ensemble approaches of machine learning. [18]

Control-flow anomalies like exchanged activities, missed activities, and duplicated activities have been used. Figure 9 shows an example of such anomalies; given our input model, these three represented cases show changes in traces when injected by these anomalies. We generated 10000 true traces as our initial dataset using Process and Log Generator 2.0, which were used to do process discovery using an Inductive miner. We have generated 500 traces of each possibility of missing, duplicated and exchanged activities in the original model, which got us 2000 traces of each anomaly.

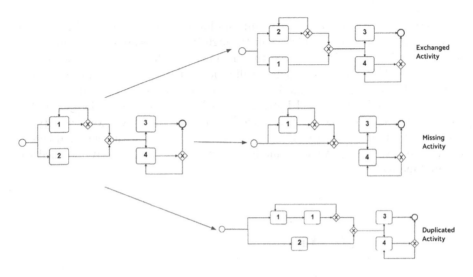

Fig. 9. Control-flow anomalies

5.3   Conformance Checking

For each anomalistic trace, we have used Token based log replay algorithm and a Trace alignment algorithm to get the fitness data along with token positions and alignments. This process gives an expected output of around 50% logs being non-conformant.

Table 1. Outcomes of statistical experiments evaluating fitness scores with various injected anomalies

Anomaly variety 1	Anomaly variety 2	p-value (Token replay)	p-value (Trace alignment)	Outcomes of Null hypothesis
Missing activities	Duplicated activities	0.398	0.423	Accepted
Missing activities	Exchanged activities	0.449	0.505	Accepted
Duplicated activities	Exchanged activities	0.373	0.427	Accepted

A paired t-test could be run for every conceivable observation pair since every fitness data point is calculated on the identical batches injected with distinct anomalies. The test findings, establishing the p-value for every test and determining if the null hypothesis was rejected or not are shown in Table 1. To prove that our hypothesis is consistent with the data, this test is done to ensure that a machine-learning model can classify the data. Furthermore, it can be observed from the table that the null hypothesis was accepted and intercepted for all pairs of data as the outcomes can distinguish with sensible confidence that any anomaly can be injected.

5.4   Bagging and Boosting Classification

Outputs of conformance-checking techniques with all anomalistic type traces are processed in a numerical format and combined to make a full dataset for the classification of anomalies. Initially, a complete dataset is divided into random test sets and train sets (70% train, 30% test). For test instance classification, applied train, test pairs to six ensemble ML algorithms: Bagging meta-estimator, Adaptive boosting(AdaBoost), Random Forest, Gradient Boosting(GBM), Extreme Gradient Boosting(XGBoost), Light Gradient Boosting(Light GBM), all adjusted with default parameters given by definitions of these classifiers in Python 3.8 and scikit-learn 1.1 package. Results are mentioned in Table 2, having evaluated each classifier's accuracy, precision and F1 score metric.

Table 2. Results obtained by ensemble approaches for classification

S.No.	Model	Token Replay			Trace Alignment		
		Accuracy	Precision	F1 Score	Accuracy	Precision	F1 Score
1	Random Forest	73.78%	73.3%	0.73	82.2%	82.5%	0.825
2	XGBoost	74.4%	83.1%	0.74	82.8%	83.6%	0.827
3	Bagging meta-estimator	74.3%	83.1%	0.74	84.3%	85.5%	0.845
4	AdaBoost	72.9%	80.0%	0.73	76.7%	80.3%	0.77
5	GBM	74.3%	83.1%	0.74	84.7%	86.3%	0.85
6	Light GBM	74.3%	83.1%	0.74	84.7%	86.3%	0.85

6   Conclusion

Resilience is always crucial for a cyber-physical system. Process Mining may also address the error prediction issue, allowing an end-user to be forewarned in advance if external events or user actions have a high likelihood of producing an error which falls under the prognostics, early warning, and scenario evaluation categories. In this work, we have investigated the usage of process mining by analyzing event log files. We explored combinations of conformance-checking methods of process mining with ensemble classification models to classify anomalies with ultimate objective of process optimization. We compared the result with KNN and C-SVC approach by [2], which achieved an accuracy of 59.56% and 62.92% and concluded that ensemble models perform better with the highest accuracy being 84.7% using trace alignment for as conformance technique with Gradient Boosting to classify anomalies. However, once a sturdy process model is built, pre-processing data methods can be refined to generate data with more information concerning the type of anomaly, and also finer ways to inject anomalies can be devised, resulting in better datasets for anomaly classification. In addition, generating process models using process mining and classifying anomalies is a step towards self-healing in cyber-physical systems. In summary, it can be claimed that adopting a system to generate consistent Event Logs across all stakeholder devices in a complex business environment can be very helpful for diagnosing faults and further opens a way for self-correcting mechanisms.

References

1. Deng, W., Liu, W., Li, Y., Zhao, T.: A petri-net-based framework for microgrid process mining. In: 2020 IEEE 4th Conference on Energy Internet and Energy System Integration (EI2), pp. 3797–3800 (2020). https://doi.org/10.1109/EI250167.2020.9346586
2. Singh, P., et al.: Using log analytics and process mining to enable self-healing in the Internet of Things. Environ. Syst. Decis. **42**, 234–250 (2022). https://doi.org/10.1007/s10669-022-09859-x
3. Bezerra, F., Wainer, J., van der Aalst, W.M.P.: Anomaly detection using process mining. In: Halpin, T., et al. (eds.) BPMDS/EMMSAD -2009. LNBIP, vol. 29, pp. 149–161. Springer, Heidelberg (2009). https://doi.org/10.1007/978-3-642-01862-6_13
4. Van der Aalst, W.M., de Medeiros, A.K.A.: Process mining and security: detecting anomalous process executions and checking process conformance. Electron. Notes Theor. Comput. Sci. **121**, 3–21 (2005). https://doi.org/10.1016/j.entcs.2004.10.013
5. Hosseini, S.M., Aghdasi, M., Teimourpour, B., Albadvi, A.: Implementing process mining techniques to analyze performance in EPC companies. Int. J. Inf. Commun. Technol. Res. **14** (2022). https://doi.org/10.52547/itrc.14.2.66
6. Fani Sani, M., van Zelst, S.J., van der Aalst, W.M.P.: Applying sequence mining for outlier detection in process mining. In: Panetto, H., Debruyne, C., Proper, H.A., Ardagna, C.A., Roman, D., Meersman, R. (eds.) OTM 2018. LNCS, vol. 11230, pp. 98–116. Springer, Cham (2018). https://doi.org/10.1007/978-3-030-02671-4_6
7. Junior, S.B., Ceravolo, P., Damiani, E., Omori, N.J., Tavares, G.M.: Anomaly detection on event logs with a scarcity of labels (2020)
8. Jagadeesh Chandra Bose, R.P., van der Aalst, W.: Trace alignment in process mining: opportunities for process diagnostics. In: Hull, R., Mendling, J., Tai, S. (eds.) BPM 2010. LNCS, vol. 6336, pp. 227–242. Springer, Heidelberg (2010). https://doi.org/10.1007/978-3-642-15618-2_17
9. https://itrust.sutd.edu.sg/itrust-labs-home/itrust-labs_epic/
10. Burattin, A.: PLG2: multiperspective processes randomization and simulation for online and offline settings (2015)
11. https://pm4py.fit.fraunhofer.de/documentation
12. Silva, P., Schukat, M.: On the use of k-nn in intrusion detection for industrial control systems. In: Proceedings of The IT&T 13th International Conference on Information Technology and Telecommunication, Dublin, Ireland, pp. 103–106 (2014)
13. Anthi, E., Williams, L., Burnap, P.: Pulse: an adaptive intrusion detection for the Internet of Things IoT (2018). https://doi.org/10.1049/cp.2018.0035
14. Pajouh, H.H., Javidan, R., Khayami, R., Dehghantanha, A., Choo, K.K.R.: A two-layer dimension reduction and two-tier classification model for anomaly-based intrusion detection in IoT backbone networks. IEEE Trans. Emerg. Top. Comput. **7**, 314–323 (2019). https://doi.org/10.1109/TETC.2016.2633228
15. Stewart, B., et al.: A novel intrusion detection mechanism for SCADA systems which automatically adapts to network topology changes. EAI Endorsed Trans. Ind. Netw. Intell. Syst. **4** (2017). https://doi.org/10.4108/eai.1-2-2017.152155

16. Hobbs, A.: The colonial pipeline hack: Exposing vulnerabilities in U.S. cybersecurity. In: Sage Business Cases. SAGE Publications Ltd, (2021). https://doi.org/10.4135/9781529789768
17. Alkhadra, R., Abuzaid, J., AlShammari, M., Mohammad, N.: Solar winds hack: in-depth analysis and countermeasures. In: 2021 12th International Conference on Computing Communication and Networking Technologies (ICCCNT), Kharagpur, India, pp. 1–7 (2021). https://doi.org/10.1109/ICCCNT51525.2021.9579611
18. Burattin, A.: Plg2: multiperspective process randomization with online and offline simulations. BPM (Demos) (2016)

Hyper-box Classification Model Using Mathematical Programming

Georgios I. Liapis and Lazaros G. Papageorgiou[✉]

The Sargent Centre for Process Systems Engineering, Department of Chemical Engineering, UCL (University College London), Torrington Place, London WC1E 7JE, UK
{georgios.liapis.20,l.papageorgiou}@ucl.ac.uk

Abstract. Classification constitutes focal topic of study within the machine learning research community. Interpretable machine learning algorithms have been gaining ground against black box models because people want to understand the decision-making process. Mathematical programming based classifiers have received attention because they can compete with state-of-the-art algorithms in terms of accuracy and interpretability. This work introduces a single-level hyper-box classification approach, which is formulated mathematically as Mixed Integer Linear Programming model. Its objective is to identify the patterns of the dataset using a hyper-box representation. Hyper-boxes enclose as many samples of the corresponding class as possible. At the same time, they are not allowed to overlap with hyper-boxes of different class. The interpretability of the approach stems from the fact that IF-THEN rules can easily be generated. Towards the evaluation of the performance of the proposed method, its prediction accuracy is compared to other state-of-the-art interpretable approaches in a number of real-world datasets. The results provide evidence that the algorithm can compare favourably against well-known counterparts.

Keywords: Mathematical programming · Data classification · Mixed integer optimisation · Hyper-box · Machine learning

1 Introduction

In classification, given a number of samples that are characterised by certain independent variables and their class membership, the aim is to identify the patterns and predict the class of a new sample based on its attributes. Scientists have proposed a plethora of classification models. Meanwhile, the incredible improvement in the last thirty years of both algorithms for mixed integer optimisation and computer hardware has led to an astonishing increase in the computational power of mixed integer optimisation solvers, as shown in [3]. Subsequently, mathematical programming based approaches became viable in the definition of a variety of machine learning methods for small and medium size datasets. Exploiting this remarkable progress, a growing body of literature focuses on mathematical programming based classifiers because they are able to combine accuracy and interpretability.

Motivated by the need to construct accurate multi-class classifiers that their decisions can be interpreted as IF-THEN rules in healthcare [16, 26] and financial management [17, 27], a single-level hyper-box approach is proposed. The extraction of an IF-THEN rule from every hyper-box demonstrates the interpretability of the approach. Meanwhile, the accuracy is highlighted by the comparison with the other approaches. The presented approach, which is inspired by the previously proposed work of Xu and Papageorgiou [24], finds the arrangement of a defined number of hyper-boxes in order to include as many samples as possible. The hyper-boxes are identified by their lower and upper bound at every attribute.

The rest of the paper is structured as follows: Sect. 2 describes mathematical programming formulations from literature that tackle classification problems. In Sect. 3, the proposed approach is presented and an illustrative example is used to demonstrate the generation of IF-THEN rules. In Sect. 4, a number of benchmark classification datasets are employed to test the performance of our proposed method against state-of-the-art interpretable approaches. Finally, conclusions are drawn in Sect. 5.

2   Related Work

Many mathematical programming formulations have been proposed that address the classification problem. Firstly, Gerhlein [12] developed a mixed integer linear programming (MILP) formulation that maximizes the number of correct classifications using a linear function for every class. Sueyoshi [22] proposed a two-stage MILP formulation that encorporates data envelopment analysis to estimate weights of a linear discriminant function by minimizing the total deviation of misclassified observations. Busygin et al. [8] formulated an optimisation based algorithm that can be used for feature selection and classification with the possibility of outlier detection.

Mathematical programming has become a useful tool for the definition of a variety of classification approaches that were solved using heuristic methods. Concerning classification trees, the widely used heuristics construct sub-optimal trees, such as CART [7] and C4.5 [20]. These greedy approaches myopically find a locally optimal split at every branch node one-at-a-time ignoring the splits at further down nodes. As a result, further down nodes apply splits that might affect generalisability. This can be tackled by a single-level training of the entire tree. In this way, the splits at each node are determined with full knowledge of all future splits.

Bertsimas and Dunn [2] presented Optimal Classification Trees (OCT), a formulation of the learning process of a classification tree as an MILP problem. The model makes decisions about the split rule at every branch node, class assignment to each leaf node, and the routing of each sample from the root node to a leaf node. The objective function is a minimisation function and contains two terms; the number of misclassified samples and the complexity of the tree. The weight of the second term is determined after tuning. Two versions of the formulation are presented; the first applies orthogonal (univariate) splits at every branch node and the second one applies oblique (multivariate) splits at every branch node. Recently, a plethora of papers relied on global exact optimisation approaches to find an optimal classification tree using mathematical programming tools [1, 6, 23].

Mathematical programming has, also, been incorporated in support vector machine (SVM) algorithms. SVM training has been modelled in a convex quadratic formulation with linear constraints that can be solved to global optimality using non-linear solvers [9]. Blanco et al. [4] proposed an SVM-based approach for multi-class classification that maximises the separation between different classes. An MILP and an MINLP version are presented, while a heuristic strategy is provided that contains dimensionality reduction and fixing of some variables. Generally, SVM is sensitive to noisy data and outliers because of its reliance on support vectors. As a result, Blanco et al. [5] presented methodologies to construct optimal support vector machine-based classifiers, which take into consideration the label noise in the dataset. They offered different alternatives based on solving MILP and MINLP that incorporate decisions on relabeling samples in the dataset.

Concerning hyper-box approaches, the first hyper-box based classifier was proposed by Simpson [21]. The algorithm generates hyperboxes, each one covering an area determined by its minimum and maximum coordinates in the N-dimensional sample space. Every hyper-box is associated with a fuzzy membership function calculating the goodness-of-fit of an input sample to a certain class. An MILP model has been proposed by Xu and Papageorgiou [24] that adopts a hyper-box representation, which extends previous work on two-dimensional plant layout [18]. In this way, the extension of the model to N dimensions creates hyper-boxes, which enclose as many samples of the corresponding class as possible. The hyper-boxes are identified by their centroid coordinate and the length on each dimension. Two constraints are included in order to ensure that there is no overlap between hyper-boxes that belong to different classes. The objective function of the mathematical model is the minimisation of the number of misclassified training samples. An iterative solution algorithm is proposed, which allows the addition of hyper-boxes in order to enclose more samples. Yang et al. [25] introduced two new proposals to improve the performance. The first one was a solution procedure that updates the sample weights during every iteration, which enforces the model to prioritise the difficult samples in the next iteration. Moreover, they introduced a data space partition method to reduce the computational cost of the proposed sample re-weighting hyper-box classifier. Finally, Üney and Türkay [28] have proposed a different formulation adopting a hyper-box representation, where boolean algebra was used to formulate the mathematical model.

3   Methodology

3.1   Problem Statement

In this work, a single-level approach is developed, in which each hyper-box is characterised by its lower and upper bound. As far as the mathematical formulation is concerned, it consists of hyper-box enclosing constraints, non-overlapping constraints and is enhanced with symmetry breaking constraints. The objective function is the minimisation of the number of misclassified samples. Overall, the problem studied can be stated as follows:

Given :

- Numerical values of S training samples with M attributes
- Classification of training samples into one of C classes
- Number of allowed hyper-boxes I per class

Determine :

- Optimal dimensions of hyper-boxes for every attribute.

So as to :

- Minimise the number of misclassified training samples

3.2   Mathematical Formulation

The indices, sets, parameters and variables associated with the model are presented below:

Indices

s Sample $(s = s_1, s_2, ..., S)$
m Attribute $(m = m_1, m_2, ..., M)$
i, j Hyper-box $(i, j = i_1, i_2, ..., I)$
c, k Class $(c, k = c_1, c_2, ..., C)$

Sets

C_s Class which sample s belongs to
I_c Set of hyper-boxes that belong to class c
S_c Set of samples that belong to class c

Parameters

A_{sm} Numerical value of sample s on attribute m
U Suitably big number
ϵ_m Minimum distance between hyper-boxes that belong to different classes on attribute m

Continuous variables

UP_{cim} Upper bound of hyper-box i of class c on attribute m
LO_{cim} Lower bound of hyper-box i of class c on attribute m

Binary variables

E_{sci} 1, if sample s is correctly classified in hyper-box i of class c; 0 otherwise.
Y_{cikjm} 1, if hyper-box i of class c is on the left of hyper-box j of class k on attribute m; 0 otherwise.

Hyper-box enclosing constraints

The first constraints are used to model whether a sample s is enclosed in a hyper-box of its class. Constraints (1) allow a sample s to be included in a hyper-box i of class c if sample value A_{sm} is higher than the lower bound of the hyper-box LO_{cim} for every attribute m. Accordingly, constraints (2) allow a sample s to be included in a hyper-box i of class c if sample value A_{sm} is lower than the upper bound of the hyper-box UP_{cim} for every attribute m. Those constraints are illustrated in Fig. 1, in which it can be seen that all samples are correctly classified in a hyper-box of their corresponding class, since their numerical value on every attribute is between lower and upper bounds of the respective hyper-box. Meanwhile, if $E_{sci} = 0$, the aforementioned constraints become redundant, since U is big enough to satisfy the constraints. Constraints (3) allow to each sample to be allocated to at most one hyper-box of its corresponding class, if correctly classified, thus avoiding possible double-counting.

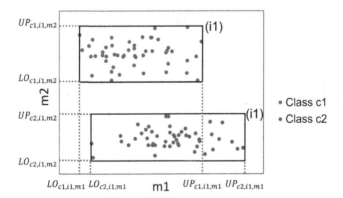

Fig. 1. Visualisation of non-overlapping constraints.

$$A_{sm} \geq LO_{cim} - U \cdot (1 - E_{sci}) \quad \forall s, c \in C_s, i \in I_c, m \qquad (1)$$

$$A_{sm} \leq UP_{cim} + U \cdot (1 - E_{sci}) \quad \forall s, c \in C_s, i \in I_c, m \qquad (2)$$

$$\sum_{i \in I_c} E_{sci} \leq 1 \quad \forall s, c \in C_s \qquad (3)$$

Non-overlapping Constraints

Hyper-boxes of different classes are prohibited from overlapping, since they express different patterns. Moreover, if there is overlapping between hyper-boxes of different classes and a sample falls inside this area, then it will be allocated to both classes. The non-overlapping is ensured by using two sets of constraints. In constraints (4), if $Y_{cikjm} = 1$, hyper-box i of class c precedes hyper-box j of class k on attribute m, because the upper bound of the first one UP_{cim} is lower than the lower bound of the second one LO_{kjm}. This is depicted in Fig. 1, in which hyper-box $i1$ of class $c2$ has upper bound on attribute $m2$, which is lower than the lower bound of hyper-box $i1$ of class $c1$ on attribute $m2$. Subsequently, $Y_{c2,i1,c1,i1,m2} = 1$ and these hyper-boxes do not overlap with each other on attribute $m2$. Thus, it is ensured that if $Y_{cikjm} = 1$ or $Y_{kjcim} = 1$, the hyper-boxes do not overlap on attribute m. This is forced by constraints (5) for every combination of hyper-boxes that belong to different classes for at least one attribute. Note that ε_m is the minimum distance between hyper-boxes that belong to different classes to prevent them from sharing the same border.

$$UP_{cim} + \varepsilon_m \leq LO_{kjm} + (U + \varepsilon_m) \cdot (1 - Y_{cikjm}) \quad \forall c, k \neq c, i \in I_c, j \in I_k, m \quad (4)$$

$$\sum_m (Y_{cikjm} + Y_{kjcim}) = 1 \quad \forall c, k < c, i \in I_c, j \in I_k \quad (5)$$

Symmetry-Breaking Constraints

Symmetry breaking constraints are added to avoid redundant equivalent solutions. More specifically, constraints (6) enforce the number of samples included in the lower indexed hyper-boxes to be higher than the number of samples included in the higher indexed hyper-boxes. In this way, some identical possible solutions are removed.

$$\sum_{s \in S_c} E_{s,c,i} \leq \sum_{s \in S_c} E_{s,c,i-1} \quad \forall c, i \in I_c, i \geq 2 \quad (6)$$

Objective Function

The objective function (7) aims to minimise the total number of misclassified samples.

$$min \sum_s \left(1 - \sum_{c \in C_s} \sum_{i \in I_c} E_{sci}\right) \quad (7)$$

The formulation that contains constraints (1) - (7) is named MHB (**Monolithic Hyper-Box model**) and its goal is to enclose as many samples as possible within the hyper-boxes.

3.3   Testing Phase

The creation of the hyper-boxes is completed during the training. During the testing phase, each testing sample is allocated to one of the hyper-boxes. The allocation happens based on the distance of the sample from the hyper-boxes [15, 24, 25]. The distance of testing sample s from hyper-box i of class c on attribute m is defined to be:

$$DIST_{scim} = max(0, A_{sm} - UP_{cim}, LO_{cim} - A_{sm}) \quad \forall s, c, i \in I_c, m \quad (8)$$

The total distance of sample s from hyper-box i of class c is given by:

$$DSI_{sci} = \sqrt{\sum_m DIST^2_{scim}} \quad \forall s, c, i \in I_c \tag{9}$$

Hence, the distances of testing samples from all hyper-boxes are calculated. Afterwards, testings samples are allocated to their nearest derived hyper-box and assigned the membership of the hyper-box.

3.4 Illustrative Example

A 2-dimensional synthetic dataset, called Moon, is created in order to demonstrate how the hyper-boxes are formed and how the rules are, afterwards, generated. Moon dataset, which is depicted in Fig. 2, was constructed with the use of *make_moons* utility of the scikit-learn library [19].

It contains 100 samples, 2 attributes and 2 classes. 50 samples belong to class $c1$ and are coloured blue and the other 50 samples belong to class $c2$ and are coloured red.

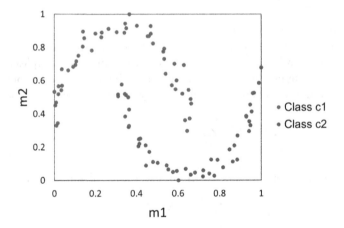

Fig. 2. Visualisation of Moon dataset.

Figure 3 illustrates the hyper-boxes generated by solving MHB for 3 hyper-boxes per class ($MHB - 3$). It is observed that 3 hyper-boxes per class are adequate to enclose all samples. As expected, there is no overlapping between hyper-boxes of different class. Meanwhile, as ensured by constraints (6), hyper-box $i1$ of each class includes the highest number of samples, followed by hyper-box $i2$ and lastly $i3$. It is worth noting that the results are post-processed, so that hyper-box lower and upper bounds for each attribute are equal to smallest and largest values, respectively, of the enclosed samples.

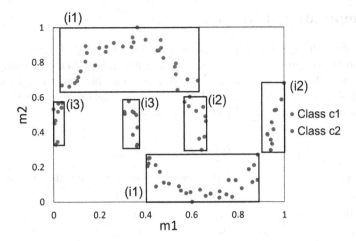

Fig. 3. Hyper-boxes created to enclose Moon's samples using $MHB - 3$.

Table 1. Bounds of each hyper-box.

Hyper-box	Class	Attribute	Lower bound	Upper bound
i1	c1	m1	0.04	0.63
		m2	0.64	1
i2	c1	m1	0.57	0.66
		m2	0.30	0.60
i3	c1	m1	0	0.04
		m2	0.33	0.57
i1	c2	m1	0.41	0.89
		m2	0	0.27
i2	c2	m1	0.93	1
		m2	0.30	0.68
i3	c2	m1	0.31	0.36
		m2	0.32	0.57

Table 1 summarises the bounds of the hyper-boxes that are formed. Using these bounds, the corresponding IF-THEN rules can easily be generated. For example, hyper-box $i1$ of class $c1$ can be interpreted as the following rule:

$$\text{IF } (0.04 \leq m1 \leq 0.63) \text{ AND } (0.64 \leq m2 \leq 1) \implies \text{Class } c1$$

Therefore, at the testing stage, if a sample satisfies both conditions of a rule, it will be assigned to the corresponding class. If it does not satisfy both conditions of any rule, it falls outside of all hyper-boxes and the distances have to be calculated in order to assign it to the nearest hyper-box, as described in Sect. 3.3.

4   Computational Results

In this section, the applicability of the proposed methodology is demonstrated by applying it to a number of datasets shown in Table 2, together with comparative analysis with literature approaches. All datasets can be downloaded from UCI machine learning repository [10] and are widely used as benchmarks to compare the performance of different classification methods.

Table 2. Datasets.

Dataset	Abbreviation	Samples	Attributes	Classes
Iris	I	150	4	3
Wine	W	178	13	3
Seeds	S	210	7	3
Glass	G	214	9	6
Thyroid New	TN	215	5	3
Heart disease Cleveland	H	297	18	5
E-coli	E	336	7	8
Ionosphere	ION	351	34	2
Data user modelling	DUM	403	5	4
Indian Liver Patient	ILP	583	10	2
Balance	B	625	4	3
Blood transfusion	BT	748	4	2
Banknote Authentication	BA	1372	4	2
Wi-fi	WF	2000	7	4
Thyroid-ANN	TA	3772	21	3

Firstly, the interpretability of MHB is demonstrated by the extraction of IF-THEN rules in a real world dataset, called Iris. It contains 150 samples with 4 attributes and they are equally distributed in 3 classes. By solving MHB, IF-THEN rules can be generated using the bounds of every attribute.

Table 3 shows the number of enclosed samples and the bounds of the hyper-boxes that are formed for Iris dataset by solving $MHB-2$. It is observed that only 2 samples are not correctly classified and they belong to class $c2$, since 48 out of 50 are correctly classified. Using the bounds, the corresponding IF-THEN rules are generated. While the procedure described in Sect. 3.3 is employed for samples that do not satisfy any of those rules. A rule generation example for hyper-box $i1$ of class $c1$ is the following:

$$\text{IF} \begin{cases} (0 \leq m1 \leq 0.417) \text{ AND} \\ (0.125 \leq m2 \leq 1) \text{ AND} \\ (0 \leq m3 \leq 0.153) \text{ AND} \\ (0 \leq m4 \leq 0.208) \end{cases} \implies \text{Class } c1$$

Table 3. Number of enclosed samples and bounds of each hyper-box for Iris dataset solving $MHB - 2$.

Hyper-box	Class	Number of enclosed samples	Attribute	Lower bound	Upper bound
i1	c1	50	m1	0	0.417
			m2	0.125	1
			m3	0	0.153
			m4	0	0.208
i1	c2	46	m1	0.167	0.750
			m2	0	0.542
			m3	0.339	0.661
			m4	0.375	0.625
i2	c2	2	m1	0.444	0.472
			m2	0.500	0.583
			m3	0.593	1
			m4	0.625	0.958
i1	c3	33	m1	0.167	0.944
			m2	0.208	0.458
			m3	0.593	1
			m4	0.667	0.958
i2	c3	17	m1	0.472	1
			m2	0.083	0.750
			m3	0.678	0.966
			m4	0.542	1

Next, the single-level hyper-box model MHB is compared with two classification tree approaches, $CART$ and OCT, which apply univariate splits. Classification trees produce solutions that can be interpreted as IF-THEN rules. Figure 4 illustrates the maximum number of rules allowed for $CART$ and OCT for different depths and for MHB for different number of hyper-boxes per class. For classification tree approaches, the maximum number of rules is equal to the number of leaf nodes, which is equal to 2^D, where D is the depth of the tree. Subsequently, a tree, whose depth is equal to $D = 2$, has 4 leaf nodes and up to 4 IF-THEN rules can be generated. The same logic applies to the rest of the depths. As far as MHB is concerned, the maximum number of rules is equal to the total number of hyper-boxes. Thus, the maximum number of rules is equal to $|I| \cdot |C|$, which means that every dataset has a different number of rules based on the number of classes. Figure 4 shows the average maximum number of rules across all examined datasets for different number of hyper-boxes per class. It is shown that classification tree approaches produce similar number of rules for depth equal to $D = 2, 3$ with hyper-box model for $I = 1, 2$ hyper-boxes per class, respectively. For larger depths $D = 4, 5$, the maximum number of rules generated by tree approaches is higher than those of MHB for $I = 3, 4$ hyper-boxes per class.

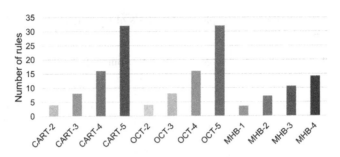

Fig. 4. Maximum number of rules allowed.

The implementation of the proposed algorithm was conducted in GAMS (General Algebraic Modeling System) [11] and the selected solver was GUROBI [13]. A time limit of 30 min was applied to every run. The other approaches, $CART$ and OCT, were implemented using Scikit-learn library [19] and InterpretableAI library [14], respectively, in order to find the corresponding testing performance. The default parameters were used, while the maximum depth of the tree was controlled. The training of CART is completed within 1 s, while the training of OCT is completed within seconds or minutes for larger datasets. It is noteworthy that a warm start solution is used during the implementation of OCT. More specifically, $CART$ is used to provide an integer-feasible solution.

The dataset is divided in training and testing subsets; 70% of the whole dataset is used for training and the rest 30% of the samples are used for the testing phase. The allocation of samples to training or testing subset is performed randomly and is repeated 15 times for every dataset, while mean prediction accuracy of the 15 iterations is reported. Note that all datasets examined undergo feature scaling in the range of [0,1]. Furthermore, the datasets that contain categorical features are converted using one-hot encoding.

Table 4 presents a comparison of average testing accuracy per dataset of $CART$ and OCT for depth $D = 2, 3$ and MHB for $I = 1, 2$ hyper-boxes per class. As explained earlier, the maximum number of rules of the approaches is similar. It is shown that $MHB - 1$ outperforms $CART - 2$ and $OCT - 2$ in 10 out of 15 datasets and it is superior on average across all datasets. The outperformance of $MHB - 1$ against $CART - 2$ and $OCT - 2$ is clear for W, G, TN and E datasets, while this is not the case for I, B and BA, in which $OCT-2$ produces the highest scores. Similarly, $MHB-2$ shows better prediction accuracy than $CART - 3$ and $OCT - 3$ on average and more specifically in 8 out of 15 examined datasets. This is easily observed in I, W, TN and H, while both classification tree approaches perform better in BT and TA.

Table 5 displays the average testing accuracy per dataset of $CART$ and OCT for depth $D = 4, 5$ and MHB for $I = 3, 4$ hyper-boxes per class. In these cases, as illustrated in Fig. 4, $CART$ and OCT produce a higher number of rules in comparison to MHB. $MHB - 3$ outperforms the other approaches for depth $D = 4$ in 7 out of 15 datasets. In terms of average performance across all datasets, $MHB - 3$ surpasses $CART - 4$ and $OCT - 4$. $MHB - 3$ prediction performance

Table 4. Testing accuracy (%) of *CART*, *OCT* for depth D=2, 3 and *MHB* for $I = 1, 2$ hyper-boxes per class.

Dataset	CART-2	CART-3	OCT-2	OCT-3	MHB-1	MHB-2
I	93.61	94.04	94.52	94.37	93.78	96.00
W	85.03	91.02	87.53	92.72	94.72	96.60
S	89.52	88.64	88.15	91.64	89.52	90.69
G	58.04	64.06	58.05	62.97	68.54	67.40
TN	90.17	91.27	89.64	90.36	92.19	94.48
H	52.98	53.17	53.40	53.48	54.08	55.36
E	76.12	80.22	79.67	82.18	84.47	85.27
ION	89.59	89.40	88.93	88.87	89.97	89.97
DUM	83.67	91.05	84.63	92.73	89.83	92.67
ILP	71.12	71.08	69.37	70.06	70.54	70.19
B	64.92	66.53	66.70	69.15	57.65	66.70
BT	76.54	76.91	74.90	76.18	76.70	76.13
BA	90.47	93.42	91.83	96.73	85.29	94.86
WF	90.87	96.51	95.78	96.73	96.48	97.51
TA	97.92	99.29	97.83	99.27	99.15	99.25
Average	**80.71**	**83.11**	**81.40**	**83.83**	**82.86**	**84.87**

exceeds significantly the other methodologies in W, E, G and B. However, it performs quite worse than $OCT-4$ in I, DUM and BA. The last comparison is between $CART-5$, $OCT-5$ and $MHB-4$, in which $MHB-4$ is superior in 10 out of 15 datasets and on average across all datasets. The only datasets that it is not the best performing approach are S, ILP, BT, BA and TA.

Apart from the previous comparison of average prediction accuracies, a scoring strategy is employed in order to evaluate the relative competitiveness of the examined approaches. For each dataset, the best performing approach is awarded 12 points, whereas the worst performing approach is awarded 1 point. The points of each methodology are averaged over the 15 datasets and the final ranking is depicted in Fig. 5.

As shown in Fig. 5, $MHB-4$ is shown to be the most accurate classification algorithm among all, achieving the highest score. Moreover, MHB has 3 out of the top 4 scores, namely $MHB-2$, $MHB-3$, $MHB-4$, while $MHB-1$ achieves to outperform $CART-3$, $CART-2$ and $OCT-2$. The results across all examined datasets show that the proposed approach MHB achieves higher testing accuracy than CART and OCT, even when using less rules.

Table 5. Testing accuracy (%) of $CART$, OCT for depth $D = 4, 5$ and MHB for $I = 3, 4$ hyper-boxes per class.

Dataset	CART-4	CART-5	OCT-4	OCT-5	MHB-3	MHB-4
I	94.41	94.15	96.15	94.37	95.26	95.11
W	91.11	91.43	93.46	94.32	96.48	96.48
S	89.72	89.53	91.22	92.06	90.48	91.75
G	64.10	64.10	68.10	68.82	71.77	72.08
TN	91.76	92.55	93.23	92.41	94.38	93.96
H	52.93	52.71	54.74	54.74	54.46	55.28
E	80.88	80.78	81.91	81.06	85.13	84.07
ION	88.52	88.33	90.94	89.93	90.48	92.25
DUM	91.25	91.31	93.22	91.90	92.33	93.33
ILP	71.09	70.98	69.94	69.71	70.35	70.61
B	67.29	70.99	69.72	73.90	72.91	76.86
BT	77.25	77.37	77.51	77.07	77.92	77.35
BA	95.42	96.98	98.43	98.12	96.61	98.09
WF	96.95	97.23	97.08	97.39	97.74	97.92
TA	99.47	99.72	99.38	99.70	99.15	99.16
Average	**83.48**	**83.88**	**85.00**	**85.03**	**85.70**	**86.29**

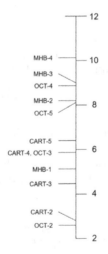

Fig. 5. Visualisation of the overall performance score of each method.

5   Concluding Remarks

This work addresses the problem of multi-class classification. A single-level approach (MHB) has been proposed, which adopts a hyper-box representation and is formulated mathematically as Mixed Integer Linear Programming (MILP) model. Key decisions of MHB involve optimal sizing and arrangement of hyper-boxes of each class in order to identify dataset patterns. The interpretability of the model is demonstrated by the IF-THEN rules generated from hyper-boxes. The proposed hyper-box model outperforms other established classification tree methodologies, namely $CART$ and OCT, in most datasets examined thus highlighting its applicability for real world datasets with high prediction accuracy, while maintaining enhanced interpretability.

Acknowledgements. Authors gratefully acknowledge the financial support from Engineering and Physical Sciences Research Council (EPSRC) under the project EP/V051008/1.

References

1. Aghaei, S., Gomez, A., Vayanos, P.: Learning optimal classification trees: strong max-flow formulations (2020). https://doi.org/10.48550/arXiv.2002.09142
2. Bertsimas, D., Dunn, J.: Optimal classification trees. Mach. Learn. **106**, 1039–1082 (2017). https://doi.org/10.1007/s10994-017-5633-9
3. Bixby, R.E.: A brief history of linear and mixed-integer programming computation. Doc. Math. **1**, 107–121 (2012)
4. Blanco, V., Japón, A., Puerto, J.: Optimal arrangements of hyperplanes for SVM-based multiclass classification. Adv. Data Anal. Classif. **14**, 175–199 (2020). https://doi.org/10.1007/s11634-019-00367-6
5. Blanco, V., Japón, A., Puerto, J.: A mathematical programming approach to SVM-based classification with label noise. Comput. Ind. Eng. **172**, 108611 (2022). https://doi.org/10.1016/j.cie.2022.108611
6. Blanquero, R., Carrizosa, E., Molero-Río, C., Morales, D.R.: Sparsity in optimal randomized classification trees. Eur. J. Oper. Res. **284**, 255–272 (2020). https://doi.org/10.1016/j.ejor.2019.12.002
7. Breiman, L., Friedman, J.H., Olshen, R.A., Stone, C.J.: Classification and Regression Trees. Taylor & Francis, Milton Park (1984). https://doi.org/10.1201/9781315139470
8. Busygin, S., Prokopyev, O.A., Pardalos, P.M.: An optimization-based approach for data classification. Optim. Meth. Softw. **22**, 3–9 (2007). https://doi.org/10.1080/10556780600881639
9. Carrizosa, E., Morales, D.R.: Supervised classification and mathematical optimization. Comput. Oper. Res. **40**, 150–165 (2013). https://doi.org/10.1016/j.cor.2012.05.015
10. Dua, D., Graff, C.: UCI machine learning repository. https://archive.ics.uci.edu/ml/index.php (2017)
11. GAMS Development Corporation: General Algebraic Model System (GAMS) (2022). Release 41.5.0, Washington, DC, USA
12. Gehrlein, W.V.: General mathematical programming formulations for the statistical classification problem. Oper. Res. Lett. 5, 299–304 (1986). https://doi.org/10.1016/0167-6377(86)90068-4
13. Gurobi Optimization, LLC: Gurobi Optimizer Reference Manual (2023). https://www.gurobi.com

14. Interpretable AI, LLC: Interpretable AI Documentation (2023). https://www.interpretable.ai
15. Maskooki, A.: Improving the efficiency of a mixed integer linear programming based approach for multi-class classification problem. Comput. Ind. Eng. **66**, 383–388 (2013). https://doi.org/10.1016/j.cie.2013.07.005
16. Müller, T.T., Lio, P.: Peclides neuro: a personalisable clinical decision support system for neurological diseases. Front. Artif. Intell. **3**, 23 (2020). https://doi.org/10.3389/frai.2020.00023
17. Nasseri, A.A., Tucker, A., Cesare, S.D.: Quantifying stockTwits semantic terms' trading behavior in financial markets: an effective application of decision tree algorithms. Expert Syst. Appl. **42**, 9192–9210 (2015). https://doi.org/10.1016/j.eswa.2015.08.008
18. Papageorgiou, L.G., Rotstein, G.E.: Continuous-domain mathematical models for optimal process plant layout. Ind. Eng. Chem. Res. **37**, 3631–3639 (1998). https://doi.org/10.1021/ie980146v
19. Pedregosa, F., et al.: Scikit-learn: machine learning in Python. J. Mach. Learn. Res. **12**, 2825–2830 (2011). https://doi.org/10.48550/arXiv.1201.0490
20. Quinlan, J.R.: Improved use of continuous attributes in c4.5. J. Artif. Intell. Res. **4**, 77–90 (1996). https://doi.org/10.1613/jair.279
21. Simpson, P.: Fuzzy min-max neural networks. I. classification. IEEE Transa. Neural Netw. **3**, 776–786 (1992). https://doi.org/10.1109/72.159066. https://ieeexplore.ieee.org/document/159066/
22. Sueyoshi, T.: Mixed integer programming approach of extended DEA-discriminant analysis. Eur. J. Oper. Res. **152**, 45–55 (2004). https://doi.org/10.1016/S0377-2217(02)00657-4
23. Verwer, S., Zhang, Y.: Learning optimal classification trees using a binary linear program formulation. In: 33rd Conference on Artificial Intelligence (2019). https://doi.org/10.1609/aaai.v33i01.33011624
24. Xu, G., Papageorgiou, L.G.: A mixed integer optimisation model for data classification. Comput. Ind. Eng. **56**, 1205–1215 (2009). https://doi.org/10.1016/j.cie.2008.07.012
25. Yang, L., Liu, S., Tsoka, S., Papageorgiou, L.G.: Sample re-weighting hyper box classifier for multi-class data classification. Comput. Ind. Eng. **85**, 44–56 (2015). https://doi.org/10.1016/j.cie.2015.02.022
26. Yoo, I., et al.: Data mining in healthcare and biomedicine: a survey of the literature. J. Med. Syst. **36**, 2431–2448 (2012). https://doi.org/10.1007/s10916-011-9710-5
27. Zibanezhad, E., Foroghi, D., Monadjemi, A.: Applying decision tree to predict bankruptcy. In: IEEE International Conference on Computer Science and Automation Engineering, vol. 4, pp. 165–169 (2011). https://doi.org/10.1109/CSAE.2011.5952826
28. Üney, F., Türkay, M.: A mixed-integer programming approach to multi-class data classification problem. Eur. J. Oper. Res. **173**, 910–920 (2006). https://doi.org/10.1016/j.ejor.2005.04.049

A Leak Localization Algorithm in Water Distribution Networks Using Probabilistic Leak Representation and Optimal Transport Distance

Andrea Ponti[1,3]([✉]), Ilaria Giordani[2,3], Antonio Candelieri[1],
and Francesco Archetti[2,4]

[1] Department of Economics, Management, and Statistics,
University of Milano-Bicocca, Milan, Italy
`andrea.ponti@unimib.it`
[2] Department of Computer Science, Systems, and Communication,
University of Milano-Bicocca, Milan, Italy
[3] OAKS s.r.l., Milan, Italy
[4] Consorzio Milano Ricerche, Milan, Italy

Abstract. Leaks in water distribution networks are estimated to account for up to 30% of the total distributed water: the increasing demand, and the skyrocketing energy cost have made leak localization and adoption even more important to water utilities. Each leak scenario is run on a simulation model to compute the resulting values of pressure and flows over the whole network. The recorded values are seen as the signature of one leak scenario. The key distinguishing element in the present paper is the representation of a leak signature as a discrete probability distribution. In this representation the similarity between leaks can be captured by a distance between their associated probability distributions. This maps the problem of leak detection from the Euclidean physical space into a space whose elements are histograms, structured by a distance between histograms, namely the Wasserstein distance. This choice also matches the physics of the system: indeed, the equations modelling the generation of flow and pressure data are non-linear. Non-linear data structure is better represented by the Wasserstein distance than by the Euclidean distance. The signatures obtained through the simulation of a large set of leak scenarios are non-linearly clustered according in the Wasserstein space using Wasserstein barycenters as centroids. As a new set of sensor measurements arrives, the related signature is associated to the cluster with the closest barycenter. The location of the simulated leaks belonging to that cluster are the possible locations of the observed leak. This new theoretical and computational framework allows a richer representation of pressure and flow data embedding both the modelling and the computational modules in the Wasserstein space whose elements are the histograms endowed with the Wasserstein distance. The computational experiments on benchmark and real-world networks confirm the feasibility of the proposed approach.

Keywords: Leak localization · Water distribution networks · Wasserstein distance

M. Sellmann and K. Tierney (Eds.): LION 2023, LNCS 14286, pp. 31–45, 2023.
https://doi.org/10.1007/978-3-031-44505-7_3

1   Introduction

1.1   Motivations

Leaks in water distribution networks are estimated to account for up to 30% of the total distributed water: this figure alone gives an idea of the positive impacts of an improvement, even relatively small in percentage, of a leak reduction. The increasing demand, driven by the growing urban population, and the skyrocketing energy cost have made leak early detection, quick localization and adoption of remedial actions ever more important to water utilities. Major obstacles are represented by the scarcity of measurements and the uncertainty in demand making the leakage localization a very challenging problem. This situation has influenced significantly not only the operational practice but also spurred the water research community towards awareness of the potential of artificial intelligence and specifically machine learning and of the importance of the synergy with machine learning community. These advances has been enabled both by new computational techniques and the growing availability of sensor of pressure and flows deployed over the water distribution network (WDN). A key concept in the present paper is the "leak scenario" characterized by the location and severity of the leak. Each scenario is simulated by the hydraulic simulator EPANET to compute the values of pressure and flows. When a possible leak is detected (e.g., with traditional methods such as Minimum Night Flow analysis) the actual pressure and flow values recorded by the sensors are compared with those of the faultless network and those obtained through simulation of all different leak scenarios. The values recorded by the sensors are seen as the features of one leak scenario and can be considered as the signature of the leak in a feature space. The signatures obtained through the simulation of all leak scenarios are clustered: as a new set of sensor measurements arrives, its feature is compared with cluster centroids in order to obtain the most similar one. The signature is then assigned to that cluster. This gives the simulated leaky pipes related to the scenarios of that cluster as potential leak locations. The overall workflow is depicted in Fig. 1.

The key distinguishing element in the present paper is the representation of a leak signature, over the simulation horizon, as a discrete probability distribution. In this representation, the similarity between leaks and between a faulty signature and the faultless one can be captured by a distance between their associated probability distributions. As probability distributions, histogram will be considered. This maps the problem of leak detection from the Euclidean physical space into a space whose elements are histograms: this space is structured by a distance between histograms, namely the Wasserstein distance also known as the optimal transport distance. The Wasserstein (WST) distance is a field of mathematics which studies the geometry of probability spaces and provides a principled framework to compare and align probability distributions. The Wasserstein distance can be traced back to the works of Gaspard Monge [11] and Lev Kantorovich [10]. The WST distance has evolved into a very rich mathematical structure whose complexity and flexibility are analyzed in a landmark

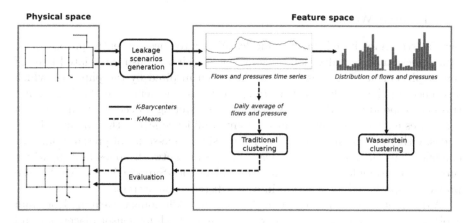

Fig. 1. General framework of the proposed approach: Wasserstein versus traditional clustering of leak-age scenarios.

volume [23] and, in the discrete domain, in the tutorial [21]. The computation of the WST distance requires the solution of a constrained linear optimization problem which can have, a very large number of variables and constraints, and can be shown to be equivalent to a min-flow problem. Recently, several specialized computational approaches have drastically reduced the computational hurdles [13]. For univariate discrete distributions, which are considered in this paper, there is a closed formula which makes the computational cost negligible. The main advantage of WST for histograms is that it is a cross binning distance, and it is not affected by different binning schemes. Moreover, the WST distance matches naturally the perceptual notion of nearness and similarity. This is not the case of commonly used distances as Kullback-Leibler (KL) and χ-square that account only for the correspondence between bins of the same index and do not use information across bins or distributions with different binning schemes, that is different supports. An important element of the WST theory is the barycenter which offers a useful synthesis of a set of distributions. The barycenter allows for a standard clustering method like k-means to be generalized to WST spaces as will be shown in Sect. 3. Several AI based strategies have been recently proposed for leak detection and localization, in some cases including the issue of sensor placement. There are mostly based on Graph Neural Networks and dictionary learning. An advantage of the Graph Neural Network (GNN) models is that they are mapped naturally on the physical network and can naturally exploit its dynamics. Moreover, most of them are purely data driven independent of the hydraulic simulation. A drawback is that neural networks used in a purely data driven approach are very expensive to train. This results in a smaller computational cost for the algorithm proposed in this paper WELL (Wasserstein Enabled Leak Localization) allowing to handle WDN's of larger size than GNN. Another important consideration is that WELL works with a much smaller number of sensors than GNN methods. The main approaches will be analyzed in Sect. 1.2.

1.2 Related Works

Several strategies based on machine learning have been recently proposed for leak detection and localization, in some cases including also the related issue of sensor placement. The present paper has been inspired by the paper [2] which also encodes the graph in a feature space whose elements are the leak signatures corresponding to different scenarios. A clustering procedure in the space of signatures enables to indicate, at different confidence levels, the potential leaky nodes. Other approaches are [24] which consider time-series of pressure data for classification and imputation as leaks of specific network nodes and [9] which propose an ensemble Convolutional Neural Network-Support Vector Machine to provide graph-based localization in water distribution systems.

A different approach is based on the concept of dictionary learning. Irofti and Stoican [8] propose a dictionary learning strategy both for sensor placement and leakage isolation. The proposed strategy is to construct a dictionary of atoms over which to project the measured pressure residuals with a sparsity constraint. Each of the measured residuals can expressed as a combination of the number of atoms. The dictionary learning approach is further developed in [7] which builds on the dictionary representation of the algorithms for sensor placement, leak detection and localization adding also a layer of graph interpolation as an input to dictionary classification. Related strategies are proposed in [20] based on pressure sensors and spatial interpolation. Romero-Ben et al. [17] use graph interpolation to estimate the hydraulic states of the complete WDN from real measurements at certain nodes of the network graph. The actual measurements together with a subset of estimated states are used to feed and train the dictionary learning scheme. Gardharsson et al. [5] aim to estimate the complete network state from data but instead of graph interpolation adopt a graph neural network approach, exploiting the topology of the physical network, in order to estimate pressures where measurements are not available. Pressure prediction errors are converted into residual signals on the edges. A similar approach has been proposed in [6] where nodal pressures are also estimated using graph neural networks. A related approach [16] encodes the data from measured nodes as images. Subsequently a clustering is performed to split the network into subnetworks and a deep neural network is used to provide the binary classification.

Morales-González et al. [12] propose simulate annealing with hyperparameter optimization for Leak Localization. In [18] the authors solve jointly sensor placement and leak localization considering the mutual information and, building on it, a measure of the relevance of a subset of nodes and a measure of redundancy. To compute these two definitions a dataset of node pressures is built covering different scenarios that consider leaks of different sizes in all nodes of the network. Soldevila et al. [19] consider jointly the leak detection and localization problems. Leaks are detected and validated analyzing statistically the inlet flow and localization is formulated as a classification problem whose computational complexity is mitigated through a clustering scheme.

The Wasserstein, a.k.a. optimal transport, distance, which is at the core of the present paper, is also considered in [1] in order to perform data driven detection and localization of leaks in industrial fluids (the example is for naphtha). The size of the network (22 nodes) and its hydraulics are quite different from the WDNs, and the method proposed is not directly relevant to the leak localization problem in WDNs. The use of WST in analyzing WDNs has been already proposed for optimal sensor placement [14] and resilience analysis [15].

1.3   Our Contributions

The main novelty of this paper is to represent the sample of pressure and flow values related to a leak scenario not through its average value but as a discrete probability distribution, specifically, a histogram. In this framework the similarity between leak scenarios is given by a distance between distributions. Among many such distances the Wasserstein distance, also known as optimal transport distance, has been used in this paper. The WST distance can be regarded as a natural extension of the Euclidean distance, lifting it into the space of probability distributions. This new theoretical and computational framework allows a richer representation of pressure and flow data embedding both the modelling and the computational modules in a space, the Wasserstein space, whose elements are the histograms, endowed with the WST distance. Moreover, also the clustering takes place in the Wasserstein space using barycenters. A result is that with respect to others AI based methods a smaller number of sensors is required.

1.4   Content Organization

The paper is organized as follows. Section 2 establishes the feature space i.e., the generation of leak scenarios. It provides the distribution representation of the leaks in the feature space. Section 3 provides the basic notions of the Wasserstein distance, its computation for the case of discrete distributions and the computation of the barycenters. Section 4 establishes the clustering algorithms in Euclidean and Wasserstein spaces. Section 5 displays the workflow of the WELL algorithm. Section 6 provides the experimental settings and results. Section 7 comments on the comparative advantages of our method and its limitations along with perspectives for future work.

2   The Wasserstein Distance

2.1   Basic Definitions

Consider the case of a discrete distribution P specified by a set of support points x_i with $i = 1, ..., m$ and their associated probabilities w_i, such that $\sum_{i=1}^{m} w_i = 1$ with $w_i \geq 0$ and $x_i \in M$ for $i = 1, ..., m$. Usually, $M = \mathbf{R}^d$ is the d-dimensional

Euclidean space where x_i are the support vectors. Therefore, P can be written as follows in Eq. (1):

$$P(x) = \sum_{i=1}^{m} w_i \delta(x - x_i) \tag{1}$$

where $\delta(\cdot)$ is the Kronecker delta. The WST distance between two distributions $P^{(1)} = \{w_i^{(1)}, x_i^{(1)}\}$ with $i = 1, ..., m_1$ and $P^{(2)} = \{w_i^{(2)}, x_i^{(2)}\}$ with $i = 1, ..., m_2$ is obtained by solving the following linear program (Eq. 2):

$$W(P^{(1)}, P^{(2)}) = \min_{\gamma_{ij} \in \mathbf{R}^+} \sum_{i \in I_1, j \in I_2} \gamma_{ij} d(x_i^{(1)}, x_j^{(2)}) \tag{2}$$

The cost of transport between $x_i^{(1)}$ and $x_j^{(2)}$, $d(x_i^{(1)}, x_j^{(2)})$ is defined by the p-th power of the norm $|x_i^{(1)}, x_j^{(2)}|$, which is usually the Euclidean distance. Two index sets can be defined as $I_1 = \{1, ..., m_1\}$ and I_2 likewise, such that:

$$\sum_{i \in I_1} \gamma_{ij} = w_j^{(2)}, \ \forall j \in I_2 \tag{3}$$

$$\sum_{j \in I_2} \gamma_{ij} = w_i^{(1)}, \ \forall i \in I_1 \tag{4}$$

Equations (3) and (4) represent the in-flow and out-flow constraints, respectively. The terms γ_{ij} are called matching weights between support points $x_i^{(1)}$ and $x_j^{(2)}$ or the optimal coupling for $P^{(1)}$ and $P^{(2)}$. The basic computation of OT between two discrete distributions involves solving a network flow problem whose computation typically scales cubically in the sizes of the measure. In the case of a one-dimensional histograms, the computation of the Wasserstein distance can be performed by a simple sorting algorithm and with the application of Eq. (5).

$$W_p(P^{(1)}, P^{(2)}) = \left(\frac{1}{n} \sum_{i}^{n} |x_i^{(1)*} - x_i^{(2)*}|^p \right)^{\frac{1}{p}} \tag{5}$$

where $x_i^{(1)*}$ and $x_i^{(2)*}$ are the sorted samples. The discrete version of the WST distance is usually called the Earth Mover Distance (EMD). For instance, when measuring the distance between grey scale images, the histogram weights are given by the pixel values and the coordinates by the pixel positions. The computational cost of optimal transport can quickly become prohibitive. The method of entropic regularization enables scalable computations, but large values of the regularization parameter can induce an undesirable smoothing effect, whereas low values not only reduce the scalability but might induce several numerical instabilities.

2.2   Wasserstein Barycenter

Under the optimal transport metric, it is possible to compute the mean of a set of empirical probability measures. This mean is known as the Wasserstein barycenter and is the measure that minimizes the sum of its Wasserstein distances to each element in that set. Consider a set of N discrete distributions, $P = \{P^{(1)}, ..., P^{(N)})\}$, with $P^{(k)} = \{(w_i^{(k)}, x_i^{(k)}) : i = 1, ..., m_k\}$ and $k = 1, ..., N$. Therefore, the associated barycenter, denoted with $\bar{P} = \{(\bar{w}_1, x_1), ..., (\bar{w}_m, x_m)\}$, is computed as follows in Eq. (6):

$$\arg\min_{P} \frac{1}{N} \sum_{k=1}^{N} \lambda_k W(P, P^{(k)}) \tag{6}$$

where the values λ_k are used to weigh the different contributions of each distribution in the computation. Without the loss of generality, they can be set to $\lambda_k = \frac{1}{N} \forall k = 1, ..., N$.

3   Wasserstein Enabled Leak Localization

3.1   Generation of Leak Scenarios

The simulation runs are performed by placing, in turn, a leak on each pipe according to EPANET specifications and varying its severity in a given range. At the end of each leakage simulation EPANET outputs pressure and flows value at each junction and pipe respectively. Only the values in correspondence of the position of monitoring devices are taken into account. The pressure and flow variations due to each simulated leak are compared to the corresponding values obtained by simulating the faultless network. Each simulated leak is stored in a dataset and represented by the pressure and flow variations (features) together with the information related to the affected pipe and the damage severity. Each monitoring point (sensor) can be represented by a matrix: rows correspond to time steps of the simulation and columns to different leakage scenarios. Let assume that there are N sensors of which n_p pressure sensors and n_f flow sensors. For each sensor (feature) and each scenario the result is a timeseries. Figure 2 displays flows and pressures values registered at sensor f_9 and sensor $p_1 1$ respectively. The color depends on the flow/pressure value with respect to the faultless network.

The EPANET simulation is performed for the 24 h horizon over 10 min intervals generating 144 observations for each monitoring point. Each entry of the matrix represents the pressure or flow (depending on the type of sensor) registered at a specific time in a specific leakage scenario. The distribution of pressures (or flows) for a leakage scenario registered in a specific monitoring point (i.e., a column of the matrices shown in Fig. 2) can be represented as a histogram: the support is given by the pressures (or flows) range (or rather the difference in pressure range with respect to the faultless network) divided into η bins and the weights are given by the number of elements falling in that bin. Figure 3 displays

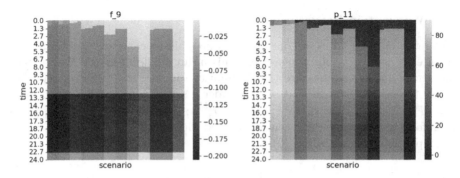

Fig. 2. Example of the matrices associated to flow sensor f_9 and pressure sensor $p_1 1$.

the histograms associated to two leakage scenarios. The upper line is related to a leak in the pipe 121 with severity 0.1 ($\ell_{121}^{0.1}$), while the bottom line is related to a leak in the pipe 31 with severity 0.3 ($\ell_{31}^{0.3}$). Two monitoring points have been considered in the example, one for flow (f_9) and one for pressure (p_{11}). The distance between the two leaks is given by the average of the Wasserstein distance of the histograms over each monitoring point.

The strategy which developed in the present paper is to normalize the sample in the range and compute a histogram whose bins correspond to the sub-intervals in the range equi-division and weight given by the elements of the sample falling in that bin. For a leak scenario, the signature is now a set of N histograms, one for each sensor. There are several options to deal with these histograms:

1. Consider the representation of a leak as one histogram with a N-dimensional support (N-dimensional histogram) with each dimension corresponding to a sensor.
2. Consider the image/heatmap given by the $N \times N$ matrix whose entries are the "distance between two sensors".
3. Consider N one-dimensional histograms.

Under any of the above options the distance between two leaks upon which we build the clustering procedure is based on the WST distance. The computational cost of the three options is very different: from very large, almost prohibitive, for the first when the number of the dimensions of the support exceeds 5, to manageable, for the second where the support is 2-dimensional. The third has a smaller cost because the distributions involved are 1-dimensional. The distance between two different leaks (different in terms of affected pipe and/or leak severity) also allows to deal with the problem of grouping similar leakage scenarios as a graph clustering task. In this paper, the application of k-means in the WST space have been considered.

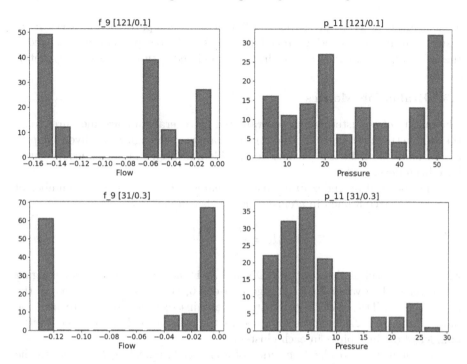

Fig. 3. Distributions of pressure (left) and flow (right) registered by sensor f_9 and p_{11} respectively. The upper line is related to the leakage scenario $\ell_{121}^{0.1}$, while the bottom line is related to the leakage scenario $\ell_{31}^{0.3}$.

3.2 Clustering in the Wasserstein Space

The concept of barycenter enables clustering among distributions in a space whose metric is the Wasserstein distance. More simply, the barycenter in a space of distributions is the analogous of the centroid in a Euclidean space. The most common and well-known algorithm for clustering data in the Euclidean space is k-means. Since it is an iterative distance-based (also known as representative-based) algorithm, it is easy to propose variants of k-means by simply changing the distance adopted to create clusters. The crucial point is that only the distance is changed, and the overall iterative two-step clustering algorithm is maintained. In the present paper the Wasserstein k-means is used, where the Euclidean distance is replaced by the Wasserstein distance and where centroids are replaced by barycenters of the distributions belonging to that cluster. As previously seen, each leakage scenario can be represented as a set of N histograms that represent flows and pressures distribution at the monitoring points. This enables the usage of a Wasserstein enabled k-means, in which the distance between two leakage is computed as the average Wasserstein distance over the N histograms associated to different sensors. This approach enables the usage of the entire distributions of pressures and flows detected during the simulation horizon instead of just considering the average values as in standard clustering approaches. To locate a leakage, the detected pressures and flows can be compared to the barycenters resulting from the clustering procedure. The set of pipes potentially damaged is the set of pipes belonging to the clusters associated with the closest barycenter.

3.3   Evaluation Metrics

The quality of a clustering can be evaluated by several standard measures (e.g., the Silhouette score, the Dunn index) which are problem agnostic. Given the particular features of the leak localization problem a specific performance measures have been developed.

The Localization Index of a cluster k, namely LI_k, refers to the number of different pipes contained in a cluster, and it is computed as:

$$LI_k = \frac{|\mathcal{P}| - |\mathcal{P}_k|}{|\mathcal{P}| - 1} \tag{7}$$

where $\mathcal{P}$ is the set of pipes of the WDN and $\mathcal{P}_k$ is the set of leaky pipes belonging to cluster k. The values of LI_k range between 0, i.e., cluster k contains all the pipes of the WDN, and 1, i.e., cluster k contains scenarios related to just one pipe. The overall LI index is obtained as the average of LI_k weighted by the number of distinct pipes in each cluster.

The Quality of Localization index of a cluster k, namely QL_k, refers to the number of scenarios related to the same (leaky) pipe (with different severities) contained in a cluster, and it is computed as:

$$QL_k = \frac{\sum_{p \in \mathcal{P}_k} \frac{n_p^k}{|\mathcal{S}|}}{|\mathcal{P}_k|} \tag{8}$$

where $\mathcal{S}$ is the set of different severity values used in the simulations and n_p^k is the number of scenarios in cluster k associated with pipe p. The values of QL_k range between 0 and 1, where 1 means that the cluster k contains all the scenarios related to the pipes in the $\mathcal{P}_k$ set. The overall QL is given by the average of QL_k. Finally, an overall index for the clustering procedure, namely QLI, can be obtained combining LI and QL as:

$$QLI = LI \times QL \tag{9}$$

4   Experimental Results

4.1   Data Resources

Three different networks (Fig. 4) have been used to test the proposed algorithm. Hanoi [22] and Anytown [4] are two benchmarks used in the literature. Hanoi is composed of 31 junctions, 1 reservoir and 34 pipes, while Any-town has 22 junctions, 1 reservoir, 2 tanks, 43 pipes and 3 pumps. Neptun [3] is the WDN of Timisoara, Romania, more specifically it is a district metered area of a large

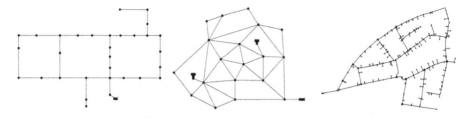

Fig. 4. The three WDN used: Hanoi (left), Anytown (center) and Neptun (right).

WDN, and it was a pilot area of the European project ICeWater. Neptun is composed of 332 junctions, 1 reservoir, 312 pipes and 27 valves.

For each of these networks five different leaks have been simulated with different severity values (ranging from 0.1 to 0.3 with a step of 0.05). A total of 170 scenarios have been simulated for Hanoi, 215 scenarios for Anytown and 1560 scenarios for Neptun. In addition, 4 sensors ($n_p = 2$, $n_f = 2$) have been considered for the networks of Hanoi and Anytown and 6 sensors ($n_p = 3$, $n_f = 3$) for the network of Neptun.

4.2   Computational Results

The proposed clustering procedure has been compared with the standard k-means. Figure 5 shows the quality of the resulting cluster in terms of QLI for different number of clusters on the three WDN considered. In the case of Hanoi and Anytown, the Wasserstein enabled clustering shows slightly better performance, in particular for smaller number of clusters. In the case of Neptun the quality of the resulting clusters by the two algorithms is comparable for $k < 30$, while k-means has slightly better performance for $k \geq 30$.

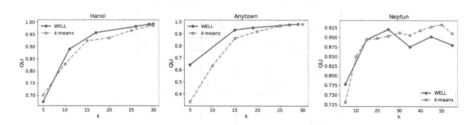

Fig. 5. QLI index over different number of clusters considering the standard k-means and the Wasserstein enabled k-means.

The advantage of WELL is particularly marked when flows and pressures have high variance over the simulation horizon. Indeed, WELL is able to consider the entire distribution of values instead of just the daily average as k-means. This is often the case of real word WDN in which WELL can offer significant improvements over standard clustering procedure.

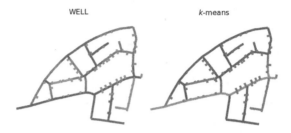

Fig. 6. On the left, the resulting clusters considering Wasserstein enabled clustering and k-means with 5 clusters. Different colors represent different cluster. On the right, barycenters obtained by the clustering procedure in the Wasserstein space on Neptun.

Furthermore, the resulting barycenters of WELL offer a signature of the leakage scenarios belonging to each cluster. This should help explain the localization of a scenario in a specific portion of the network by giving the typical behavior of different clusters. Figure 6 shows an example of the clustering results while Fig. 7 shows an example of the barycenter in the case of Neptun with 5 clusters. For example, each histogram on the first row (c_0) is obtained as the Fréchet mean (i.e., barycenter) of the histograms associated to the leak scenarios belonging to the cluster c_0.

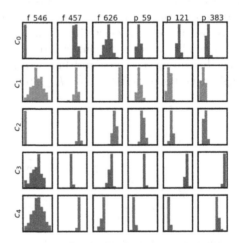

Fig. 7. Barycenters obtained by the clustering procedure in the Wasserstein space on Neptun.

Finally, to analyze the predictive capabilities of WELL, a test set has been built with the same procedure described in Sect. 3.1 but varying the severity values. Table 1 shows the resulting accuracy in the test set.

Table 1. Computational results in terms of prediction accuracy for the two algorithms and the three WDN.

Network	k	k-means	WELL					
			n = 5	n = 15	n = 25	n = 35	n = 45	n = 55
Hanoi	5	**1.000**	0.941	0.971	**1.000**	0.971	0.971	**1.000**
	15	**1.000**		0.941	0.941	0.971	0.941	0.971
	25	**1.000**			0.971	**1.000**	**1.000**	**1.000**
	35	**1.000**				0.971	0.971	0.941
	45	**1.000**					0.941	0.941
	55	**1.000**						0.941
Anytown	5	**1.000**	0.791	0.907	**1.000**	**1.000**	**1.000**	**1.000**
	15	**1.000**		0.488	0.581	0.837	0.721	0.884
	25	**1.000**			**1.000**	0.860	**1.000**	**1.000**
	35	**1.000**				0.860	0.860	**1.000**
	45	**1.000**					**1.000**	**1.000**
	55	**1.000**						**1.000**
Neptun	5	**0.996**	0.877	0.972	0.985	0.991	0.984	0.986
	15	**0.993**		0.954	0.973	0.980	0.976	0.989
	25	**0.981**			0.917	0.929	0.929	0.933
	35	0.940				0.950	0.959	**0.969**
	45	0.944					0.946	**0.954**
	55	**0.939**						0.928

5   Conclusions, Limitations, and Perspectives

The approach based on the distributional representation of the leak in the feature space is methodologically sound and offers a good computational performance. WELL can also work with a relatively smaller number of sensors than other methods. The WST distance offers a more efficient exploration than other methods.

A limitation can be related to the computational cost of the WST distance: in the 1-dimensional approximation used in this paper the cost is negligible. Using a multivariate representation, the computation can become prohibitive already for 5 sensors. The perspectives are theoretical: indeed, WST theory is extremely rich and new theoretical tools could further improve the performance. WELL seems to be a natural candidate when the WDN, as is still the case, has few sensors working properly. Another advantage of the distributional representation and the WST distance is that it matches naturally the perceptual notion of nearness and similarity. Moreover, the barycenter offers a useful synthesis of a set of distributions as shown in the discussion of Fig. 7.

References

1. Arifin, B., Li, Z., Shah, S.L., Meyer, G.A., Colin, A.: A novel data-driven leak detection and localization algorithm using the kantorovich distance. Comput. Chem. Eng. **108**, 300–313 (2018)
2. Candelieri, A., Conti, D., Archetti, F.: A graph based analysis of leak localization in urban water networks. Procedia Eng. **70**, 228–237 (2014)
3. Candelieri, A., Soldi, D., Archetti, F.: Cost-effective sensors placement and leak localization-the neptun pilot of the icewater project. J. Water Supply Res. Technol. AQUA **64**(5), 567–582 (2015)
4. Farmani, R., Walters, G.A., Savic, D.A.: Trade-off between total cost and reliability for anytown water distribution network. J. Water Resour. Plan. Manag. **131**(3), 161–171 (2005)
5. Garðarsson, G.Ö., Boem, F., Toni, L.: Graph-based learning for leak detection and localisation in water distribution networks. IFAC-PapersOnLine **55**(6), 661–666 (2022)
6. Hajgató, G., Gyires-Tóth, B., Paál, G.: Reconstructing nodal pressures in water distribution systems with graph neural networks. arXiv preprint arXiv:2104.13619 (2021)
7. Irofti, P., Romero-Ben, L., Stoican, F., Puig, V.: Data-driven leak localization in water distribution networks via dictionary learning and graph-based interpolation. In: 2022 IEEE Conference on Control Technology and Applications (CCTA), pp. 1265–1270. IEEE (2022)
8. Irofti, P., Stoican, F.: Dictionary learning strategies for sensor placement and leakage isolation in water networks. IFAC-PapersOnLine **50**(1), 1553–1558 (2017)
9. Kang, J., Park, Y.J., Lee, J., Wang, S.H., Eom, D.S.: Novel leakage detection by ensemble CNN-SVM and graph-based localization in water distribution systems. IEEE Trans. Industr. Electron. **65**(5), 4279–4289 (2017)
10. Kantorovich, L.V.: On the translocation of masses. In: Doklady Akademii Nauk USSR (NS), vol. 37, pp. 199–201 (1942)
11. Monge, G.: Mémoire sur la théorie des déblais et des remblais. Mem. Math. Phys. Acad. Royale Sci. 666–704 (1781)
12. Morales-González, I., Santos-Ruiz, I., López-Estrada, F.R., Puig, V.: Pressure sensor placement for leak localization using simulated annealing with hyperparameter optimization. In: 2021 5th International Conference on Control and Fault-Tolerant Systems (SysTol), pp. 205–210. IEEE (2021)
13. Peyré, G., Cuturi, M., et al.: Computational optimal transport: with applications to data science. Found. Trends® Mach. Learn. **11**(5–6), 355–607 (2019)
14. Ponti, A., Candelieri, A., Archetti, F.: A wasserstein distance based multiobjective evolutionary algorithm for the risk aware optimization of sensor placement. Intell. Syst. Appl. **10**, 200047 (2021)
15. Ponti, A., Candelieri, A., Giordani, I., Archetti, F.: Probabilistic measures of edge criticality in graphs: a study in water distribution networks. Appl. Netw. Sci. **6**(1), 1–17 (2021)
16. Romero, L., Blesa, J., Puig, V., Cembrano, G., Trapiello, C.: First results in leak localization in water distribution networks using graph-based clustering and deep learning. IFAC-PapersOnLine **53**(2), 16691–16696 (2020)
17. Romero-Ben, L., Alves, D., Blesa, J., Cembrano, G., Puig, V., Duviella, E.: Leak localization in water distribution networks using data-driven and model-based approaches. J. Water Resour. Plan. Manag. **148**(5), 04022016 (2022)

18. Santos-Ruiz, I., López-Estrada, F.R., Puig, V., Valencia-Palomo, G., Hernández, H.R.: Pressure sensor placement for leak localization in water distribution networks using information theory. Sensors **22**(2), 443 (2022)
19. Soldevila, A., Boracchi, G., Roveri, M., Tornil-Sin, S., Puig, V.: Leak detection and localization in water distribution networks by combining expert knowledge and data-driven models. Neural Comput. Appl. 1–21 (2022)
20. Soldevila, A., Fernandez-Canti, R.M., Blesa, J., Tornil-Sin, S., Puig, V.: Leak localization in water distribution networks using Bayesian classifiers. J. Process Control **55**, 1–9 (2017)
21. Solomon, J., Rustamov, R., Guibas, L., Butscher, A.: Wasserstein propagation for semi-supervised learning. In: International Conference on Machine Learning, pp. 306–314. PMLR (2014)
22. Vasan, A., Simonovic, S.P.: Optimization of water distribution network design using differential evolution. J. Water Resour. Plan. Manag. **136**(2), 279–287 (2010)
23. Villani, C., et al.: Optimal Transport: Old and New, vol. 338. Springer, Heidelberg (2009). https://doi.org/10.1007/978-3-540-71050-9
24. Wang, Z., Oates, T.: Imaging time-series to improve classification and imputation. arXiv preprint arXiv:1506.00327 (2015)

Fast and Robust Constrained Optimization via Evolutionary and Quadratic Programming

Konstantinos I. Chatzilygeroudis[(✉)] and Michael N. Vrahatis

Computational Intelligence Laboratory (CILab), Department of Mathematics,
University of Patras, 26110 Patras, Greece
costashatz@upatras.gr, vrahatis@math.upatras.gr

Abstract. Many efficient and effective approaches have been proposed
in the evolutionary computation literature for solving constrained optimization problems. Most of the approaches assume that both the objective function and the constraints are black-box functions, while a few of
them can take advantage of the gradient information. On the other hand,
when the gradient information is available, the most versatile approaches
are arguably the ones coming from the numerical optimization literature.
Perhaps the most popular methods in this field are sequential quadratic
programming and interior point. Despite their success, those methods
require accurate gradients and usually require a well-shaped initialization
to work as expected. In the paper at hand, a novel hybrid method, named
UPSO-QP, is presented that is based on particle swarm optimization and
borrows ideas from the numerical optimization literature and sequential
quadratic programming approaches. The proposed method is evaluated
on numerous constrained optimization tasks from simple low dimensional
problems to high dimensional realistic trajectory optimization scenarios,
and showcase that is able to outperform other evolutionary algorithms
both in terms of convergence speed as well as performance, while also
being robust to noisy gradients and bad initialization.

Keywords: Constrained Optimization · Particle Swarm
Optimization · Quadratic Programming

1 Introduction and Related Work

Constraint Optimization Problems (COP) appear in many diverse research fields
and applications, including, among others, structural optimization, engineering
design, VLSI design, economics, allocation and location problems, robotics and
optimal control problems [9,11,17,28]. All these real-world problems are typically represented by a mathematical model that can contain both binary and

This work was supported by the Hellenic Foundation for Research and Innovation
(H.F.R.I.) under the "3rd Call for H.F.R.I. Research Projects to support Post-Doctoral
Researchers" (Project Acronym: NOSALRO, Project Number: 7541).

continuous variables, and a set of linear and non-linear constraints, ranging from simple, low dimensional numerical functions to high-dimensional noisy estimates.

The *Numerical Optimization* (NumOpt) literature [11,17] has given us a wide range of powerful tools to tackle those problems. Methods such as *Feasible Direction* (FD) [2], *Generalized Gradient Descent* (GGD) [18], *Interior Point* (IP) [27] and *Sequential Quadratic Programming* (SQP) [17] are able to solve very effectively COP problems even in high dimensions as well as to tackle problems for cases where the objective function or any of the constraints are non-convex. Despite their success, those methods require the availability of the exact gradient values of both the objective and constraints functions, which can be difficult to have in several real-world situations where only approximations are available. Moreover, it is well known that in practice those methods require good initialization to achieve reasonable convergence rates. On the other hand, many *Evolutionary Algorithms* (EAs) have been proposed for solving COP problems [4–7,12,21,24] and have been applied with success to many COP problems, even in cases where the corresponding function values are corrupted by noise. Most of the methods do not require the knowledge of the gradients [12,21], while some of them attempt to improve convergence or performance by using the gradient information [5]. Nevertheless, EAs are known to need many functions evaluations to be able to find high performing solutions, and can often fail to find the global optimum.

In the paper at hand, we take inspiration both from the numerical optimization and the evolutionary computation literature and propose UPSO-QP, a novel approach for solving COPs that attempts to merge the two fields. In particular, UPSO-QP is based on *Particle Swarm Optimization* (PSO) [20,22] and borrows some ideas from SQP. We extensively evaluate UPSO-QP in many scenarios ranging from low-dimensional noiseless settings to high-dimensional non-convex problems, and showcase that UPSO-QP is able to outperform other evolutionary algorithms both in terms of convergence speed as well as performance, while also being robust to noisy gradients and bad initialization.

The rest of the paper is organized as follows. In Sect. 2 the problem formulation and a brief presentation of the required background material are presented. In Sect. 3 a detailed description of the proposed method is provided, while in Sect. 4 experimental results are presented. The paper ends in Sect. 5 with some concluding remarks.

2   Problem Formulation and Background Material

We aim at solving the following problem:

$$\operatorname*{argmin}_{x \in \mathbb{R}^N} f(x),$$
$$\text{s.t.} \quad h_i(x) = 0,$$
$$g_j(x) \geqslant 0, \tag{1}$$

where $x \in \mathbb{R}^N$ is the optimization variable, $f : \mathbb{R}^N \to \mathbb{R}$ is the objective function, $h_i : \mathbb{R}^N \to \mathbb{R}$ with $i = 1, 2, \ldots, N_{\text{eq}}$ are the equality constraints, and $g_j : \mathbb{R}^N \to \mathbb{R}$ with $j = 1, 2, \ldots, N_{\text{ineq}}$ are the inequality constraints.

2.1 Particle Swarm Optimization

Particle Swarm Optimizers (PSO) [20,22] are evolutionary algorithms that are inspired by the aggregating behaviors of populations. PSO algorithms consist of a swarm of particles that are randomly positioned in the search space and communicate with their neighbors. Each particle performs objective function evaluations and updates its position as a function of previous evaluations of its neighborhood and the whole swarm. The two main strategies are *local PSO* (PSO-LS) and *global PSO* (PSO-GS). In the local strategy, each particle has memory of the its best position ever visited, and the single best position ever visited by its neighbors. In the global strategy, particles maintain memory of their personal best position and the best position ever found by the whole swarm.

More formally, a predetermined set of M particles are initialized at random positions in the search space. The position of particle q at initialization (first iteration, $k = 0$) is denoted by $x_q(0) \in \mathbb{R}^N$. Additionally, each particle is provided with a random velocity vector denoted $v_q(0) \in \mathbb{R}^N$. At each iteration $k+1$, particles update their positions with the following equations:

$$x_q(k+1) = x_q(k) + v_q(k+1). \tag{2}$$

In the *local PSO strategy*, the velocity is given by $l_q(k+1) = \chi \left[v_q(k) + c_1 r_1 \left(x_q^b(k) - x_q(k) \right) + c_2 r_2 \left(x_q^{lb}(k) - x_q(k) \right) \right]$, while in the *global PSO strategy* $g_q(k+1) = \chi \left[v_q(k) + c_1 r_1 \left(x_q^b(k) - x_q(k) \right) + c_2 r_2 \left(x^b(k) - x_q(k) \right) \right]$. $x_q^b(k)$ denotes the best position visited by particle i from initialization and up to time k, and similarly, $x_q^{lb}(k)$ and $x^b(k)$ denote the best position ever found by the neighbors of i and the whole swarm respectively. Scalars c_1, c_2 are user defined parameters and $r_1, r_2 \in [0,1]$ are randomly generated numbers, while χ is a user defined parameter similar to what learning rate is in gradient descent. The *Unified Particle Swarm Optimizer* (UPSO) [21] is an algorithm that combines the behaviors of the local and global PSO strategies:

$$v_q(k+1) = u\, g_q(k+1) + (1-u)\, l_q(k+1), \tag{3}$$

where $u \in [0,1]$ is a user defined parameter. Notice that if $u = 0$, UPSO coincides with the local strategy, PSO-LS, whereas, if $u = 1$, UPSO coincides with PSO-GS. In that sense, UPSO gives the "best of both worlds" by allowing the user to achieve superior performance with the fine-tuning of a single parameter.

2.2 Sequential Linear Quadratic Programming

The main intuition of *Sequential Linear Quadratic Programming* (SLQP) is to tackle the problem of Eq. (1) by splitting it into easier subproblems that are iteratively solved. In particular, the problem at each iteration is split into to two phases: a) the *Linear Programming* (LP) phase, and b) the *Equality Quadratic Programming* (EQP) phase. Before delving more into the details of SLQP, we can first see that the Lagrangian of Eq. (1) is as follows:

$$\mathcal{L}(\boldsymbol{x}, \boldsymbol{\lambda}, \boldsymbol{\mu}) = f(\boldsymbol{x}) - \sum_i \lambda_i h_i(\boldsymbol{x}) - \sum_j \mu_j g_j(\boldsymbol{x})$$

$$= f(\boldsymbol{x}) - \boldsymbol{\lambda}^\top \boldsymbol{h}(\boldsymbol{x}) - \boldsymbol{\mu}^\top \boldsymbol{g}(\boldsymbol{x}), \tag{4}$$

where $\boldsymbol{h}(\cdot)$ and $\boldsymbol{g}(\cdot)$ are the stacked versions of the constraints.

In the first phase of SLQP, the problem is linearized around the current estimate $\boldsymbol{x}_k$ and an LP is formulated as follows:

$$\operatorname*{argmin}_{\boldsymbol{p} \in \mathbb{R}^N} f(\boldsymbol{x}_k) + \nabla f(\boldsymbol{x}_k)^\top \boldsymbol{p},$$

$$\text{s.t.} \quad h_i(\boldsymbol{x}_k) + \nabla h_i(\boldsymbol{x}_k)^\top \boldsymbol{p} = 0,$$

$$g_j(\boldsymbol{x}_k) + \nabla g_j(\boldsymbol{x}_k)^\top \boldsymbol{p} \geqslant 0,$$

$$\|\boldsymbol{p}\|_\infty \leqslant \Delta_k^{\mathrm{LP}}, \tag{5}$$

where $f(\boldsymbol{x}_k)$ can be omitted from the optimization since it is constant, the solution of the problem is defined as $\boldsymbol{x}_k^{\mathrm{LP}} = \boldsymbol{x}_k + \boldsymbol{p}^{\mathrm{LP}}$, and Δ_k^{LP} is a trust-region radius in order to make the problem bounded.

Once the above problem is solved, we define the *Active Sets*, $\mathcal{A}_k^{\mathrm{eq}}$ and $\mathcal{A}_k^{\mathrm{ineq}}$, and the *Violating Sets*, $\mathcal{V}_k^{\mathrm{eq}}$ and $\mathcal{V}_k^{\mathrm{ineq}}$ to be the sets where the constraints are equal to zero and where the constraints are violated respectively.

In the second phase of SLQP, we define the following EQP problem:

$$\operatorname*{argmin}_{\boldsymbol{p} \in \mathbb{R}^N} f(\boldsymbol{x}_k) + \frac{1}{2}\boldsymbol{p}^\top \nabla_{\boldsymbol{x}\boldsymbol{x}}^2 \mathcal{L}_k \, \boldsymbol{p} + \left(\nabla f(\boldsymbol{x}_k) + \alpha_k \sum_{i \in \mathcal{V}_k^{\mathrm{eq}}} \gamma_i \nabla h_i(\boldsymbol{x}_k)\right.$$

$$\left. + \alpha_k \sum_{j \in \mathcal{V}_k^{\mathrm{ineq}}} \gamma_j \nabla g_j(\boldsymbol{x}_k)\right)^\top \boldsymbol{p},$$

$$\text{s.t.} \quad h_i(\boldsymbol{x}_k) + \nabla h_i(\boldsymbol{x}_k)^\top \boldsymbol{p} = 0, i \in \mathcal{A}_k^{\mathrm{eq}},$$

$$g_j(\boldsymbol{x}_k) + \nabla g_j(\boldsymbol{x}_k)^\top \boldsymbol{p} = 0, j \in \mathcal{A}_k^{\mathrm{ineq}},$$

$$\|\boldsymbol{p}\|_2 \leqslant \Delta_k^{\mathrm{EQP}}, \tag{6}$$

where $\nabla_{\boldsymbol{x}\boldsymbol{x}}^2 \mathcal{L}_k$ is the Hessian of the Lagrangian over the optimization variables $\boldsymbol{x}$ evaluated at the current estimate $(\boldsymbol{x}_k, \boldsymbol{\lambda}_k, \boldsymbol{\mu}_k)$, γ_i, γ_j are the algebraic signs of the i-th or j-th violated constraint, α_k is a penalty factor, and Δ_k^{EQP} is a trust-region radius in order to make the problem bounded. Practical implementations include line search, techniques for updating the penalty factors, estimating the Hessian instead of computing it, and trust-region radii as well as introducing slack variables to make the sub-problems always feasible (linearization can yield infeasible problems). For more details, we refer the interested reader to [17] and the references therein.

3   The Proposed UPSO-QP Approach

In the paper at hand, we combine the *Unified PSO* (UPSO) with *Sequential Linear Quadratic Programming* (SLQP). The intuition lies in the fact that SLQP is among the "strongest" nonlinear optimizers in the literature and practical applications, while UPSO is effective in black-box settings including constrained optimization [19,21]. The goal of our approach is to "fuse" the robustness and ease of usage of PSO methods with the convergence properties of SLQP methods. To this end, we propose a new hybrid algorithm, called UPSO-QP, that is based on UPSO, but also borrows ideas from SLQP. UPSO-QP follows the general UPSO framework, but we make some alternations to greatly improve its convergence when gradient (possibly imprecise or noisy) information is available.

3.1   Local QP Problems

First, we add a procedure to take advantage of gradient information of the objective and constraint functions. In particular, each particle q with probability $r_{\mathrm{qp}} \in [0,1]$ will solve the following QP problem:

$$\operatorname*{argmin}_{p \in \mathbb{R}^N} \frac{1}{2} p^\top p + \nabla f\big(x_q(k)\big)^\top p,$$

$$\text{s.t.} \quad h_i\big(x_q(k)\big) + \nabla h_i\big(x_q(k)\big)^\top p = 0,$$

$$g_j\big(x_q(k)\big) + \nabla g_j\big(x_q(k)\big)^\top p \geqslant 0,$$

$$\|p\|_\infty \leqslant v_{\max}, \tag{7}$$

where $v_{\max}$ is the maximum allowed velocity for each particle. This problem is inspired by the LP phase of SLQP (and in general by the SQP literature) with the added quadratic cost. This problem, similar to Eq. (5), can be infeasible because of the linearization. Instead of adding slack variables to ensure the feasibility of the problem (or other similar "tricks" from the numerical optimization literature), we take a practical approach, give the QP solver a fixed iteration budget and take the solution it has achieved so far even if infeasible. If the problem is infeasible, most QP solvers will converge to the least squares solution of the problem. So we expect to get a least squares approximation if the linearization yields infeasibility. In any case, we assume the solution returned by the QP problem to be $v_q^{\mathrm{qp}}(k+1)$, while we denote the update from UPSO as $v_q^{\mathrm{pso}}(k+1)$. The final velocity for each particle q is computed as:

$$v_q(k+1) = \alpha_{\mathrm{qp}}\, v_q^{\mathrm{qp}}(k+1) + \big(1 - \alpha_{\mathrm{qp}}\big)\, v_q^{\mathrm{pso}}(k+1), \tag{8}$$

for a user defined parameter $\alpha_{\mathrm{qp}} \in [0,1]$. In this paper, we use the ProxQP solver [1] to solve the QP problems.

3.2   UPSO for Constrained Optimization

Apart from moving into the "right" direction, we also need a method for comparing particles. This is important since the "best" particle (either in the neighborhood or globally) is crucial for UPSO's performance. We follow [19] and we augment the objective function with a penalty function:

$$\tilde{f}(\boldsymbol{x}) = f(\boldsymbol{x}) + H(\boldsymbol{x}), \tag{9}$$

where

$$
\begin{aligned}
H(\boldsymbol{x}) &= h(k)\, P(\boldsymbol{x}), \\
P(\boldsymbol{x}) &= \sum_{i} \theta\big(\mathrm{cv}_i(\boldsymbol{x})\big)\, \mathrm{cv}_i(\boldsymbol{x})^{\gamma(\mathrm{cv}_i(\boldsymbol{x}))} + \sum_{j} \theta\big(\mathrm{cv}_j(\boldsymbol{x})\big)\, \mathrm{cv}_j(\boldsymbol{x})^{\gamma(\mathrm{cv}_j(\boldsymbol{x}))}, \\
\mathrm{cv}_i(\boldsymbol{x}) &= \big|h_i(\boldsymbol{x})\big|, \\
\mathrm{cv}_j(\boldsymbol{x}) &= \big|\min\{0,\, g_j(\boldsymbol{x})\}\big|, \\
h(k) &= k\sqrt{k}.
\end{aligned}
\tag{10}
$$

We use the same functions $\theta(\cdot)$ and $\gamma(\cdot)$ as in [19].

3.3   Considerations

The main idea is that solving the problem in Eq. (5) will push each particle to follow the local linearized approximation of the original problem. This approximation is very effective close to the actual solution, while it can be bad in far away regions. The main intuition behind this merging of UPSO with SLQP is that this local approximation will generally move the particles closer to the solution, while UPSO can compensate for inaccuracies of those approximations. Moreover, this problem is solved individually by each particle and thus solved in many different locations of the search space simultaneously. In this way, we increase the probability that one of the initial conditions will be in a good region to enable convergence to the global solution. Moreoever, we get an *implicit averaging* [13,25] effect that helps UPSO "see through" the noise and inaccuracies and converge to a better optimum. Additionally, the constraint $\|\boldsymbol{p}\|_{\infty} \leqslant v_{\max}$ in Eq. (5) ensures that we stay in regions where the local linearization is expected to be true, and thus produce well behaved search velocities.

In smooth and well-behaved objective and constraint functions, the solution of Eq. (5) will provide strong directions towards the optimal solution, and thus accelerating the convergence of UPSO. In noisy, discontinuous and/or non-convex problems, the solution of Eq. (5) will at least provide an approximate direction towards minimizing the constraint violation (least squares solution) that can help UPSO converge faster.

The parameter r_{qp} is used to handle the trade-off between effectiveness and wall time performance. The bigger the value more particles will solve the QP and thus we get better approximations of the search landscape. At the same time, this means that we solve more QP problems that can increase the wall time significantly. On the other hand, the parameter α_{qp} is used to specify how much we want to trust the solution of the QP problem. In smooth and noiseless functions, we should set $\alpha_{qp} \approx 1$ since the solution of the QP will most likely provide a good search direction. On the contrary, in noisy or non-convex problems we should decrease this value, as the QP estimate can be less accurate and even misleading. Overall, one can change the behavior of the solver by setting the appropriate values to these parameters.

4   Experiments

We extensively evaluate the effectiveness of UPSO-QP with multiple experiments and comparing to strong baselines. We aim at answering the following questions:

a) How does UPSO-QP perform in well-defined numerical constrained optimization problems? How does it compare to other evolutionary algorithms? How does it compare to state of the art SQP and IPM methods? We will answer those questions in Sect. 4.1.
b) How does UPSO-QP handle problems with noisy values and gradients? How does it compare in this domain to state of the art SQP and IPM methods? We will answer those questions in Sect. 4.2.
c) How does UPSO-QP operate on realistic high-dimensional constrained optimization problems? How sensitive it is in well-shaped initialization? How does it compare to state of the art SQP and IPM methods? We will answer those questions in Sect. 4.3.

In the subsequent sections, we compare the following algorithms:

1) UPSO-QP — custom implementation in C++ of our approach[1].
2) UPSO augmented with a penalty function for constrained optimization as in [19] (UPSO-Pen)—we use our own custom C++ implementation.
3) UPSO augmented with a penalty function and gradient-based repair technique as in [5] (UPSO-Grad) — we use our own custom C++ implementation.
4) Sequential Least Squares Programming (SLSQP) — this is an SQP approach as implemented in [14] and it is closely related to SLQP as described above[2].
5) A Primal-Dual Interior Point Algorithm as described in [27] (Ipopt) — this is a state-of-the-art IPM method widely used in practice[3].

We have carefully chosen the algorithms to compare UPSO-QP to in order to be able to highlight the main properties of our proposed method. In particular, UPSO-Pen does not have access to any gradient information and the penalty function technique is one of the most widely used in the evolutionary computation community. UPSO-Grad is an evolutionary method that takes advantage

---

[1] The code is available at https://github.com/NOSALRO/algevo.
[2] We use the implementation provided by scipy.
[3] We use the C++ implementation provided by the Ipopt library.

of the gradient information of the constraints in order to improve performance and has been shown to have superior performance over other evolutionary techniques [3]. Lastly, SLSQP and Ipopt are two of the most versatile and widely used numerical optimization algorithms.

4.1 Numerical Constrained Optimization Problems

In the first set of experiments, we select two low-dimensional numerical constrained optimization problems with known optimal solutions. This will give us the ability to extensively test and compare UPSO-QP with other methods, both from the evolutionary computation literature as well as the numerical optimization one. For each problem/algorithm pair we run 20 replicates with different initial conditions. For all the evolutionary algorithms, we used $M = 40$ particles, with 10 neighborhoods, $\chi = 0.729, c_1 = c_2 = 2.05$ and $u = 0.5$. For UPSO-QP, we also set $r_{qp} = 0.5$ and $\alpha_{qp} = 1$.

Problem 1. For $f : \mathbb{R}^2 \rightarrow \mathbb{R}$ [11],

$$f(\boldsymbol{x}) = (x_1 - 2)^2 + (x_2 - 1)^2,$$
$$h_1(\boldsymbol{x}) = x_1 - 2x_2 + 1 = 0,$$
$$g_1(\boldsymbol{x}) = -0.25 x_1^2 - x_2^2 + 1 \geqslant 0.$$

The best known optimal feasible solution is $f(\boldsymbol{x}^*) = 1.3934651$.

Problem 2. For $f : \mathbb{R}^6 \rightarrow \mathbb{R}$ [15],

$$f(\boldsymbol{x}) = -10.5 x_1 - 7.5 x_2 - 3.5 x_3 - 2.5 x_4 - 1.5 x_5 - 10 x_6 - 0.5 \sum_{i=1}^{5} x_i^2,$$

$$g_i(\boldsymbol{x}) = x_i \geqslant 0, \ i = 1, 2, \ldots, 5,$$
$$g_{i+5}(\boldsymbol{x}) = 1 - x_i \geqslant 0, \ i = 1, 2, \ldots, 5,$$
$$g_{11}(\boldsymbol{x}) = x_6 \geqslant 0,$$
$$g_{12}(\boldsymbol{x}) = 6.5 - 6 x_1 - 3 x_2 - 3 x_3 - 2 x_4 - x_5 \geqslant 0,$$
$$g_{13}(\boldsymbol{x}) = 20 - 10 x_1 - 10 x_3 - x_6 \geqslant 0.$$

The best known optimal feasible solution is $f(\boldsymbol{x}^*) = -213$.

Results. The results showcase that UPSO-QP outperforms all other evolutionary baselines and always converges to the optimal solution faster (*cf.* Fig. 1, 2). Moreover compared to SLSQP and IPopt, UPSO-QP is able to achieve the same level of accuracy while also having comparable total wall time measurements.

4.2 Constrained Optimization with Noisy Functions Values

In this section, we will solve the same problems as above but we will add noise in different ways to showcase the robustness of UPSO-QP.

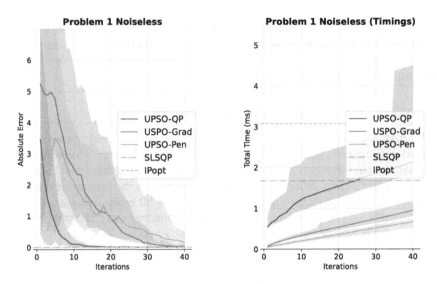

Fig. 1. Results for Problem 1. Solid lines are the median over 20 replicates and the shaded regions are the regions between the 5-th and 95-th percentiles.

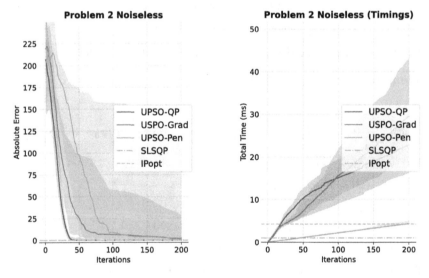

Fig. 2. Results for Problem 2. Solid lines are the median over 20 replicates and the shaded regions are the regions between the 5-th and 95-th percentiles.

Impact of Noise. The impact of imprecise information with respect to the values of the objective or constraint function can be studied and analyzed by simulating the imprecise values using, for instance, the following approach [20, 22]. Information about the function values is obtained in the form of $f^\eta(x)$ which determines an approximation to the true function value of the objective function $f(\boldsymbol{x})$, contaminated by a small amount of noise η. To this end, the function values are obtained, for the case of the additive noise, as follows [8, p.40]: $f^{\eta}(\boldsymbol{x}) = f(\boldsymbol{x}) + \eta$. For the case of the multiplicative noise, the function values are obtained as follows: $f^{\eta}(\boldsymbol{x}) = f(\boldsymbol{x})(1 + \eta)$, where η is a Gaussian noise term with zero mean and standard deviation σ, $\eta \sim \mathcal{N}\left(0, \sigma^2\right)$, that determines the noise strength.

Experiments. In order to showcase the effectiveness of UPSO-QP, we run each algorithm/problem pair with different noise settings: 3 noise levels for additive noise and 3 noise levels for multiplicative noise. For each problem, we select different levels in order for the noise to have an effect in performance. We also inject noise in both the objective functions and all the constraint functions. For each distinct scenario we run 20 replicates with different initial conditions. The results showcase that UPSO-QP is clearly outperforming all the other evolutionary algorithms and SLSQP (*cf.* Fig. 3, 4, 5, 6). Moreover, UPSO-QP performs as par with Ipopt in most cases and outperforms it in scenarios with big noise. We used the same hyper-parameters as in the previous section.

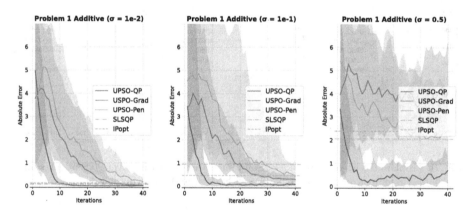

Fig. 3. Problem 1 with additive noise. Solid lines are the median over 20 replicates and the shaded regions are the regions between the 5-th and 95-th percentiles.

4.3   Evaluation on High Dimensional Problems

In this section, we will highlight the effectiveness of UPSO-QP in high dimensional and realistic examples. In particular, we will use two examples of the *Trajectory Optimization* (or Optimal Control) problem (TO) [10, 16, 26, 28]. This type of problems tend to be quite high dimensional while also having many constraints and being sensitive to good initialization. The simplest formulation of TO problems is defined as:

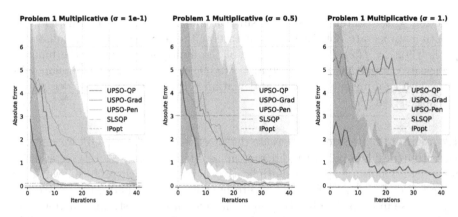

Fig. 4. Problem 1 with multiplicative noise. Solid lines are the median over 20 replicates and the shaded regions are the regions between the 5-th and 95-th percentiles.

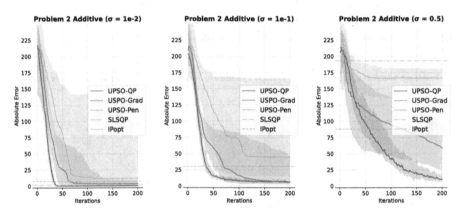

Fig. 5. Problem 2 with additive noise. Solid lines are the median over 20 replicates and the shaded regions are the regions between the 5-th and 95-th percentiles.

$$\underset{s_1,\ldots,s_L,u_1,\ldots,u_{L-1}}{\text{argmin}} \sum_{l}^{L-1} C(s_l, u_l) + C_{\text{final}}(s_L),$$

$$\text{s.t.} \quad \text{Dyn}(s_l, u_l, s_{l+1}) = 0, \tag{11}$$

where s_l is the state at time l, u_l is the action taken at time l, $C(\cdot, \cdot)$, $C_{\text{final}}(\cdot)$ define the cost functions, while $\text{Dyn}(\cdot)$ defines the dynamics equations of the system. We can additionally add more constraints depending on the problem (*e.g.* bounds on the variables). In essence, the above formulation assumes that we discretize the continuous signal at L points and enforcement all the constraints only at those points. More advanced formulations assume piece-wise polynomials and enforce the constraints on arbitrary points [28,29]. Overall, the optimization searches for the states and actions that respect the dynamics equations and minimize the costs.

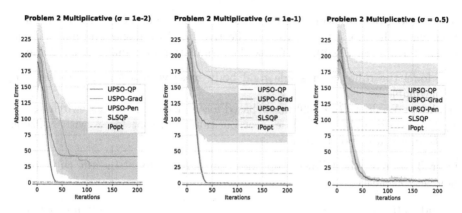

Fig. 6. Problem 2 with multiplicative noise. Solid lines are the median over 20 replicates and the shaded regions are the regions between the 5-th and 95-th percentiles.

Double Integrator. The first example that we will use is the *Double Integrator* (DI) system [23]. The DI's state is defined as $s = \{x, \dot{x}\} \in \mathbb{R}^2$, while the actions are defined as $u = \{\ddot{x}\} \in \mathbb{R}$. The dynamics equations of motion are given by:

$$s_{l+1} = \begin{bmatrix} 1 & dt \\ 0 & 1 \end{bmatrix} s_l + \begin{bmatrix} \frac{1}{2}dt^2 \\ dt \end{bmatrix} u_l, \qquad (12)$$

where dt is the time-step of integration in seconds. In this particular setup, the system starts at $s_1 = \{1, 0\}$ and has to reach at $s_L = \{0, 0\}$ while minimizing the magnitude of the actions taken. $C_{\text{final}}(s_l) = \frac{1}{2}s_l^\top s_l$, and $C(s_l, u_l) = C_{\text{final}}(s_l) + \frac{1}{2}0.1\,u_l^\top u_l$. We use $dt = 0.1$ and $L = 51$ steps. *The total dimensions of the optimization variables is 151 (i.e. $x \in \mathbb{R}^{151}$), while the total number of equality constraints are 102 (i.e. $h(\cdot) \in \mathbb{R}^{102}$).* We will use this example to compare UPSO-QP to other evolutionary methods. The results showcase that UPSO-QP is able to solve the problem and converge rapidly to the optimal solution, while the other evolutionary baselines struggle at finding a good solution (*cf.* Fig. 7). This is mainly because of the dimensionality of problem and we would need to perform an extensive hyper-parameter search to make them competitive. On the contrary, UPSO-QP is able to take advantage of the

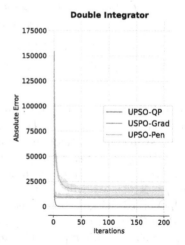

Fig. 7. Double Integrator Results. Solid lines are the median over 20 replicates and the shaded regions are the regions between the 5-th and 95-th percentiles.

local linearizations in the search space and converge quickly to the optimal value. We used the same hyper-parameters as in the previous section.

Monopod Locomotion. Here we take one example of TO for legged locomotion, where the task is to generate an effective gait for a monopod robot walking on flat terrain (*cf.* Fig. 8). We follow the formulation of Winkler *et al.* [29] and:

a) Model the robot as a single rigid body mass with a leg that its mass is negligible;
b) Adopt Winkler *et al.* [29] phase-based formulation for contact switching;
c) Parameterize the body pose, foot positions and foot forces with multiple Hermite cubic polynomials.

Overall, we have the following optimization problem (omitting the cubic polynomials for clarity):

$$
\begin{aligned}
\text{find} \quad & r(t),\ r : \mathbb{R} \to \mathbb{R}^3, && \text{(Body positions)} \\
& \theta(t),\ \theta : \mathbb{R} \to \mathbb{R}^3, && \text{(Body Euler angles)} \\
& p(t),\ p : \mathbb{R} \to \mathbb{R}^3, && \text{(Foot position)} \\
& f(t),\ f : \mathbb{R} \to \mathbb{R}^3, && \text{(Foot force)} \\
\text{s.t.} \quad & \mathrm{srbd}(r,\theta,p,f) = \{\ddot{r},\ddot{\theta}\}, && \text{(Dynamics)} \\
& \{r(0),\theta(0)\} = \{r_{\text{init}},\theta_{\text{init}}\}, && \text{(Initial State)} \\
& \{r(T),\theta(T)\} = \{r_{\text{goal}},\theta_{\text{goal}}\}, && \text{(Goal State)} \\
& p(t) \in \mathcal{B}\big(r(t),\theta(t)\big), && \text{(Bounds wrt body)} \\
& \dot{p}(t) = 0,\ \text{for } t \in \text{Contact}, && \text{(No slip)} \\
& p(t) \in \mathcal{T},\ \text{for } t \in \text{Contact}, && \text{(Contact on terrain)} \\
& f(t) \in \mathcal{F},\ \text{for } t \in \text{Contact}, && \text{(Pushing force/friction cone)} \\
& f(t) = 0,\ \text{for } t \notin \text{Contact}, && \text{(No force in air)}
\end{aligned}
\tag{13}
$$

In this particular setup, the monopod starts at pose $r(0) = \{0,0,0.5\}$, $\theta(0) = 0$, and has to reach $r(T) = \{1,0,0.5\}$, $\theta(T) = 0$ in $T = 2\,s$, while it is allowed for 3 swing phases (foot in the air). *The total dimensions of the optimization variables is 339 (i.e. $x \in \mathbb{R}^{339}$). The total number of equality constraints are 291 (i.e. $h(\cdot) \in \mathbb{R}^{291}$), and the total number of inequality constraints are 225 (i.e. $g(\cdot) \in \mathbb{R}^{225}$).* We will use this example to compare against Ipopt and evaluate whether UPSO-QP can be more robust to the initial solution guess. Here for UPSO-QP we used $M = 400$ particles, with 20 neighborhoods, $\chi = 0.729, c_1 = c_2 = 2.05$, $u = 0.5$, $r_{\text{qp}} = 0.005$ and $\alpha_{\text{qp}} = 1$.

The problem we are trying to solve in this section is highly non-linear, non-convex with many "bad" local optima that the optimization can be trapped around and not able to get away. For this reason and in order to test the robustness of the algorithms to the initial solution guess, we take a well-shaped initialization and add to each variable Gaussian noise $\eta \sim \mathcal{N}\left(0,\sigma^2\right)$. We vary σ from 0 to 1. This way we can have a meaningful comparison, while also getting reasonable convergence. We ran 10 replicates per experiment with different initialization parameters. The results showcase that both algorithms are able to find

Fig. 8. Monopod: an example solution using UPSO-QP. The shaded "ghost" robot is the target. The visualizations (1–7) are snapshots at time intervals. A video of the optimized behavior is available at https://youtu.be/ZnDs8wc96eM.

the optimal solution (*cf.* Fig. 8) 100% of the time up to perturbation of $\sigma = 0.7$. For $\sigma = 0.8$ and $\sigma = 0.9$, UPSO-QP is always able to find the optimal solution, while Ipopt struggles and does not find the solution even after 5000 iterations. For $\sigma = 1$, UPSO-QP rarely (1/10 runs) finds the optimal solution before 2000 iterations. The results verify that UPSO-QP keeps the effectiveness of numerical optimization methods, while being more robust to bad initialization.

5 Concluding Remarks

We have proposed UPSO-QP, a novel algorithm that effectively combines the evolutionary and numerical optimization literature, and solves general COPs. UPSO-QP is able to keep convergence rates/wall-time similar to the analytical methods, while being robust to noisy measurements and bad initialization similar to EAs. Overall, UPSO-QP is getting the "best of both of worlds". There needs to be more investigation in which problems/scenarios the effect of the linearization part of Eq. (8) is dominant, and in which ones the PSO part dominates.

References

1. Bambade, A., et al.: PROX-QP: yet another quadratic programming solver for robotics and beyond. In: RSS 2022-Robotics: Science and Systems (2022)
2. Beck, A., Hallak, N.: The regularized feasible directions method for nonconvex optimization. Oper. Res. Lett. **50**(5), 517–523 (2022)
3. Cantú, V.H., et al.: Constraint-handling techniques within differential evolution for solving process engineering problems. Appl. Soft Comput. **108**, 107442 (2021)
4. Chatzilygeroudis, K., Cully, A., Vassiliades, V., Mouret, J.-B.: Quality-diversity optimization: a novel branch of stochastic optimization. In: Pardalos, P.M., Rasska-zova, V., Vrahatis, M.N. (eds.) Black Box Optimization, Machine Learning, and No-Free Lunch Theorems. SOIA, vol. 170, pp. 109–135. Springer, Cham (2021). https://doi.org/10.1007/978-3-030-66515-9_4
5. Chootinan, P., et al.: Constraint handling in genetic algorithms using a gradient-based repair method. Comput. Oper. Res. **33**(8), 2263–2281 (2006)

6. D'Angelo, G., Palmieri, F.: GGA: a modified genetic algorithm with gradient-based local search for solving constrained optimization problems. Inf. Sci. **547**, 136–162 (2021)

7. Elsayed, S.M., Sarker, R.A., Mezura-Montes, E.: Particle swarm optimizer for constrained optimization. In: 2013 IEEE Congress on Evolutionary Computation, pp. 2703–2711. IEEE (2013)

8. Elster, C., Neumaier, A.: A method of trust region type for minimizing noisy functions. Computing **58**, 31–46 (1997)

9. Floudas, C.A., Pardalos, P.M.: A Collection of Test Problems for Constrained Global Optimization Algorithms, vol. 455. Springer, Heidelberg (1990). https://doi.org/10.1007/3-540-53032-0

10. Hargraves, C.R., Paris, S.W.: Direct trajectory optimization using nonlinear programming and collocation. J. Guid. Control. Dyn. **10**(4), 338–342 (1987)

11. Himmelblau, D.M., et al.: Applied Nonlinear Programming. McGraw-Hill, New York (2018)

12. Jain, H., Deb, K.: An evolutionary many-objective optimization algorithm using reference-point based nondominated sorting approach, part ii: handling constraints and extending to an adaptive approach. IEEE Trans. Evol. Comput. **18**(4), 602–622 (2013)

13. Jin, Y., Branke, J.: Evolutionary optimization in uncertain environments - a survey. IEEE Trans. Evol. Comput. **9**, 303–317 (2005)

14. Kraft, D.: A software package for sequential quadratic programming. German Research and Testing Institute for Aerospace (1988)

15. Michalewicz, Z.: Genetic Algorithms + Data Structures = Evolution Programs, 3rd edn. Springer, Heidelberg (1996). https://doi.org/10.1007/978-3-662-03315-9

16. Murray, D., Yakowitz, S.: Differential dynamic programming and newton's method for discrete optimal control problems. J. Optim. Theory Appl. **43**(3), 395–414 (1984)

17. Nocedal, J., Wright, S.: Numerical Optimization, 2nd edn., pp. 497–528. Springer, New York (2006). https://doi.org/10.1007/978-0-387-40065-5

18. Norkin, V.I.: Generalized gradients in dynamic optimization, optimal control, and machine learning problems. Cybern. Syst. Anal. **56**(2), 243–258 (2020). https://doi.org/10.1007/s10559-020-00240-x

19. Parsopoulos, K.E., Vrahatis, M.N.: Particle swarm optimization method for constrained optimization problems. In: Intelligent technologies-theory and application: New trends in intelligent technologies, vol. 76, pp. 214–220. IOS Press (2002)

20. Parsopoulos, K.E., Vrahatis, M.N.: Recent approaches to global optimization problems through particle swarm optimization. Nat. Comput. **1**, 235–306 (2002)

21. Parsopoulos, K.E., Vrahatis, M.N.: Unified particle swarm optimization for solving constrained engineering optimization problems. In: Wang, L., Chen, K., Ong, Y.S. (eds.) ICNC 2005. LNCS, vol. 3612, pp. 582–591. Springer, Heidelberg (2005). https://doi.org/10.1007/11539902_71

22. Parsopoulos, K.E., Vrahatis, M.N.: Particle Swarm Optimization and Intelligence: Advances and Applications. Information Science Publishing (2010)

23. Rao, V.G., Bernstein, D.S.: Naive control of the double integrator. IEEE Control Syst. Mag. **21**(5), 86–97 (2001)

24. Sun, Y., et al.: A particle swarm optimization algorithm based on an improved deb criterion for constrained optimization problems. PeerJ Comput. Sci. **8**, e1178 (2022)

25. Tsutsui, S., Ghosh, A.: Genetic algorithms with a robust solution searching scheme. IEEE Trans. Evol. Comput. **1**, 201–208 (1997)

26. Von Stryk, O., Bulirsch, R.: Direct and indirect methods for trajectory optimization. Ann. Oper. Res. **37**(1), 357–373 (1992)
27. Wächter, A., Biegler, L.T.: On the implementation of an interior-point filter line-search algorithm for large-scale nonlinear programming. Math. Program. **106**(1), 25–57 (2006)
28. Wensing, P.M., et al.: Optimization-based control for dynamic legged robots. arXiv:2211.11644 (2022)
29. Winkler, A.W., Bellicoso, C.D., Hutter, M., Buchli, J.: Gait and trajectory optimization for legged systems through phase-based end-effector parameterization. IEEE Robot. Autom. Lett. **3**(3), 1560–1567 (2018)

Bayesian Optimization for Function Compositions with Applications to Dynamic Pricing

Kunal Jain[1($\boxtimes$)], K. J. Prabuchandran[2], and Tejas Bodas[1]

[1] International Institute of Information Technology, Hyderabad, Hyderabad, India
kunal.jain@research.iiit.ac.in, tejas.bodas@iiit.ac.in
[2] Indian Institute of Technology, Dharwad, Dharwad, India
prabukj@iitdh.ac.in

Abstract. Bayesian Optimization (BO) is used to find the global optima of black box functions. In this work, we propose a practical BO method of function compositions where the form of the composition is known but the constituent functions are expensive to evaluate. By assuming an independent Gaussian process (GP) model for each of the constituent black-box function, we propose Expected Improvement (EI) and Upper Confidence Bound (UCB) based BO algorithms and demonstrate their ability to outperform not just vanilla BO but also the current state-of-art algorithms. We demonstrate a novel application of the proposed methods to dynamic pricing in revenue management when the underlying demand function is expensive to evaluate.

Keywords: Bayesian optimization · revenue maximization · function composition · dynamic pricing and learning

1 Introduction

Bayesian Optimization (BO) is a popular technique for optimizing expensive-to-evaluate black-box functions. Such a function might correspond to the case where evaluating it can take up to hours or days, which for example, is the case in re-training massive deep learning models with new hyper-parameters [29]. In some cases, functions can be financially costly to evaluate, such as drug testing [22] or revenue maximization [21]. In such black-box optimization problems, one often has a fixed budget on the total number of function evaluations that can be performed. For example, one typically has a budget on the computational capacity spent in the hyper-parameter tuning of neural networks [25]. In such cases, BO proves to be very effective in providing a resource-conserving iterative procedure to query the objective function and identify the global optima [24].

Prabuchandran K.J. was supported by the Science and Engineering Board (SERB), Department of Science and Technology, Government of India for the startup research grant 'SRG/2021/000048'.

The key idea in BO is to use a surrogate Gaussian Process (GP) model [23] for the black-box function, which is updated as and when multiple function evaluations are performed. To identify the location (in the domain) of the following query point, an acquisition function (a function of the surrogate model) is designed and optimized. The design of the acquisition function depends not only on the application in mind but also on the trade-off between exploration and exploitation that comes naturally with such sequential optimization problems. Some popular acquisition functions used in the literature are the Expected Improvement (EI), Probability of Improvement (PI), Knowledge Gradient (KG) and Upper Confidence Bound (UCB) [7,9,16].

In this work, we consider Bayesian optimization of composite functions of the form $g(x) = h(f_1(x), f_2(x), \ldots f_M(x))$ where functions f_i are expensive black-box functions while h is known and easy to compute. More specifically, we are interested in maximizing g and identifying its optima. A vanilla BO approach can be applied to this problem, ignoring the composite structure of the function [9, 16]. In this approach, one would build a GP posterior model over the function g based on previous evaluations of $g(x)$ and then select the next query point using a suitable acquisition function computed over the posterior. However, such a vanilla BO approach ignores the available information about the composite nature of the functions, which we show can easily be improved upon. In this work, we model each constituent function f_i using an independent GP model and build acquisition functions that use the known structure of the composition. Our algorithms outperform the vanilla BO method as well as the state-of-art method [1] in all test cases and practical applications that we consider. Our algorithms are also more practical and less computationally intensive than the methods proposed in [1].

Note that function compositions arise naturally in the real world. One such example is the revenue maximization problem based on the composition of price and demand function. Another example could be the optimization of the F1 score in classification problems that can be seen as a composition of precision and recall metrics [18]. A key novelty of our work lies in the application of our BO algorithms to dynamic pricing problems in revenue management. To the best of our knowledge, ours is the first work to perform dynamic pricing for revenue optimization using Bayesian optimization methods.

1.1 Related Work

Optimization of function compositions has been studied under various constraints such as convexity, inexpensive evaluations and derivative information [4,14,28]. The scope of these work are somewhat restrictive and differ from our key assumption that the constituent functions in the composition are black-box functions and are relatively expensive to evaluate. Our work is closely related to Astudillo and Frazier [1] who optimize black-box function compositions of the form $g(x) = h(f(x))$ using Bayesian Optimization. In this work, the constituent function $f(x)$ is expensive to evaluate and is modelled as a Multi-Ouput GP.

This work was further improved by Maddox et al. [20] using Higher Order Gaussian Process (HOGP). These work assume that the member functions in the composition are correlated and dedicates a significant amount of computational power to capturing these correlations. They propose an EI-based acquisition function for estimating the value of g using a MOGP over the functions f. The calculation of this acquisition function requires them to compute the inverse of the lower Cholesky factor of the covariance matrix, which is a computationally expensive task (runtime increases with order $\mathcal{O}(N^3)$ where N is the size of the covariance matrix), especially when optimizing high dimensional problems.

Our work differs from them in that we consider a composition of multiple constituent functions with single output and assume an independent GP model for each such constituent. This results in significantly lesser computational requirements and faster iterations. Our work focuses on practical deployments of the technique, as showcased in Sect. 4. We also propose a UCB-based algorithm where the problem of matrix inversion does not arise. The UCB-based algorithm allows the user to trade-off between exploration and exploitation during iterations, making it more practical for our use cases. Finally, our key contribution lies in applying the proposed methods to dynamic pricing problems, a brief background of which is discussed in the next subsection.

1.2 Dynamic Pricing and Learning

Dynamic pricing is a phenomenon where the price for any commodity or good is changed to optimize the revenue collected. Consider the scenario of a retailer with a finite inventory, finite time horizon and a known probabilistic model for the demand. On formulating this as a Markov decision problem, it is easy to see that a revenue optimal pricing policy would be non-stationary, resulting in different optimal prices for the same inventory level at different time horizons. In this case, the dynamic nature of pricing is a by-product of finite inventory and horizon effects. See [21] for more details.

Now consider a second scenario of a retailer with an infinite horizon and infinite inventory, trying to find the optimal price for his product. Assuming that the underlying probabilistic demand model is unknown to the retailer, this becomes a simultaneous demand learning and price optimization problem. To learn the underlying demand function, the retailer is required to probe or explore the demand for the product at various prices, and use the information gathered to converges to an optimal price over time. Clearly, in this setting, uncertainty in the demand process naturally leads to exploration in the price space, resulting in the dynamic nature of the pricing policy. See [6] for more details. This second scenario (black-box demand function) is of particular interest to us, and we apply our BO algorithms in this setting, something that has not been done before.

Dynamic pricing with learning is a traditional research topic in Operations Research with a long history (see [13] for a historical perspective). Lobo and Boyd [15] introduced an exploration-exploitation framework for the demand pricing problem, which balances the need for demand learning with revenue maximization. The problem has been studied under various conditions, such as limited [2] and unlimited inventory [17], customer negotiations [19], monopoly [10,12], limited price queries [11] etc. One recurrent theme in these work is to assume a parametric form for the demand function in terms of price and other exogenous variables. Reinforcement learning (RL) methods are then used to simultaneously learn the unknown parameters of the demand function and set prices that have low regret. See, for example, Broder and Rusmevichientong [8] where linear, polynomial and logic demand function models have been assumed. To the best of our understanding, these assumptions on the demand function are rather over-simplified and are typically made for technical convenience. Further, an RL method suited for a particular demand model (say linear demand) may not work when the ground truth model for the demand is different (say, logit model). There have been recent models which try to avoid this issue by modelling the demand as a parameterized random variable (here price is a parameter) but end up making similar convenient assumptions on the parametric form of the mean or variance of the demand, see [5] and references therein.

To keep the demand function free from any specific parametric form, in this work, we assume that it is a black-box expensive to evaluate function and instead model it using a Gaussian process. The revenue function can be expressed as a composition of the price and the demand and this allows us to apply our proposed methods (of Bayesian optimization for function composition) to the dynamic pricing and learning problem.

1.3 Contributions and Organization

The following are the key contributions of our work:

1. We propose novel acquisition functions cEI and cUCB for Bayesian Optimization for function compositions. These acquisition functions are based on EI and UCB acquisition functions for vanilla BO and are less compute intensive and faster to run through each iteration as compared to state-of-art algorithm for function composition proposed in [1].
2. We assume independent GPs to model the constituent functions and this allows for possible parallelization of the posterior update step.
3. As a key contribution of this work, we propose to use BO based dynamic pricing algorithms to optimize the revenue. To the best of our knowledge, we are the first, to use BO for learning the optimal price when the demand functions are expensive to evaluate or are black-box in nature.
4. We consider various revenue maximization scenarios, obtain the revenue function as a composition of price and demand and illustrate the utility of our algorithms in each of these setting.

The following is the organization of the rest of the paper. Section 2 formally describes the problem statement and Sect. 3 describes our proposed algorithms; Sect. 4 details our experimental results; and finally, we conclude in Sect. 5.

2   Problem Description

We begin by describing the problem of BO for composite functions in Subsect. 2.1. In Subsect. 2.2 we describe the dynamic pricing problem and model the revenue function as a function composition to which BO methods for composite functions can be applied.

2.1   BO for Function Composition

We consider the problem of optimizing $g(\mathbf{x}) = h(f_1(\mathbf{x}), f_2(\mathbf{x}), \dots, f_M(\mathbf{x}))$ where $g : \mathcal{X} \to \mathrm{R}$, $f_i : \mathcal{X} \to \mathrm{R}$, $h : \mathrm{R}^M \to \mathrm{R}$ and $\mathcal{X} \subseteq \mathrm{R}^d$. We assume each f_i is a black-box expensive-to-evaluate continuous function while h is known and cheap to evaluate given the values of f_i. The optimization problem that we consider is

$$\max_{\mathbf{x} \in \mathcal{X}} h(f_1(\mathbf{x}), f_2(\mathbf{x}), \dots, f_M(\mathbf{x})). \tag{1}$$

We want to solve Problem 1 in an iterative manner where in the nth iteration, we can use the previous observations $\{\mathbf{x}_i, f_1(\mathbf{x}_i), \dots, f_M(\mathbf{x}_i)\}_{i=1}^{n-1}$ to request a new observation $\{\mathbf{x}_n, f_1(\mathbf{x}_1), \dots, f_M(\mathbf{x}_n)\}$.

A vanilla BO algorithm applied to this problem would first assume a prior GP model on g, denoted by $\mathcal{GP}(\mu(\cdot), K(\cdot, \cdot))$ where μ and K denote the mean and covariance function of the prior model. Given some function evaluations, an updated posterior GP model is obtained. A suitable acquisition function, such as EI or PI can be used, to identify the next query point. For example, in the $n + 1^{\text{th}}$ update round, one would first use the n available observations $(g(\mathbf{x}_1), g(\mathbf{x}_2), \dots, g(\mathbf{x}_n))$ to update the GP model to $\mathcal{GP}(\mu^{(n)}(\cdot), K^{(n)}(\cdot, \cdot))$ where $\mu^{(n)}(\cdot)$ is the posterior mean function and $K^{(n)}(\cdot, \cdot)$ is the posterior covariance function, see [23] for more details. The acquisition function then uses this posterior model to identify the next query location $\mathbf{x}_{n+1}$. In doing so, vanilla BO ignores the values of the member functions in the composition h.

BO for composite function, on the other hand, takes advantage of the available information about h, and its easy-to-compute nature. Astudillo and Frazier [1] model the constituent functions of the composition by a single multi-output function $\mathbf{f}(\mathbf{x}) = (f_1(\mathbf{x}), \dots, f_M(\mathbf{x}))$ and then model the uncertainty in $\mathbf{f}(\mathbf{x})$ using a multi-output Gaussian process to optimize $h(\mathbf{f}(\mathbf{x}))$. Since the prior over f is modelled as a MOGP, the proposed method tries to capture the correlations between different components of the multi-output function $\mathbf{f}(\mathbf{x})$. Note that the proposed EI and PI-based acquisition functions are required to be computed using Monte Carlo sampling. Furthermore, a sample from the posterior distribution is obtained by first sampling an n variate normal distribution, then scaling it by the lower Cholesky factor and then centering it with the mean of the posterior GP. Two problems arise due to this: 1. Such simulation based averaging approach increases the time complexity of the procedure linearly with the number of samples taken for averaging and 2. calculation of the lower Cholesky factor increases the function's time complexity cubically with the number of data points. These factors render the algorithm unsuitable, particularly for problems with large number of member functions or for problems with large dimensions.

To alleviate these problems, in this work, we model the constituent functions using independent GPs. This modelling approach allows us to train GPs for each output independently and hence the posterior GP update can be parallelized. We propose two acquisition functions, cEI which is based on the EI algorithm and cUCB, which is based on the GP-UCB algorithm [26]. Our cEI acquisition function is similar in spirit to the EI-CF acquisition function of [1] but is less computationally intensive owing to the independent GP model. Since we have independent one dimensional GP model for each constituent function, sampling points from the posterior GP does not require computing the Cholesky factor (and hence matrix inversion), something that is needed in the case of high-dimensional GP's of [1]. This greatly reduces the complexity of the MC sampling steps of our algorithm (see Sect. 3 for more details). However, the cEI acquisition function still suffers from the drawback of requiring Monte Carlo averaging. To alleviate this problem, we propose a UCB based acquisition function that uses the current mean plus scaled variance of the posterior GP at a point as a surrogate for the constituent function at that point. As shown by Srinivas et al. [26], while the mean term in the surrogate guides exploitation, it is the variance of the posterior GP at a point that allows for suitable exploration. The scaling of the variance term is controlled in such a way that it balances the trade off between exploration and exploitation. In Sect. 4, we illustrate the utility of our method, first for standard test functions and then as an application to dynamic pricing problem. Our algorithms, especially the cUCB one, outperforms not only vanilla BO but also those proposed in Astudillo and Frazier [1].

2.2 Bayesian Optimization for Dynamic Pricing

We consider Bayesian Optimization for two types of revenue optimization problems. The first problem optimizes the revenue per customer where customers are characterized by their willingness-to-pay distribution (which is unknown). In the second problem, we assume a parametric demand model (the functional form is assumed to be unknown) and optimizes the associated revenue.

In the first model, we assume that an arriving customer has an associated random variable, V, with complimentary cumulative distribution function $\bar{F}$, indicating its maximum willingness to pay for the item sold. For an item on offer at a price p, an arriving customer purchases it with the probability

$$d(p) := \bar{F}(p) = \Pr\{V \geq p\}. \tag{2}$$

In this case, when the product is on offer at a price p, the revenue per customer $r(p)$ is given by $r(p) = p\bar{F}(p)$. The revenue function is a composition of the price and demand or purchase probability and we assume that the distribution of the purchase probability i.e., F is not known and also expensive to estimate. One could perform a vanilla BO algorithm by having a GP model on $r(p)$ itself. However to exploit the known nature of the revenue function, we will apply our function composition method by instead having a GP on $\bar{F}(p)$ and demonstrate its superiority over vanilla BO.

In the second model, we assume that the true demand $d(p)$ for a commodity at price p has a functional form. This forms the ground truth model that governs the demand, but we assume that the functional form for this demand is not known to the manager optimizing the revenue. In our experiments, we assume linear, logit, Booth and Matyas functional forms for the demand (see Sect. 4 for more details.) Along similar lines, one could build more sophisticated demand models to account for external factors (such as supply chain issues, customer demographics or inventory variables), something that we leave for future explorations.

Note that we make some simplifying assumptions about the retail environment in these two models and our experiments. We assume a non-competitive monopolistic market with an unlimited supply of the product and no marginal cost of production. However, these assumptions can easily be relaxed by changing the ground truth demand model appropriately, which are used in the experiments to reflect these aspects. The fact that we use a GP model as a surrogate for the unknown demand model offers it the ability to model a diverse class of demand functions under diverse problem settings. We do not discuss these aspects further but focus on the following simple yet meaningful experimental examples that one typically encounters in revenue management problems.

In the following, **p** denotes the price vector:

1. **Independent demand model:** A retailer supplies its product to two different regions whose customer markets behave independently from each other. Thus, the same product has independent and different demand functions (and hence different optimal prices) in different geographical regions and under such black-box demand models for the two regions (d_1, d_2). The retailer is interested in finding the optimal prices, leading to the optimization of the following function: $g(\mathbf{p}) = p_1 d_1(p_1) + p_2 d_2(p_2)$.

2. **Correlated demand model:** Assume that a retailer supplies two products at prices p_1 and p_2 and the demand for the two products is correlated and influenced by the price for the other product. Such a scenario can be modelled by a revenue function of the form $g(\mathbf{p}) = p_1 d_1(\mathbf{p}) + p_2 d_2(\mathbf{p})$. Consider the example where the prices of business and economy class tickets can influence the demand in each segment. Similarly, the demand for a particular dish in a fast food chain might be influenced by the prices for other dishes.

3. **Identical price model:** In this case, the retailer is compliant with having a uniform price across locations. However, the demand function across different locations could be independent at each of these locations, leading to the following objective function: $g(p) = p d_1(p) + p d_2(p)$. This scenario can be used to model different demand functions for different population segments in their age, gender, socio-economic background, etc.

3   Proposed Method

As discussed in Sect. 2, we propose that instead of having a single GP model over g, we have M different GP models over each constituent function in the composition. Each prior GP model will be updated using GP regression whenever the observations of the constituent functions are available. A suitably designed acquisition function would then try to find the optimal point when the constituent functions should all be evaluated at, in the next iteration. For ease of notation, we use the shorthand $h(\{f_i(\mathbf{x})\})$ to denote $h(f_1(\mathbf{x}), \ldots, f_M(\mathbf{x}))$ in the subsequent sections.

3.1   Statistical Model and GP Regression

Let $f_i^{1:n}$, $i \in \{1, 2, \ldots, M\}$ denote the function evaluations of the member functions at locations $\{\mathbf{x}_1, \mathbf{x}_2, \ldots, \mathbf{x}_n\}$ denotes as $\mathbf{x}^{1:n}$. In the input space $\mathcal{X} \subset \mathbb{R}^d$, let $\mathcal{GP}(\mu_i^{(n)}, k_i^{(n)})$ be the posterior GP over the function f_i where $\mu_i^{(n)} : \mathcal{X} \to \mathbb{R}$ is the posterior mean function, $k_i^{(n)} : \mathcal{X} \times \mathcal{X} \to \mathcal{R}$ is the positive semi-definite covariance function and the variance of the function is denoted by $\sigma_i^{(n)}(\mathbf{x})$. The superscript n is used to denote the fact that the posterior update accounts for n function evaluations made till now. For each such GP, the underlying prior is a combination of a constant mean function $\mu_i \in \mathbb{R}$ and the squared exponential function k_i

$$k_i(\mathbf{x}, \mathbf{x}') = \sigma^2 \exp\left(-\frac{(\mathbf{x} - \mathbf{x}')^T(\mathbf{x} - \mathbf{x}')}{2l_i^2}\right)$$

The kernel matrix K_i is then defined as

$$\mathbf{K}_i := \begin{bmatrix} k_i(\mathbf{x}_1, \mathbf{x}_1) & k_i(\mathbf{x}_1, \mathbf{x}_2) & \ldots & k_i(\mathbf{x}_1, \mathbf{x}_n) \\ \vdots & \vdots & \vdots & \vdots \\ k_i(\mathbf{x}_n, \mathbf{x}_1) & k_i(\mathbf{x}_n, \mathbf{x}_2) & \ldots & k_i(\mathbf{x}_n, \mathbf{x}_n) \end{bmatrix}$$

and with abuse of notation define $\mathbf{K} = \mathbf{K} + \lambda^2 I$ (to account for noise in the function evaluations). The posterior distribution on the function $f_i(\mathbf{x})$ at any input $\mathbf{x} \in \mathcal{X}$ [23] is given by

$$P(f_i(\mathbf{x})|\mathbf{x}^{1:n}, f_i^{1:n}) = \mathcal{N}(\mu_i^{(n)}(\mathbf{x}), \sigma_i^{(n)}(\mathbf{x}) + \lambda^2), \quad i \in \{1, 2, \ldots, M\} \text{ where}$$

$$\mu_i^{(n)}(\mathbf{x}) = \mu^{(0)} + \mathbf{k}_i^T \mathbf{K}_i^{-1}(f_i^{1:n} - \mu_i(\mathbf{x}^{1:n})) \text{ and } \sigma_i^{(n)}(\mathbf{x}) = k_i(\mathbf{x}, \mathbf{x}) - \mathbf{k}_i^T \mathbf{K}_i^{-1} \mathbf{k}_i$$

$$\mathbf{k}_i = [k_i(\mathbf{x}, \mathbf{x}_1) \quad \ldots \quad k_i(\mathbf{x}, \mathbf{x}_n)]$$

3.2   cEI and cUCB Acquisition Functions

For any fixed point $\mathbf{x} \in \mathcal{X}$, we use the information about the composition function h to estimate g by first estimating the value of each member function at $\mathbf{x}$. However, this is not a straightforward task and needs to be performed in a way

---

Algorithm 1. cEI: Composite BO using EI based acquisition function

---

Require: $T \leftarrow$ Budget of iterations
Require: $h(\cdot), f_1(\cdot), \ldots, f_M(\cdot) \leftarrow$ composition and member functions
Require: $\mathbf{X} = \{\mathbf{x}_1, \ldots, \mathbf{x}_s\} \leftarrow s$ starting points
Require: $\mathbf{F} = \{(f_1(\mathbf{x}), \ldots, f_M(\mathbf{x}))\}_{\mathbf{x} \in \mathbf{X}} \leftarrow$ function evaluations at starting points
 1: **for** $n = s+1, \ldots, s+T$ **do**
 2:      **for** $i = 1, \ldots, M$ **do**
 3:          Fit model $\mathcal{GP}(\mu_i^{(n)}(\cdot), K_i^{(n)}(\cdot, \cdot))$ using evaluations of f_i at points in $\mathbf{X}$
 4:      **end for**
 5:      Find new point $\mathbf{x}_n$ by optimizing cEI$(\mathbf{x}, L)$ (defined below)
 6:      Get $(f_1(\mathbf{x}_n), \ldots, f_M(\mathbf{x}_n))$
 7:      Augment the data $(f_1(\mathbf{x}_n), \ldots, f_M(\mathbf{x}_n))$ into $\mathbf{F}$ and update $\mathbf{X}$ with $\mathbf{x}_n$
 8: **end for**
 9: **function** cEI$(\mathbf{x}, L)$
10:      **for** $l = 1, \ldots, L$ **do**
11:          Draw M samples $Z_{(l)} \sim \mathcal{N}_M(0_M, I_M)$
12:          Compute $\alpha^{(l)} := \{h(\{\mu_i^{(n)}(\mathbf{x}) + \sigma_i^{(n)}(\mathbf{x})Z_i^{(l)}\}) - g_n^*\}^+$
13:      **end for**
14:      **return** $E_n(\mathbf{x}) = \frac{1}{L}\Sigma_{l=1}^L \alpha^{(l)}$
15: **end function**

---

similar to the vanilla EI acquisition using Monte Carlo sampling. We propose to use the following acquisition function, that we call as cEI.

$$E_n(\mathbf{x}) = \mathbb{E}_n \left[h(\{\mu_i^{(n)}(\mathbf{x}) + \sigma_i^{(n)}(\mathbf{x})Z_i\}) - g_n^* \right]^+ \tag{3}$$

where Z is drawn from an M-variate normal distribution and g_n^* is the best value observed so far. This acquisition function is similar to EI as we subtract the best observation, g^*, so far and only consider negative terms to be 0. Assuming independent GPs over the functions allows constant time computation of the variance at $\mathbf{x}$. However, since each function f_i is being considered an independent variable with mean $\mu_i^{(n)}(\cdot)$ and variance $\sigma_i^{(n)}(\cdot)$, the calculation of $E_n(\boldsymbol{x})$ does not have a closed form and thus, the expectation needs to be evaluated empirically with sampling. Algorithm 1 provides the complete procedure for doing BO with this acquisition function.

To alleviate this complexity in estimating the acquisition function, we propose a novel UCB-style acquisition function. This function estimates the value of each member function using the GP priors over them and controls the exploration and exploitation factor with the help of the hyperparameter λ_n:

$$U_n(\mathbf{x}) = h(\{\mu_i^{(n)}(\mathbf{x}) + \beta_n \sigma_i^{(n)}(\mathbf{x})\}) \tag{4}$$

Algorithm 2 gives the complete details for using this acquisition function. The user typically starts with a high value for β to promote exploration and reduces iteratively to exploit the low reget regions it found. For our experiments, we start with $\beta = 1$ and exponentially decay it in each iteration by a factor of 0.99.

Algorithm 2. cUCB: Composite BO using UCB based acquisition function
___

Require: $T \leftarrow$ Budget of iterations
Require: $h(\cdot), f_1(\cdot), \ldots, f_M(\cdot) \leftarrow$ composition and member functions
Require: $\mathbf{X} = \{\mathbf{x}_1, \ldots, \mathbf{x}_s\} \leftarrow s$ starting points
Require: $\mathbf{F} = \{(f_1(\mathbf{x}), \ldots, f_M(\mathbf{x}))\}_{\mathbf{x} \in \mathbf{X}} \leftarrow$ function evaluations at starting points
Require: $\beta \leftarrow$ Exploration factor
  1: **for** $n = s+1, \ldots, s+T$ **do**
  2:     **for** $i = 1, \ldots, M$ **do**
  3:         Fit model $\mathcal{GP}(\mu_i^{(n)}(\cdot), K_i^{(n)}(\cdot, \cdot))$ using evaluations of f_i at points in $\mathbf{X}$
  4:     **end for**
  5:     Find new point $\mathbf{x}_n$ suggested by the composition function using Eq. 4
  6:     Get $(f_1(\mathbf{x}_n), \ldots, f_M(\mathbf{x}_n))$
  7:     Augment the data $(f_1(\mathbf{x}_n), \ldots, f_M(\mathbf{x}_n))$ into $\mathbf{F}$ and update $\mathbf{X}$ with $\mathbf{x}_n$
  8:     Update β
  9: **end for**
___

4   Experiments and Results

In this section, we compare the results of our cUCB and cEI algorithms with Vanilla EI, Vanilla UCB and the state-of-the-art BO for composite functions (BO-CF) [1] using HOGP [20] in terms of loss in regret and runtime of the algorithms. We first compare our methods on 3 test functions and then move on to show their applications to three different pricing scenarios. Our code is available here.

Our algorithms are implemented with the help of the BoTorch framework [3] and use the APIs provided by them to declare and fit the GP models. We assume noiseless observations for our results in this section, and the same results can be obtained when we add Gaussian noise to the problem with a fixed mean and variance. We start with the same 10 initial random points and run our BO algorithms for 70 iterations. We use a system with 96 Intel Xeon Gold 6226R CPU @2.90 GHz and 96 GB of memory shared between the CPUs.

We compare the performance of different algorithms based on the log of mean minimum regret till each iteration, averaged over 100 runs. In a single BO run, the regret at iteration i in the kth run is defined as $l_i^k = g^* - g(\mathbf{x}_i)$ where g^* is the global maximum of the objective function. The minimum regret at iteration i in the kth run is defined as $m_i^k = \min_{1 \leq j \leq i} l_j^k$ and the final metric at iteration i averaged across 100 runs is calculated as $r_i = \log_{10} \left(\frac{1}{100} \sum_{k=1}^{100} m_i^k \right)$.

4.1   Results on Test Functions

Langermann Function: To express this function as a function composition, we consider each outer iteration of the Langermann function to be a separate constituent function, that is, $f_i(\mathbf{x}) = \exp\left(-\frac{1}{\pi} \sum_{j=1}^{d} (x_j - A_{ij})^2 \right)$

$\cos\left(\pi \sum_{j=1}^{d}(x_j - A_{ij})\right)$. The composition for this will be $h(\{f_i(\mathbf{x})\}) = \sum_{i=1}^{m} c_i f_i(\mathbf{x})$ with $d = 2$, $c = (1, 2, 5, 2, 3)$, $m = 5$, $A = ((3,5),(5,2),(2,1),(1,4),(7,9))$ and domain $\mathcal{X} = [0,10]^2$. Note that the terms differ only in the columns of hyperparameter A for different member functions and thus, should have a high covariance.

Dixon-Price Function: In this function, we take the term associated with each dimension of the input to be a separate constituent, that is, $f_1(\mathbf{x}) = (x_1 - 1)^2$ and $f_i(\mathbf{x}) = i(2x_i^2 - x_{i-1})^2$. The composition for this function will be $h(\{f_i(\mathbf{x})\}) = \sum_{i=1}^{d} f_i(\mathbf{x})$ with $d = 5$ and domain $\mathcal{X} = [-10,10]^d$. Since only consecutive terms in this function share one variable, the member functions do have a non-zero covariance that will be much lower than the above function.

Ackley Function: Here, we build a more complex composition function by considering the terms in the exponents as the member functions, resulting in

$$f_1(\mathbf{x}) = \sqrt{\frac{1}{d}\sum_{i=1}^{d} x_i^2} \text{ and } f_2(\mathbf{x}) = \frac{1}{d}\sum_{i=1}^{d}\cos(cx_i)$$

and $h(\mathbf{x}) = -a\exp(-bf_1(\mathbf{x})) - \exp(f_2(\mathbf{x})) + a + \exp(1)$ with $d = 5$, $a = 20$, $b - 0.2$, $c = 2\pi$ and domain $\mathcal{X} = [-32.768, 32.768]^d$.

Results: Figure 1a, 1b and 2a compares the results of different algorithms. Vanilla EI and UCB algorithms do not consider the composite nature of the function while BO-CF and our methods use the composition defined above. Even with the high covariance between the members in Langermann function, cUCB outperforms BO-CF while the cEI algorithm has a similar performance level. However, when that covariance reduces in the Dixon-Price function, the cEI

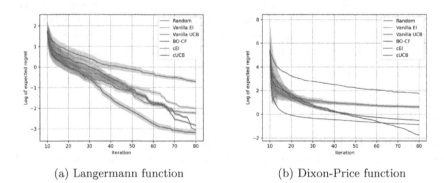

(a) Langermann function   (b) Dixon-Price function

Fig. 1. Log regret for test functions with a composite nature

algorithm performs better than BO-CF while the cUCB algorithm significantly outperforms it. Figure 2a shows that our algorithms work well with complicated composition functions as well and both, cUCB and cEI, outperform BO-CF.

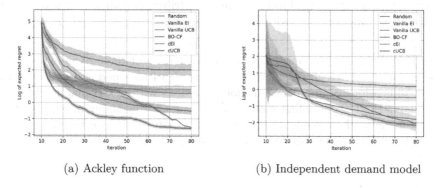

(a) Ackley function   (b) Independent demand model

Fig. 2. Log of expected regret for pricing tasks

4.2 Results for Demand Pricing Experiments

We now test our approach on the demand models discussed in Sect. 2.2.

Independent Demand Model: Recall that in this model, we allow the price at each location to be different and model the demand in a region to depend only on the price therein. We consider 4 regions where each region has a parametric demand functions and randomly chosen parameters. Particularly, we assume that two regions have a logit demand function $(d(p) = \frac{e^{-z_1-z_2 p}}{1+e^{-z_1-z_2 p}})$ with $z_1 \in [1.0, 2.0]$, $z_2 \in [-1.0.1.0]$ and the other two regions have a linear demand function $(d(p) = z_1 - z_2 p)$ with $z_1 \in [0.75, 1]$ and $z_2 \in [2/3, 0.75]$. The domain for this model is $\mathcal{X} = [0,1]^4$ and the composition is The composition function looks $h(d_1(\mathbf{p}), d_2(\mathbf{p}), d_3(\mathbf{p}), d_4(\mathbf{p})) = \sum_{i=1}^{4} p_i d_i(p_i)$.

Correlated Demand Model: In this example, we assume that different products are for sale at different prices, and there is a certain correlation between demand for different products via their prices. We consider the case of 2 products where one of the product has a demand function governed by the Matyas function [27]. We assume that the demand for the second product is governed by the Booth function [27]. More specifically, the constituent functions d_1 and d_2 are $d_1(\mathbf{p}) = 8(100 - \texttt{Matyas}(\mathbf{p}))$, $d_2(\mathbf{p}) = 1154 - \texttt{Booth}(\mathbf{p})$ where Matyas function is defined as $\texttt{Matyas}(\mathbf{x}) = 0.26(x_1^2 + x_2^2) - 0.48 x_1 x_2$ and Booth function is defined as $\texttt{Booth}(\mathbf{x}) = (x_1 + 2x_2 - 7)^2 + (2x_1 + x_2 - 5)^2$. The domain of the problem is $\mathcal{X} = [0,10]^2$ and finally composition for this will be $h(d_1(\mathbf{p}), d_2(\mathbf{p})) = p_1 d_1(\mathbf{p}) + p_2 d_2(\mathbf{p})$.

Identical Price Model: In this example, we assume that a commodity is sold for same price at two different regions but the willingness to pay variable for customers in the two regions is different. We assume that the willingness to pay distribution in one region follows exponential distribution with $\lambda = 5.0$. n the other region, this is assumed to be a gamma distribution with $\alpha = 10.0, \beta = 10.0$ The resulting function composition is given by : $h(d_1(p), d_2(p)) = pd_1(p) + pd_2(p)$.

Results: Figures 2b, 3a and 3b compare the results of these dynamic pricing models for the different BO algorithms. Our algorithms perform well even in higher dimensions of input and member functions with cUCB marginally out-performing BO-CF in the first model. cUCB matches the minimum regret in the second model and converges to it much faster than BO-CF. In the case of the third model, having independent GP's performs better than BO-CF with cEI.

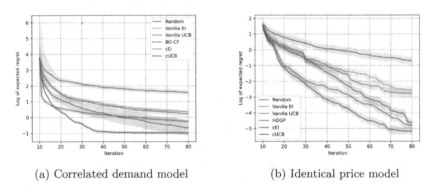

(a) Correlated demand model   (b) Identical price model

Fig. 3. Log of expected regret for pricing tasks

Table 1. Run-time for 70 iterations across algorithms in seconds

Task	EI	UCB	cEI	cUCB	HOGP
Langermann function	1.74	1.73	9.71	9.34	36.30
Dixon-Price function	1.70	1.69	14.47	12.71	41.72
Ackley function	1.63	1.55	11.56	11.37	47.91
Independent demand model	2.05	2.03	7.85	7.12	19.12
Correlated demand model	3.91	3.29	9.63	9.19	53.01
Identical price model	2.55	2.38	7.27	7.81	24.37

4.3 Runtime Comparisons with State of the Art

Along with the performance of our algorithms being superior in terms of regret, the methodology of training independent GPs is significantly faster in terms of run time. As shown in Table 1, both of our algorithms are between 3 to 4 times faster than BO-CF using HOGP on average and their run time increases linearly with the number of member functions in the composition when compared to vanilla EI and UCB. By not having to compute the inverse matrix for estimating the lower Cholesky factor of the covariance matrix, we gain large improvements in run time. The elimination of inverse matrix computation while estimating with the help of lower Cholesky factor of the covariance matrix results in the large improvement in run time over BO-CF. Also note that our UCB variant is marginally faster than the EI variant as well due to the elimination of MC sampling in the process.

5 Conclusion

In this work, we have proposed EI and UCB based BO algorithms, namely cEI and cUCB for optimizing functions with a composite nature. We further apply our algorithms to the revenue maximization problem and test our methods on different market scenarios. We show that our algorithms, particularly cUCB, outperforms vanilla BO as well as the current state of the art BO-CF algorithm. Our algorithms are computationally superior because they do not require multiple Cholesky decompositions as required in the BO-CF algorithm.

As part of future work, we would like to provide theoretical bounds on cumulative regret for the proposed algorithms. We would also like to see the applicability of the proposed algorithms in hyper-parameter tuning for optimizing F1 score. It would also be interesting to propose BO algorithms for an extended model wherein the member functions can be probed independently from each other at different costs.

References

1. Astudillo, R., Frazier, P.: Bayesian optimization of composite functions. In: Chaudhuri, K., Salakhutdinov, R. (eds.) Proceedings of the 36th International Conference on Machine Learning. Proceedings of Machine Learning Research, vol. 97, pp. 354–363. PMLR (2019)
2. Babaioff, M., Dughmi, S., Kleinberg, R., Slivkins, A.: Dynamic pricing with limited supply. ACM Trans. Econ. Comput. **3**(1), 1–26 (2015)
3. Balandat, M., et al.: BoTorch: a framework for efficient Monte-Carlo Bayesian optimization. In: Advances in Neural Information Processing Systems, vol. 33 (2020)
4. Barber, R.F., Sidky, E.Y.: MOCCA: mirrored convex/concave optimization for nonconvex composite functions. J. Mach. Learn. Res. **17**(144), 1–51 (2016)
5. den Boer, A.V., Zwart, B.: Simultaneously learning and optimizing using controlled variance pricing. Manage. Sci. **60**(3), 770–783 (2014)

6. den Boer, A.V., Zwart, B.: Dynamic pricing and learning with finite inventories. Oper. Res. **63**(4), 965–978 (2015)
7. Brochu, E., Cora, V.M., de Freitas, N.: A tutorial on Bayesian optimization of expensive cost functions, with application to active user modeling and hierarchical reinforcement learning (2010)
8. Broder, J., Rusmevichientong, P.: Dynamic pricing under a general parametric choice model. Oper. Res. **60**(4), 965–980 (2012)
9. Candelieri, A.: A gentle introduction to Bayesian optimization. In: 2021 Winter Simulation Conference (WSC), pp. 1–16 (2021)
10. Chen, Y.M., Jain, D.C.: Dynamic monopoly pricing under a Poisson-type uncertain demand. J. Bus. **65**(4), 593–614 (1992)
11. Cheung, W.C., Simchi-Levi, D., Wang, H.: Technical note—dynamic pricing and demand learning with limited price experimentation. Oper. Res. **65**(6), 1722–1731 (2017)
12. Crapis, D., Ifrach, B., Maglaras, C., Scarsini, M.: Monopoly pricing in the presence of social learning. Manage. Sci. **63**(11), 3586–3608 (2017)
13. den Boer, A.V.: Dynamic pricing and learning: historical origins, current research, and new directions. Surv. Oper. Res. Manage. Sci. **20**(1), 1–18 (2015)
14. Drusvyatskiy, D., Paquette, C.: Efficiency of minimizing compositions of convex functions and smooth maps. Math. Program. **178**(1–2), 503–558 (2018)
15. Elreedy, D.F., Atiya, A.I., Shaheen, S.: A novel active learning regression framework for balancing the exploration-exploitation trade-off. Entropy **21**(7), 651 (2019)
16. Frazier, P.I.: A tutorial on Bayesian optimization (2018)
17. Harrison, J.M., Keskin, N.B., Zeevi, A.: Bayesian dynamic pricing policies: learning and earning under a binary prior distribution. Manage. Sci. **58**(3), 570–586 (2012)
18. Injadat, M., Salo, F., Nassif, A.B., Essex, A., Shami, A.: Bayesian optimization with machine learning algorithms towards anomaly detection. In: 2018 IEEE Global Communications Conference (GLOBECOM), pp. 1–6 (2018)
19. Kuo, C.W., Ahn, H.S., Aydin, G.: Dynamic pricing of limited inventories when customers negotiate. Oper. Res. **59**(4), 882–897 (2011)
20. Maddox, W., Balandat, M., Wilson, A.G., Bakshy, E.: Bayesian optimization with high-dimensional outputs. In: Beygelzimer, A., Dauphin, Y., Liang, P., Vaughan, J.W. (eds.) Advances in Neural Information Processing Systems (2021)
21. Phillips, R.L.: Pricing and Revenue Optimization. Stanford University Press (2021)
22. Pyzer-Knapp, E.O.: Bayesian optimization for accelerated drug discovery. IBM J. Res. Dev. **62**(6), 2:1–2:7 (2018)
23. Rasmussen, C.E., Williams, C.K.I.: Gaussian Processes for Machine Learning. The MIT Press (2005)
24. Scotto Di Perrotolo, A.: A theoretical framework for bayesian optimization convergence. Master's thesis, KTH, Optimization and Systems Theory (2018)
25. Snoek, J., Larochelle, H., Adams, R.P.: Practical Bayesian optimization of machine learning algorithms. In: Pereira, F., Burges, C., Bottou, L., Weinberger, K. (eds.) Advances in Neural Information Processing Systems, vol. 25. Curran Associates, Inc. (2012)
26. Srinivas, N., Krause, A., Kakade, S.M., Seeger, M.W.: Information-theoretic regret bounds for gaussian process optimization in the bandit setting. IEEE Trans. Inf. Theory **58**(5), 3250–3265 (2012)
27. Surjanovic, S., Bingham, D.: Virtual library of simulation experiments: test functions and datasets. https://www.sfu.ca/~ssurjano. Accessed 7 Feb 2023

28. Woodworth, B.E., Srebro, N.: Tight complexity bounds for optimizing composite objectives. In: Lee, D., Sugiyama, M., Luxburg, U., Guyon, I., Garnett, R. (eds.) Advances in Neural Information Processing Systems, vol. 29. Curran Associates, Inc. (2016)
29. Wu, J., Chen, X.Y., Zhang, H., Xiong, L.D., Lei, H., Deng, S.H.: Hyperparameter optimization for machine learning models based on Bayesian optimization. J. Electron. Sci. Technol. **17**(1), 26–40 (2019)

A Bayesian Optimization Algorithm for Constrained Simulation Optimization Problems with Heteroscedastic Noise

Sasan Amini$^{(\boxtimes)}$ and Inneke Van Nieuwenhuyse

Flanders Make@UHasselt, Data Science Institute, Hasselt University,
Hasselt, Belgium
sasan.amini@uhasselt.be

Abstract. In this research, we develop a Bayesian optimization algorithm to solve expensive, constrained problems. We consider the presence of heteroscedastic noise in the evaluations and thus propose a new acquisition function to account for this noise in the search for the optimal point. We use stochastic kriging to fit the metamodels, and we provide computational results to highlight the importance of accounting for the heteroscedastic noise in the search for the optimal solution. Finally, we propose some promising directions for further research.

Keywords: Bayesian optimization · Constrained problems ·
Heteroscedastic noise · Stochastic Kriging · Barrier function

1 Introduction

In optimization problems, the decision maker may not only be concerned about optimizing the primary performance measure of interest (the *objective function*) but also about secondary performance measures that need to satisfy pre-specified thresholds (i.e., *constraints*). These thresholds can be externally imposed (e.g., in the case of governmental regulations) or be internally linked to the optimization (e.g., output quality indicators). Examples of constrained problems are abundant in the literature,; see for example, [16,19], and [11]. Often, the objective and constraints (hereafter referred to as the *outputs*) in such problems cannot be expressed as closed-form expressions of the decision variables (the *inputs*); instead, they can only be evaluated numerically, using physical or computer experiments. Problems of this type are referred to as *black box optimization* problems and simulation optimization is often used for optimizing such problems [1]. Unfortunately, many approaches in this field rely on population-based heuristics; consequently, they may require many function evaluations before converging to the optimal solution. This causes a problem in settings where the experiments

This study is supported by the Special Research Fund (BOF) of Hasselt University (grant number: BOF19OWB01), and the Flanders Artificial Intelligence Research Program (FLAIR).

are expensive (because of long computation times, high monetary costs, or haz-ardous experiments), as the total experimental budget is then usually too limited to obtain high-quality results.

In recent years, Bayesian optimization (BO) [4] has become relatively popular in the literature for solving expensive black-box optimization problems. In BO, a computationally inexpensive metamodel, usually a Gaussian Process Regression (GPR) model [21], is estimated based on a small set of input-output observa-tions; the metamodel information is then used to sequentially select new inputs to simulate using an *acquisition function*, in view of converging to the optimal solution within a small number of iterations (i.e., a small number of additional experiments). GPR and BO are terms that are commonly used in the Computer Science and Engineering fields; in the OM/OR field, GPR is commonly referred to as Kriging, and Bayesian Optimization as (Efficient) Global Optimization [13,21]. Ordinary Kriging models tend to be most popular here; these models assume, though, that the outputs are observed without noise. Yet, in many real-life simulation optimization problems, the output observations are *noisy*. While *re-sampling* is a common strategy to reduce the noise on the observed outputs (both in the evolutionary approaches [3] and in some BO papers, such as [16]), it can only be implemented to a limited extent in settings with a small bud-get and an expensive simulator: indeed, more replications at already observed inputs will decrease the available budget for observing new points, which poten-tially affects the performance of the algorithm [22]. Additionally, real-life sim-ulation optimization problems are often *constrained*. The current BO literature on *stochastic* constrained settings is relatively scarce (see, e.g., [6,9,11,16], and [24]), and assumes that the noise is *homogenous*, i.e., that it is independent of the input location. Yet, in many real-life settings, the noise is *heterogenous*: see, e.g., [12] and [7].

In this research, we focus on constrained problems of the following type:

$$\min_{\mathbf{x}} \mathbb{E}[f(\mathbf{x}, \xi_{\mathbf{x}})]$$

$$s.t. \ \prod_{k=1}^{K} Pr\left(c_k(\mathbf{x}, \xi_{\mathbf{x}}) \leq 0\right) \geq p_0, \tag{1}$$

$$k = 1, 2, ..., K$$

where $f : D \to \mathbb{R}$ denotes a scalar-valued objective function, and $c_k : D \to \mathbb{R}$ denotes a scalar-valued constraint function. The objective and the constraint functions are expensive black-box functions, and their evaluations are perturbed by heterogeneous random noise (denoted by $\xi_{\mathbf{x}}$). The user-defined parameter p_0 denotes the minimum required *joint* probability of feasibility for *all* constraints (as is common in the literature [6,24], we assume that the constraints are inde-pendent, hence the multiplication operator). We propose a Bayesian optimization algorithm that accounts for the heterogenous noise in the metamodel (Stochastic Kriging, [2]), and that leverages the model information in the acquisition func-tion. We show that these aspects enable the algorithm to converge to solutions that are closer to the true optimum, without the need for (intense) re-sampling.

The actual acquisition function is inspired by the barrier function approach of [19], as further detailed in Sect. 3.

To the best of our knowledge, this research proposes the first algorithm specifically designed for constrained problems with heteroscedastic noise. We use an instructive example from the literature to show that both the use of Stochastic Kriging and the actual form of the acquisition function matter for the performance of the algorithm. Our results also highlight the need for further research on the identification of the final optimal solution in noisy settings.

The remainder of the paper is organized as follows. Section 2 provides an introduction to Bayesian optimization and Stochastic Kriging. Section 3 describes the proposed algorithm in more detail. Section 4 explains the numerical experiments designed to test the performance of the proposed algorithm. The results are discussed in Sect. 5, while Sect. 6 presents the conclusions, along with some promising directions for further research.

2   Bayesian Optimization (BO) and Stochastic Kriging (SK): Notation and Terminology

In what follows, we first outline the basics of a general BO algorithm (Subsect. 2.1); next, we provide some basic theory and notation related to Stochastic Kriging (Subsect. 2.2).

2.1   Bayesian Optimization (BO)

The general steps of a BO algorithm are summarized in Fig. 1. The optimization procedure starts with the evaluation of an initial set of points with good space-filling properties (e.g., obtained by latin hypercube sampling, LHS). The size of this initial set is commonly set equal to $10d$ [17], or to $11d - 1$ [13]), or $\frac{(d+1)(d+2)}{2}$ [25], where d denotes the number of input dimensions of the problem. The metamodel is then trained with this initial set to model the expensive function(s).

Gaussian Processes Regression (GPR) is the most common type of metamodel in the Bayesian optimization literature. A GPR model is not only able to predict the outcomes at unobserved solutions, but also yields an estimate of the uncertainty on these predictions. The algorithm leverages this information in an *infill criterion* or *acquisition function* to decide which input vector to sample next. Different acquisition functions have been put forward, such as Expected Improvement [18], Knowledge Gradient [5], and Entropy Search [10]. After evaluating (i.e., simulating) this new infill point, the new input/output information is used to update the metamodel. The algorithm continues to sequentially add extra infill points, each time updating the metamodels, until the stopping criterion (e.g., a maximum available computational budget) has been reached, after which the algorithm identifies the optimal solution.

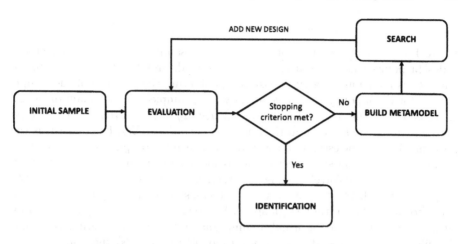

Fig. 1. General steps in a Bayesian Optimization algorithm

2.2 Stochastic Kriging (SK)

A Kriging model essentially models a distribution over functions, such that the joint distribution at any finite set of points is a multivariate Gaussian distribution. Such a model is fully defined by its mean function and its covariance function (also referred to as *kernel*). The parameters of these are estimated from the input/output observations in the dataset, e.g. by means of maximum likelihood estimation. Once the model has been fit to the data, it provides a cheap/fast way to generate predictions of the function at non-observed input locations; importantly, it also provides an estimate of the uncertainty (i.e., the mean squared error or MSE) of these predictions. Details about Ordinary Kriging can be found in [14]. Ordinary Kriging assumes that the output observations are *deterministic*. Consequently, it interpolates between the available output observations: at observed input/output locations, the predictor coincides with the observed output, and the MSE of the predictor is zero.

This is no longer the case in Stochastic Kriging [2]. For a given output function and an arbitrary design point x_i, *Stochastic Kriging* represents the observed objective value $\hat{f}_r(x_i)$ in the rth replication as Eq. 2.

$$\hat{f}_r(x_i) = \beta_0 + M(x_i) + \epsilon_r(x_i) \tag{2}$$

The first two terms of Eq. 2 are equivalent to Ordinary Kriging. The term $\epsilon_r(x_i)$ is referred to as *intrinsic uncertainty*, as it is the uncertainty inherent in stochastic simulation. The intrinsic uncertainty is (naturally) independent and identically distributed across replications, having mean 0 and variance $\tau^2(x_i)$ at any arbitrary point x_i. Note that the model allows for heterogeneous noise, implying $\tau^2(x_i)$ need not be constant throughout the design space.

The Stochastic Kriging prediction at an arbitrary location $\mathbf{x}$ (whether observed or not) is given by:

$$\mu^{sk}(\mathbf{x}) = \beta_0 + \Sigma_M(\mathbf{x}, .)^T [\Sigma_M + \Sigma_\epsilon]^{-1} (\bar{\mathbf{f}} - \beta_0 \mathbf{1}_p) \tag{3}$$

where $\bar{\mathbf{f}}$ is the $p \times 1$ vector containing all the sample means obtained at the p design points that have already been observed, and $\mathbf{1}_p$ is a $p \times 1$ vector of ones. Σ_ϵ is a diagonal matrix of size $p \times p$, showing the variance of the sample means (for the p points that have already been sampled) on the main diagonal. Σ_ϵ thus reflects the so-called *intrinsic uncertainty* of the system, as it is caused by the (heterogenous) noise in the output observations. Σ_M, by contrast, is the $p \times p$ matrix with the estimated covariance between the outputs of each couple of already sampled points; this covariance is modeled using a *covariance* function that assumes spatial correlation, i.e., it assumes that the correlation between the outputs at two distinct input locations increases as the two input locations are closer to each other in the design space. We refer to the original publication by [2] for more details on the usual assumptions of this covariance function, and the different types of spatial correlation models (or *kernels*) that can be used. Analogously, the notation $\Sigma_M(\mathbf{x}, .)$ is the $p \times 1$ vector containing the covariances between the point under study, and the p already sampled points. The mean squared error (MSE) of this predictor is given by:

$$\sigma^{sk^2}(\mathbf{x}) = \Sigma_M(\mathbf{x}, \mathbf{x}) - \Sigma_M(\mathbf{x}, .)^T [\Sigma_M + \Sigma_\epsilon]^{-1} \Sigma_M(\mathbf{x}, .) + \frac{\Gamma^T \Gamma}{\mathbf{1}_p^T [\Sigma_M + \Sigma_\epsilon]^{-1} \mathbf{1}_p} \tag{4}$$

where $\Gamma = 1 - \mathbf{1}_p^T [\Sigma_M + \Sigma_\epsilon]^{-1} \Sigma_M(\mathbf{x}, .)$. The essential difference between Stochastic Kriging and Ordinary Kriging is the presence of Σ_ϵ in the expressions for the predictor and the MSE. In the absence of noise, Σ_ϵ disappears from the equations, and the predictor and MSE expressions boil down to the expressions of an Ordinary Kriging model.

3 Proposed Algorithm

Our algorithm uses a barrier function approach. In such approach, the constraints in the original problem are replaced by a penalty term in the objective function: this term penalizes the objective when its value approaches the boundaries of the feasible region and should thus ensure that solutions outside this feasible area are avoided. To develop our algorithm, we build on a recent study by [19], which compares different BO approaches for *deterministic* constrained problems using barrier functions. To avoid harsh discontinuities in the penalty function, the authors propose a log barrier function. For a derivation of the function and the subsequent alternative infill criteria proposed, we refer to the original publication; from the results, we select the following infill criterion as the basis for our (stochastic) alternatives:

$$\alpha(\mathbf{x}) = EI^{ok}(\mathbf{x}) + \sigma_f^{ok^2}(\mathbf{x}) \sum_{k=1}^{K} \left(\log\left(-\mu_{c_k}^{ok}(\mathbf{x}) \right) + \frac{\sigma_{c_k}^{ok^2}(\mathbf{x})}{2\mu_{c_k}^{ok^2}(\mathbf{x})} \right) \tag{5}$$

$$EI^{ok}(\mathbf{x}) = [f_{min} - \mu^{ok}(\mathbf{x})]\,\Phi\left(\frac{f_{min} - \mu^{ok}(\mathbf{x})}{\sigma^{ok}(\mathbf{x})}\right) + \sigma^{ok}(\mathbf{x})\phi\left(\frac{f_{min} - \mu^{ok}(\mathbf{x})}{\sigma^{ok}(\mathbf{x})}\right)$$

(6)

The notation $EI^{ok}(\mathbf{x})$ [13] refers to the Expected Improvement at input location $\mathbf{x}$, calculated by Ordinary Kriging information. In Eq. 6, f_{min} represents the best mean objective function value evaluated so far, Φ denotes the normal cumulative distribution and ϕ denotes the normal probability density function.

A straightforward adaptation of this function to the stochastic setting can then be obtained by replacing the Ordinary Kriging estimates with Stochastic Kriging estimates:

$$\alpha'(\mathbf{x}) = EI^{sk}(\mathbf{x}) + \sigma_f^{sk^2}\sum_{k=1}^{K}\left(\log\left(-\mu_{c_k}^{sk}(\mathbf{x})\right) + \frac{\sigma_{c_k}^{sk^2}(\mathbf{x})}{2\mu_{c_k}^{sk^2}(\mathbf{x})}\right)$$

(7)

$$EI^{sk}(\mathbf{x}) = [f_{min} - \mu^{sk}(\mathbf{x})]\,\Phi\left(\frac{f_{min} - \mu^{sk}(\mathbf{x})}{\sigma^{sk}(\mathbf{x})}\right) + \sigma^{sk}(\mathbf{x})\phi\left(\frac{f_{min} - \mu^{sk}(\mathbf{x})}{\sigma^{sk}(\mathbf{x})}\right)$$

(8)

where EI^{sk} stands for *expected improvement* calculated using Stochastic Kriging information. Yet, as previous research in [20], and [7] has shown that *modified expected improvement (MEI)* improves the search (compared to the traditional EI acquisition function) in stochastic settings, we opt to use the following criterion in our proposed algorithm:

$$\alpha''(\mathbf{x}) = MEI(\mathbf{x}) + \sigma_f^{sk^2}\sum_{k=1}^{K}\left(\log\left(-\mu_{c_k}^{sk}(\mathbf{x})\right) + \frac{\sigma_{c_k}^{sk^2}(\mathbf{x})}{2\mu_{c_k}^{sk^2}(\mathbf{x})}\right)$$

(9)

$$MEI(\mathbf{x}) = [\mu^{sk}(x_{min}) - \mu^{sk}(\mathbf{x})]\,\Phi\left(\frac{\mu^{sk}(x_{min}) - \mu^{sk}(\mathbf{x})}{\sigma^{ok}(\mathbf{x})}\right)$$
$$+\sigma^{ok}(\mathbf{x})\phi\left(\frac{\mu^{sk}(x_{min}) - \mu^{sk}(\mathbf{x})}{\sigma^{ok}(\mathbf{x})}\right)$$

(10)

Algorithm 1 outlines the proposed algorithm. We start with an initial LHS design consisting of $11d-1$ design points. We then run a fixed number of replications per design point, yielding a sample mean for the outcome of each function (objective and constraints), and an estimate of the variance on these sample means. The SK model is fit to the observations, and this model information is used in step 4 to select the next design point to sample (i.e., the point that maximizes the infill criterion in Eq. 9). After simulating the infill point, the meta-models are updated, and the algorithm continues to add extra infill points until the computational budget is depleted. It then proceeds to identify the optimal solution. We follow the most common approach (proposed in, e.g., [6,11,16]), which suggests identifying the optimal solution as the one that attains the *best predicted value* for the objective function across the search space (subject to all the constraints being satisfied with joint estimated probability larger than p_0).

Algorithm 1. SK/MEI

1: Construct the initial design set using a latin hypercube sample,
2: Replicate at each initial design point and update the set of sampled points,
3: Fit a Stochastic Kriging metamodel to the observed values,
4: Search and select the infill point with the highest α'',
5: Replicate at the selected point and update the set of sampled points,
6: If the computational budget is depleted, go to step 7; otherwise, return to step 4.
7: Return the solution that attains the *best predicted value* for the objective function
   while satisfying the minimum required probability of feasibility (p_0).
---

4   Numerical Experiments

In this section, we design experiments to assess the performance of the proposed algorithm. As, to the best of the authors' knowledge, there is no competing algorithm in the literature for constrained problems with *heteroscedastic noise* on all outputs, we benchmark our SK/MEI algorithm against two variants (see Table 1): (1) OK/EI: uses Ordinary Kriging to build surrogates and Expected Improvement as the first term in the acquisition function (i.e., the original algorithm proposed in [19]), (2) SK/EI: uses Stochastic Kriging to build surrogates and EI as the first term in the acquisition function. We assess the performance of the algorithms on the test problem proposed in [8]; this test problem has been widely used in the subsequent constrained BO literature (see e.g., [19], and [15]). It is formulated as follows:

$$\min f(x_1, x_2) = x_1 + x_2 \tag{11}$$

$$s.t. \ c_1(x_1.x_2) = \frac{3}{2} - x_1 - 2x_2 - \frac{1}{2}\sin\left[2\pi(x_1^2 - 2x_2)\right]$$

$$c_2(x_1.x_2) = -\frac{3}{2} + x_1^2 + x_2^2$$

Table 1. Algorithms' specifications

Algorithm	Surrogate	Acquistion function
OK/EI	Ordinary Kriging	$\alpha(\mathbf{x})$ (see Eq. 5)
SK/EI	Stochastic Kriging	$\alpha'(\mathbf{x})$ (see Eq. 7)
SK/MEI	Stochastic Kriging	$\alpha''(\mathbf{x})$ (see Eq. 9)

All functions (objective and two constraints) are perturbed by additive, heterogeneous Gaussian noise $\left(\epsilon_j(\mathbf{x}) \sim \mathcal{N}(0, \tau_j(\mathbf{x}))\right)$ using the approach proposed in [12], where $j = 1, 2, 3$. Each algorithm starts with an initial LHS design consisting of 21 points; it then sequentially selects 120 additional infill points to evaluate (simulating $n = 5$ replications at each point). To facilitate the optimization of the acquisition function in each iteration, we discretize the search space into a large but finite set of points (the so-called *candidate set*) (as common in the literature; see, e.g., [7] and [12]). We set the size of this candidate set equal to 1000 times the number of input dimensions (hence, 2000 for the problem considered). The algorithm can then simply evaluate the acquisition function value in all unvisited alternatives, and choose the alternative with the highest acquisition function value as the next infill point. At the end of each algorithm, we require a joint probability of feasibility of at least 0.9 ($p_0 = 0.9$) for the identification of the estimated optimum.

We repeat this experiment 25 times for each algorithm, yielding 25 macroreplications. Each macroreplication starts from a different initial design; this initial design is the same across all algorithms, to allow for a fair comparison. The candidate set remains the same across all algorithms, and across all macroreplications. Figure 2 shows the details of this candidate set, where solutions that are truly feasible (i.e., in the deterministic problem shown in Eq. (11)) are shown in green, while infeasible solutions are shown in red. In total, among 2000 solutions, we have 933 feasible and 1067 infeasible solutions. The black star represents the global optimum within this discretized set of solutions. Evidently, since we treat the problem as a black box, and the output functions are disturbed by noise, the three algorithms have no insight into this "true" problem. Instead, they sequentially "learn" about the different output functions, during the iterations of the algorithm, by using the different acquisition functions.

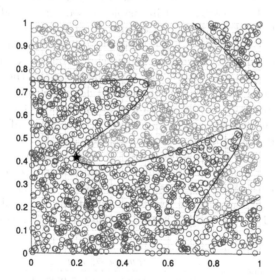

Fig. 2. Candidate solution set for the test problem proposed in [8], showing feasible solutions in green, and infeasible solutions in red).

5 Results

In this section, we discuss the results of the experiments. The three plots in the left column of Fig. 3 represent the 25 final optima, identified by the three algorithms respectively, in the 25 macro-replications. The OK/EI algorithm clearly fails to handle the problem; the resulting solutions are scattered across the solution space, and potentially far from the true optimal solution (shown as the green point in the plots.) SK/EI provides much more consistent results already, most of which are located in the immediate neighborhood of the true optimal solution. This remarkable improvement is solely due to the use of Stochastic Kriging instead of Ordinary Kriging, highlighting the importance of accounting for the intrinsic uncertainty in the metamodel. Finally, SK/MEI returns the most consistent and close solutions to the true global optimum.

This is shown even more clearly in the three plots in the right column of Fig. 3, where the horizontal axis shows, for the same 25 final optima, the *Euclidean distance (Δd, in the solution space)* to the true global optimum, whereas the vertical axis shows the *difference in the true objective value (Δf;* note that positive Δf values indicate final optima that have a true objective that is *better* than the global optimum, which indicates that these optima are, in reality, infeasible). The further from the origin, the worse the solution quality. Clearly, SK/MEI provides the best solutions for this problem.

Part of the difference in performance is due to the fact that the OK/EI and SK/EI algorithms do not succeed in recognizing the feasible versus infeasible areas of the search space to a sufficient extent. As an illustration, Fig. 4 shows the areas identified as feasible and infeasible (based on the calculated values for the joint probability of feasibility) at the end of the algorithm, for an arbitrary macroreplication. Clearly, OK/EI fails to correctly identify the borders of the feasible region. SK/EI performs significantly better (thanks to the use of SK), but is still outperformed by SK/MEI. Table 2 shows the mean values of some main performance metrics at the end of 25 macro-replications. The first column shows the number of feasible solutions wrongly identified as infeasible(*False Negative*), while the second column represents the number of infeasible solutions wrongly identified as feasible (*False Positive*). We also calculate *precision, recall,* and *F1-Score*. These measures are used in the information retrieval domain to measure how well a model retrieves the data, and are defined as follows. More details on these performance metrics can be found in [23]. As evident from the results, SK/MEI outperforms the two other algorithms w.r.t. all metrics. Both the use of SK, and the use of MEI in the acquisition function, contribute to the improvement.

$$precision = \frac{true\ positives}{true\ positives + false\ positives} \qquad (12)$$

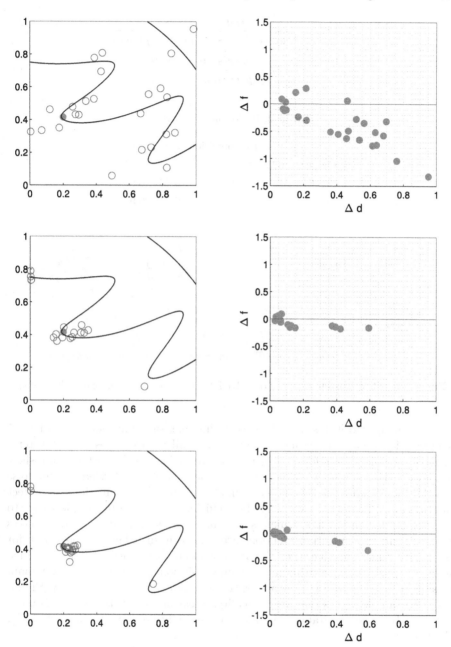

Fig. 3. Final optimal solutions obtained from 25 macro-replications (Left column) and Deviation from the optimal solution in the solution space and objective space (Right column): OK/EI(top row), SK/EI(center row), and SK/MEI(bottom row)

$$recall = \frac{true\ positives}{true\ positives + false\ negatives} \qquad (13)$$

$$F1 = 2 \times \frac{precision \times recall}{precision + recall} \qquad (14)$$

Table 2. Performance metrics

	FN	FP	precision	recall	F1
OK/EI	675	53	0.86	0.27	0.41
SK/EI	102	42	0.94	0.89	0.91
SK/MEI	50	8	0.98	0.94	0.96

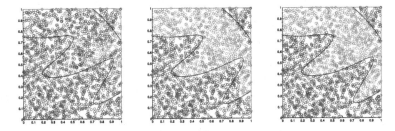

Fig. 4. Identified feasible region; OK/EI(left), SK/EI(center), and SK/MEI(right)

The three algorithms also show very different sampling behavior. Figure 5 shows which points are actually sampled as infill points, across the 120 iterations of each algorithm, for an arbitrary macroreplication. OK/EI does not show any systematic sampling pattern; it is, in fact, constantly misled by the noise (as this algorithm treats the noisy estimates of the output functions as perfect information). The other two algorithms efficiently manage to sample closer to the feasible region borders, which is desirable in this problem. Indeed, it shows that the barrier acquisition functions are successful in balancing the search for better goal function values (i.e., the first term in the acquisition functions, which tends to "push" the algorithm into the infeasible zone), and the desire to remain feasible (the second term in the acquisition functions). Note that also SK/EI and SK/MEI sample points that are, in reality, infeasible; yet, this behavior is not a problem. It is even desirable in view of (sequentially) learning the boundaries of the feasible region.

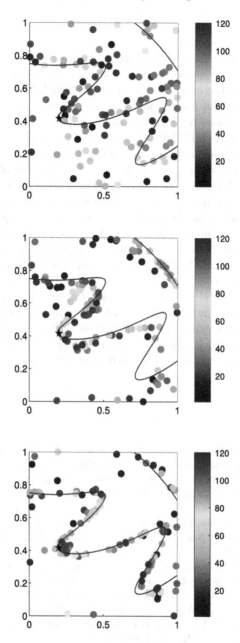

Fig. 5. Order of 120 sampled points by OK/EI(top), SK/EI(middle), and SK/MEI(bottom)

6 Conclusion

This research proposes a Bayesian optimization algorithm for constrained problems with heteroscedastic noise. The results show that using Stochastic Kriging significantly improves the optimization of systems with heteroscedastic noise: it significantly enhances the recognition of feasible versus infeasible areas of the search space, and leads to final optima that are closer to the true global optimum. Replacing EI with MEI further helps to improve the results; however, its impact is not as significant as the impact of using SK.

Last but not least, we want to highlight that, in spite of the encouraging results obtained by SK/MEI, a significant problem remains: in the identification step, the true global optimum was never identified as the final optimum, even though the algorithm has actually sampled this true global optimum during the search in 21 of the 25 macroreplications. This happens because of two main reasons: (1) sometimes the algorithm wrongly classifies the true optimum as an infeasible point, or (2) the algorithm classifies the true optimum as feasible, but some other point has (incorrectly) estimated to be better in terms of predicted goal value. These results show that, in a noisy setting, the current approach for identifying the final optimum is too simplistic: potential errors in the estimation of the probability of feasibility, and in the estimation of the goal value, are ignored. Further research is required on incorporating the uncertainty on these outcomes in the identification step, to further improve the algorithms for use in noisy problems.

References

1. Amaran, S., Sahinidis, N.V., Sharda, B., Bury, S.J.: Simulation optimization: a review of algorithms and applications. Ann. Oper. Res. **240**(1), 351–380 (2016)
2. Ankenman, B., Nelson, B.L., Staum, J.: Stochastic kriging for simulation metamodeling. Oper. Res. **58**(2), 371–382 (2010). https://doi.org/10.1287/opre.1090.0754
3. Fieldsend, J.E., Everson, R.M.: The rolling tide evolutionary algorithm: a multiobjective optimizer for noisy optimization problems. IEEE Trans. Evol. Comput. **19**(1), 103–117 (2014)
4. Frazier, P.I.: Bayesian optimization. In: Recent advances in optimization and modeling of contemporary problems, pp. 255–278. Informs (2018)
5. Frazier, P.I., Powell, W.B., Dayanik, S.: A knowledge-gradient policy for sequential information collection. SIAM J. Control. Optim. **47**(5), 2410–2439 (2008)
6. Gelbart, M.A., Snoek, J., Adams, R.P.: Bayesian optimization with unknown constraints. arXiv preprint arXiv:1403.5607 (2014)
7. Gonzalez, S.R., Jalali, H., Van Nieuwenhuyse, I.: A multiobjective stochastic simulation optimization algorithm. Eur. J. Oper. Res. **284**(1), 212–226 (2020)
8. Gramacy, R.B., et al.: Modeling an augmented Lagrangian for blackbox constrained optimization. Technometrics **58**(1), 1–11 (2016)
9. Gramacy, R.B., Lee, H.K.: Optimization under unknown constraints. In: Proceeding of the ninth Bayesian Statistics International Meeting, pp. 229–256. Oxford University Press (2011)

10. Hennig, P., Schuler, C.J.: Entropy search for information-efficient global optimization. J. Mach. Learn. Res. **13**(6), 1809–1837 (2012)
11. Hernández-Lobato, J.M., Gelbart, M.A., Adams, R.P., Hoffman, M.W., Ghahramani, Z.: A general framework for constrained Bayesian optimization using information-based search. J. Mach. Learn. Res. **17**(1), 1–53 (2016)
12. Jalali, H., Van Nieuwenhuyse, I., Picheny, V.: Comparison of kriging-based algorithms for simulation optimization with heterogeneous noise. Eur. J. Oper. Res. **261**(1), 279–301 (2017)
13. Jones, D.R., Schonlau, M., Welch, W.J.: Efficient global optimization of expensive black-box functions. J. Global Optim. **13**(4), 455–492 (1998)
14. Kleijnen, J.P.: Kriging metamodeling in simulation: a review. Eur. J. Oper. Res. **192**(3), 707–716 (2009)
15. Kleijnen, J.P., Van Nieuwenhuyse, I., van Beers, W.: Constrained optimization in simulation: efficient global optimization and karush-kuhn-tucker conditions (2021)
16. Letham, B., Karrer, B., Ottoni, G., Bakshy, E.: Constrained Bayesian optimization with noisy experiments. Bayesian Anal. **14**(2), 495–519 (2019)
17. Loeppky, J., Sacks, J., Welch, W.: Choosing the sample size of a computer experiment: a practical guide. Technometrics **51**(4), 366–376 (2009)
18. Močkus, J.: On Bayesian methods for seeking the extremum. In: Marchuk, G.I. (ed.) Optimization Techniques 1974. LNCS, vol. 27, pp. 400–404. Springer, Heidelberg (1975). https://doi.org/10.1007/3-540-07165-2_55
19. Pourmohamad, T., Lee, H.K.: Bayesian optimization via barrier functions. J. Comput. Graph. Stat. **31**(1), 74–83 (2022)
20. Quan, N., Yin, J., Ng, S.H., Lee, L.H.: Simulation optimization via kriging: a sequential search using expected improvement with computing budget constraints. IIE Trans. **45**(7), 763–780 (2013)
21. Rasmussen, C.E., Williams, C.: Gaussian Processes for Machine Learning. MIT press, Cambridge (2006)
22. Rojas-Gonzalez, S., Van Nieuwenhuyse, I.: A survey on kriging-based infill algorithms for multiobjective simulation optimization. Comput. Oper. Res. **116**, 104869 (2020)
23. Sammut, C., Webb, G.I.: Encyclopedia of Machine Learning. Springer, New York (2011)
24. Ungredda, J., Branke, J.: Bayesian optimisation for constrained problems. arXiv preprint arXiv:2105.13245 (2021)
25. Zeng, Y., Cheng, Y., Liu, J.: An efficient global optimization algorithm for expensive constrained black-box problems by reducing candidate infilling region. Inf. Sci. **609**, 1641–1669 (2022)

Hierarchical Machine Unlearning

HongBin Zhu[1], YuXiao Xia[1], YunZhao Li[1], Wei Li[2], Kang Liu[3(✉)],
and Xianzhou Gao[2]

[1] Big Data Center, State Grid of China, Beijing 100012, China
[2] State Grid Smart Grid Research Institute, Shanghai 200016, China
[3] State Key Laboratory for Novel Software Technology, Department of Computer
Science and Technology, Nanjing University, Nanjing 200016, China
`mg20330035@smail.nju.edu.cn`

Abstract. In recent years, deep neural networks have enjoyed tremendous success in industry and academia, especially for their applications in visual recognition and natural language processing. While large-scale deep models bring incredible performance, their massive data requirements pose a huge threat to data privacy protection. With the growing emphasis on data security, the study of data privacy leakage in machine learning, such as machine unlearning, has become increasingly important. There have been many works on machine unlearning, and other research has proposed training several submodels to speed up the retraining process, by dividing the training data into several disjoint fragments. When the impact of a particular data point in the model is to be removed, the model owner simply retrains the sub-model containing this data point. Nevertheless, current learning methods for machine unlearning are still not widely used due to model applicability, usage overhead, etc. Based on this situation, we propose a novel hierarchical learning method, Hierarchical Machine Unlearning (HMU), with the known distribution of unlearning requests. Compared with previous methods, ours has better efficiency. Using the known distribution, the data can be partitioned and sorted, thus reducing the overhead in the data deletion process. We propose to train the model using the hierarchical data set after partitioning, which further reduces the loss of prediction accuracy of the existing methods. It is also combined with incremental learning methods to speed up the training process. Finally, the effectiveness and efficiency of the method proposed in this paper are verified by multiple experiments.

Keywords: Data privacy · Machine unlearning · Data deletion

1 Introduction

Currently, a large amount of data is used in machine learning. For example, the dataset for training medical diagnosis models contains a large amount of personal

Supported by science and technology project of Big Data Center of State Grid Corporation of China, "Research on Trusted Data Destruction Technology for Intelligent Analysis" (No. SGSJ0000AZJS2100107) and the National Key Research and Development Program of China under Grants 2020YFB1005900.

patient information; the dataset for recommendation systems has users' usage history on the Internet. The use of large amounts of data raises many questions about data security and privacy.

Data poison attack is a very common type of attack that compromises the security of a model [1]. Specifically, an attacker carefully designs the input data model for the training process to disrupt the true distribution of the data and cause the model to obtain the wrong output. Once the model receives a poisoning attack, the accuracy of the model will be reduced, and the output will be biased towards the direction chosen by the attacker. For a system trained on user data, an attacker can inject malicious data simply by creating a user account. Such data poisoning attacks force us to rethink the meaning of system security [2].

On the other hand, there is numerous research on data privacy leakage in machine learning models, among which the Membership Inference Attack refers to the attacker who reverses the training set of a model by analyzing the published machine learning model [3]. For example, an attacker can obtain the privacy information of a target by determining whether the target is in the dataset of a certain type of disease diagnosis model. Since the Membership Inference Attack does not require a specific model structure, but only calls the machine learning model interfaces provided by large Internet companies on the Web, it poses a great danger to privacy, and many users want the model trainer to remove their own data in the dataset and erase the influence of their personal data in the trained model. In response to data poisoning attacks, previous research has proposed to improve the robustness of the model to prevent training data poisoning. However, the various mechanisms previously used, including Differential Privacy, are not perfectly implemented for user requests to remove data and its influence. Therefore, researchers have started to use Machine Unlearning to satisfy the user's right to be forgotten [4].

The most straightforward way to forget learning data is to remove the samples requested to be forgotten from the original training data set and retrain the model from scratch. However, when the dataset size is large and frequent forgotten requests occur, retraining from scratch incurs a very high computational overhead. To reduce the overhead in the data forgetting process, the SISA (or Shared, Isolated, Sliced, Aggregated) training method has been proposed. In this method, how to completely forget the requested data is the key to the design of the forgettable training method. For the data unlearning request from the user, the data influence in the existing model has to be removed, and a new model with the same output distribution is obtained as the model obtained by retraining from scratch; on the other hand, the computational overhead of the unlearning process is an important metric in the design of machine untraining mechanism. Compared to retraining, unlearning must reduce computational overhead, achieve system redeployment more quickly, and obtain higher availability.

Based on the above analysis, the existing machine learning forgettable training methods broadly have the following problems:

- Most existing machine unlearning methods, which are not very applicable, need to be adapted by security professionals for specific models. For example, the machine learning model transformed into a statistical query approach proposed by Cao et al. is difficult to apply to more complex learning models, such as neural network models. Such untraining methods are difficult to be used on a large scale.
- Converting a regular model to a forgettable model can affect the usability of the model. In the SISA model, there is a 3%–15% loss in the accuracy of the original model depending on the data set, the number of slices, the complexity of the model, and other settings.
- The existing machine unlearning methods take a lot of time. Although machine unlearning can save some time compared to retraining from scratch, it still adds a large time overhead that limits the application of machine unlearning in various types of systems.

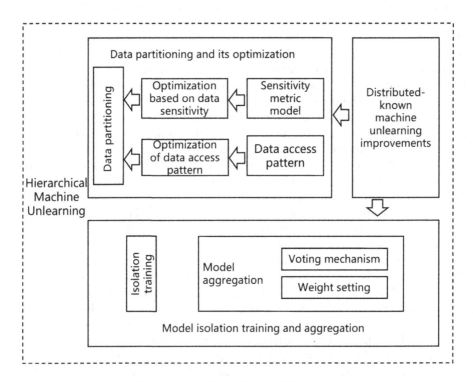

Fig. 1. Hierarchical Machine Unlearning

To address the above problems, this paper proposes a hierarchical machine unlearning training method(see Fig. 1). On the one hand, through the analysis of known data unlearning request probability, the original dataset is divided into low-unlearning-probability data and high-unlearning-probability data to reduce the computational overhead of the unlearning process; on the other hand, this paper achieves a compromise between model availability and unlearning time by fixing the parameters between model layers. In general, the contributions of this paper include the following aspects:

1. To address the problem of low applicability of the existing forgettable training methods, we reduce the cost of the unlearning process. To achieve this aim, we design a novel dataset partitioning method and model isolation learning strategy without introducing a complex transformation process.
2. Optimize data grouping. For data grouping with differentiated distribution, the original data set is divided into low-forgetting-probability data and high-forgetting-probability data, followed by optimization based on data sensitivity, hierarchical design of the model, and optimization of the access pattern of the data, thus reducing the additional overhead.
3. Based on the proposed method, the optimization effects of array grouping, and data access patterns are verified through theoretical analysis and relevant experiments. The results show that the method proposed in this paper can reduce the overhead of the data-forgetting process while improving the model prediction effect.

Section 2 of this paper provides a review of related work; Sect. 3 presents the background knowledge and relevant theoretical foundations applied; Sect. 4 presents the proposed hierarchical machine learning forgettable training method; Sect. 5 gives the time overhead analysis of the proposed method; Sect. 6 presents the experimental design and results analysis; Sect. 7 concludes and perspectives the work of this paper.

2   Related Work

Machine unlearning aims at eliminating the impact of data in the model with low computational overhead and without completely retraining the model from scratch. The concept of machine unlearning was first introduced by Cao and Yang in 2015. It is an application of the right to be forgotten proposed by the European Union in machine learning.

Cao and Yang [4] proposed to transform the learning algorithm into a summation form after statistical query learning to decompose the dependencies between the training data. To remove a data point, the model owner simply removes the transformation of this data point from the summation that depends on this instance. However, Cao and Yang's algorithm is not applicable to learning algorithms that cannot be transformed into summation forms, such as neural networks.

Ginart et al. [5] studied the machine unlearning method for the K-Means clustering algorithm, which focuses on proposing an efficient learning algorithm for data deletion. Specifically, their algorithm constructs data deletion as an online problem and gives a time-efficient analysis of the optimal deletion efficiency. A data deletion operation can be defined as the random variables $A(D_{-i})$ and $R_A(D, A(D), i)$, which are distributionally equivalent for all D and i. But this technique cannot be generalized to other machine learning models. Ginart et al. pointed out that data unlearning in machine learning should be related to the model training intermediate process, inspiring that we can reduce the time of data forgetting by recording the intermediate process of model training.

Therefore, Bourtoule et al. [6] proposed a more general training method SISA (Shared, Isolated, Sliced, and Aggregated) for deep learning. The main idea of SISA is to divide the training data into several disjoint fragments, each of which trains a sub-model [6]. To remove a particular data point, the model owner simply retrains the submodel containing this instance. To further speed up the unlearning process, they propose to split each shard into several slices and store intermediate model parameters as each shard updates the model. However, the SISA method cannot make good use of unlearning probability information when applied to scenarios where the data unlearning probability is known. The special slicing approach proposed by Bourtoule et al. for scenarios where the data forgetting probability is known reduces the accuracy of model prediction as well as the usability of the model.

Another machine unlearning research aims to verify that model owners comply with data deletion requests. Sommer et al. proposed a backdoor-based approach. The main idea is to allow data owners in a Machine Learning as a Service (MLaaS) scenario to plant backdoors in their data and then train the model. When the data owner later requests the deletion of their data, they can verify that their data has been deleted by checking the success rate of the backdoor.

There are also many related works on privacy preservation for machine learning models, such as federation learning and differential privacy [8]. However, most of these approaches use cryptographic tools to protect data privacy or make data indistinguishable rather than perform efficient data deletion.

In response to the research of related work, there are no efficient machine unlearning methods in scenarios where the probability of data unlearning requests is known.

3   Preliminary

3.1   Machine Unlearning

Consistent with the SISA, in our work, the participants in machine unlearning consist of an honest service provider S and a set of users U. The service provider may be a large organization that collects information about its users; and each user $u \in U$ provides its data d_u to the service provider, forming the service provider's dataset D. The users have the right to request that the impact of their own data be removed from the dataset and the model from which the data is generated, and the service provider needs to provide the users with credible evidence proving that the data has been removed.

In machine learning, given a data set D, a target model M that better fits the data will be obtained by training in the hypothesis space. If we add new data d_u to the original data set D to obtain a new data set D', we can train on the new data set D' to obtain a new model M'. Since there is no effective method to evaluate the impact of the new data d_u on the model parameters, it is difficult to eliminate it in M' without setting up save points in advance. To assure the data provider that the effect of the data has been removed, the most convincing way to obtain the model is to retrain the model after removing specific data points from the dataset. The retraining results in a model that provides a strong privacy guarantee to the data provider. The goal of machine unlearning is to provide training as well as data unlearning methods equivalent to retraining.

Theorem 1. *Machine Unlearning. Let $D = d_i : i \in U$ denote the data set collected from the set of users U, and $D' = D \cup d_u$ denote the data set that contains forgotten data d_u. Let D_M denote the distribution of the model with training on D' and unlearning d_u using the mechanism M; and let D_{real} denote the distribution of the model obtained by the training mechanism M on D. If the two distributions D_M and D_{real} are equivalent, then M satisfies unlearning.*

Although the goal of machine unlearning is very clear, i.e., to remove the influence of certain data from the dataset in the existing model, the unlearning task is easy to implement due to the stochastic nature of machine learning algorithms, among other reasons. Existing unlearning methods all suffer from various problems, and the ideal unlearning method should satisfy the following requirements.

1. **Understandability.** Since the underlying retraining approach is well understood and implemented, arbitrary forgetting learning algorithms need to be easy to understand and simple to apply and correct for nonexperts.
2. **Usability.** If a large amount of data needs to be removed or if more typical data points must be removed, it is understandable that the accuracy of the model decreases. Even retraining the model from scratch can cause a drop in accuracy when data is destroyed. A better forgetting learning method should be able to control this degradation to a level similar to retraining the model.
3. **Provability.** Just like retraining, unlearning should be able to prove that the unlearning data points no longer have an impact on the model parameters. In addition, this proof should be concise and not require help from experts.
4. **Applicability.** For a good unlearning method, any machine learning model should be usable, and independent of the complexity or other properties of the model.
5. **Limited unlearning time.** Unlearning methods should be faster than retraining in all cases.
6. **No additional overhead.** Any usable unlearning method should not introduce additional computational overhead into the training process of the original complex computational model.

3.2 PAC Learning

The goal of a machine learning algorithm is to learn a mapping from a sample space X to a labeled space Y. According to PAC (Probably Approximately Correct) learning, a mapping c is said to be a target concept if for a mapping c

such that for any sample $(x, y), c(x) = y$ holds. The learning algorithm L knows nothing about the target concept but can describe the distribution of the data by accessing a known data set D. The set of all possible concepts considered by the learning algorithm L is called the "hypothesis space", denoted by H. Any possible value $h \in H$ in the hypothesis space is called a hypothesis. For a given dataset D, the learning algorithm L obtains a hypothesis h that is as close as possible to the target concept c by solving the objective function.

In the solving process of the machine learning algorithm, the randomness of the results comes from two main sources: the randomness in the training process and the randomness in the learning algorithm.

Randomness in the Training Process: Given a dataset D, it is usually necessary to first draw small batches of data randomly from the dataset, and the order in which the data are drawn varies from one training to another. In addition, training is usually parallel without explicit synchronization, meaning that the random data acquisition order of the parallel training process can make the training uncertain.

Randomness in the Learning Algorithms: Intuitively, the purpose of a learning algorithm is to find the optimal hypothesis h in a vast hypothesis space H. Usually, this hypothesis is defined by setting a fixed weight of parameters to the learning model. PAC learning considers the hypothesis h, which is as close as possible to the target concept c, as one of the many hypotheses that minimize the empirical risk. However, commonly used optimization functions, such as stochastic gradient descent, can only converge to one of several local minima for any convex loss function. Coupled with the randomness involved in training, it is very challenging to obtain the same final hypothesis h using the same learning algorithm for the same dataset D.

Due to the randomness of machine learning, it is difficult to quantify the impact of certain data points on the model, and it is difficult to remove them from the final model.

4   Hierarchical Machine Unlearning

In this paper, we propose a novel machine unlearning method, Hierarchical Machine Unlearning (HMU), to improve the efficiency of the SISA and achieve a trade-off between unlearning speed and prediction accuracy in the unlearning process. By analyzing the known unlearning probability of data, the original dataset is divided into S pieces of low-unlearning-probability data and S pieces of high-unlearning-probability data. After data partitioning, isolated learning models are first built using the low-unlearning-probability data, and then the high- unlearning-probability data are introduced into the corresponding models using incremental learning. Finally, the output model is obtained by an aggregation method. The structure of data processing and model training process in our model is shown in Fig. 2.

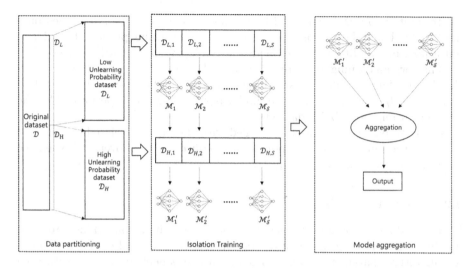

Fig. 2. The structure of Hierarchical Machine Unlearning, including data partitioning, isolation training and model aggregation

As with the SISA, we do not make assumptions about the specific algorithm of the machine learning models used, and whether these isolated models are homogeneous or heterogeneous. This is because partitioning the dataset for models of arbitrary structure makes no difference other than making the dataset accessible to the model smaller. And incremental learning can be applied to any machine learning algorithm that has iterative learning, which includes gradient descent class algorithms such as logistic regression and neural networks.

We still maintain the isolated training property of the SISA: no data are exchanged between the different isolated models during the iterative training process. Since the individual isolated models are trained on a smaller dataset only, this may reduce the model fitting ability, which we will perform in the experimental section in Chapter VI.

Assuming that the server in the scenario is honest, our unlearning algorithm can also provide data deletion guarantees when the server performs according to the designed algorithm.

4.1   Data Partitioning

Assuming that the data unlearning probability is known, all the data are sorted by the unlearning probability. Then, the original dataset D is divided into low unlearning probability dataset D_L and high unlearning probability dataset D_H, and $D_L \cap D_H = $ while $D_L \cup D_H = D$. The percentage of low unlearning probability data is $P_L = \frac{|D_L|}{|D|}$, and the percentage of high unlearning probability data is $P_H = \frac{|D_H|}{|D|}$.

To limit the impact generated by the unlearning data, the partitioned data needs to be shared. For the ordered data with low unlearning probability and high unlearning probability, they are divided into S equal-sized shards respectively.

The low unlearning probability dataset D_L should be partitioned into S shards, and for any different shards $D_{L,i}$ and $D_{L,j}$, they have $|D_{L,i}| = |D_{L,j}|$ and $D_{L,i} \cap D_{L,j} = \emptyset$. Similarly, the similar partitioning is also performed for the high unlearning probability dataset D_H.

By partitioning the dataset, combined with subsequent isolated training, we can limit the influence of data points, and reduce the time overhead when the server receives an unlearning request.

4.2 Isolation Training

In our work, the training of each pair of low and high unlearning probability shards is performed in isolation. In contrast, in a typical distributed training algorithm, parameter updates from each component are shared with each other, and the training iterations of the model are based on these joint parameter updates. Although isolated training affects the generalization ability of the whole model, this problem can be effectively avoided when the number of shards is appropriate [6]. Isolation is a subtle but powerful structure that provides concrete, provable, and intuitive guarantees for data deletion.

Since we train the model separately on each pair of low and high-unlearning-probability data shards, this limits the influence of unlearning data points on the overall model. The low-unlearning-probability data can affect at most the data in the same shard and the data in the corresponding high-unlearning-probability data shard, while the high-unlearning-probability data can only affect the data in the same shard. Finally, when the request for data unlearning arrives, we only need to retrain the affected part of the model. Since the shards are much smaller than the entire training dataset, this reduces the retraining time to achieve data deletion.

By further slicing the shared data and fixing the model parameter in incremental training, we can make the time overhead even lower.

In each shard, the data is further divided into smaller slices. Also, incremental training is performed on the slices and the model parameters are saved before introducing a new slice to reduce the retraining time.

For a shard d, the specific incremental training procedure on d can be performed by uniformly dividing it into R disjoint data slices d_i for $i \in \{1, ..., R\}$ such that $\cap_{i \in [R]} d_i = \emptyset$ and $\cup_{i \in [R]} d_i = d$ as follows:

1. Import the target model, randomly initialize the parameters, or initialize the parameters using the saved model, and train the model using the data slice d_1 to obtain model M_1 and save the model parameters.
2. On the basis of M_1, train model using $d_1 \cup d_2$ to obtain M_2 and save the model parameters.
3. Similarly, in step R, use $\cup_{i \in [R]} d_i$ to train model M_{R-1} and obtain model M_R and save model M_R as the output model M in this stage.

For any low unlearning probability data shard $d_{l,k}$, the corresponding model M_k can be obtained by incremental training and the parameters of the model are saved for the next training step.

For a model M_k trained with low unlearning probability data shard $D_{L,k}$, it can be further trained with the corresponding high unlearning probability data shard $D_{H,k}$. After the incremental training, we get the modified model M_k'.

In some scenarios, the number of high unlearning probability data may be much smaller than the number of low unlearning probability data, and we can improve the model prediction accuracy and speed up the model training by freezing some parameters of M_k model.

4.3   Model Aggregation

By fine-tuning the models using high unlearning probability data shard, S mutually isolated models can be obtained. Since the datasets for training S models are of the same size, these models have similar performance, it can avoid the decrease of prediction accuracy, when multiple learners are aggregated.

Common aggregation strategies are averaging, voting, and stacking methods. In machine unlearning, the aggregation strategy needs to have the following properties: the aggregation strategy should not involve training data, otherwise the aggregation process itself will have to be unlearning in some cases. Therefore, we can use the majority voting aggregation, where the overall prediction is the majority vote of all constituent models. It is not difficult to find that changing the aggregation strategy can improve or reduce the accuracy of the overall model. In practical application scenarios, we can choose properly from the candidate aggregation strategies according to the scenario-specific requirements.

5   Time Overhead

Experimentally, the number of data points affected by the unlearning process and the unlearning time are proportional, so we can analyze the impact of each parameter on the retraining time by analyzing the number of samples affected by the unlearning process.

Since we divide the dataset into two parts, low unlearning probability dataset D_L and high unlearning probability dataset D_H, so the retraining time overhead can be computed differently when the unlearning data appearing in D_L or appearing in D_H.

Suppose there are a total of K unlearning requests, the probability that the i-th unlearning request occurs in D_L is denoted as p_L, and the probability that it occurs in D_H is denoted as p_H, according to the dataset partitioning rule, $p_L << p_H$.

If the i-th unlearning request occurs in D_H, the expected number of data points to be affected is

$$\sum_{j=0}^{i-1} \binom{i-1}{j} (\frac{p_H}{S})^j (1 - \frac{p_H}{S})^{i-j-1} (\frac{D_H}{S} - 1 - j).$$

If the i-th unlearning request occurs in D_L, the expected upper bound on the number of data points to be affected is

$$\sum_{j=0}^{i-1} \binom{i-1}{j} (\frac{p_L}{S})^j (1 - \frac{p}{S})^{i-j-1} (\frac{D_L}{S} - 1 - j) + \frac{D_H}{S}.$$

Combining above two equations, the expected upper bound on the number of all data points in each unlearning request is

$$p_H \frac{D_H}{S} + p_L (\frac{D_H}{S} + \frac{D_L}{S}).$$

The SISA model has an upper bound on the number of samples expected in each forgetting request of $\frac{|D|}{S}$, and since $|D| = |D_H| + |D_L|$ and [8] has been shown that $|D_H| << |D|$, the hierarchical machine unlearning algorithm has a large improvement in unlearning time compared to the SISA model.

6 Experiment

Our experiment is designed to test the performance of our HMU training method in the scenario where the service provider has the information about the nature of the distribution of the unlearning request. We perform some studies to answer the following questions:

1. What is the impact of data partitioning on accuracy and retraining time?
2. Does HMU training have a better prediction accuracy than SISA training?
3. Does HMU training improve the retraining time?

The information of the datasets we used is shown in Table 1. The datasets we selected are diverse in terms of the total number of samples, data dimensionality, and samples per category. This allows us to explore tasks of different complexity, where the MNIST dataset and the Purchase dataset are simple, while the SVHN [10] dataset and CIFAR-10 [11] are relatively complex.

As in previous work related to machine unlearning, in the Purchase dataset, we select the 600 most purchased items as category attributes. SVHN is a real-world image dataset for machine learning and object recognition algorithms with minimal requirements for data pre-processing and formatting. The CIFAR-10 dataset consists of 60000 32 × 32 colour images in 10 classes, with 6000 images per class. There are 50000 training images and 10000 test images. All these datasets are commonly used in machine learning.

Table 1. Dataset Information

Dataset	Dimensionality	Size	#Class
Purchase	600	250000	2
SVHN	$32 \times 32 \times 3$	604833	10
CIFAR-10	$32 \times 32 \times 3$	60000	10

Table 2. DNN Model Features

Dataset	Model Architecture
Purchase	2 FC layers
SVHN	Wide ResNet-1-1
CIFAR-10	ResNet-18

We choose three different model structures in our experiments, and the specific model information is shown in Table 2, which includes various deep neural networks with different numbers of hidden layers and different layer sizes.

For each of the three different datasets, we assume that the data unlearning probability distribution is exponential.

First, we test the impact of data partitioning on accuracy and retraining time. In our experiment, we choose a different ratio of the size of low-unlearning-probability data to that of high-unlearning-probability data.

In Fig. 3, the prediction accuracy and retraining time of the output model are shown as the ratio of P_L and P_H values. We can find that by adjusting the data partitioning, we can speed up the retraining of the model without significantly reducing the prediction accuracy.

Then, we test the prediction accuracy of the trained models using the SISA method and our method with the same number of shards.

Figures 4 and 5 show the comparison of the prediction accuracy of the proposed hierarchical machine unlearning training method with the SISA training method for purchase data and SVHN data under different shards counting scenarios, respectively.

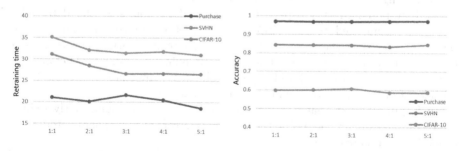

Fig. 3. Retraining Time and Accuracy Variation Depending on Different Data Partitioning when the number of shards is 6.

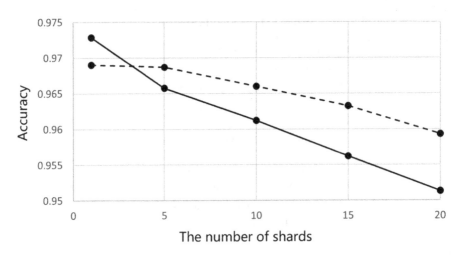

Fig. 4. Purchase Dataset Model Prediction Accuracy with Different Numbers of Shards when $P_L : P_H = 4 : 1$.

The prediction accuracy comparison shows that the prediction accuracy of the SISA method is higher when the number of shards is 1, i.e., when no data shard is used, because in the hierarchical machine forgetting learning method, even if the number of shards is 1, the data are partitioned according to the unlearning probability, which reduces the fitting ability of the model; when the number of shards is greater than 1, in both sets of experiments mentioned above, SISA shows a more significant decrease in prediction accuracy.

This is because the SISA training method uses a smaller amount of data with high forgetting probability to train a partially isolated model, and obtains a partially weak learner, which affects the prediction accuracy of the overall model through model aggregation, and this decrease becomes more obvious with the increase in the number of shards.

In contrast, in our method, since each isolated model dataset is of the same size and has similar learning ability, there is no drop in accuracy due to the imbalance in learning ability.

According to the analysis in Sect. 5, the hierarchical machine unlearning method proposed in this paper affects fewer data points in the data deletion phase compared to the SISA training method under the condition of $|D_H| << |D_L|$, since the experiments simulate the realistic situation with a small amount of high-unlearning-probability data. By fixing the parameter before fine-tuning the models using high-unlearning-probability data shares, we can achieve a trade-off of retraining time and prediction accuracy.

In Fig. 6, we show that our proposed HMU model can significantly reduce the retraining time by increasing the number of layers of frozen parameters with a small loss in model accuracy.

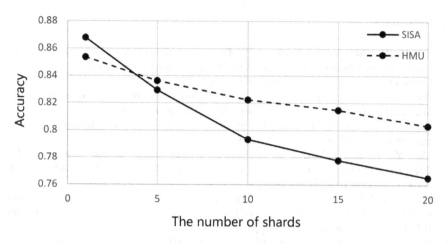

Fig. 5. SVHN Dataset Model Prediction Accuracy with Different Numbers of Shards when $P_L : P_H = 4 : 1$.

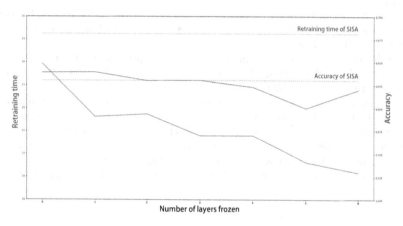

Fig. 6. Performance of HMU and SISA models on the CIFAR-10 dataset with varying numbers of frozen layers when the number of shards is 4 and $P_L : P_H = 4 : 1$.

7   Conclusion

In this paper, we propose an improvement of the SISA machine unlearning training method. When data unlearning requests with different distributions are known, the original dataset is first divided into low-unlearning-probability data and high-unlearning-probability data. Based on the divided data, multiple isolation models are trained, and the two types of data are put into the training of the models in an incremental learning manner. Through experiments, we demonstrate that our scheme improves the prediction accuracy and reduces the time of data deletion compared to the SISA method.

References

1. Biggio, B., Nelson, B., Laskov, P.: Poisoning attacks against support vector machines. In: Proceedings of the 29th International Conference on International Conference on Machine Learning, Edinburgh, Scotland (2012)
2. Steinhardt, J., Koh, P.W.W., Liang, P.S.: Certified defenses for data poisoning attacks. In: Advances in Neural Information Processing Systems, vol. 30 (2017)
3. Shokri, R., Stronati, M., Song, C., Shmatikov, V.: Membership inference attacks against machine learning models (2017)
4. Cao, Y., Yang, J.: Towards making systems forget with machine unlearning. In: 2015 IEEE Symposium on Security and Privacy, 01 May 2015. IEEE (2015). https://doi.org/10.1109/sp.2015.35
5. Ginart, A., Guan, M., Valiant, G., Zou, J.Y.: Making AI forget you: data deletion in machine learning. In: Proceedings of the 33rd International Conference on Neural Information Processing Systems, p. Article 316 (2019)
6. Bourtoule, L., et al.: Machine unlearning. In: 2021 IEEE Symposium on Security and Privacy (SP), 01 May 2021. IEEE (2021). https://doi.org/10.1109/sp40001.2021.00019
7. Gupta, V., Jung, C., Neel, S., Roth, A., Sharifi-Malvajerdi, S., Waites, C.: Adaptive machine unlearning. arXiv:2106.04378. https://ui.adsabs.harvard.edu/abs/2021arXiv210604378G
8. Wei, K., et al.: Federated learning with differential privacy: algorithms and performance analysis. IEEE Trans. Inf. Forensics Secur. **15**, 3454–3469 (2020). https://doi.org/10.1109/TIFS.2020.2988575
9. Bertram, T., et al.: Five years of the right to be forgotten. In; Proceedings of the 2019 ACM SIGSAC Conference on Computer and Communications Security, 06 Nov 2019. ACM (2019). https://doi.org/10.1145/3319535.3354208
10. Netzer, Y., Wang, T., Coates, A., Bissacco, A., Wu, B., Ng, A.: Reading digits in natural images with unsupervised feature learning. In: NIPS (2011)
11. Krizhevsky, A.: Learning multiple layers of features from tiny images (2009)

Explaining the Behavior of Reinforcement Learning Agents Using Association Rules

Zahra Parham[1]([✉]), Vi Tching de Lille[2], and Quentin Cappart[1]

[1] Ecole Polytechnique de Montréal, Montreal, Canada
{zahra.parham,quentin.cappart}@polymtl.ca
[2] StockholmSyndrome.ai, Montreal, Canada
vitching@stockholmsyndrome.ai

Abstract. Deep reinforcement learning algorithms are increasingly used to drive decision-making systems. However, there exists a known tension between the efficiency of a machine learning algorithm and its level of explainability. Generally speaking, increased efficiency comes with the cost of decisions that are harder to explain. This concern is related to *explainable artificial intelligence*, which is a hot topic in the research community. In this paper, we propose to explain the behaviour of a black-box sequential decision process, built with a deep reinforcement learning algorithm, thanks to standard data mining tools, i.e. association rules. We apply this idea to the design of *playing bots*, which is ubiquitous in the video game industry. To do so, we designed three agents trained with a deep Q-learning algorithm for the game *Street Fighter Turbo II*. Each agent has a specific playing style and the data mining algorithm aims to find rules maximizing the lift, while ensuring a minimum threshold for the confidence and the support. Experiments show that association rules can provide insights on the behavior of each agent, and reflect their specific playing style. We believe that this work is a next step towards the explanation of complex models in deep reinforcement learning.

Keywords: Association Rules · Explainable Reinforcement Learning

1 Introduction

Multiplayer video games refer to video games that involve more than one person playing together at the same time, either as a team (cooperative game) or as opponents (competitive game). In such games, the interactions between players are of the utmost importance and must be carefully designed in order to make the game enjoyable. However, ensuring and maintaining proper interactions all throughout the playing session is a hard goal to achieve in practice. First, enough people must be available to play the game, and second, people must remain active until the end of the game. Besides, the more people are involved in the game, the more difficult it is to ensure these goals. For instance, each session of the massively-played game *Leagues of Legends* is roughly about 30 min and

M. Sellmann and K. Tierney (Eds.): LION 2023, LNCS 14286, pp. 107–120, 2023.
https://doi.org/10.1007/978-3-031-44505-7_8

involves 10 players, split into 2 teams. Having only one player leaving the game deteriorates general satisfaction, especially for the impacted team.

A natural solution to this issue is to integrate artificial agents, dedicated to mimic the behaviour of human players. Such agents are commonly referred to as *bots*. When a person is missing or leaves the game before the end, the bot will replace the player. Despite the simplicity of this idea, building believable and fun-to-play bots is a non-trivial task: they must be developed specifically for each game and their behaviour in the game must remain realistic for the other players. An additional asset is to be able to replace the player with a bot having a similar playing style as the replaced player in order to smooth the transition. Albeit possible and already used by big video game companies, building efficient and human-like bots are generally beyond the range of independent studios with limited resources. It is why an innovative way to program bots should be designed. The requirements are as follows: the approach should be (1) *generic*, meaning that it must be possible to use the approach for different games, (2) *credible* as it should mimic a human behaviour, and (3) *transparent*, in the sense that a developer must be able to understand the rationale behind the actions of the bot and to re-calibrate it if required.

In another context, *reinforcement learning* [26] has been successfully applied to various kinds of video games in combination with *deep learning* [15], *imitation learning* [20], and *league-style training*, such as for Dota 2 [4], Minecraft [9], or Doom [13]. The idea is to let the bot play the game, reward it when appropriate actions are performed, and use this reward as feedback to train the agent. Once trained, the agent can then be used for new sessions of the game. Provided that the learning was successful, good performances from the bot are expected. There exist in the literature a plethora of learning algorithms that can be used in this context. Notable examples are DQN [19], PPO [25], DDPG [17], or SAC [10]. From an industrial point of view, the main benefit of this approach is that the training algorithm is generic, only the definition of the environment changes from one game to another. This directly ensures the first requirement about the genericity of the approach. However, the requirements on the credibility and transparency remain unaddressed. Broadly speaking, these concerns are related to *explainable artificial intelligence* [6]. It must be possible to understand and to trace why specific predictions are performed by a model. By doing so, the confidence that we can have in the model is increased. This aspect is critical for many real applications, such as in healthcare [21]. Although widely studied for supervising learning approaches [5], there are fewer methods dedicated to explainability in reinforcement learning [23], and even less that are applied on the video game industry [18].

Based on this context, we propose to use data mining tools, such as *association rules* [2,22], in order to provide meaningful information that can be used to infer explanations about decisions carried out by reinforcement learning agents. The reason behind this choice is that the explanations provided by association rules are tightly related to expert knowledge injected in traditional bots. It is done as follows: when the agent is deployed on a game, it generates samples consisting of a set of observations and of the decision that it has carried out. The

(a) Action: jumping (Player 1 - red).   (b) Action: punching (Player 2 - yellow).

Fig. 1. Illustrations of Street Fighter Turbo II game.

proposed idea is to use association rule mining tools in order to detect which components of the observations often involve specific decisions. Using this tool, we propose to find rules maximizing the *lift*, while ensuring a minimum threshold for the *confidence* and the *support*. Such metrics, common in data mining, are defined subsequently. Briefly, this turns in finding rules that are the most significant while ensuring constraints on their occurrences in the set of observations. Then, we can deduce that these observations are features that often trigger the decision. We evaluate this idea on the 2-players competitive game *Street Fighter Turbo II*. We trained three agents for this game. Each of them is characterized by a specific playing style (aggressive, defensive, and balanced) and has been trained accordingly. The mined rules that we computed from millions of samples obtained from each agent show that we can discriminate each agent by its playing style and thus explain their specific behavior.

The paper is organized as follows. The next section presents the nature of the game and the preprocessing steps that have been done. Then, Sect. 3 formally describes the reinforcement learning environment. The subsequent section introduces the reinforcement learning agent interacting with the environment. The rule extraction methodology is then explained in Sect. 5. Finally, Section 6 presents the rules that we obtained for each agent.

2   Case Study: Street Fighter Turbo II

Street Fighter II Turbo is a competitive fighting game released by Capcom for arcades in 1992. Briefly, the game features two opponents. The goal of each player is to deplete the health of the opponent before the timer expires. The winner is the surviving player or, in case of timeout, the player having the most remaining health. To do so, each player can perform a variety of actions, such as moving forward, moving backward, jumping, crouching, kicking the enemy, etc. Illustrations of the game interface with two actions are proposed in Fig. 1.

From the point of view of a human player, actions are performed thanks to a predefined combination of keys on a keyboard. By limiting the combination to at most 2 keys as proposed in [7], we consider 21 different actions. They are summarized in Table 1. The raw environment of any video game is the visual frames displayed to the player, i.e., a grid of pixels. Although such an input can be successfully leveraged by deep learning architectures, i.e., thanks to a convolutional neural network [16], it does not give an input that is understandable by humans. For such a reason, the first step is to pre-process the visual frames and to translate them into a set of high-level features. To do so, we used *BizHawk* emulator[1] to obtain low-level information located in the RAM and related to a specific state of the game. Among the many features available in the RAM, we have extracted few of them, with the possible values they can take. A summary of them is proposed in Table 2. As we can see, we have 15 different features, 12 of them are related to a specific player, 2 of them relates to both players, and the last one (remainingTime) is a general status of the game.

Table 1. List of available actions.

Action name	Description
movingForward	The agent walks in the forward direction
movingBackward	The agent walks in the backward direction
jumping	The agent jumps straight up
jumpingForward	The agent jumps in the forward direction
jumpingBackward	The agent jumps in the backward direction
jumpingWithKicking	The agent jumps in the forward while kicking the opponent
neutralJumpingStrong	The agent punches medium while jumping
farStandingRoundhouse	The agent kicks hard while standing far
farStandingFierce	The agent punches hard while standing far
farStandingJab	The agent punches light while standing far
farStandingShort	The agent kicks light while standing far
farStandingForward	The agent kicks high side while standing
crouchingShort	The agent kicks low while crouching
crouchingForward	The agent kicks with good reach while crouching
crouchingStrong	The agent punches medium crouching
crouchingJab	The agent punches light while crouching
crouchingFierce	The agent punches hard while crouching
crouchingRoundhouse	The agent kicks hard while crouching
sitDown	The agent sits down in place
sitBackward	The agent walks in the backward direction while siting down
idling	The agent stays in place without doing any action

3 Definition of the Environment

The first step in reinforcement learning is to define an *environment* as a *Markov Decision Process* (MDP). Briefly, let $\langle S, A, T, R \rangle$ be a tuple representing a deter-

[1] https://github.com/TASEmulators/BizHawk.

Table 2. Summary of the observations used to create the environment.

Observation name	Domain	Description
isMoving(p_1)	$\{0, 1\}$	Indicate if the player p_1 is currently moving
isMoving(p_2)	$\{0, 1\}$	Indicate if the player p_2 is currently moving
isCrouching(p_1)	$\{0, 1\}$	Indicate if the player p_1 is currently crouching
isCrouching(p_2)	$\{0, 1\}$	Indicate if the player p_2 is currently crouching
isJumping(p_1)	$\{0, 1\}$	Indicate if the player p_1 is currently jumping
isJumping(p_2)	$\{0, 1\}$	Indicate if the player p_2 is currently jumping
horizontalCoord(p_1)	$[0, 498]$	The horizontal coordinate of player p_1
horizontalCoord(p_2)	$[0, 498]$	The horizontal coordinate of player p_2
verticalCoord(p_1)	$[0, 204]$	The vertical coordinate of player p_1
verticalCoord(p_2)	$[0, 204]$	The vertical coordinate of player p_2
health(p_1)	$[0, 176]$	The remaining health of player p_1
health(p_2)	$[0, 176]$	The remaining health of player p_2
horizontalDelta(p_1, p_2)	$[0, 189]$	The horizontal distance between player p_1 and p_2
verticalDelta(p_1, p_2)	$[0, 158]$	The vertical distance between player p_1 and p_2
remainingTime	$[0, 99]$	The time until the end of the game (99 s in total)

ministic and fully observable environment, where S is the set of states, A is the set of actions that an agent can perform inside the environment, $T : S \times A \rightarrow S$ is the transition function leading the agent to another state, and $R : S \times A \rightarrow \mathbb{R}$ is a function rewarding (or penalizing) the realization of an action $a \in A$ from a state $s \in S$. The behaviour of an agent is defined by a policy $\pi : S \rightarrow A$, indicating the action to be performed on a specific state. The goal of an agent is to learn a policy maximizing the accumulated reward during its lifetime, defined as a sequence of states $s_t \in S$ with $t \in \{1, \ldots, \theta\}$. This is commonly referred to as an *episode*. The final state s_θ is referred to as the *terminal state* and is commonly reached when a halting condition is reached. This formalization is common in any task related to reinforcement learning [26]. The model we have designed for *Street Fighter II Turbo* is as follows:

State A state $s \in S$ is defined as a vector $\langle x_1, \ldots, x_{15} \rangle$ of 15 features. It corresponds to the 15 observations summarized in Table 2. A state is terminal when one of these three conditions is fulfilled: (1) the health of the first player is depleted, i.e., health(p_1) $= 0$, (2) the health of the second player is depleted, i.e., health(p_2) $= 0$, or (3) when the timer is exceeded, i.e., remainingTime $= 0$.

Action An action $a \in A$ simply corresponds to an available action proposed in Table 1. There are then 21 possible actions that the agent can perform inside the environment.

Transition The transition function updates the current state s_t according to the action performed at the time t. The definition of the transition directly relies on the game mechanisms as executed by the emulator. For instance, assuming that farStandingJab is an action performed by the first player and that deals 40 damages to the opponent, the state information s_{t+1} related to health(p_2) gets the value health(p_2) − 40. Internally, the transition between two states corresponds to 7 visual frames.

Reward The goal of the reward is to encourage the agent to perform actions that will lead it to win the game. A simple way to define the reward is to only give a positive value when the agent wins the game, and a negative value when it loses it. This reward signal is defined in Eq. (1). It indicates that an action a performed at a state s is positively rewarded if the next state is a terminal state corresponding to a victory for the first player. On the other hand, it is negatively rewarded (i.e. punished) in case of a defeat.

$$R^{\text{final}}(s_t, a) = \begin{cases} 1 & \text{if health}(p_1) > \text{health}(p_2) \wedge \text{isTerminal}(s_{t+1}) \\ -3 & \text{else if health}(p_1) < \text{health}(p_2) \wedge \text{isTerminal}(s_{t+1}) \\ 0 & \text{otherwise} \end{cases} \quad (1)$$

The values of 1 and −3 have been calibrated manually. The drawback is that non-zero rewards are collected only at the end of an episode. This yields the *sparse reward* issue, which is known to complicate the training process [24]. We tackle this issue by introducing an intermediate reward, that can be collected in the middle of an episode. This is also known as a *reward shaping* method. The idea is to evaluate the impact of an action on the remaining health of both players. Intuitively, each health point depleted from the opponent will be rewarded, and each health point that is inflicted will be punished. Let $\delta_t^{\text{health}}(p) = \text{health}_{t+1}(p) - \text{health}_t(p)$ be the difference in the health for the player p between state s_{t+1} and s_t, the intermediate reward is defined in Eq. (2).

$$R^{\text{intermediate}}(s_t, a) = \alpha^{\text{win}} \delta_t^{\text{health}}(p_2) - \alpha^{\text{lose}} \delta_t^{\text{health}}(p_1) \quad (2)$$

On this equation, α^{win} is a positive coefficient giving incentive to the agent to deplete the health of the opponent, and α^{lose} a second coefficient giving it incentive to not lose health. Based on both equations, the reward function used in our model is as follows.

$$R(s_t, a) = R^{\text{intermediate}}(s_t, a) + R^{\text{final}}(s_t, a) \quad (3)$$

4   Learning Algorithm

The learning algorithm relies on a *deep Q-learning* approach [19]. Briefly, the idea is to estimate the quality of taking an action a from a state s. This estimation is referred to as a *Q-value* and is obtained thanks to a trained deep neural network. In our case, we used a fully-connected neural network of two hidden layers of 64 neurons each together with a ReLU activation [8]. The output is a real value for each action, corresponding to the *Q-value*. Once trained, the agent policy consists in always selecting the action that has the best *Q-value*.

Three agents have been designed in this work. Each of them has been trained between 2,000,000 and 3,000,000 time steps. This corresponds to around 10,000 game sessions and 15 h of training time on a Intel(R) Xeon(R) CPU @ 2.30 GHz CPU and a Tesla P100-PCIE-16 GB GPU using Adam optimizer [14]. Additionally, the game consists of three episodes (i.e. three rounds), but we only consider the first one in each game session. The playing style of the opponent is the one implemented by the game developers and natively integrated in the game. Finally, we would like to point out that building the most efficient agent is not the motivation of this paper. In contrast, our goal was to build only a decent agent and to show that its behaviour can be successfully explained thanks to association rules.

5   Explanation with Association Rules

This section describes the methodology we used to extract information explaining the behaviour of the trained agents. We do it by means of association rules [1]. Briefly, the idea is to extract relevant implications from a large database T of transactions. In our context, each transaction is defined as a set $\{I_1, \ldots, I_n\}$ of n items together with a specific item Y. An association rule is defined as an implication of the form $I \rightarrow Y$, where $I = \{I_j, \ldots, I_k\}$ is a subset of existing items. The goal is to find association rules that are the most frequent in T. Provided with this information, we can then infer that the item Y is often obtained when the items I are also present. Three metrics are commonly used for determining how relevant a rule is. There are as follows:

1. The *support* indicates how frequent a rule is. It is computed as the ratio between the transactions containing both I and Y with the total number of transactions in T. It is formalized in Eq. (4).

$$\text{Support}(I \rightarrow Y) = \frac{\#\{t \in T \mid I \in t \wedge Y \in t\}}{\#\{t \in T\}} \qquad (4)$$

2. The *confidence* is the ratio between the transactions containing both I and Y, with the transactions containing only I. It is formalized in Eq. (5).

$$\text{Confidence}(I \rightarrow Y) = \frac{\#\{t \in T \mid I \in t \wedge Y \in t\}}{\#\{t \in T \mid I \in t\}} \qquad (5)$$

3. The *lift* measures the performance at predicting the presence of both Y and I in a transaction against a random prediction. A lift of 1 indicates that the probabilities of occurrence of both I and Y are independent. In such a case, no relevant rule can be drawn involving those items. This measure is formalized in Eq. (6), where $\text{Support}(I)$ is the support of I in the database. Intuitively,

a lift of 2 shows that the Y of the corresponding rule is twice more likely to be present compared to the average.

$$\text{Lift}(I \rightarrow Y) = \frac{\text{Support}(I \cup Y)}{\text{Support}(I) \times \text{Support}(Y)} \tag{6}$$

In our case, we opted to find rules maximizing the lift, while ensuring a minimum support threshold of 0.01 and a minimum confidence threshold of 0.01 as well. The choice of maximizing the lift has been done arbitrarily and depending on the application, another metric could be considered. An additional analysis where we maximize the confidence instead is also proposed. There exist many algorithms in the literature for finding association rules, such as *Apriori* [2] or its variants [27, 28].

To obtain a database of transactions, we mapped the state of the reinforcement learning environment (i.e., observations from Table 2) with the $\{I_1, \ldots, I_n\}$ items, and the actions that are taken (i.e., actions from Table 1) with the Y item. We collected such information by letting the trained agents play the game 1,000 times. It roughly gives a database $\mathcal{T}$ between 100,000 and 200,000 transactions for each agent. It means that the agent performs between 100 and 200 actions per game. One difficulty that arose is that standard association rule mining algorithms assume that the items have categorical values. It is not the case of some observations of the environment, such as the remaining health of a player ($\text{health}(p)$). Although alternative algorithms exist in the literature for such a situation [11], we selected the option to discretize each numerical observation into a set of meaningful categories. The main reason is that categorical data are easier to interpret. A summary of this discretization is proposed in Table 3. For instance, the first discretization shows that we have a category rightOf if the horizontal coordinate of player p_1 is higher than the horizontal coordinate of player p_2. The rule extraction has been carried out in R using `arules` package[2] and was executed in less than 30 s for each agent.

Table 3. Summary of the discretization performed on the observations.

Modified observation	Domain	Categorical domain
horizontalCoord(p_1)	$[0, 498]$	rightOfP2 : hCoord(p_1) > hCoord(p_2)
horizontalCoord(p_2)		leftOfP2 : hCoord(p_1) < hCoord(p_2)
horizontalDelta(p_1, p_2)	$[0, 189]$	close : $[0, 63]$, middle : $[64, 126]$, far : $[127, 189]$
verticalCoord(p)	$[0, 204]$	jumping : $[0, 191]$, standing : $[192, 204]$
health(p)	$[0, 176]$	low : $[0, 58]$, medium : $[59, 117]$, high : $[118, 176]$
verticalDelta(p_1, p_2)	$[0, 158]$	Not used (redundant with verticalCoord)
remainingTime	$[0, 99]$	Not used (not player-dependant)

---

[2] https://github.com/mhahsler/arules/.

6   Analysis of the Rules Obtained

This section presents the best rules that we have been able to extract from the agents we trained. Three agents are presented: a *balanced*, a *defensive*, and an *aggressive* one. Their differences relates to the reward function that has been used to define the environment. Detailed information is proposed subsequently.

Rules for a Balanced Agent

The first agent we trained uses the reward function defined in Eq. (3) with $\alpha^{win} = \alpha^{lose} = 1$. Intuitively, this agent has an equal incentive to protect its health and to deplete the health of the opponent, hence its qualification of being balanced. The progression of the reward during the training phase is illustrated in Fig. 2. Then, Table 4 shows the top-5 rules obtained, sorted by their lift, while enforcing a minimum threshold of 0.01 for the support and the confidence. Interestingly, we can see that the agents perform both aggressive (jumpingWithKicking) and defensive (jumpingForward) actions with a relatively similar lift value. For instance, an interpretation of the first rule is that the agent is likely to jump toward the enemy when it is not too far (horizontalDelta(p_1, p_2) = middle) and on the left side of the opponent (horizontalCoord(p_1) = leftOfP2). The result is to land behind the opponent, which is a common tactical move for this game. Table 5 proposes a similar analysis when maximizing the confidence instead, while enforcing the support to be more than 0.01 and the lift more than 2. As we can observe, similar rules are obtained.

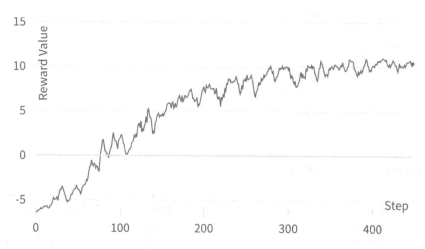

Fig. 2. Evolution of the reward during the training for the balanced agent.

Table 4. Top-5 rules obtained for the balanced agent (lift measure).

Rule antecedent (I)	Rule consequent (Y)	Support	Confidence	Lift
health(p_1) = high, isJumping(p_1) = 1, verticalCoord(p_2) = standing, health(p_2) = high	crouchingStrong	0.01	0.36	15.61
horizontalDelta(p_1, p_2) = middle, isMoving(p_2) = 1,isCrouching(p_2) = 0, horizontalCoord(p_1) = leftOfP2	jumpingForward	0.01	0.20	12.19
isCrouching(p_2) = 1, health(p_1) = high, health(p_2) = high, horizontalCoord(p_1) = rightOfP2, horizontalDelta(p_1, p_2) = middle	movingBackward	0.01	0.97	5.89
verticalCoord(p_2) = standing, isMoving(p_2) = 1	jumpingBackward	0.01	0.02	1.35
health(p_2) = high, isMoving(p_2) = 1, health(p_1) = high	jumpingWithKicking	0.02	0.98	9.76

Table 5. Top-5 rules obtained for the balanced agent (confidence measure).

Rule antecedent (I)	Rule consequent (Y)	Support	Confidence	Lift
verticalCoord(p_1) = jumping, health(p_1) = high, health(p_2) = high	jumpingWithKicking	0.02	0.98	9.76
isCrouching(p_1) = 0, verticalCoord(p_1) = standing health(p_1) = high, health(p_2) = medium, horizontalDelta(p_1, p_2) = middle, horizontalCoord(p_1) = rightOfP2	movingBackward	0.01	0.97	5.89
isJumping(p_2) = 1, isJumping(p_1) = 1, health(p_2) = high, isMoving(p_2) = 1, horizontalCoord(p_1) = leftOfP2	farStandingFierce	0.01	0.78	4.27
isMoving(p_1) = 1, isStanding(p_1) = 1 isJumping(p_2) = 1, health(p_1) = high, health(p_2) = high, horizontalDelta(p_1, p_2) = close, horizontalCoord(p_1) = leftOfP2	idling	0.02	0.69	4.62
isCrouching(p_1) = 1, isJumping(p_2) = 1, health(p_2) = low, horizontalDelta(p_1, p_2) = close,	sitDown	0.01	0.59	9.27

Rules for a Defensive Agent

As our goal is to assess whether association rules can find relevant rules explaining the behaviour of a reinforcement learning agent, we trained another agent, which has incentive to protect its health. This is done by setting $\alpha^{\text{win}} = 0$ and $\alpha^{\text{lose}} = 1$. Intuitively, the agent does not receive reward anymore if it hits the opponent, but it still gets punished if it loses health. Provided that our hypothesis is correct, the top rules should be related to more defensive actions. They are summarized in Table 6. A first observation is that the rules obtained are highly different than the ones related to the balanced agent. The first top-rule (crouchingFierce) is an in-between aggressive/defensive action and is obtained with a high lift value. The second top-rule (movingBackward) is a purely defensive action. This shows a playing style more defensive than the balanced agent.

Rules for an Aggressive Agent

Following the same idea, we performed the same analysis on an aggressive agent. It has been implemented by setting $\alpha^{win} = 1$ and $\alpha^{lose} = 0$. Intuitively, it still receives rewards when it hits the opponent, but it is not punished anymore when it loses health. Provided that our hypothesis is correct, the top rules should be related to more aggressive actions. They are summarized in Table 7. Compared to the defensive agent, roughly the same rules are obtained but within a different importance order. For instance, the second top-rule is now an attack instead of a defensive move. Another offensive move (jumpingWithKicking) also gains importance (lift of 12.3 instead of 5.73), showing that the agent is, as intended, more aggressive than the defensive one.

Table 6. Top-5 rules obtained for the defensive agent (lift measure).

Rule antecedent (I)	Rule consequent (Y)	Support	Confidence	Lift
isMoving(p_1) = 1, health(p_2) = medium, isCrouching(p_2) = 0, horizontalDelta(p_1, p_2) = close, isJumping(p_1) = 1, verticalCoord(p_2) = standing	crouchingFierce	0.01	0.60	31.10
isCrouching(p_1) = 0, isMoving(p_1) = 0, health(p_2) = high, isJumping(p_1) = 0, health(p_1) = high, horizontalDelta(p_1, p_2) = middle, horizontalCoord(p_1) = rightOfP2	movingBackward	0.01	0.97	19.27
isMoving(p_1) = 1, isJumping(p_1) = 1, horizontalDelta(p_1, p_2) = close, health(p_1) = high, verticalCoord(p_2) = jumping	crouchingJab	0.01	0.40	16.00
isCrouching(p_1) = 1, isMoving(p_1) = 0, health(p_2) = low, isCrouching(p_2) = 0, isMoving(p_2) = 1	sitDown	0.01	0.20	7.03
isCrouching(p_1) = 0, health(p_2) = medium, horizontalDelta(p_1, p_2) = close, health(p_1) = high, verticalCoord(p_2) = jumping, isMoving(p_2) = 1	jumpingWithKicking	0.01	0.10	5.73

Table 7. Top-5 rules obtained for the aggressive agent (lift measure).

Rule antecedent (I)	Rule consequent (Y)	Support	Confidence	Lift
isMoving(p_1) = 1, health(p_2) = medium, isJumping(p_1) = 1, isCrouching(p_2) = 0, health(p_1) = high, verticalCoord(p_2) = standing, horizontalDelta(p_1, p_2) = close	crouchingFierce	0.01	0.76	39.67
isMoving(p_1) = 1, isJumping(p_1) = 1, health(p_1) = high, horizontalDelta(p_1, p_2) = close, verticalCoord(p_2) = jumping, health(p_2) = medium, horizontalCoord(p_1) = leftOfP2	crouchingJab	0.01	0.50	20.29
isCrouching(p_1) = 0, isMoving(p_1) = 0, health(p_2) = high, isJumping(p_1) = 0, health(p_1) = high, horizontalDelta(p_1, p_2) = middle, horizontalCoord(p) = rightOf	movingBackward	0.01	0.97	19.28
isCrouching(p_1) = 0, health(p_2) = medium, health(p_1) = high, horizontalDelta(p_1, p_2) = close, verticalCoord(p_2) = jumping, isMoving(p_2) = 1, isJumping(p_1) = 0, horizontalCoord(p) = leftOf	jumpingWithKicking	0.01	0.94	12.3
isCrouching(p_1) = 1, isMoving(p_1) = 0, isCrouching(p_2) = 0, isMoving(p_2) = 1	sitDown	0.01	0.10	3.96

7   Conclusion and Future Work

Deep reinforcement learning is increasingly considered for driving decision making systems. However, the trade-off between the efficiency of a model and its explainability level is a challenge, which is critical for numerous applications. In this paper, we proposed to use association rules to explain the decisions performed by an agent trained by a deep reinforcement learning algorithm. We proposed an application, on *StreetFighter Turbo II* video game, and trained three agents, each with a specific playing style. The results obtained show that the playing style of an agent has an impact on the rules obtained and on their rank. This directly corroborates the hypothesis that association rules can be a relevant tool to explain the behaviour of a black-box decision process, such as one obtained by a deep reinforcement learning algorithms. Although our application is for the video game industry, the approach proposed is generic and could be considered for other applications of reinforcement learning. However, an important limitation is that the input observations must be intrinsically explainable. It is not always the case, especially when the inputs are a grid of pixels. Targeting this limitation is an interesting line of future work. We also point out that other data mining algorithms, such as classification rule learners [12] or subgroup discovery [3], could be also considered for this task.

References

1. Agrawal, R., Imieliński, T., Swami, A.: Mining association rules between sets of items in large databases. SIGMOD Rec. **22**(2), 207–216 (1993). https://doi.org/10.1145/170036.170072
2. Agrawal, R., Srikant, R., et al.: Fast algorithms for mining association rules. In: Proceedings of 20th International Conference on Very Large Data Bases, VLDB, vol. 1215, pp. 487–499. Citeseer (1994)
3. Atzmueller, M.: Subgroup discovery. Wiley Interdisc. Rev. Data Min. Knowl. Discovery **5**(1), 35–49 (2015)
4. Berner, C., et al.: Dota 2 with large scale deep reinforcement learning. CoRR abs/1912.06680 (2019). http://arxiv.org/abs/1912.06680
5. Burkart, N., Huber, M.F.: A survey on the explainability of supervised machine learning. J. Artif. Intell. Res. **70**, 245–317 (2021)
6. Došilović, F.K., Brčić, M., Hlupić, N.: Explainable artificial intelligence: a survey. In: 2018 41st International Convention on Information and Communication Technology, Electronics and Microelectronics (MIPRO), pp. 0210–0215. IEEE (2018)
7. Fletcher, A.: How we built an AI to play Street Fighter II - can you beat it? https://medium.com/gyroscopesoftware/how-we-built-an-ai-to-play-street-fighter-ii-can-you-beat-it-9542ba43f02b. Accessed 18 Nov 2022
8. Glorot, X., Bordes, A., Bengio, Y.: Deep sparse rectifier neural networks. In: Proceedings of the Fourteenth International Conference on Artificial Intelligence and Statistics, pp. 315–323. JMLR Workshop and Conference Proceedings (2011)
9. Guss, W.H., et al.: The minerl competition on sample efficient reinforcement learning using human priors. CoRR abs/1904.10079 (2019). http://arxiv.org/abs/1904.10079
10. Haarnoja, T., Zhou, A., Abbeel, P., Levine, S.: Soft actor-critic: off-policy maximum entropy deep reinforcement learning with a stochastic actor. In: International Conference on Machine Learning, pp. 1861–1870. PMLR (2018)
11. Hong, T.P., Kuo, C.S., Chi, S.C.: Mining association rules from quantitative data. Intell. Data Anal. **3**(5), 363–376 (1999)
12. Jovanoski, V., Lavrač, N.: Classification rule learning with APRIORI-C. In: Brazdil, P., Jorge, A. (eds.) EPIA 2001. LNCS (LNAI), vol. 2258, pp. 44–51. Springer, Heidelberg (2001). https://doi.org/10.1007/3-540-45329-6_8
13. Kempka, M., Wydmuch, M., Runc, G., Toczek, J., Jaskowski, W.: Vizdoom: a doom-based AI research platform for visual reinforcement learning. CoRR abs/1605.02097 (2016). http://arxiv.org/abs/1605.02097
14. Kingma, D.P., Ba, J.: Adam: a method for stochastic optimization. arXiv preprint arXiv:1412.6980 (2014)
15. LeCun, Y., Bengio, Y., Hinton, G.: Deep learning. Nature **521**(7553), 436–444 (2015)
16. LeCun, Y., Bengio, Y., et al.: Convolutional networks for images, speech, and time series. Handb. Brain Theory Neural Netw. **3361**(10), 1995 (1995)
17. Lillicrap, T.P., et al.: Continuous control with deep reinforcement learning. arXiv preprint arXiv:1509.02971 (2015)
18. Madumal, P., Miller, T., Sonenberg, L., Vetere, F.: Explainable reinforcement learning through a causal lens. In: Proceedings of the AAAI Conference on Artificial Intelligence, vol. 34, pp. 2493–2500 (2020)
19. Mnih, V., et al.: Playing atari with deep reinforcement learning. arXiv preprint arXiv:1312.5602 (2013)

20. Osa, T., et al.: An algorithmic perspective on imitation learning. Found. Trends® Rob. **7**(1–2), 1–179 (2018)
21. Pawar, U., O'Shea, D., Rea, S., O'Reilly, R.: Explainable AI in healthcare. In: 2020 International Conference on Cyber Situational Awareness, Data Analytics and Assessment (CyberSA), pp. 1–2. IEEE (2020)
22. Peake, G., Wang, J.: Explanation mining: post hoc interpretability of latent factor models for recommendation systems. In: Proceedings of the 24th ACM SIGKDD International Conference on Knowledge Discovery & Data Mining, pp. 2060–2069 (2018)
23. Puiutta, E., Veith, E.M.: Explainable reinforcement learning: a survey (2020)
24. Riedmiller, M., et al.: Learning by playing solving sparse reward tasks from scratch. In: International Conference on Machine Learning, pp. 4344–4353. PMLR (2018)
25. Schulman, J., Wolski, F., Dhariwal, P., Radford, A., Klimov, O.: Proximal policy optimization algorithms. arXiv preprint arXiv:1707.06347 (2017)
26. Sutton, R.S., Barto, A.G.: Reinforcement Learning: An Introduction. MIT press, Cambridge (2018)
27. Wu, H., Lu, Z., Pan, L., Xu, R., Jiang, W.: An improved apriori-based algorithm for association rules mining. In: 2009 Sixth International Conference on Fuzzy Systems and Knowledge Discovery, vol. 2, pp. 51–55. IEEE (2009)
28. Yuan, X.: An improved apriori algorithm for mining association rules. In: AIP Conference Proceedings, vol. 1820, p. 080005. AIP Publishing LLC (2017)

Deep Randomized Networks for Fast Learning

Richárd Rádli$^{(\boxtimes)}$ and László Czúni

University of Pannonia, Veszprém, Hungary
{radli.richard,czuni.laszlo}@mik.uni-pannon.hu

Abstract. Deep learning neural networks show a significant improvement over shallow ones in complex problems. Their main disadvantage is their memory requirements, the vanishing gradient problem, and the time consuming solutions to find the best achievable weights and other parameters. Since many applications (such as continuous learning) would need fast training, one possible solution is the application of sub-networks which can be trained very fast. Randomized single layer networks became very popular due to their fast optimization while their extensions, for more complex structures, could increase their prediction accuracy. In our paper we show a new approach to build deep neural models for classification tasks with an iterative, pseudo-inverse optimization technique. We compare the performance with a state-of-the-art backpropagation method and the best known randomized approach called hierarchical extreme learning machine. Computation time and prediction accuracy are evaluated on 12 benchmark datasets, showing that our approach is competitive in many cases.

Keywords: extreme learning machines · classification · optimization

1 Introduction

While deep neural networks (DNNs) are very successful in machine learning tasks their training is time demanding and thus often requires special hardware and software solutions such as parallelization on CPUs, GPUs, TPUs, or on a cluster of computers [1,9,14]. Since their task-solving performance increases with their depth and width the required memory and computing power can reach implementation limits not talking about the problems of vanishing gradients and sub-optimal solutions. Circuit complexity theory deals with the size and depth of networks to compute a function, for example wide residual networks [26] could reach better results by balancing between depth and width but there are also many technical aspects to be considered. Wider networks allow many multiplications to be computed in parallel, whilst deeper networks require more sequential computations (since they depend on each other from layer to layer). For more efficient design, Google created the family of networks called EfficientNet [23] where a compound scaling method is used to scale the depth (number of layers),

M. Sellmann and K. Tierney (Eds.): LION 2023, LNCS 14286, pp. 121–134, 2023.
https://doi.org/10.1007/978-3-031-44505-7_9

the width (number of kernels in a layer), and resolution (size of input image) of the network.

In our paper we introduce a different approach relying on the early observation that random weights can solve some problems very efficiently [11,19,20]. The drawback of these initial approaches was their shallowness, the limitation to a single hidden layer, what was later resolved by multilayer representational learning [6,25] or by ensemble of networks [3,7].

Contrary to these methods, we propose a multiple phase dense network building strategy (we call it multiple phase dense randomized neural network, MP-DRNN), which shows good accuracy for classification tasks at fast training speed. Our purpose is to be able to build up and train dense deep networks more efficiently than with random initialization and backpropagation training. The extension of our approach to other types of networks (f.e. convolutional) is for future work. The main contributions of our article are:

- We show how to build deep randomized structures in consecutive phases, where it is guaranteed that the classification performance will not decrease as the depth of the network grows. (There is no vanishing gradient problem, and the best solution, already found, is always kept after each phase.)
- While our classification accuracy outperformed the popular Adam optimizer, with almost identical deep structures, in the majority of test sets, our computation time is typically a fraction of it.
- The generated structure's weights are determined in separate phases without applying back propagation. The determination of weights can also be sped up by GPUs but in our tests we used only CPUs.
- We compare our approach also to two of the most popular hierarchical random networks H-ELM [24] and H-LR21-ELM [15] by accuracy and training time on 12 benchmark datasets.

Our paper does not aim to reach the highest possible accuracy for some specific datasets. We used simple dense connections, no convolutional layers and no residual connections were applied between the layers, necessary to reach the best accuracy in most cases. While our method can be directly used for the building of top-level classification layers over well-proven (f.e. convolutional) backbone networks [18], there is more work to generalize our idea to reach the best optimal results utilizing special connection structures. The relevance of our results can also be supported by the facts that convolutional networks can be considered as special cases of fully connected ones and moreover, in many tasks, there is no spatially limited relation between input features (calling for convolution). It is encouraging to notice that the proposed optimization outperformed the popular Adam optimizer [12] in many test cases, running on the same dense networks, typically at the fraction of time. In future much effort should be taken to apply similar ideas for other types of networks (e.g. convolutional).

In the next section we overview the most closely related neural network models which show somehow similar structure and apply similar training approaches. In Sect. 3 first we introduce the main steps of our approach then the datasets are described followed by empirical evaluations on them. The different alternatives, to design connections and weights, are analysed in Sect. 5.2 while conclusions and further work are described in Sect. 6.

2   Related Articles

While in the previous section we mentioned different approaches to tackle the problems coming from the size and complexity of DNNs, now we are going into the details of how random neural networks (RNNs) can be utilized to attack some of these problems.

It is well known that randomizing weights in neural networks can result in improved accuracy. Published three decades ago in [20], the input data of a single (hidden)-layer feed-forward neural network (SLFN) were weighted with random numbers. Then, in the output layer, a fully connected network with bias was applied, where its weights could be calculated by solving a linear set of equations by numerical methods. Later, in [19], the random vector functional links network (RVFL) was proposed, where beside the input patterns, their randomized version is also generated and fed to the output using the standard weighted connections. Both architectures showed good prediction performance in several experiments. While the so called extreme learning machines (ELM) [11] have subtle variations to these randomized networks (no direct connection between inputs and outputs like RVFL, other usage of bias than in [20]) they became more popular recently, in spite of reports that RVFL showed better performance than ELM in some cases [22]. For a wider review on neural networks with random weights, we propose to read [5].

All those solutions had good results for some problems and their training time was very short (thanks to the non-iterative weight-setting mechanism), however, they had shallow structures without hierarchy, needed for more complex tasks. To increase the depth of ELMs, multilayer ELMs (ML-ELM) were introduced in [6] using a sequence of basis transformations where the new basis vectors were generated by autoencoders (AE) called ELM-AE. The encoder network $E : \mathbb{R}^h \to \mathbb{R}^d$, the latent space $\mathbb{R}^d$, and the decoder network $D : \mathbb{R}^d \to \mathbb{R}^h$ form a computation mechanism for the compact representation of input information by $\mathbf{z}$:

$$\hat{\mathbf{x}} = D(E(\mathbf{x})) = D(\mathbf{z}), \tag{1}$$

where $\mathbf{x}$ is the input data, $\mathbf{z}$ is the latent information, and $\hat{\mathbf{x}}$ is the output of the ELM-AE network (approximated input). Randomness is playing role as the weights and the bias of the encoder is randomly generated. These random weights were chosen to be orthogonal in [6] but in later variations [24] this condition was omitted. The weights of the decoder were computed as in [11], and then the inverted weights of the decoder were used as basis vectors for the hidden layer representation of input data. After a few such steps (2 in the published article) the obtained representation is fed to a normal ELM.

While the ML-ELM approach was tested only in one dataset (MNIST) with good results, a slightly modified version, called hierarchical ELM (H-ELM) [24], was evaluated on several different datasets and could outperform previous variants. Beside leaving the orthogonality condition for the random weights of the ELM-AE, the l_2 norm was replaced by l_1, and the FISTA algorithm [2] was used to compute the output weights instead of the matrix inversion technique. Details of experiments are given in Sect. 5. While H-ELM and its variant (introduced later as H-LR21-ELM) has generally the best accuracy so far (in the discussed context) there are several questions not answered yet: how many ELM-AE blocks are necessary to achieve the best results, how regularization constant C (see later Eq. 8) is determined.

The multilayer randomized solution for representational learning introduced in [25] was also named ML-ELM, but it applies a different idea than [6]. One of the purposes of this approach (and target of the tests) was efficient dimension reduction. Here, instead of using single hidden neurons as in the basic ELM feature mapping, this method employs sub-network nodes that consist of multiple general hidden neurons. It is worth noting that a node in the basic ELM feature mapping can be considered as a special case of the sub-network node. The number of general nodes and output dimension is independent, but the number of hidden neurons in each general node should equal the dimension of outputs.

In [16] a multi-modal extreme learning machine was introduced with local receptive fields on RGB and depth images (Multi-Modal Local Receptive Field Extreme Learning Machine: MM-LRF-ELM) for object recognition. While it maintains ELMs' advantages of training efficiency, the MM-LRF-ELM is relatively shallow. Since it is adopted for images, feature extraction is randomized by the convolutional kernels themselves while the final classification weights are computed by matrix inversion as in [11].

As for many kinds of classifiers, ensemble learning is also effective for ELMs. Examples for this are in [4] (voting-based ELM, V-ELM), in [7] (hierarchical ensemble of ensembles, H-ELM-E), or in [13], where a trained combiner is used to integrate components for hyperspectral image classification. In [3] a hierarchical ensemble of ELMs (HE-ELM) is proposed where to encourage the diversity of component ELMs with two strategies followed: the sparse connection to component ELMs and feature bagging. To achieve representation learning, in contrast to the previous shallow ensembles, it applies multiple layer feature representation. Unfortunately, as the number of ELMs is increasing the time complexity is getting worse (but unfortunately not discussed in [3]).

Finally, we are making a note about the hierarchical approach called H-LR21-ELM [15], where a specific combination of L_{21} norm based loss function and regularization is used to improve the robustness and the sparsity of H-ELMs. While in most optimizations the L_2 norm is used, we have already seen [24] that the L_1 norm can improve results. The proposed L_{21} norm in [15] is claimed to make the H-ELM model less sensitive to outliers and impulsive noises that are pervasive in real-world data, moreover, the L_{21} regularization can learn the most-relevant sparse representations to reduce the intrinsic complexity of models.

We can summarize that while (shallow) random networks has the advantage of high speed training and good accuracy for some datasets, deeper structures should be implemented for high accuracy on more complex problems. H-ELMs and H-LR21-ELMs have good results but their parameter settings is not clear and the sequence of AEs, used in them, might loose important information as focusing mainly on data compression (reconstructability). Other solutions are still shallow ([16]) or require set of networks ([4,7,13]). For the better understanding of the ELM, H-ELM, and our proposal, we explain their mechanisms in details, evaluate test results, and propose future developments in the following parts of our article.

3   The Proposed MP-DRNN Method

Our proposed method can be considered as a global optimization approach where both randomly generated and optimally computed weights are applied in dense layers built one after the other in consecutive phases as in differential evolution techniques (DE).

Differential evolution, first introduced by [21], is an evolutionary computation method to optimize a problem iteratively trying to improve a candidate solution considering a cost function. DE methods are commonly known as metaheuristics as they make few or no assumptions about the problem and can scan very large spaces of candidate solutions. In general, DE optimizes a problem by maintaining a population of candidate solutions and creating new candidate solutions by combining existing ones and also keeping best one(s) without the need for gradients.

Unfortunately, maintaining whole networks as members of a population would be very memory and time demanding (similarly to ensemble approaches we would like to avoid). Instead, we propose to add new neurons to the structure, hoping that useful new features can be extracted with the help of them. The number of these new neurons and their weights are discussed in Sect. 3.2 and Sect. 5. The overview of our approach can be drawn by two kinds of phases in four main steps:

1. Create a single ELM network and compute its output weights as the first iteration. This is the first approximation for the solution. See Fig. 1 for illustration.
2. Add new neurons to extend the output layer (which now becomes a hidden layer) of the previous phase and add new output neurons at the top level becoming the new output layer. See Fig. 2 for illustration.
3. Compute the weights for the new output layer with the matrix inversion technique (see Eq. 7).
4. Repeat Steps 2–3 as further phases.

More detailed description of these steps is given in Algorithm 1. The advantages of this mechanism is:

– In Phase 1 any type of ELM can be utilized. In our experiments we implemented the plain old ELM [11].

- As in new extension phases we always keep the previous cells and connections, the extended network should have identical or improved accuracy. Since the added extra weights are computed by the evaluation of all training data (as in previous phases) in worst case scenario zero weights are given to new hidden neurons and ones for the previous output neurons (i.e. the output is simply copied).
- The distribution of numbers of neurons can be set arbitrarily, and even pruning can be applied to get a sparser network (not implemented in our article).

Algorithm 1. Building and training of MP-DRNN

Input: Training set $(\mathbf{x}_i, \mathbf{y}_i)$ **Output:** MP-DRNN structure with weights

1: Randomly generate hidden node weights
2: Calculate the hidden layer output $\mathbf{H}$
3: Calculate the Moore-Penrose generalized inverse $\mathbf{H}^\dagger$
4: Calculate the weight matrix $\mathbf{W} := \mathbf{H}^\dagger \mathbf{T}$
5: Calculate values of target vectors t_O^1
6: **for** (i=1; i$\leq$ n; i=i+1) **do**
7:     Copy the outputs neurons t_O^i from the previous phase, and add Gaussian noise $\mathcal{N}(\mu, \sigma^2)$ to their weights
8:     Add more neurons $H_L^{i+1} rnd$ with random weights
9:     Concatenate t_O^i, $t_O^i \delta$ and $H_L^{i+1} rnd$
10:     Calculate the $i+1^{th}$ hidden layer output
11:     Calculate the Moore-Penrose generalized inverse $\mathbf{H}^\dagger$
12:     Calculate the weight matrix $\mathbf{W} := \mathbf{H}^\dagger \mathbf{T}$
13: **end for**

3.1 The Initial Phase of MP-DRNN

First, we formalize the basic idea behind basic SLFN random networks according to [11]. Consider a set of N distinct training samples $(\mathbf{x}_i, \mathbf{y}_i)$, $i = 1, ..., N$. Then, a SLFN with L hidden neurons has the following output equation:

$$\mathbf{t}(\mathbf{x}_i) = \sum_{j=1}^{L} \mathbf{w}_j \phi(\mathbf{r}_j \mathbf{x}_i + b_j), \tag{2}$$

where ϕ is an activation function, $\mathbf{r}_j$ are the random and fixed-input weight vectors, b_j are the biases, $\mathbf{w}_j$ are the output weight vectors to be tuned, and $\mathbf{t}$ is the target vector (the outputs for the different classes). $\mathbf{r}_i^1$ and $\mathbf{w}_i^1$ in Fig. 1 correspond to the weights here since the upper index will denote the first phase of the algorithm. In practice, closed-form solutions can be used to find $\mathbf{w}_j$ in a matrix form. Thus, we can shorten the equation using matrices as:

$$\mathbf{T} = \mathbf{HW}, \tag{3}$$

as the outputs of all hidden neurons are gathered into the matrix $\mathbf{H}$:

$$\mathbf{H} = \begin{bmatrix} \phi(\mathbf{r}_1\mathbf{x}_1 + b_1) & \cdots & \phi(\mathbf{r}_L\mathbf{x}_1 + b_L) \\ \vdots & \ddots & \vdots \\ \phi(\mathbf{r}_1\mathbf{x}_N + b_1) & \cdots & \phi(\mathbf{r}_L\mathbf{x}_N + b_L) \end{bmatrix}, \tag{4}$$

given $\mathbf{W} = \left(\mathbf{w}_1^T \cdots \mathbf{w}_L^T\right)^T$, and $\mathbf{T} = \left(\mathbf{t}_1^T \cdots \mathbf{t}_C^T\right)^T$.

A unique solution for this system can be given by using the Moore–Penrose generalized inverse (pseudoinverse) [17] of the matrix H, denoted as $H^\dagger$. From Eq. 3:

$$\mathbf{W} = \mathbf{H}^{-1}\mathbf{T}. \tag{5}$$

To find the "best fit" (least squares) solution to the system of linear equations the pseudoinverse is computed [17]:

$$\mathbf{H}^\dagger = (\mathbf{H}^T\mathbf{H})^{-1}\mathbf{H}^T. \tag{6}$$

Finally, we get the weights:

$$\mathbf{W} := \mathbf{H}^\dagger\mathbf{T}. \tag{7}$$

Alternatively, to apply regularization on $\mathbf{W}$:

$$\mathbf{H}^\dagger = \begin{cases} (C^{-1}\mathbf{1} + \mathbf{H}^T\mathbf{H})^{-1}\mathbf{H}^T & \text{if N>L} \\ \mathbf{H}^T(C^{-1}\mathbf{1} + \mathbf{H}\mathbf{H}^T)^{-1} & \text{if N<L} \end{cases} \tag{8}$$

where C is a scalar regularization constant for the control of magnitude of weights [10] and $\mathbf{1}$ is the identity matrix.

At this point we've arrived to the first phase of the MP-DRNN. In further steps we will extend this network at its end layer as specified in the following.

3.2   Extension Phases

Following the general ideas of differential evolution, we have chosen three strategies to add new hidden neurons (denoted as Hidden layer 2 in Fig. 2):

1. $t_O^1\delta$: clones of output neurons of phase one by perturbating their weights with Gaussian noise with 0 mean. The purpose of these neurons is to test the parameter space around the optimal weights computed in the previous phase by Eq. 7.
2. H_L^2rnd: random neurons for random scanning of parameter space.
3. output neurons (with t_O^2 output values) to represent class labels.

In our baseline approach the sum of new neurons, of consecutive phases, is equally distributed. In another variant (Exp. ort. - also evaluated in Sect. 5) the number of neurons were exponentially decreasing through the phases (following the

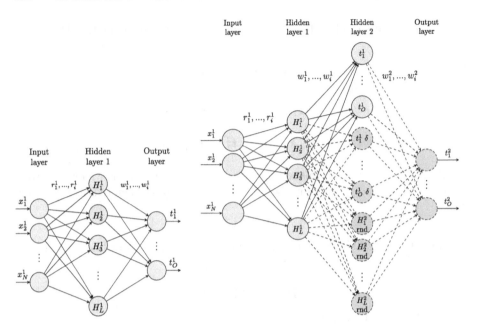

Fig. 1. First phase of MP-DRNN is equivalent to a basic ELM.

Fig. 2. Second phase of MP-DRNN. Lime coloured nodes denote first phase neurons, while light blue nodes denote the second phase neurons. Dotted lines indicate the new weights as incremental structures added to phase one. (Color figure online)

general pattern of fully connected DNNs). Since we apply random numbers for weights in each phase we are still following ELM universal approximation theory [11] while building deep hierarchical networks.

4   Datasets

All of the utilized benchmark datasets are the same as those used in [15]. Most of them are from the UCI Machine Learning Repository. We included six image datasets and six classic real-world datasets as classification problems. The datasets were divided into train and test sets the same way as in [15]. Before feeding the data to the network, we normalized the datasets to the [0, 1] scale. Detailed description of the datasets can be seen in Table 1. This table presents a summary of the benchmark datasets used in our study, including the number of instances, the number of features, and the data type. The data type column indicates whether the dataset is image-based or feature-based.

Table 1. Main numerical characteristics of the datasets.

Dataset	#Features	#Train	#Test	#Classes	#Data type
Connect4	42	50,000	17,577	3	Feature
Isolet	617	6,238	1,559	26	Feature
Letter	16	10,500	9,500	26	Image
MNIST	784	60,000	10,000	10	Image
Fashion-MNIST	784	60,000	10,000	10	Image
Musk2	166	3,000	3,598	2	Feature
Optdigits	64	3,822	1,797	10	Feature
Page-blocks	10	4,385	1,100	5	Feature
Segmentation	19	1,733	577	7	Image
Shuttle	9	29,834	26,936	7	Feature
USPS	256	7,291	2,007	10	Image
YaleB	1024	1,680	734	38	Image

5   Evaluations and Further Studies

For the qualitative evaluation of our proposed method we used three competitive networks as references:

1. **H-ELM** networks with main parameters specified in [15]. The implementation available at [8] was used for the running time evaluations. We left the implementation intact, only set the number of neurons in each layer and tuned C (which is the L2 penalty of the last layer ELM) and s (the scaling factor) in order to obtain the accuracy published in the original article, since these parameters were not published. By tuning these parameters we could not reproduce all testing accuracy values as published in [15] thus for accuracy we use their published numbers.
2. Our main rival is the **H-LR21-ELM** network [15]. Again, the above parameters were not specified for each of the dataset, therefore we relied on the claimed results of [15].
3. Fully-connected neural network (**FCNN**) with 3 layers and the same number of neurons as the other networks. The hyper-parameters were the following: the network was trained until convergence, we used the ADAM optimizer with learning rate set to 10^{-3}. Batch size was selected to 128 in all cases. The main purpose with this network is to get impressions how successfully the iterative backpropagation (BP) could be substituted with the randomized approach.

The evaluation covered testing accuracy, as well as training times; H-ELM was implemented in MATLAB, while the other two methods were written in Python. The authors of [15] used three layers for both H-ELM and H-LR21-ELM, while we have implemented our network with various number of layers for each dataset (for details see Table 2). With the MP-DRNN, the activation

Table 2. Number and distribution of neurons at different layers for the tested models.

Dataset	H-ELM/H-LR21-ELM				MP-DRNN Base method					
	N1	N2	N3	Sum	N1	N2	N3	N4	N5	Sum
Connect4	400	200	2000	2600	866	866	866	–	–	2598
Isolet	800	400	3000	4200	1050	1050	1050	1050	–	4200
Letter	100	50	500	650	130	130	130	130	130	650
MNIST	1000	5000	10000	16000	5333	5333	5333	–	–	15999
Fashion-MNIST	1000	5000	10000	16000	5333	5333	5333	–	–	15999
Musk2	400	200	2000	2600	520	520	520	520	520	2600
Optdigits	100	50	500	650	162	162	162	162	–	648
Page-blocks	100	50	500	650	130	130	130	130	130	650
Segmentation	100	50	500	650	216	216	216	–	–	648
Shuttle	10	50	500	650	216	216	216	–	–	648
USPS	400	200	2000	2600	1300	1300	–	–	–	2600
YaleB	800	400	3000	4200	840	840	840	840	840	4200

functions were Leaky ReLU (slope was set to 0.2), except for the last layer, where it was sigmoid. (For H-ELM and H-LR21-ELM, the authors applied the sigmoid activation function.) Our initial MP-DRNN model (referred to as MP-DRR Base Model) did not use regularization nor scaling, had the same number of neurons in each layer. The only hyper-parameter is the number of layers.

The leaky ReLU was chosen for our method because empirical tests showed that it produced better results than using only sigmoid activation functions in all layers. Furthermore, it is worth noting that computation time of Leaky ReLU is less expensive. Specification of the software and hardware environment: AMD Ryzen 5 5600X CPU, 32 GB DDR4 RAM, Windows 11, MATLAB R2022a, Python 3.8.0 and NumPy 1.22.1.

5.1 Evaluation of Reference Methods and Our Base Model

The testing accuracy and training times of the different methods are shown in Fig. 3 and in Table 3 with best values in bold. For MP-DRNN we repeated the experiments for 5 times, and the averaging results were computed for comparison. FCNN was trained until convergence, while the results of H-ELM and H-LR21-ELM (also average values) originate from [15]. Considering accuracy, as summarized in the bottom line of Table 3, H-LR21-ELM won in most cases, while our base model (MP-DRNN BM) was second with 4 wins.

5.2 Improvements for Building MP-DRNN Models

As further investigations we made experiments with the following alternative configurations modifying the MP-DRNN base network:

Table 3. Classification accuracy and training times of our base and reference methods. Bold values are for the best testing accuracy. Training time of H-LR21-ELM is not given since we used the data from [15] and the different computer platforms doesn't allow fair comparisons.

Dataset	H-ELM		H-LR21-ELM		FCNN		MP-DRNN BM	
	Testing accuracy	Training time [s]	Testing accuracy	Training time [s]	Testing accuracy	Training time [s]	Testing accuracy	Training time [s]
Connect4	68.01	2.968	69.41	N.A	**79.21**	43.477	74.38	12.822
Isolet	95.41	1.625	**95.89**	N.A	94.55	15.684	95.29	6.2847
Letter	87.83	0.115	**88.07**	N.A	81.08	4.343	83.76	0.899
MNIST	**98.87**	93.837	98.76	N.A	97.94	855.580	97.78	404.060
Fashion-MNIST	89.78	42.523	**89.96**	N.A	87.49	852.654	88.64	407.072
Musk2	98.21	0.317	99.29	N.A	99.97	2.669	**100**	2,0236
Optdigits	97.48	0.052	97.78	N.A	97.16	2.207	**97.79**	0,3693
Page-blocks	96.71	0.051	97.06	N.A	96.18	3.299	**97.43**	0.384
Segment	94.75	0.025	95.38	N.A	96.01	6.088	**96.24**	0.292
Shuttle	99.02	0.294	99.02	N.A	**99.54**	9.544	99.44	1,277
USPS	97.68	0.659	**98.09**	N.A	95.71	15.666	94.56	4.663
YaleB	98.34	1.339	**99.17**	N.A	89.78	32.353	95.53	0.751
Number of wins	1		5		2		4	

Table 4. Classification results of the different methods on the 12 datasets. Bold values are for the best accuracy.

Dataset	H-LR21-ELM		Exp-Ort		Exp-Ort-C		Large Exp-Ort-C		
	Test. acc.	Trn. time [s]	Test. acc.	Trn. time [s]	Test. acc.	Trn. time [s]	Test. acc.	Trn. time [s]	Increase
Connect4	69.41	N.A	75.37	15.33	76.24	7.589	–	–	–
Isolet	95.89	N.A	95.93	8.967	**96.01**	7.027	–	–	–
Letter	88.07	N.A	83.92	0.783	84.20	0.559	**91.3**	2.1	×3
MNIST	98.76	N.A	97.89	500.131	97.91	400.176	97.94	869.786	×1.5
Fashion-MNIST	**89.96**	N.A	88.75	505.423	88.82	404.152	88.9	870.41	×1.5
Musk2	99.29	N.A	99.9	2.765	**100**	1.409	–	–	–
Optdigits	97.48	N.A	98.3	0.371	**98.45**	0.281	–	–	–
Page-blocks	97.06	N.A	96.58	0.433	**97.66**	0.377	–	–	–
Segment	95.38	N.A	96.43	0.423	**96.80**	0.206	–	–	–
Shuttle	99.02	N.A	99.49	1.14	**99.57**	0.752	–	–	–
USPS	**98.09**	N.A	94.61	5.287	94.70	2.237	–	–	–
YaleB	**99.17**	N.A	94.67	0.965	95.67	0.613	–	–	–
Number of wins	3		0		6		1		

1. Exp-Ort: To follow the general pattern of deep networks, instead of equally distributed number of neurons, their number was decreasing exponentially from layer to layer (the sum of neurons did not change). Besides, to increase the uniformity of random weights, only half of the originally determined random weights (H_L^2rnd) were generated, while the other half was generated as their orthogonal vectors.

2. Exp-Ort-C: In another case we modified the Exp. Ort. model by adding the C regularization factor and tried to find its optimal value (similarly as competitor techniques H-ELM and H-LR32-ELM do).

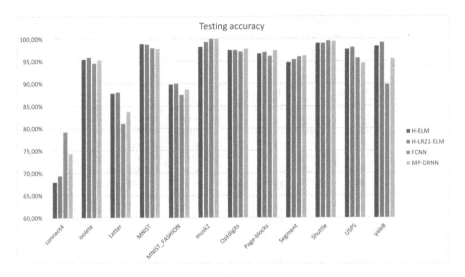

Fig. 3. Classification accuracy of our base and reference methods.

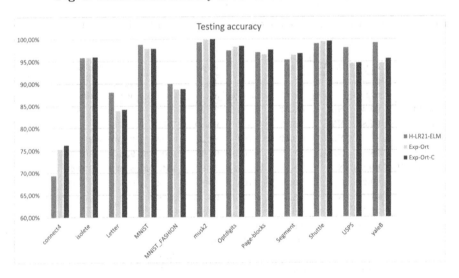

Fig. 4. Classification results of the different methods on the 12 datasets

3. Large Exp-Ort-C: Finally, for some datasets, we increased the number of neurons in the hidden layers, in order to test whether we can achieve higher testing accuracy.

Results can be seen in Fig. 4 and in Table 4. The performance of these alternatives showed little variations, but in general, each modification resulted in improved accuracy, over-performing competitors in many cases. While we think that the comparisons of H-LR21-ELM and Exp-Ort-C is fair (since both used hand tuned parameter C) the Large Exp-Ort-C variant is just to show the limitations of the current state of our approach. The training time of the MP-DRNN technique is between the FCNN and the H-ELM implementations while our classification accuracy is almost always higher than the FCNN. It looks obvious, that H-ELM and H-LR21-ELM have advantages in case of datasets with large dimensions (such as the images of MNIST, Fashion-MNIST, USPS, and YaleB - see Table 1), reasoned by the good feature compression abilities of autoencoders on highly correlated data. This implicates to use H-ELM or H-LR21-ELM (at least) in the first phase of our future implementations of MP-DRNN.

6   Conclusions and Future Work

In our article a new building approach for dense hierarchical randomized DNNs was introduced. Its training is significantly (circa 5 times in average) faster than backpropagation while its classification is more accurate in most test cases (at the same number of neurons). In more than half of the test cases our solution called Exp-Ort-C MP-DRNN could outperform H-ELM and H-LR21-ELM while the version without the need of fine tuning the regularization parameter (Exp-Ort MP-DRNN) could win for datasets with low dimensions (in 5 cases).

We think that our approach can already be used as the final classification layers of larger backbone networks (similarly to the approach in [18]) but after using AEs for the settings of some of the random weights (similarly to H-ELM) we can improve the accuracy for high-dimensional (correlated) data such as images. As more future work we are to increase the sparsity of the network by pruning, extend the approach to convolutional layers, and investigate the usability of the L21 norm.

Acknowledgements. We acknowledge the financial support of the Hungarian Scientific Research Fund grant OTKA K-135729. We are grateful to the NVIDIA corporation for supporting our research with GPUs obtained by the NVIDIA Hardware Grant Program.

References

1. Awan, A.A., Jain, A., Anthony, Q., Subramoni, H., Panda, D.K.: HyPar-Flow: exploiting MPI and Keras for scalable hybrid-parallel DNN training with tensorflow. In: Sadayappan, P., Chamberlain, B., Juckeland, G., Ltaief, H. (eds.) ISC High Performance 2020. LNCS, vol. 12151, pp. 83–103. Springer, Cham (2020). https://doi.org/10.1007/978-3-030-50743-5_5
2. Beck, A., Teboulle, M.: A fast iterative shrinkage-thresholding algorithm for linear inverse problems. SIAM J. Imag. Sci. **2**(1), 183–202 (2009)
3. Cai, Y., Liu, X., Zhang, Y., Cai, Z.: Hierarchical ensemble of extreme learning machine. Pattern Recogn. Lett. **116**, 101–106 (2018)
4. Cao, J., Lin, Z., Huang, G.B., Liu, N.: Voting based extreme learning machine. Inf. Sci. **185**(1), 66–77 (2012)
5. Cao, W., Wang, X., Ming, Z., Gao, J.: A review on neural networks with random weights. Neurocomputing **275**, 278–287 (2018)

6. Chamara, L., Zhou, H., Huang, G.B., Vong, C.M.: Representational learning with extreme learning machine for big data. IEEE Intell. Syst. **28**(6), 31–34 (2013)
7. Cvetković, S., Stojanović, M.B., Nikolić, S.V.: Hierarchical ELM ensembles for visual descriptor fusion. Inf. Fusion **41**, 16–24 (2018)
8. Hierarchical ELM MATLAB source codes. https://www.extreme-learning-machines.org
9. Han, J., Xu, L., Rafique, M., Butt, A.R., Lim, S.H.: A quantitative study of deep learning training on heterogeneous supercomputers. Oak Ridge National Lab. (ORNL), Oak Ridge, TN (United States) (2019)
10. Huang, G.B., Zhou, H., Ding, X., Zhang, R.: Extreme learning machine for regression and multiclass classification. IEEE Trans. Syst. Man. Cybern. Part B (Cybern.) **42**(2), 513–529 (2011)
11. Huang, G.B., Zhu, Q.Y., Siew, C.K.: Extreme learning machine: theory and applications. Neurocomputing **70**(1–3), 489–501 (2006)
12. Kingma, D.P., Ba, J.: Adam: a method for stochastic optimization (2014). arXiv preprint arXiv:1412.6980
13. Ksieniewicz, P., Krawczyk, B., Woźniak, M.: Ensemble of extreme learning machines with trained classifier combination and statistical features for hyperspectral data. Neurocomputing **271**, 28–37 (2018)
14. Lee, S., Nirjon, S.: SubFlow: a dynamic induced-subgraph strategy toward real-time DNN inference and training. In: 2020 IEEE Real-Time and Embedded Technology and Applications Symposium (RTAS), pp. 15–29. IEEE (2020)
15. Li, R., Wang, X., Song, Y., Lei, L.: Hierarchical extreme learning machine with L21-norm loss and regularization. Int. J. Mach. Learn. Cybern. **12**(5), 1297–1310 (2021)
16. Liu, H., Li, F., Xu, X., Sun, F.: Multi-modal local receptive field extreme learning machine for object recognition. Neurocomputing **277**, 4–11 (2018)
17. Moore, E.H.: On the reciprocal of the general algebraic matrix. Bull. Am. Math. Soc. **26**, 394–395 (1920)
18. Nagy, A.M., Czúni, L.: Classification and fast few-shot learning of steel surface defects with randomized network. Appl. Sci. **12**(8), 3967 (2022)
19. Pao, Y.H., Park, G.H., Sobajic, D.J.: Learning and generalization characteristics of the random vector functional-link net. Neurocomputing **6**(2), 163–180 (1994)
20. Schmidt, W.F., Kraaijveld, M.A., Duin, R.P.: Feed forward neural networks with random weights. In: International Conference on Pattern Recognition, pp. 1. IEEE Computer Society Press (1992)
21. Storn, R.: On the usage of differential evolution for function optimization. In: Proceedings of North American Fuzzy Information Processing, pp. 519–523. IEEE (1996)
22. Suganthan, P.N., Katuwal, R.: On the origins of randomization-based feedforward neural networks. Appl. Soft Comput. **105**, 107239 (2021)
23. Tan, M., Le, Q.: EfficientNet: rethinking model scaling for convolutional neural networks. In: International Conference on Machine Learning, pp. 6105–6114. PMLR (2019)
24. Tang, J., Deng, C., Huang, G.B.: Extreme learning machine for multilayer perceptron. IEEE Trans. Neural Netw. Learn. Syst. **27**(4), 809–821 (2015)
25. Yang, Y., Wu, Q.J.: Multilayer extreme learning machine with subnetwork nodes for representation learning. IEEE Trans. Cybern. **46**(11), 2570–2583 (2015)
26. Zagoruyko, S., Komodakis, N.: Wide residual networks (2016). arXiv preprint arXiv:1605.07146

Generative Models via Optimal Transport and Gaussian Processes

Antonio Candelieri[1]([✉]) [iD], Andrea Ponti[1,2] [iD], and Francesco Archetti[1] [iD]

[1] University of Milano-Bicocca, Milan 20126, Italy
{antonio.candelieri,andrea.ponti,francesco.archetti}@unimib.it
[2] OAKS srl, Milan 20126, Italy

Abstract. Generative models have recently gained a renewed interest due to their success in the development of new real-life applications, such as artificial intelligence generated images, texts, audios. The most recent and successful approaches combine neural network learning and Optimal Transport theory, exploiting the so-called *transportation map/plan* to generate a new element of a domain starting from an element of a different one, while preserving statistical properties of the data generation processes of the two domains. Although effective, the Neural Optimal Transport (NOT) approach is largely computationally expensive – due to the training of two nested deep neural networks – and requires *injecting* additional noise to improve generative properties. In this paper we present an alternative method, based on Gaussian Process (GP) regression, which overcomes these limitations. Contrary to a neural model, a GP is *probabilistic*, meaning that, for a given input, it provides both a prediction and the associated uncertainty. Thus, the generative properties are, by design, guaranteed by sampling the generated element around the prediction and depending on the uncertainty. Results on both toy examples and a dataset of images are provided to empirically demonstrate the benefits of the proposed approach.

Keywords: Generative models · Optimal Transport · Gaussian Process

1 Introduction

A generative model learns and provides a probabilistic representation of the underlying data generation process, given an available set of data. In the Machine Learning (ML) community, generative models have been used for many decades [22], along with the *discriminative* ones. The main difference is that discriminative methods only model the decision boundary between different classes. Formally, given a *training set* $\mathbf{X}$ with the associated *labels* $\mathbf{Y}$, a discriminative model just learns a possible representation of the *conditional probability* $p(\mathbf{Y}|\mathbf{X})$, while a generative model captures the *joint probability* $p(\mathbf{X}, \mathbf{Y})$, or just $p(\mathbf{X})$ if no labels are provided (i.e., unsupervised learning).

© The Author(s), under exclusive license to Springer Nature Switzerland AG 2023
M. Sellmann and K. Tierney (Eds.): LION 2023, LNCS 14286, pp. 135–149, 2023.
https://doi.org/10.1007/978-3-031-44505-7_10

Both generative and discriminative models can solve the same tasks, but only generative models can generate new data coherent with the approximated underlying data generation process. Generating new data consists in sampling from the learned joint probability, whichever is the type of data. This is the main motivation of the increasing adoption of generative models for the development of new real-life applications, such as artificial intelligence generated images, texts, audios [17,21,24,29,34,39,45,46].

Generative Adversarial Networks (GANs), first proposed in [14], are the most widely adopted generative models, nowadays [1]. Motivations of their success are (*i*) a highly accurate data generation and (*ii*) a more efficient training than their precursors, specifically Variational Auto-Encoders and Boltzmann machines.

However, it is well-known that GAN training could be unstable [4]. This has recently led to the so-called Wasserstein GANs (WGANs) [5,9,20,36], where the Wasserstein distance between the distributions of the training data and the labels is taken as loss function of the WGAN training process.

Almost all the proposed WGANs use just the value of the Wasserstein distance (aka Optimal Transport cost) while ignoring the associated Optimal Transport (OT) map/plan, that is the function *translating* elements from the *source* to the *target* domain (i.e., translating $\mathbf{X}$ into $\mathbf{Y}$). Only recently, it has been demonstrated that OT plan itself can be used as a generative model [15,35]. Following this idea, Neural Optimal Transport (NOT) [6,28] has been recently proposed to compute OT plans via Deep Neural Networks (DNNs).

Although NOT overcomes some limitations of other methods computing OT plans/maps – such as, poor scalability of Input Convex Neural Networks to large-scale problems [2] and difficulties in sampling from the plan in the case of entropy regularized OT methods [15,38] – it could result computationally expensive (due to the training of two nested DNNs) and it could also lead to *fake* plans which are not optimal, an issue explained and addressed in [27]. Although it has been proposed as a method for solving the OT problem, the actual value of NOT is its capability to model the OT plan and exploit it to generate new data, coherent with $\mathbf{Y}$, starting from new data coherent with $\mathbf{X}$.

Contributions. The main contributions of this paper can be summarized as follows:

– Our focus is on learning a model of the OT plan, instead of computing it (as in NOT). We propose a generative model strategy, based on Gaussian Process regression, which is an effective and efficient alternative to neural networks.
– Indeed, the proposed approach can generalize over few data, drastically reducing the computational burden for model training (i.e., DNNs requires a lot of data to be trained).
– As a probabilistic model, the proposed approach naturally deals with noise and uncertainty, and does not require *noise outsourcing* – basically noise injection – which is, instead, at the core of NOT.
– Empirical results on both toy examples and an image generation task provides evidence of the benefits of the approach.

Related Works. Most of the relevant works have been already quoted; here the most relevant are recalled. The basic idea of learning an OT plan and exploiting it for generating new data has been presented in [15, 35]. All the works related to WGANs are out of scope: they use OT theory for training WGAN, but do not compute neither learn/use the OT plan. On the contrary, NOT [6, 27, 28] is the only method – at our knowledge – learning OT plan and exploit it as a generative model itself. Finally, an overview on generative models – not only neural – can be found in [1].

Although many research works can be found, linking OT and GP, their aims are significantly different from this paper. For instance, in [7, 13] GP regression over distributional inputs is considered, using the regularized OT distance between inputs instead of the Euclidean distance. In [31, 32], the computation of the distance between two GP models – considered as two distributions of functions – is addressed, with proposed methods based on OT.

Organization of the Paper. The rest of the paper is organized as follows: Sect. 2 summarizes the methodological background about the Optimal Transport theory (i.e., the Wasserstein distance) and the Gaussian Process (GP) regression, that is the alternative modelling strategy proposed in this paper to learn and generalize OT plans (readers with knowledge about these two topics can skip the section). Sect. 3 presents the proposed algorithm and remarks the differences with other approaches, especially NOT. Section 4 and Sect. 5 report the experimental settings and the results, respectively. Finally, conclusions are given in Sect. 6, remarking achievements, limitations, and perspectives.

2 Background

2.1 Optimal Transport

In this section we summarize the key notions of Optimal Transport (OT) theory used in the paper. For a more detailed review, we refer to [33, 42].

Consider two continuous probability distributions, namely $\mathcal{P}$ and $\mathcal{Q}$, representing the data generation processes into two spaces, respectively $\mathcal{X} \in \mathbb{R}^d$ and $\mathcal{Y} \in \mathbb{R}^d$. Moreover, assume there exists a function, $c : \mathcal{X} \times \mathcal{Y} \to \mathbb{R}$, giving the cost for moving (probability) mass from a *source* location, $\mathbf{x} \in \mathcal{X}$, towards a *target* location, $\mathbf{y} \in \mathcal{Y}$, with the aim to match $\mathcal{P}$ with $\mathcal{Q}$. Computing the Wasserstein distance between $\mathcal{P}$ and $\mathcal{Q}$, under the cost function c, means searching for an OT *map* minimizing the overall transportation cost:

$$\mathcal{W}_c(\mathcal{P}, \mathcal{Q}) = \inf_{T_\# \mathcal{P} = \mathcal{Q}} \int_{\mathcal{X}} c\left(\mathbf{x}, T(\mathbf{x})\right) d\mathcal{P}(\mathbf{x}) \tag{1}$$

where $T_\#$ is called *push-forward operator* associated to the *map* $T : \mathcal{X} \to \mathcal{Y}$. The solution of (1) is named OT map and denoted with T^*. This formulation of the OT problem is known as Monge's primal, it is not symmetric and does not allow for mass splitting, thus there might no exists a map satisfying $T_\# \mathcal{P} = \mathcal{Q}$.

Successively, the so-called Kantorovitch formulation, was proposed, allowing for mass splitting:

$$\mathcal{W}_c(\mathcal{P}, \mathcal{Q}) = \inf_{\pi \in \Pi(\mathcal{P}, \mathcal{Q})} \int_{\mathcal{X} \times \mathcal{Y}} c(\mathbf{x}, \mathbf{y}) \, d\pi(\mathbf{x}, \mathbf{y}) \qquad (2)$$

In this formulation an OT *plan* is searched for, denoted with $\pi^* \in \Pi(\mathcal{P}, \mathcal{Q})$, where $\Pi(\mathcal{P}, \mathcal{Q})$ is the set of all the possible joint-probability distributions having $\mathcal{P}$ and $\mathcal{Q}$ as marginals. If π^* is of the form $[Id_{\mathcal{X}}, T^*]_{\#}\mathcal{P}$, for some T^*, then T^* minimizes (1) and π^* is called *deterministic*, otherwise it is called *stochastic* ($Id_{\mathcal{X}}$ denotes the identity function on the space $\mathcal{X}$).

Duality. The Kantorovich formulation is a constrained convex minimization problem and it can be naturally transformed into the associated dual formulation, which is a constrained concave maximization problem:

$$\mathcal{W}_c(\mathcal{P}, \mathcal{Q}) = \sup_{(f,g)} \int_{\mathcal{X}} f(x) d\mathcal{P}(x) + \int_{\mathcal{Y}} g(y) d\mathcal{Q}(y)$$

$$s.t.$$

$$f(x) + g(y) \le c(x, y). \qquad (3)$$

The two continuous function $f(x)$ and $g(x)$ are also called *Kantorovich potentials*. In this paper we will specifically consider the case $c(x, y) = \|x - y\|^2$, usually denoted with $\mathcal{W}_2$ and named Wasserstein-2, which allow us to exploit the Brenier Theorem (i.e., Theorem 2.1 in [33]).

Brenier Theorem. *In the case $\mathcal{X} = \mathcal{Y} = \mathbb{R}^d$ and $c(x, y) = \|x - y\|^2$, and with at least one of the two probability distributions – let say $\mathcal{P}$ – having a density with respect to the Lebesgue measure, then the OT plan π^* is unique and is of the form $[Id_{\mathcal{X}}, T^*]_{\#}\mathcal{P}$. Furthermore, the optimal Monge's map T^* is uniquely defined as the gradient of a convex function $\varphi(x)$, such that $T^*(x) = \nabla\varphi(x)$, where $\varphi(x)$ is the unique (up to an additive constant) convex function such that $[\nabla\varphi(x)]_{\#}\mathcal{P} = \mathcal{Q}$. This convex function is related to the Kantorovich potential as $\varphi(x) = \frac{\|x\|^2}{2} - f(x)$.*

Thus, the theorem guarantees, under its assumptions, the existence and the uniqueness of both the OT plan π^* and the OT map T^*: it is also known as Monge-Kantorovich equivalence. It is important to remark that, otherwise, the OT plan π^* is generally not unique (Remark 2.3 in [33]).

2.2   Gaussian Process Regression

This section briefly summarizes the GP regression method [19,43]. A GP can be though as a collection of random variables, any finite number of which have a joint Gaussian distribution. Two functions completely define a GP, they are the *mean* and the *covariance* function, respectively denoted with $\mu : \mathcal{X} \to \mathbb{R}$ and $k : \mathcal{X} \times \mathcal{X} \to \mathbb{R}$.

From a ML point of view, GP regression is a *kernel method* [37], with the covariance function defined by any valid kernel function (e.g., Squared Exponential, Matérn, Exponential, etc.), providing structural assumptions about the shape of the function to be learned. The GP model is trained by tuning the kernel's hyperparameters depending on the training data, via Maximum Log-likelihood Estimation (MLE) or Maximum A Posteriori (MAP). Consequently, the two GP functions becomes conditioned on the training data, according to the following two equations:

$$\mu(\mathbf{x}) = \mathbf{k}(\mathbf{x}, \mathbf{X}) \left[\mathbf{K} + \lambda^2 \mathbf{I} \right]^{-1} \mathbf{y} \tag{4}$$

$$\sigma^2(\mathbf{x}) = k(\mathbf{x}, \mathbf{x}) - \mathbf{k}(\mathbf{x}, \mathbf{X}) \left[\mathbf{K} + \lambda^2 \mathbf{I} \right]^{-1} \mathbf{k}(\mathbf{X}, \mathbf{x}) \tag{5}$$

where $\mathbf{X} = \left\{ \mathbf{x}^{(i)} \right\}_{i=1:N}$ and $\mathbf{y} = \left\{ y^{(i)} \right\}_{i=1:N}$ denote, respectively, the training data (i.e., $\mathbf{x}^{(i)} \in \mathcal{X}$) and the associated labels (i.e., $y^{(i)} \in \mathbb{R}$); $k(x, x)$ is the kernel function whose hyperparameters are tuned depending on $(\mathbf{X}, \mathbf{y})$; $\mathbf{K}$ is an $N \times N$ matrix (i.e., kernel matrix) with entries $\mathbf{K}_{ij} = k(\mathbf{x}^{(i)}, \mathbf{x}(j))$ and $\mathbf{x}^{(i)}, \mathbf{x}^{(j)} \in \mathcal{X}$; $\lambda^2 \in \mathbb{R}$ is a term used to avoid ill-conditioning in the matrix inversion as well as dealing with *noisy* data (i.e., different labels for the same input); $\mathbf{k}(\mathbf{x}, \mathbf{X})$ is a row vector with elements $\mathbf{k}_i = k(\mathbf{x}, \mathbf{x}^{(i)})$. Finally, just for completeness, $\mathbf{k}(\mathbf{X}, \mathbf{x})$ is the transposed of $\mathbf{k}(\mathbf{x}, \mathbf{X})$.

While $\mu(\mathbf{x})$ provides the prediction for any input $\mathbf{x} \in \mathcal{X}$, $\sigma^2(\mathbf{x})$ provides the associated *uncertainty*. This property makes a GP a *probabilistic* model, contrary to other *deterministic* kernel methods, such as Support Vector Regression [37] (at least its original formulation). As a probabilistic model, a GP offers generative properties *by-design*, allowing to sample multiple predictions for a certain point $\mathbf{x} \in \mathcal{X}$ according to a Gaussian distribution with mean $\mu(\mathbf{x})$ and covariance $\sigma^2(\mathbf{x})$, that is $\hat{y} \sim \mathcal{N}(\mu(\mathbf{x}), \sigma^2(\mathbf{x}))$.

With respect to the interplay between ML and optimization, one of the most successful application of GP regression is Bayesian Optimization [3,10–12,18], but also optimal control and Reinforcement Learning [8,16,25,40].

Multi-output GP. Although it was born as a single-output kernel-based regression method, GP modelling has been extended to the multi-output case, that is $\mathbf{Y} = \left\{ \mathbf{y}^{(i)} \right\}_{i=1:N}$ with $\mathbf{y}^{(i)} = \mathbf{f}(\mathbf{x}^{(i)})$ and $\mathbf{f} : \mathcal{X} \to \mathbb{R}^b, b > 1$, a vector-valued function [30]. This is the specific setting considered in our experiments: all the case studies are characterized by $\mathcal{X} = \mathcal{Y} = \mathbb{R}^d$, with $d > 1$. It is important to remark that adopting separate GP models – one for each scalar-valued component $f_l(\mathbf{x})$ of $\mathbf{f}(\mathbf{x}) = \left[f_1(\mathbf{x}), \ldots, f_l(\mathbf{x}), \ldots f_d(\mathbf{x}) \right]$ – is usually more convenient than using a single multi-output model [12,41,44].

Preliminary experiments (not reported due to limitations on the length of the paper) confirmed that a pool of single-output GP models is more effective and computationally efficient than a single multi-output GP, also in the setting addressed in this paper. Therefore, we have used a pool of separate GP models such that the resulting vector-valued predictive mean and variance are given by $\boldsymbol{\mu}(\mathbf{x}) = [\mu_1(\mathbf{x}), \ldots, \mu_d(\mathbf{x})]$ and $\boldsymbol{\sigma}^2(\mathbf{x}) = [\sigma_1^2(\mathbf{x}), \ldots, \sigma_d^2(\mathbf{x})]$, with every $\mu_l(\mathbf{x})$ and $\sigma_l^2(\mathbf{x})$ conditioned to $\mathbf{X}$ and $\mathbf{Y}_l = \left\{ y_l^{(i)} \right\}_{i=1:N}$.

More simply, the l-th GP model in the pool is aimed at predicting the value of the l-th coordinate of the data point generated from $\mathbf{x}$, that is $\mu_l(\mathbf{x})$, along with the associated predictive uncertainty $\sigma_l^2(\mathbf{x})$. From a computational point of view, separation of the GPs allows to exploit the *embarrassingly parallel* nature of the entire pool learning. In any case, we simply refer to the pool of GPs as multi-output GP in the rest of the paper.

3   Learning and Generalizing Optimal Transport Maps

Assume two sets of data are given, namely $\mathbf{X} = \left\{ \mathbf{x}^{(i)} \right\}_{i=1:N}$ and $\mathbf{Y} = \left\{ \mathbf{y}^{(i)} \right\}_{i=1:N}$, with $\mathbf{x}^{(i)} \in \mathcal{X}$ and $\mathbf{y}^{(i)} \in \mathcal{Y}$. Denote with $\mathcal{P}$ and $\mathcal{Q}$ the associated data generation processes, such that $\mathbf{X} \sim \mathcal{P}$ and $\mathbf{Y} \sim \mathcal{Q}$.

We consider the setting $\mathcal{X} = \mathcal{Y} = \mathbb{R}^d$ and $c(x, y) = \|\mathbf{x} - \mathbf{y}\|^2$ for which the Brenier Theorem (Sect. 2.1) holds. Thus, we know that an OT plan π^* exists and is unique, and it is of the form $\pi^* = [Id_{\mathcal{X}}, T^*]_\# \mathcal{P}$, with T^* a unique OT map.

It is important to remark that our approach – differently from NOT – is not aimed at computing the OT plan, but just to efficiently learn a generative model representing it. To achieve this goal, any suitable solver (preferably one implementing a primal-dual method [26]) can be used to obtain the OT map, T^*, between $\mathbf{X}$ and $\mathbf{Y}$. This OT plan is the ground truth for learning the multi-output GP providing $\boldsymbol{\mu}(\mathbf{x}) \approx T^*(\mathbf{x})$.

The main advantage of our approach is that the OT plan T^* is computed just once, for every $\mathbf{X}$, $\mathbf{Y}$ pair. On the other hand, NOT aims, itself, at computing the OT plan, but it requires to compute, several times, a so-called *empirical estimator* of the $\mathcal{W}_2$ distance, used as loss function for training two nested DNNs.

In the following, a sketch of the proposed algorithm is reported.

Algorithm.

- **INPUT:** a source dataset $\mathbf{X} \sim \mathcal{P}$ and a target dataset $\mathbf{Y} \sim \mathcal{Q}$, such that $|\mathbf{X}| = |\mathbf{Y}| = N$.

- **STEP 1:** Obtain the OT map T^* associated to the available *source* and *target* datasets, respectively $\mathbf{X}$ and $\mathbf{Y}$. Specifically, we use a primal-dual OT solver to obtain T^*.
- **STEP 2:** Denote with $\mathcal{J}$ the index set $\mathcal{J} = \left\{ j : T^*(\mathbf{x}^{(i)}) = \mathbf{y}^{(j)} \right\}$. Thus, $\mathcal{J}$ represents the optimal matching of each source data to a target one.

- **STEP 3:** Order $\mathbf{Y}$ according to $\mathcal{J}$ and obtain $\widetilde{\mathbf{Y}} = \{\widetilde{\mathbf{y}}^{(i)}\}$, with $\widetilde{\mathbf{y}}^{(i)} = \mathbf{y}^{(\mathcal{J}_i)}$. This means, more simply, that $\widetilde{\mathbf{y}}^{(i)} = T^*(\mathbf{x}^{(i)})$.
- **STEP 4:** Train a multi-output GP depending on $\mathbf{X}$ and $\widetilde{\mathbf{Y}}$. As previously mentioned (Sect. 2.2), as a multi-output GP model we will use a pool of d independent GPs, whose predictive mean and variance are respectively denoted with $\mu_l(\mathbf{x})$ and $\sigma_l^2(\mathbf{x})$, with $l = 1, \dots, d$.

- **OUTPUT:** a multi-output GP (as a pool of separate single-output GPs) generalizing the OT as follows $\boldsymbol{\mu}(\mathbf{x}) \approx T^*(\mathbf{x})$.

4 Experimental Setting

Our experimental setting is taken from NOT papers [6, 28]. More specifically, we consider two different 2-dimensional examples, showing the generative properties of our approach on simple and easy-to-understand problems (Sect. 4.1). Then, we consider an image generation task on a well-known dataset of handwritten single digits (Sect. 4.2). All the experiments refer to the so-called **one-to-one translation** task, that is the generation of an element of the target domain starting from an element of the source one.

GP's Kernel. For all the experiments and for all the GP models, we have used the exponential kernel:

$$k(\mathbf{x}, \mathbf{x}') = \sigma_f^2 \, e^{-\frac{|\mathbf{x}-\mathbf{x}'|}{\ell}} \tag{6}$$

where $\sigma_f^2 \in \mathbb{R}$ and $\ell \in \mathbb{R}^d$ (i.e., anisotropic kernel) are the kernel's hyperparameters tuned via MLE. The resulting predictive mean function will be continuous but not continuously differentiable. Preliminary analysis has empirically demonstrated that smoother kernels (i.e., Matérn 3/2, Matérn 5/2, and Squared Exponential) lead to worse results.

Computational Setting. The approach has been developed in R and the code is accessible for free on github[1]. All the experiments have been performed on an Intel(R) Core(TM) i7-7700HQ CPU at 2.80 GHz (4 physical cores, 8 virtual cores), 16 GB of RAM, Microsoft Windows 10. GP models in the pools have been trained in parallel by using 7 out of the 8 (virtual) cores available.

4.1 Toy 2D Examples

As preliminary experiments, we have considered the following two 2-dimensional toy examples:

- the source distribution, $\mathcal{P}$, is a Gaussian while the target distribution, $\mathcal{Q}$, is a Gaussian-mixture with 8 components. We considered $|\mathbf{X}| = |\mathbf{Y}| = 1000$ and

---

[1] https://github.com/acandelieri/GenOTGP_LION17.git.

then generated 100 new target points as the translation of as many source points sampled from $\mathcal{P}$.

- the source distribution, $\mathcal{P}$, is a Gaussian while the target distribution, $\mathcal{Q}$, is a Swiss Roll. We considered $|\mathbf{X}| = |\mathbf{Y}| = 1000$ and then generated 200 new target points as the translation of as many source points sampled from $\mathcal{P}$.

4.2 Image Generation

The second experiment uses the handwritten single digit dataset named USPS[2]. Every image is codified with 16×16 grey-scale pixels (i.e., with numerical precision 10^{-6}). The aim is to create a generative model for one-to-one translation of each digit into a different one. This means that $10 \times 9 = 90$ one-to-one translation tasks have been performed.

Pre-processing. Each pixel has a value ranging from -1 (black) to 1 (white); we have scaled all the values into the range $[0, 1]$ and inverted the color scale (i.e., 0 for white and 1 for black), just to have black digits on a white background (for a more clear visualization). Every image is then represented as a 256-dimensional vector with components in $[0, 1]$.

Training-Generation Splitting. A different number of images is available for each handwritten digit: we have decided to limit the number to 500 images each. Every one-to-one translation task uses a training set $\mathbf{X}$ (i.e., images of the source digit i) and a label set $\mathbf{Y}$ (i.e., images of the target digit $j \neq i$) of 300 images each, that is $|\mathbf{X}| = |\mathbf{Y}| = 300$. The remaining 200 images of the source digit, denoted with $\mathbf{X}_{GEN}$, are used to generate as many images of the target digit.

5 Results

5.1 Results on Toy 2D Examples

This section summarizes the results on the toy 2-dimensional examples.

Gaussian to Gaussian-Mixture. Figure 1 shows the sets of data sampled from the source (on the left) and the target (in the middle) distributions, that is $\mathbf{X} \sim \mathcal{P}$ and $\mathbf{Y} \sim \mathcal{Q}$, with $\mathcal{P}$ and $\mathcal{Q}$ as defined in Sect. 4.1. Finally, the OT map, as computed by the primal-dual OT solver, is shown (on the right). The multi-output GP trained on the OT map had $RMSE = 0$ (perfect interpolation).

The data generation accuracy of the multi-output GP can be qualitatively appreciated in Fig. 2, showing: the source data sampled from $\mathcal{P}$ (on the left), its translation towards $\mathcal{Q}$ (in the middle), and the overlap between the generated (aka transported, translated) data and the expected distribution (on the right). To quantify the quality of the approximated map (i.e., $\boldsymbol{\mu}(x) \approx T^*(\mathbf{x})$), we have considered the generated points, namely $\mathbf{Y}_{GEN} = \boldsymbol{\mu}(\mathbf{X}_{GEN})$, as a sample

---

[2] https://paperswithcode.com/dataset/usps (last access: 2023–02–06).

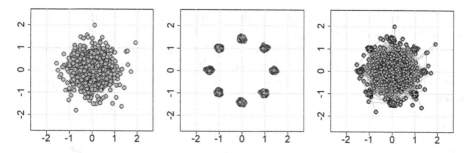

Fig. 1. Learning one-to-one translation from Gaussian distribution to Gaussian mixture: (left) source data $\mathbf{X} \sim \mathcal{P}$, (middle) target data $\mathbf{Y} \sim \mathcal{Q}$ (with $|\mathbf{X}| = |\mathbf{Y}| = 1000$), and (right) Optimal Transport map, T^*, from $\mathbf{X}$ to $\mathbf{Y}$.

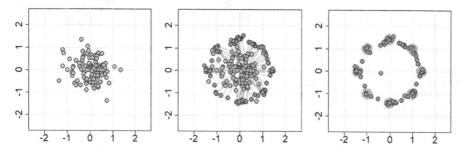

Fig. 2. Generating one-to-one translation from Gaussian to Gaussian mixture: (left) a sample of source data, $\mathbf{X}_{GEN} \sim \mathcal{P}$, with $|\mathbf{X}_{GEN}| = 100$, (middle) generated (aka transported, translated) data, and (right) qualitative matching between generated data (in green) and target distribution $\mathcal{Q}$ represented through a sample of 10000 data (blue shaded areas). (Color figure online)

from the target distribution $\mathcal{Q}$ and computed the actual OT map between $\mathbf{X}_{GEN}$ and $\mathbf{Y}_{GEN}$ – denoted with $T^*_{GEN}(\mathbf{x})$ – through the primal-dual OT solver. Then, the costs associated to the approximated and the actual (i.e., optimal) map, respectively denoted with $\mathcal{W}_2^\mu$ and $\mathcal{W}_2^{T^*}$ are computed, and their percentage difference, specifically $\Delta_{\mathcal{W}_2} = 100 \big(\mathcal{W}_2^{T^*} - \mathcal{W}_2^\mu \big) / \mathcal{W}_2^{T^*}$, is finally used as a quality measure. It is important to notice that this measure is related to the Frechét Inception Distance (FID) usually adopted to assess the quality of images created by generative models like GANs and WGANs [23, 26]. Specifically, FID is the $\mathcal{W}_2^2$ between two Gaussian distributions associated to generated and real images, respectively. On the other hand, our $\Delta_{\mathcal{W}_2}$ is more focused on evaluating if the optimality of $\mu(\mathbf{x})$ still holds with data generation. Thus, a value of $\Delta_{\mathcal{W}_2}$ close to 0% means that the matching between each new source point and its translation is approximately the same of that provided by T^* as computed from scratch.

For the Gaussian to Gaussian mixture example we obtained $\Delta_{\mathcal{W}_2} = 0.04\%$.

Gaussian to Swiss Roll. Figure 3 shows the sets of data sampled from the source (on the left) and the target (in the middle) distributions, that is $\mathbf{X} \sim \mathcal{P}$ and $\mathbf{Y} \sim \mathcal{Q}$, with $\mathcal{P}$ and $\mathcal{Q}$ as defined in Sect. 4.2. Finally, the OT map, as computed by the primal-dual OT solver, is shown (on the right). Also in this case, the multi-output GP trained on the OT map had $RMSE = 0$ (perfect interpolation).

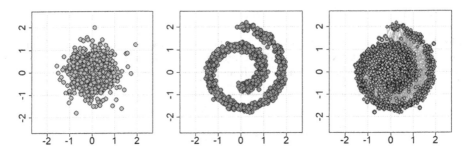

Fig. 3. Learning one-to-one translation from Gaussian to Swiss Roll: (left) source data $\mathbf{X} \sim \mathcal{P}$, (center) target data $\mathbf{Y} \sim \mathcal{Q}$ (with $|\mathbf{X}| = |\mathbf{Y}| = 1000$), and (right) Optimal Transport map from $\mathbf{X}$ to $\mathbf{Y}$.

The data generation accuracy of the multi-output GP can be qualitatively appreciated in Fig. 4, showing: the source data sampled from $\mathcal{P}$ (on the left), its translation towards $\mathcal{Q}$ (in the middle), and the overlap between the generated (aka transported, translated) data and the expected distribution (on the right).

For the Gaussian to Swiss Roll example we obtained $\Delta_{\mathcal{W}_2} = 0.03\%$.

In both the cases, the very small values of $\Delta_{\mathcal{W}_2}$ remark that the proposed approach is less prone – or almost free – to learn *fake OT plans*, an issue arising in NOT and recently addressed in [27].

5.2   Results on Image Generation

In this section we summarize the most relevant results for the image generation task. Figure 5 shows an example of one-to-one translation with digit 1 chosen as source and translated into the remaining nine digits, separately. This means that the algorithm described in Sect. 3 has been performed nine times, leading to as many multi-output GPs, each one used to provide a specific one-to-one translation towards a certain digit. It is important to remark that this is just an example: all the generated data, that is translations from 200 images of each digit to as many images of each other digit, can be viewed for free on github[3], under the folder "generated_digits".

---

[3] https://github.com/acandelieri/GenOTGP_LION17.git.

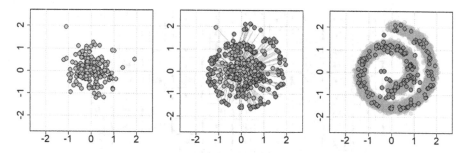

Fig. 4. Generating one-to-one translation from Gaussian to Swiss Roll: (left) a sample of source data, $\mathbf{X}_{GEN} \sim \mathcal{P}$, with $|\mathbf{X}_{GEN}| = 200$, (middle) generated (aka transported,translated) data, and (right) qualitative matching between generated data (in green) and target distribution $\mathcal{Q}$ represented through a sample of 10000 data (blue shaded area). (Color figure online)

Fig. 5. One-to-one translation of a 1 digit's image (from $\mathbf{X}_{GEN}$) into any other digit. Every target digit is generated by a separate multi-output GP model.

We have empirically noticed that all the images of the digit 1, into the dataset, are very similar. This *"lack of variability"* can represent an issue when the target digit to generate has, instead, a high variability. In the following, we show the generation of 200 images of the digit 4 (high variability) starting from as many images of the digit 1 (low variability), in Fig. 6, and as many images of the digit 2 (high variability), in Fig. 7.

Although the images of the digit 4 are quite good in the two cases, taking the digit 1 as the source digit leads to generate 4 digits very similar one to each other, and also more blurred than those generated starting form the images of the digit 2. This empirically confirms our suspect that a low variability of the source could affect the quality of the translation.

From a computational point of view, training the multi-output GP model, on 300 pairs of handwritten digit images, and then using it for generating 200 new images, requires 5 min, on average. Thus, the computational cost is negligible if compared against training times reported in papers using neural networks, especially if time needed for searching for the most effective neural architecture and network's hyperparameters is also taken into account.

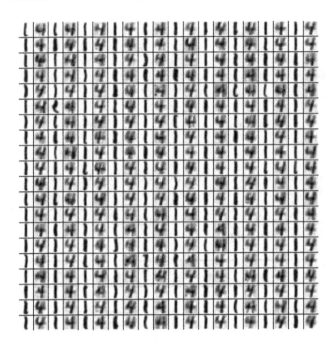

Fig. 6. One-to-one translation of 200 images of digit 1 into as many target digits 4 (source digit on the left and generated target digit on the right).

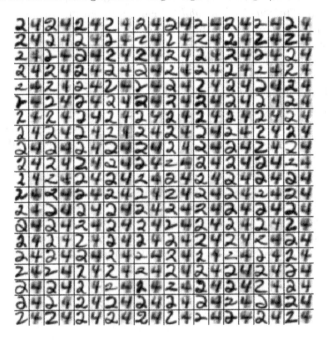

Fig. 7. One-to-one translation of 200 images of digit 2 into as many target digits 4 (source digit on the left and generated target digit on the right).

6   Conclusions

The proposed GP regression based generative model can effectively and efficiently learn and represent OT plans, and is a valid and convenient alternative to neural networks, especially in the small-data setting. Indeed, DNNs require large datasets to be trained, while GP can generalize over few examples (while it scales poorly on large datasets). Moreover, a pool of separate GPs can be (embarassingly) trained in parallel, with a significant lower computational time with respect to DNN learning. Although an image generation task has been considered, our opinion is that the proposed approach cannot replace – and it is not the aim – the current GANs and WGANs: we are more interested into exploiting statistically sound generative properties of our method in learning-and-optimization settings such as multi-task and transfer/meta learning. Another ongoing research is aimed at extending the approach to also solve the OT problem – as in NOT – further than generalizing an OT plan provided by an OT solver.

References

1. Aggarwal, A., Mittal, M., Battineni, G.: Generative adversarial network: an overview of theory and applications. Int. J. Inf. Manage. Data Insights **1**(1), 100004 (2021)
2. Amos, B., Xu, L., Kolter, J.Z.: Input convex neural networks. In: International Conference on Machine Learning, pp. 146–155. PMLR (2017)
3. Archetti, F., Candelieri, A.: Bayesian Optimization and Data Science. Springer, Cham (2019). https://doi.org/10.1007/978-3-030-24494-1
4. Arjovsky, M., Bottou, L.: Towards principled methods for training generative adversarial networks. In: International Conference on Learning Representations
5. Arjovsky, M., Chintala, S., Bottou, L.: Wasserstein generative adversarial networks. In: International Conference on Machine Learning, pp. 214–223. PMLR (2017)
6. Asadulaev, A., Korotin, A., Egiazarian, V., Burnaev, E.: Neural optimal transport with general cost functionals. arXiv preprint arXiv:2205.15403 (2022)
7. Bachoc, F., Béthune, L., Gonzalez-Sanz, A., Loubes, J.M.: Gaussian processes on distributions based on regularized optimal transport. In: International Conference on Artificial Intelligence and Statistics, pp. 4986–5010. PMLR (2023)
8. Berkenkamp, F., Schoellig, A.P., Krause, A.: Safe controller optimization for quadrotors with Gaussian processes. In: 2016 IEEE International Conference on Robotics and Automation (ICRA), pp. 491–496. IEEE (2016)
9. Biau, G., Sangnier, M., Tanielian, U.: Some theoretical insights into Wasserstein GANs. J. Mach. Learn. Res. **22**(1), 5287–5331 (2021)
10. Candelieri, A.: A gentle introduction to Bayesian optimization. In: 2021 Winter Simulation Conference (WSC), pp. 1–16. IEEE (2021)
11. Candelieri, A., Perego, R., Archetti, F.: Green machine learning via augmented gaussian processes and multi-information source optimization. Soft Comput., 1–13
12. Candelieri, A., Ponti, A., Archetti, F.: Fair and green hyperparameter optimization via multi-objective and multiple information source Bayesian optimization. arXiv preprint arXiv:2205.08835 (2022)

13. Candelieri, A., Ponti, A., Archetti, F.: Gaussian process regression over discrete probability measures: on the non-stationarity relation between Euclidean and Wasserstein squared exponential kernels. arXiv preprint arXiv:2212.01310 (2022)
14. Courville, A., Bengio, Y.: Generative adversarial nets. In: Advance in Neural (2014)
15. Daniels, M., Maunu, T., Hand, P.: Score-based generative neural networks for large-scale optimal transport. Adv. Neural. Inf. Process. Syst. **34**, 12955–12965 (2021)
16. Deisenroth, M.P., Fox, D., Rasmussen, C.E.: Gaussian processes for data-efficient learning in robotics and control. IEEE Trans. Pattern Anal. Mach. Intell. **37**(2), 408–423 (2013)
17. Dong, C., et al.: A survey of natural language generation. ACM Comput. Surv. **55**(8), 1–38 (2022)
18. Frazier, P.I.: Bayesian optimization. In: Recent Advances in Optimization and Modeling of Contemporary Problems, pp. 255–278. Informs (2018)
19. Gramacy, R.B.: Surrogates: Gaussian Process Modeling, Design, and Optimization for the Applied Sciences. Chapman and Hall/CRC, London (2020)
20. Gulrajani, I., Ahmed, F., Arjovsky, M., Dumoulin, V., Courville, A.C.: Improved training of Wasserstein GANs. In: Advances in Neural Information Processing Systems, vol. 30 (2017)
21. Haidar, M.A., Rezagholizadeh, M.: TextKD-GAN: text generation using knowledge distillation and generative adversarial networks. In: Meurs, M.-J., Rudzicz, F. (eds.) Canadian AI 2019. LNCS (LNAI), vol. 11489, pp. 107–118. Springer, Cham (2019). https://doi.org/10.1007/978-3-030-18305-9_9
22. Harshvardhan, G., Gourisaria, M.K., Pandey, M., Rautaray, S.S.: A comprehensive survey and analysis of generative models in machine learning. Comput. Sci. Rev. **38**, 100285 (2020)
23. Heusel, M., Ramsauer, H., Unterthiner, T., Nessler, B., Hochreiter, S.: GANs trained by a two time-scale update rule converge to a local Nash equilibrium. In: Advances in Neural Information Processing Systems, vol. 30 (2017)
24. Huang, F., Guan, J., Ke, P., Guo, Q., Zhu, X., Huang, M.: A text GAN for language generation with non-autoregressive generator (2021)
25. Jaquier, N., Rozo, L., Calinon, S., Bürger, M.: Bayesian optimization meets Riemannian manifolds in robot learning. In: Conference on Robot Learning, pp. 233–246. PMLR (2020)
26. Korotin, A., Li, L., Genevay, A., Solomon, J.M., Filippov, A., Burnaev, E.: Do neural optimal transport solvers work? a continuous Wasserstein-2 benchmark. Adv. Neural. Inf. Process. Syst. **34**, 14593–14605 (2021)
27. Korotin, A., Selikhanovych, D., Burnaev, E.: Kernel neural optimal transport. arXiv preprint arXiv:2205.15269 (2022)
28. Korotin, A., Selikhanovych, D., Burnaev, E.: Neural optimal transport. arXiv preprint arXiv:2201.12220 (2022)
29. Li, B., Qi, X., Lukasiewicz, T., Torr, P.: Controllable text-to-image generation. In: Advances in Neural Information Processing Systems, vol. 32 (2019)
30. Liu, H., Cai, J., Ong, Y.S.: Remarks on multi-output Gaussian process regression. Knowl.-Based Syst. **144**, 102–121 (2018)
31. Mallasto, A., Feragen, A.: Learning from uncertain curves: the 2-Wasserstein metric for Gaussian processes. In: Advances in Neural Information Processing Systems, vol. 30 (2017)
32. Masarotto, V., Panaretos, V.M., Zemel, Y.: Procrustes metrics on covariance operators and optimal transportation of Gaussian processes. Sankhya A **81**, 172–213 (2019)

33. Peyré, G., Cuturi, M., et al.: Computational optimal transport: with applications to data science. Found. Trends® Mach. Learn. **11**(5–6), 355–607 (2019)
34. Ramesh, A., et al.: Zero-shot text-to-image generation. In: International Conference on Machine Learning, pp. 8821–8831. PMLR (2021)
35. Rout, L., Korotin, A., Burnaev, E.: Generative modeling with optimal transport maps. arXiv preprint arXiv:2110.02999 (2021)
36. Salimans, T., Zhang, H., Radford, A., Metaxas, D.: Improving GANs using optimal transport. arXiv preprint arXiv:1803.05573 (2018)
37. Scholkopf, B., Smola, A.J.: Learning with Kernels: Support Vector Machines, Regularization, Optimization, and Beyond. MIT press, Cambridge (2018)
38. Seguy, V., Damodaran, B.B., Flamary, R., Courty, N., Rolet, A., Blondel, M.: Large-scale optimal transport and mapping estimation. arXiv preprint arXiv:1711.02283 (2017)
39. Singh, N.K., Raza, K.: Medical image generation using generative adversarial networks: a review. Health Inf.: Comput. Perspect. Healthc., 77–96 (2021)
40. Sui, Y., Gotovos, A., Burdick, J., Krause, A.: Safe exploration for optimization with gaussian processes. In: International Conference on Machine Learning, pp. 997–1005. PMLR (2015)
41. Svenson, J., Santner, T.: Multiobjective optimization of expensive-to-evaluate deterministic computer simulator models. Comput. Stat. Data Anal. **94**, 250–264 (2016)
42. Villani, C.: Topics in Optimal Transportation, vol. 58. American Mathematical Soc. (2021)
43. Williams, C.K., Rasmussen, C.E.: Gaussian Processes for Machine Learning, vol. 2. MIT press Cambridge, MA (2006)
44. Zhan, D., Cheng, Y., Liu, J.: Expected improvement matrix-based infill criteria for expensive multiobjective optimization. IEEE Trans. Evol. Comput. **21**(6), 956–975 (2017)
45. Zhang, H., Xie, L., Qi, K.: Implement music generation with GAN: a systematic review. In: 2021 International Conference on Computer Engineering and Application (ICCEA), pp. 352–355. IEEE (2021)
46. Zhu, Y., et al.: Quantized GAN for complex music generation from dance videos. In: Avidan, S., Brostow, G., Cisse, M., Farinella, G.M., Hassner, T. (eds.) ECCV 2022. LNCS, vol. 13697, pp. 182–199. Springer, Cham (2022). https://doi.org/10.1007/978-3-031-19836-6_11

Real-World Streaming Process Discovery from Low-Level Event Data

Franck Lefebure, Cecile Thuault, and Stephane Cholet[✉]

Softbridge Technology, 97122 Baie-Mahaut, GP, France
`stephane.cholet@softbridge.fr`

Abstract. New perspectives on monitoring and optimising business processes have emerged from advances in process mining research, as well as techniques for their discovery, conformance checking and enhancement. Many identify process discovery as the most challenging task, as it is the first in line, and as the methods deployed could impact the effectiveness of subsequent analysis. More and more companies are ready to use process mining products, but not at any cost. They require them to be highly reliable, fast and to introduce a limited overload on their resources (including human). In a real-word setting, specific business constraints add to common issues, such as high data frequency and asynchrony. This also adds to the complexity of real-world processes and of process mining itself. Mainstream studies propose to use a finite set of high-level and business-oriented event logs, where key attributes (i.e., case, activity and timestamp) are known, and apply unscaled discovery techniques to produce control-flow process models. In this research, we propose an original approach we have designed and deployed to mine processes of businesses. It features fully streamed and real-time techniques to mine low-level technical event data, where key attributes do not exist and have to be forged. We will focus on the scope of process discovery, and expose our adoption of an organizational perspective, driven by (semi-) unsupervised discovery, streaming and scaling features.

Keywords: Streaming process mining · Unsupervised process discovery · Real time · Scaling process mining · Organizational perspective

1 Introduction

Process mining is an emerging field of research that lies at the intersection of data science and process science [3]. Recent developments have brought to light the importance of process supervision in companies, and techniques have been developed for process discovery, conformance checking, and enhancement.

Process discovery is a challenging task, as it is the first step towards process mining, and because the effectiveness of subsequent analysis depends on the methods used. The raw material of process discovery comes from event logs, typically stored in data sources (e.g., files or databases). These logs are generated

M. Sellmann and K. Tierney (Eds.): LION 2023, LNCS 14286, pp. 150–164, 2023.
https://doi.org/10.1007/978-3-031-44505-7_11

by various software systems during process operations. Each event in the log contains information about a case, an activity and a timestamp, plus some contextual attributes such as an involved resource. Thus, a case can be represented as an ordered sequence of events, which we will later refer to as a *trace*. There are several algorithms available in the literature for discovering processes from traces, some of which are discussed in Sect. 3. However, common issues such as missing data or model reliability must be addressed in any situation.

Companies are increasingly interested in using process mining products, but not at any cost [11]. To be successful, a process mining solution must demonstrate its ability to quickly and effectively improve processes with minimal effort. In a real-world setting, there is a need for both streaming and real-time processing. While streaming aims to mine events as they arrive – likely unordered, rather than using a finite set of data, real-time aims to minimize the delay between the process operation and its discovery in the product. Besides, very large amounts of traces can be produced, and companies also expect deployed solutions to mine as much of their new gold as possible, still in a reduced time; hence introducing the need for scaling. Furthermore, processes in a company are not static: a process could have variants at a given time period but it could as well evolve over time. These changes may or may not be known by all process actors, but will always be visible from the produced logs. Unsupervised discovery techniques can address this aspect, but they are infrequently used in both literature and industry [3,14].

Explainability is an important consideration when discovering processes. Most techniques rely on traces produced by high-level components, where events already have necessary attributes, context and structure. Such techniques allow discovery of processes and their issues but do not explain them at a lower level. For example, if a bottleneck is discovered, it would be possible to understand its origin at the same level by applying dedicated analysis, such as decision trees. But it would be difficult to explain the root cause of the bottleneck, which is located at a lower level, likely in the back-end side of the information system.

In this paper, we aim to provide a comprehensive overview of our process discovery techniques, which have been designed and deployed in real-world settings to meet industrial requirements for streaming, real-time, and scaling. In Sect. 2, we provide context regarding the use of process mining in the real world, with the example of a company. In Sect. 3, we propose an overview of the literature on process mining, including perspectives and streaming process mining. In Sect. 4, we present the techniques we have developed and deployed to mine processes in companies and meet industrial requirements regarding streaming, real-time and scaling. In Sect. 5, we describe a setup where our solution is currently deployed and being used. Eventually, we conclude and suggest future directions for our work in Sect. 6.

2 Context

2.1 Naming Conventions

The concept of a digital enterprise is often compared to that of a complex system [5], where various elements interact with each other and with their environment to produce an optimal result. In particular, a company can be modeled at three levels of organization, as proposed in [13]: the information system, the business processes, and the customer journeys, as shown in Fig. 1. The information system represents the hardware and software infrastructure of the company, while business processes represent a set of interconnected tasks aimed at achieving organizational goals. Customer journeys represent the interactions that customers may have with the company.

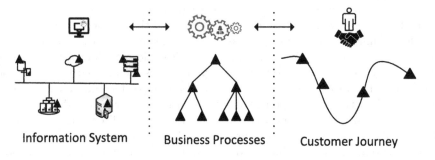

Fig. 1. Company modeling with three layers, according to [13]. The symbol ▲ represents traces, activities and customer interactions related to a given process.

In this paragraph, we introduce several important naming conventions. In the context of business processes, A company can serve its customers through different *channels*, each of which features several *touchpoints* or front-end applications. *Roles* are often used to group functional aspects [1], and a touchpoint can initiate one or more business processes. These processes are triggered in the front-end and drive an *application chain* over an *application domain* (see Fig. 2), with each application proposing a set of *services* that can exchange information through *service interfaces*. The back-end of the application chain is generally unknown to customers and front-end actors, who may assume that initiating a process from a touchpoint is sufficient to complete it. Mostly, any service or application that is not a touchpoint will be considered to belong to the back-end of the application chain. Also, it is worth noting that when a service from a back-end application performs a tasks, it will likely not let an exploitable service event log for common process discovery algorithms, because it will not feature a case and an activity.

2.2 A Company's Expectations and Related Challenges

Although process mining is not yet widely adopted in companies, it appears from a recent survey [11] that they have high expectations for this technology. The survey found that 77% of companies expect process mining products to optimize their processes, and 57% expect to gain more transparency in their execution. Cost reduction ranked third, while several other hot topics such as automatization, monitoring, standardization, and conformance checking also made the top ten. These business-oriented expectations imply scientific challenges for the field of process mining. In this subsection, we attempt to make a soft bridge between identified business expectations and the scientific challenges they unveil.

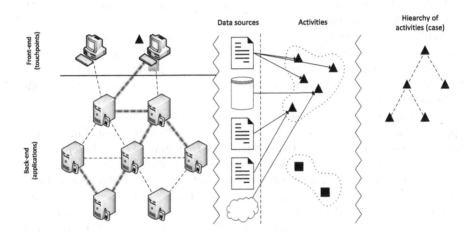

Fig. 2. Processes are triggered from a touchpoint and describe an application chain along its operation (highlighted in grey). Services store execution logs in various data sources, which can be used to further create activities. Ultimately, activities are gathered and can be organized (e.g., hierarchically) to form of a case. Activities ▲ and ■ belong to two distinct processes. On the figure, only the process with ▲ activities is illustrated.

Companies are complex due to their numerous interconnected components, making it difficult for an individual to have a complete understanding of the entire organization. One expectation of process mining is to model processes in accordance with the existing *enterprise architecture* to facilitate better comprehension and enhance operations. This challenge requires adopting an appropriate mining perspective, such as the organizational perspective.

Business processes are rather evolving entities than static and seamless workflows that never change. Although most algorithms have the capability to discover new variants of processes that have already been observed, processes must first be explicitly monitored to be observed. For instance, the addition or deletion of services to an application may go unnoticed. Also, if a process is modified after services have been added, related activities may remain unobserved. The expectation is that services and processes can be discovered without the need for supervision, and with minimal configuration. This challenge can be referred to as the unsupervised discovery challenge.

A company's complexity is further compounded by the fact that it involves actors with varying profiles and backgrounds, such as marketing-oriented, back-office, or executive. While these different profiles may be interested in the same set of processes, they often have different expectations regarding the level of analysis (typically, technical or business) or the ergonomics of the process mining solution (typically, dashboard or expert mode). The usability of the process mining solution is thus a crucial expectation that needs to be addressed. Although event data is often too much refined to offer technical insights, it can be used to build high-level process models. Therefore, a challenge in process mining is to build multi-level processes from technical data that can accommodate the diverse needs of different actors.

While it is true that expectations for industrial process mining can vary based on factors such as a company's goals, industry, or size, we posit that the subset of expectations outlined previously can be considered fundamental. Nonetheless, to delve further into the challenges faced by *industrial process mining*, we offer additional insights.

The complexity of information systems can present both a challenge and an opportunity for process mining. These systems are often distributed and involve numerous distinct and unrelated components that may operate independently of each other, although they do work together. As a result, log data is produced from multiple sources, in different formats, and with no set chronology. Given the fast-paced, competitive nature of modern business, it is crucial that data is collected and analyzed in real time from streaming sources rather than static files. Achieving this requires advanced tools and technologies that can handle the distributed and dynamic nature of information systems.

A crucial aspect of process mining is the ability to analyze information systems in real-time, facilitating both proactive and corrective strategies through the use of alerts and automated decision layers. However, it is not enough to identify where a problem is located in a process: it's better to know why. For this reason, the ability to explain at the lower level, e.g., technical, why did a process instance failed can also be considered as an expectation. The answer might come from a traditional root-cause analysis or from the research of a process (or cross-process) pattern that explains the issue.

It makes almost no doubt that process mining will become a "new normal" in the next couple years. Although most techniques remain focused on so-called *lab data*, researchers are aware of the challenges brought by industrial process mining [7]. In this section we have discussed some of these challenges, and in the next one we will give necessary background about the state of the art in process mining.

3   Related Work

Several running challenges for process mining to overcome have been identified in the literature [1,14]. In addition, extensive reviews have been published [1,3,7], addressing several aspects of the field. In this section, we cover selected aspects of process mining that relate to our contribution.

3.1   A Brief Overview of Process Mining

In the last decade, process mining has been structured and formalized, allowing researchers to uncover many challenges related to the discipline. The raw material of process mining are event logs produced by the components of an information system. Let L denote a multi-set of traces $[\sigma_1, \ldots, \sigma_m]$ where a single trace is a sequence of events such as $\sigma = \langle a_1, a_2, \ldots, a_n \rangle$. In most approaches, three core attributes are necessary to each event: case (process instance), activity (task) and timestamp. A log featuring the core attributes is an *event log* and is suitable for process mining tasks. Other contextual attributes can provide information about a process execution, and can also be used to analyze or model processes from different perspectives. Logs that do not have the core attributes, for instance technical logs, cannot be directly used to build activities and cases. Table 1 shows example logs from both categories.

Table 1. Examples of two (unrelated) log files. On the left, an event log featuring the three core attributes. On the right, a real-world technical log with a timestamp and various information for each entry.

Case	Activity	t	Log line contents
pizza-17	create base	18:10	pubsub; prj-api-pubsub; PROCESSED; 2023-04-13T
pizza-17	add tomato	18:17	13:46:04Z; 2023-04-13T13:46:04Z; 077dd8d4-ac8e- 96bf-f19804a62ab; 74636743906476; STATUS=EXECUTED;
pizza-18	add tomato	18:18	HTTP_CODE=204; PENDING_ORDER_ID=20230329-
pizza-17	add cheese	18:18	24475001; CID=4560897; NB_MODIFIED=4

Process discovery involves the task of learning a process model from a multi-set of traces L, where each trace is a sequence of events with core attributes such as case, activity, and timestamp. The Alpha algorithm [2] is one of the widely used and extended process discovery algorithms, but several advanced algorithms such as region-based [8], Split Miner and Log Skeleton have also been proposed. Each algorithm has its own strengths and weaknesses, and their choice depends on various factors such as process complexity, available data, and process mining goals. While region-based miners are suitable for situations with duplicate activities, Split Miner is known for its ability to balance the fitness and precision of models. From the literature, it is clear that there is no one-size-fits-all approach to process mining, and the selection of a suitable method should consider the aforementioned factors.

One commonly used approach in process mining is to view the events in a trace σ as forming a partial order $\prec$ with respect to their occurrence time, which is known as the control-flow perspective [1]. This perspective enables the identification of the main process backbone by ordering the activities according to the partial order. In the context of a business environment, the control-flow perspective is well suited as it allows for the examination of dependencies between well-defined entities. However, other perspectives such as the case, information, and application perspectives are also useful for analyzing processes from different angles.

3.2 The Organizational Perspective

An organizational perspective can be adopted in order to derive the organizational structure of an information system [15]. While control-flow analysis focuses on the sequencing of activities, the organizational perspective emphasizes the discovery of relationships between the components of an entity. These components can be individuals, and in this case, a social network can be inferred and analyzed to understand how people work together, identify who is responsible for most problems, and detect bottlenecks in the workflow. Alternatively, components can be distinct business services, which can be used to study the effectiveness of their collaboration. These approaches have been investigated in an industrial context [14].

An alternative approach is to map the components of an information system to its applications, interfaces, and processes. This perspective offers a comprehensive view of the system, which would be nearly impossible to obtain by a single person. By applying suitable techniques, the entire system can be reconstructed, providing insights into the behavior and dependencies of its different components. This approach is particularly relevant in complex and dynamic systems, where interactions between components are often intricate and difficult to unravel.

3.3 Streaming Process Discovery

Streaming process mining refers to techniques applied to the analysis of data streams, rather than static and finite log files [3]. A typical use case that motivates the use of streaming techniques is the need to understand active processes, as opposed to those that have already completed. This perspective also qualifies for further proactive or corrective strategies that may be activated to optimize processes.

Streaming process mining comes with a set of interesting challenges which have been underlined in the literature [3]. First of all, data can be handled either in real-time, incrementally or online. Real time systems can be hard, soft or firm, based on the emphasis put on the delay between the availability of the input and the production of the output. Also, by definition an event stream is not bound, and it is not possible to store the entire stream. As a consequence, produced models and analysis might be considered as incomplete most of time. Thus, the question of *data approximation* for a given process has to be raised. Backtracking over stream is prohibited (each data point is processed only once), and stream rate can fluctuate over time. These peculiarities are thoroughly discussed in [6].

Another major challenge relates to so-called *data unorderness.* In a real-world setting, with a distributed environment, logs are produced at independent frequencies by applications. As a result, the order in which events arrive is often different than the order in which they occur. In [4], the authors propose to gather events in a windowed fashion to periodically reorder them. In background, a directly follows graph is maintained to feature a control-flow process model.

A well know algorithm for streaming process discovery is the Heuristic Miner with Lossy Counting (HM-LC) [6]. It employs a dependency measure to evaluate the causal dependency between two activities. Based on a threshold, a dependency graph is maintained for activities that qualify, while other activities are treated as noise.

Eventually, it is worth noting that only a few studies address the issue of deleting or updating activities from a stream (see [9] for an example), which brings to light the complexity of the task. Also, contributions on scaling such techniques remain scarce.

The literature of process mining is rich. Concepts are getting structured and use cases in various fields show the importance of mining processes, and not just data. In the next section, we propose our contribution to process mining and expose some details about a set of production-grade techniques that are actively deployed and used.

4   Contribution

4.1   Supervision of a Whole Application Domain

Most approaches require to explicitly configure the process mining tool to monitor specific processes. Produced event logs are then used for further process discovery, conformance checking or other mining tasks. Although it seems obvious, it's worth noting that processes that are not monitored will not be discovered. The same applies at the service and application levels. For instance, interfaces between services that are out of the scope of the supervised processes will remain uncontrolled.

To avoid having a fragmented view of the information system, we have designed a method that only requires to configure an application domain (as defined in Sect. 2.1), rather than specific processes. It implies that all services and served processes are monitored by default. An interesting example of the need of such an implementation is the addition of a service, which is quite common. Whenever a service is added, it will be monitored and discovered with no prior configuration. The same also applies for processes and variants.

4.2   Unsupervised and Streaming Process Discovery

The proposed framework (later referred to as *the streamer*) is designed with prominent streaming capabilities implemented all along the process discovery. This design is particularly well-suited for the distributed nature of contemporary information systems, where logs and applications may be spread across multiple locations and media, such as databases and files. The core concept behind the streaming approach is the continuous construction of entities as data becomes available, in contrast to processing a finite dataset that is entirely available at once. This characteristic also enables real-time monitoring and analysis, as process models are constructed iteratively.

The streamer is designed with four steps, namely from logs to rawdata[1], from rawdata to activities, from activities to case and from cases to processes. As data flows through the successive steps, it is analyzed and aggregated to create an always up-to-date process model.

From Logs to Rawdata. The first step collects log entries from various sources to build *rawdata*, a data structure that represents a fragment of activity. A rawdata contains as much information as possible about an execution, mainly contextual attributes, a timestamp and various identifiers. These identifiers are technical identifiers which can be found in multiple log entries from distinct services.

Log information can either be used to update or create rawdata. This step achieves an important compression work by only pushing relevant information from the logs to the rawdata. It is important to note that at some point during the step, a rawdata may not yet contain core attributes. The main reason, besides infrastructure reasons, is that necessary information is not contained in a unique location. Several log files have to be processed, at distinct moments, to build a rawdata that could move ahead through the pipeline. It is part of the step to gather into a single data structure several information fragments.

From Rawdata to Activities. As stated earlier, the framework monitors a full application domain instead of explicitly configured processes. Thus, processes are automatically discovered and built, by correlating rawdata and activities to unveil their organizational setup.

When a rawdata features enough information, it is pushed to a pool where it will be used either to update or create an activity. As a consequence of the large number of monitored services, a very high volume of technical log transactions has to be analyzed. The second step of the streamer can be seen as a correlation step, where two or more rawdata can be correlated and joined into a single event. Correlation is made possible thanks to a set of rules that are checked against data. For confidentiality reasons, we cannot provide a detailed explanation about how the rules are built, maintained and checked. However, as a result of their application, a rawdata is either turned into an activity or used to populate an existing activity.

---

[1] By convention, the plural form of the word is unchanging.

From Activities to Cases. The framework mines processes with an organizational perspective. Thus, timestamps are barely used to compute the activities' hierarchy. The second step can be understood as a correlation and linking step.

We use a correlation function on the activities, previously populated from rawdata. Prior to mapping an activity to a parent, a data-driven learning phase has to be conducted by using the attributes found in the activities to determine their hierarchy. It is then possible to organize two activities into an execution tree. In such a situation, the activities are attached together to form a case fragment, which is pushed to the following step. A notable benefit from this step is the capacity of building a documented map of the information system without any prior configuration. This contributes to the organizational perspective and allows for multiple level root-cause analysis.

More details about this step are given in Subsects. 4.3 and 4.5.

From Cases to Processes. *The streamer* intends to provide real-time process mining to users, by producing accurate metrics and process models at any time for any discovered process.

A process model can be seen as a structure that holds for all of the variants of a process. Computation of metrics over a process model requires to traverse each case (through each activity). This computation would be overly complex to handle if it was realized continuously, i.e., each time a case is updated. To alleviate this complexity, we have developed an approach that consists in computing the models per contiguous temporal slices over time.

As explained early in the paper, all necessary information to build a case becomes arbitrarily available over time, e.g., the last activity can become available before the first one, which excludes building the case in a handy or deterministic order. As a corollary, we postulate that newer cases are updated more frequently than older ones - this is also true for rawdata and activities. Thus, we compute process model chunks containing variants over slices of time. The most recent chunk models last seen variants of the process from now backwards to a point in the past, called $chunk_t$. As time moves ahead, variants aged older than a given period fall into another chunk, $chunk_{t-1}$. Chunks are built for all hours of the present day, then daily, weekly and monthly. As a consequence, when a case is updated with new data, only the corresponding process chunk is fully recomputed. The whole process model, comprising all of the process chunks, can always be up to date.

An interesting challenge existing in all steps is that sufficient information to achieve the step's goal becomes arbitrarily available on the time line. For example, in the second step, an activity may remain isolated until the previous step provides enough information to associate the activity with a case. To alleviate this issue, we have implemented an exponential backoff.

4.3   Control

The order in which data arrives is not predictable. Given the distributed aspect of information systems, events may become visible in a order that is different than the order in which they have happened. To illustrate the issue, we consider the second step (from activities to cases), although the control strategy applies to all steps.

To build a hierarchical tree of activities, we basically need to know which service called another service. Let $s1$ and $s2$ be two services where $s1 \to s2$ denotes $s1$ making a call to $s2$. During a process instance, $s_i(\cdot) = a_i$ denotes an activity of a case. Let $\succ$ be a partial order over activities such that $s_i \to s_j \implies a_i \succ a_j$. If we note $t_{obs}(a)$ the moment on the timeline where a was observed, then $s_i \to s_j \not\implies t_{obs}(a_i) \succ t_{obs}(a_j)$. Thus, it is not possible to rely on the observation moment to build the hierarchical tree of activities over the call order. When an activity is observed for the first time, it is queued and the algorithm tries to find and update its case hierarchy, as explained in Sect. 4.2. If no suitable case fragment is found, the activity is used to build a new case fragment. Else, if the activity was itself updated during the search, it is re-queued (see Subsect. 4.5).

We have implemented an exponential backoff re-queuing strategy over each step, so that any rawdata, activity, case or process that cannot be handled immediately for some reason is re-queued to be handled later. Let Q be a queue stack with $Q = \langle q_0, q_1, \ldots, q_n \rangle$. Each queue q has a backoff parameter b controlling its delivery rate, such as q_i has a backoff parameter of $b_i = 2^i$. In this configuration, elements in q_0 are handled immediately. More generally, an element c in a stack with n queues is handled after an expected time $E(c)$ in seconds:

$$E(c) = \frac{1}{n+1} \sum_{i=0}^{n} 2^i = \frac{2^n - 1}{n+1}$$

Exponential backoff is a widely used technique in computer science, and is appreciated for providing an efficient way to manage collisions. In our running example, a collision occurs when an activity cannot be inserted into a case because. The expectation of the strategy is that delaying a treatment that has failed will lead to success in a future attempt, probably because sufficient information would have become available in the meantime – allowing the correlation function to produce an acceptable output. With ten queues in the stack ($n = 9$), an element would wait an average of 46.45 s. In practice, this delay is much shorter.

4.4   Scaling

In the process mining literature, the term *scaling process mining* is often used to describe the expansion of process mining to cover larger scopes [3]. One such example is the transition from analyzing a few processes to monitoring an entire information system. This type of scaling can be viewed as functional scaling. Another type of scaling which is more rarely discussed is technical scaling. This aspect becomes critical in the context of streaming process mining, where data ingestion and process monitoring are performed in real time.

The presented framework, *the streamer*, is scaling-capable on both functional and technical aspects. As discussed earlier, a whole application domain is put under control, and processes that operate under the defined domain are discovered automatically. The user can then navigate in a process population and chose to explore a specific process. This approach seems essential to scale functionally, as it would be tedious to configure processes individually. From a more technical point of view, the scaling capability of a solution depends on both the algorithms and the infrastructure that supports them. Each step described in Subsect. 4.2 can be multi-threaded in order to scale with the rate and volume of data. To alleviate issues regarding deadlocks and optimize performance, we have developed a custom optimistic locking mechanism over the steps.

4.5 Optimistic Locking Mechanism

In this subsection, we keep the running example of step two (from activities to cases, paragraph Sect. 4.2). Linking an activity to a case from a stream of data may fail, e.g., if the structure of the case was updated in the mean time. To optimize the linking success, we have implemented an optimistic locking mechanism over the steps.

When the case is read and found to be eligible to have a new activity bounded, a version number is stored. At writing time, if the version number has changed, the activity is re-queued and the step will attempt to bound it to a case later, with respect to the control strategy exposed in Sect. 4.3. Optimistic locking avoids hard locking records (a.k.a pessimistic locking) and deadlocks related issues, while preserving coherence of data.

The production-grade techniques explained in this section have proved effective in real life situations, where data streams at high rates and volumes. In the next section, we provide a use case where the techniques have been deployed.

5 Deployment

So far, we have given details about a set of techniques developed in the so-called *streamer*, a framework for streaming and scaling process mining. This framework is a core back-end component of a larger software, Data Explorer, commercialized by Softbridge Technology [12], a process mining company. In this section, we provide details about a deployment of Data Explorer (including *the streamer*) in a client company.

Data Explorer has been used for five years on a daily basis by several users from various profiles. The client company is in the telecommunications industry and has an online store, as well as over a hundred physical stores. They also manage both internal and external call centers. Its customers are both professionals and individuals who buy terminals, sign up for fiber or 4G subscriptions, buy paid options, receive free credits, complaint, terminate contracts, etc. These elements are a few examples of real business processes for the company.

An overview of the supervised application domain is provided in Table 2. A total of 456 processes have been discovered in an unsupervised manner. It would have been highly tedious and time-consuming to configure each process individually. Instead, after a minimal configuration step (see Subsect. 4.1), *the streamer* allows unsupervised discovery of processes supported by the defined application domain.

Table 2. Details regarding the application domain supervised by the deployed solution. *The streamer* mines 456 processes and their variants, providing both data ingestion and analysis in real time.

Item	Description
Application domain	84 applications among 7 channels
Log locations	33 distinct data sources (files, databases, etc.)
Business processes	456 business processes (not including variants)
Data volume	Monthly average over 10Tio

Given its high complexity, the cost of this project was established to € 300.000, plus an undisclosed annual fee. In Table 3, we provide some examples of cost cuts that have been observed and used by the client company to evaluate its ROI (Return On Interest). From its adoption in 2018 to today, the ROI of Data Explorer has been evaluated to € 1.243.680 on a three-year period, starting with as high as € 214.560 in the first year (annual fee included). Customer satisfaction has also been significantly improved, as shown by the huge reduction factor on its related cost (x17).

Table 3. Examples of cost cuts realized after the adoption of Softbridge Technology's process mining product. After the adoption of the product in 2018, more than € 730.000 were saved in annual complaint costs, and a seven-figure amount was saved overall. Proposed metrics are annual averages, computed from 2015 to 2018 for the column Before, and from 2019 to 2023 for the column After.

	Before	After	Reduction factor
Annual customer interactions	4 800 000	4 800 000	
Failed interactions	240.000 (5%)	48.000 (0.5%)	x5
Average diagnosis time	2 h	30 min	x4
Diagnosis costs (€ 38/h)	€ 729.600	€ 36.480	x20
Annual complaints cost	€ 777.600	€ 46.080	x17

The streamer introduces a very limited overload on the information system, as its operations are read-only. In the presented use case, the solution was deployed on premises. In nominal situation, a process becomes visible in the framework less than ten minutes after its creation in real life. This delay also applies to updates in a process, for instance when a new event occurs. Both technical and business teams can take appropriate action in a reduced time after a problem has occurred. It helps minimize or eliminate negative impacts on business value and customer experience.

The deployment phase of the system at the client company required minimal human resources and initially focused on monitoring a limited set of sale processes on a single touchpoint. This decision was driven by billing issues that couldn't be resolved using traditional manual methods. Initially, there was skepticism among stakeholders that activities could be linked without prior knowledge of their originating process and context. The logs that were collected by the client company before the definitive adoption of Data Explorer were convenient to mine processes, but with minimal information. The client company then made a strategic decision to augment the logs, in order to obtain highly documented processes, and to expand the scope of process mining by broadening the range of applications and business processes to be monitored.

6 Conclusion

Process mining is a research area located at the intersection of data science and process science. Lately, a huge number of studies and use cases have proved that it could help improve operations (see Sects. 2 and 4). While mainstream studies focus on offline process mining, attempts have been made regarding streaming process mining, and have contributed to provide a better understanding of its necessity (see Subsect. 3.3). In this study, we have proposed our contribution to the field by presenting our framework, *the streamer*, a process mining product.

The streamer meets some important criteria expected in the industry to provide business value, as presented in Sect. 2. Both data ingestion and analysis are performed from a stream rather than static data, thus achieving real-time monitoring of processes. In distributed information systems, log entries are asynchronously written to data sources. The streamer continuously ingests data from these sources after a lightweight configuration phase, enabling unsupervised discovery of all processes operating under the defined domain without the need for per-process configuration.

Real-time process discovery and monitoring is achieved with updates becoming visible in the product in less than ten minutes after the operation was performed in reality. By utilizing low-level technical log transactions, processes can be explored at multiple levels, including technical and business levels, enabling precise mapping of the information system. This approach facilitates root-cause analysis, which can capture cross-process patterns instead of single activities. These real-time and streaming operations provide tangible improvements of business value while being grounded in strong scientific and technical principles.

We have adopted an organizational perspective to fit our multilevel process discovery. Indeed, instead of presenting flow of events, we organize processes just like they exist in their context. In addition to providing an always up-to-date vision, this perspective contributes to faster problem auditing and resolution.

There are still many challenges left for process mining. As a future work direction, we have started to develop a conformance checking feature that would as well work in a streaming context.

In 2020, a study quoted that real-time monitoring of processes is an important process intelligence capability not yet commercially available [10]. Recently in 2022, the authors of [3] stated that real-time data ingestion is rarely observed in the industry, and real-time analysis is still to be achieved. Although discreet, the framework of which we have exposed some of the functionalities is commercially available since 2017 and effectively proposes both real-time data ingestion and real-time analysis [12].

References

1. van der Aalst, W.M.P., Weijters, A.J.M.M.: Process mining: a research agenda. Comput. Ind. **53**, 231–244 (2004). https://doi.org/10.1016/j.compind.2003.10.001
2. van der Aalst, W.M.P., Weijters, A.J.M.M., Maruster, L.: Workflow mining: discovering process models from event logs. IEEE Trans. Knowl. Data Eng. **16**(9), 1128–1142 (2004). https://doi.org/10.1109/TKDE.2004.47
3. van der Aalst, W.M.P., Carmona, J.: Process Mining Handbook, vol. 448. Springer, Cham (2022). https://doi.org/10.1007/978-3-031-08848-3
4. Awad, A., Weidlich, M., Sakr, S.: Process mining over unordered event streams. In: 2020 2nd International Conference on Process Mining (ICPM), pp. 81–88 (2020). https://doi.org/10.1109/ICPM49681.2020.00022
5. von Bertalanffy, L., Hofkirchner, W., Rousseau, D.: General System Theory: Foundations, Development, Applications. George Braziller, Incorporated (2015)
6. Burattin, A., Sperduti, A., van der Aalst, W.M.P.: Control-flow discovery from event streams. In: Proceedings of the IEEE Congress on Evolutionary Computation, CEC 2014, Beijing, China, 6–11 July 2014, pp. 2420–2427. IEEE (2014). https://doi.org/10.1109/CEC.2014.6900341
7. Burratin, A.: Process Mining Techniques in Business Environments, vol. 207. Springer, Cham (2015). https://doi.org/10.1007/978-3-319-17482-2
8. Darondeau, P.: Deriving unbounded Petri nets from formal languages. In: Sangiorgi, D., de Simone, R. (eds.) CONCUR 1998. LNCS, vol. 1466, pp. 533–548. Springer, Heidelberg (1998). https://doi.org/10.1007/BFb0055646
9. Gama, J., Aguilar-Ruiz, J.S., Klinkenberg, R.: Knowledge discovery from data streams. In: Intelligent Data Analysis (2009)
10. Modi, A., Shahdeo, U.: Process mining state of the market. Technical report. EGR-2020-38-R-3808, Everest Group (2020)
11. Muller, X., Lhoste, P.: Adoption du process mining et facteurs de réussite. Technical report, Deloite (2022)
12. Softbridge Technology, a Process Mining Company (2015). https://softbridge-technology.com. Accessed 07 Feb 2023
13. Thuault, C., Cholet, S.: Understanding and mastering complex systems. Technical report. SB-WP-2022-9, Softbridge Technology, Baie-Mahault, GP, France (2022)
14. van der Aalst, W., et al.: Business process mining: an industrial application. Inf. Syst. **32**(5), 713–732 (2007). https://doi.org/10.1016/j.is.2006.05.003
15. Zhao, W., Zhao, X.: Process mining from the organizational perspective. In: Wen, Z., Li, T. (eds.) Foundations of Intelligent Systems. AISC, vol. 277, pp. 701–708. Springer, Heidelberg (2014). https://doi.org/10.1007/978-3-642-54924-3_66

Robust Neural Network Approach to System Identification in the High-Noise Regime

Elisa Negrini[1]([⊠])[ID], Giovanna Citti[2], and Luca Capogna[3]

[1] Department of Mathematics, University of California Los Angeles, Los Angeles, USA
enegrini@ucla.edu
[2] Department of Mathematics, University of Bologna, Bologna, Italy
giovanna.citti@unibo.it
[3] Department of Mathematical Sciences, Smith College, Northampton, USA
lcapogna@smith.edu

Abstract. We present a new algorithm for learning unknown governing equations from trajectory data, using a family of neural networks. Given samples of solutions $x(t)$ to an unknown dynamical system $\dot{x}(t) = f(t, x(t))$, we approximate the function f using a family of neural networks. We express the equation in integral form and use Euler method to predict the solution at every successive time step using at each iteration a different neural network as a prior for f. This procedure yields M-1 time-independent networks, where M is the number of time steps at which $x(t)$ is observed. Finally, we obtain a single function $f(t, x(t))$ by neural network interpolation. Unlike our earlier work, where we numerically computed the derivatives of data, and used them as target in a Lipschitz regularized neural network to approximate f, our new method avoids numerical differentiations, which are unstable in presence of noise. We test the new algorithm on multiple examples in a high-noise setting. We empirically show that generalization and recovery of the governing equation improve by adding a Lipschitz regularization term in our loss function and that this method improves our previous one especially in the high-noise regime, when numerical differentiation provides low quality target data. Finally, we compare our results with other state of the art methods for system identification.

Keywords: Deep Learning · System Identification · Network Regularization

1 Introduction

System identification refers to the problem of building mathematical models and approximating governing equations using only observed data from the system. Governing laws and equations have traditionally been derived from expert

M. Sellmann and K. Tierney (Eds.): LION 2023, LNCS 14286, pp. 165–178, 2023.
https://doi.org/10.1007/978-3-031-44505-7_12

knowledge and first principles, however in recent years the large amount of data available resulted in a growing interest in data-driven models and approaches for automated dynamical systems discovery. The applications of system identification include any system where the inputs and outputs can be measured, such as industrial processes, control systems, economic data and financial systems, biology and the life sciences, medicine, social systems, and many more (see [3] for more examples of applications).

In this work we train a family of neural networks to learn from noisy data a nonlinear and potentially multi-variate mapping f, right-hand-side of the differential equation:

$$\dot{x}(t) = f(t, x) \tag{1}$$

The trained network can then be used to predict the future system states.

In general, two main approaches can be used to approximate the function f with a neural network. The first approach aims at approximating the function f directly, like we did in our previous paper [11], which we refer to as *splines method*. In this work, inspired by the work of Oberman and Calder in [12], we use a Lipschitz regularized neural network to approximate the RHS of the ODE (1), directly from observations of the state vector $x(t)$. The target data for the network is made of discrete approximations of the velocity vector $\dot{x}(t)$, which act as a prior for f. One limitation of this approach is that, in order to obtain accurate approximations of the function f, one needs to obtain reliable target data, approximations of the velocity vector, from the observations of $x(t)$. This proved to be hard when a large amount of noise was present in the data. The second approach aims at approximating the function f implicitly by expressing the differential equation (1) in integral form and enforcing that the network that approximates f satisfies an appropriate update rule. This is the approach used in [18], which we refer to as *multistep method*, where the authors train the approximating network to satisfy a linear multistep method. An advantage of this approach over the previous one is that the target data used to train the multistep network is composed only of observations of the state vector $x(t)$. However, noise in the observations of $x(t)$ can still have a strong impact on the quality of the network approximation of f.

In this work we build on the second approach and introduce a new idea to overcome the limitations of the methods mentioned above. Similarly to the multistep method, we express the differential equation in integral form and train the network that approximates f to satisfy Euler update rule (with minimal modifications one can use multistep methods as well). This implicit approach overcomes the limitations of the splines method, whose results were strongly dependent on the quality of the velocity vector approximations used as target data. Differently than the multistep method, our proposed approach is based on a Lipschitz regularized family of neural networks and it is able to overcome the sensitivity to noise. Later on we compare these methods and other methods for system identification with our proposed approach and show that in the high-noise setting our method produces more accurate results thanks to the use of Lipschitz regularization and multiple networks.

The rest of the paper is organized as follows: Sect. 2 outlines relevant efforts in the field of system identification. In Sect. 3 we describe in detail our proposed method. Section 4 describes and discusses the experimental results. Finally conclusion and future research directions are outlined in Sect. 5

2   Related Works

In recent years many methods have been proposed for data-driven discovery of nonlinear differential equations. Commonly used approaches are sparse regression, Gaussian processes, applied Koopmanism and dictionary based approaches, among which neural networks. Sparse regression approaches are based on a user-determined library of candidate terms from which the most important ones are selected using sparse regression [4, 20–22]. These methods provide interpretable results, but they are sensitive to noise and require the user to choose an "appropriate" sets of basis functions a priori. In contrast, since neural networks are universal approximators our method allows to accurately recover very general and complex RHS functions even when no information on the target function is available. Identification using Gaussian processes places a Gaussian prior on the unknown coefficients of the differential equation and infers them via maximum likelihood estimation [16,17,19]. The Koopman approach is based on the idea that non linear system identification in the state space is equivalent to linear identification of the Koopman operator in the infinite-dimensional space of observables. Since the Koopman operator is infinite-dimensional, in practice one computes a projection of the Koopman operator onto a finite-dimensional subspace of the observables. This approximation may result inaccurate in presence of noise and has proven challenging in practical applications [5,9,10]. In contrast our proposed method is able to overcome the sensitivity to noise thanks to the use of Lipschitz regularization and multiple networks. Since neural networks are universal approximators, they are a natural choice for nonlinear system identification: depending on the architecture and on the properties of the loss function, they can be used as sparse regression models, they can act as priors on unknown coefficients or completely determine an unknown differential operator [2,6,7,11,13–15,18]. Our method is part of this category, but adds to the existing literature thanks to the use of multiple networks and Lipschitz Regularization. Moreover, since our proposed method is based on weak notion of solution using integration it can be used to reconstruct both smooth and non-smooth RHS functions. This is especially an advantage over models that rely on the notion of classical solution like the splines method [11] and make it an extremely valuable approach when working with real-world data.

3   Proposed Method

In this section we describe the architecture used in the experiments.

Our goal is to approximate a vector-valued RHS $f(t, x)$ of a system of differential equations $\dot{x}(t) = f(t, x)$, directly from discrete noisy observations of the state vector $x(t) \in \mathbb{R}^d$. We propose to do so using a neural network architecture composed of two blocks: the *target data generator* and the *interpolation network*. See Algorithm 1 for the full architecture algorithm.

The Target Data Generator: The target data generator is a family of neural networks whose goal is to produce reliable velocity vector approximations which will be used as target data for the interpolation network. The data is selected as follows: given time instants $t_1, \ldots, t_M$ and K trajectories, define

$$x_i(t_j) \in \mathbb{R}^d, \quad i = 1, \ldots, K, \quad j = 1, \ldots, M$$

to be an observation of the state vector $x(t)$ at time t_j for trajectory i. For each time instant t_j, $j = 1, \ldots, M - 1$ we train a feed forward neural network $N_j(x(t_j))$ to approximate the velocity vector $\dot{x}(t)$ at time instant t_j. Indicating by θ^j the network parameters, the loss function L_j used for training forces each N_j to satisfy Euler update rule and it is defined as:

$$L_j(\theta^j) = \frac{1}{K} \sum_{i=1}^{K} \| \Delta t \, N_j(x_i(t_j), \theta^j) + x_i(t_j) - x_i(t_{j+1}) \|_2^2 \quad j = 1 \ldots, M - 1$$

Once the networks N_j are trained, they collectively provide a discrete approximation of the velocity vector on the full time domain, which we indicate by $\widetilde{\dot{x}(t)}$.

The Interpolation Network: The interpolation network N_{int} is a Lipschitz regularized neural network (as defined in [11]) which takes as input a time t and an observation of the state vector $x(t)$ and uses as target data the approximation of the velocity vector $\widetilde{\dot{x}(t)}$ given by the target data generator (this acts as a prior for the unknown function $f(t, x)$). Once trained the interpolation network N_{int} provides an approximation of the RHS function f on its domain, that is $N_{int}(t, x) \approx f(t, x)$. The loss function $L(\theta_{int})$ minimized to train the interpolation network contains two terms. The first one is the Mean Squared Error (MSE) between the network output and the target data: this forces the network predictions to be close to the observed data. The second term is a a Lipschitz regularization term which forces the Lipschitz constant of the network N_{int} to be small (the Lipschitz constant is a proxy for the size of the network's gradient):

$$L(\theta_{int}) = MSE\left(\widetilde{\dot{x}(t)}, N_{int}(t, x(t); \theta_{int}) \right) + \alpha \text{Lip}(N_{int}).$$

Here $\alpha > 0$ is a regularization parameter and $\text{Lip}(N_{int})$ is the Lipschitz constant of the network N_{int}. Inspired by the work of Jin et al. in [8] the Lipschitz constant of network is computed as:

$$\text{Lip}(N_{int}) \leq \|W_{int}^1\|_2 \|W_{int}^2\|_2 \ldots \|W_{int}^L\|_2. \tag{2}$$

Since this is an explicit constraint on the network weights, the computational cost for this approximation is low. This is in contrast with the approximation we used in our previous paper [11], (Section 3.1) based on Rademacher's theorem, which requires to compute the gradient of the network at each iteration.

Algorithm 1. Full Architecture Algorithm

Require: $t = (t_j)$, $x(t) = (x_i(t_j))$, $i = 1, \ldots, K$, $j = 1, \ldots, M$

Train Target data generator

for $j = 1 : M - 1$ **do**

$\quad \Theta_j \leftarrow \text{argmin}_{\Theta_j} MSE(\Delta t \, N_j(x(t_j); \theta^j) + x(t_j), \, x(t_{j+1}))$

end for

Obtain $\widetilde{x(t)} := (N_j(x(t_j)))$, $j = 1, \ldots, M - 1$

Train Interpolation Network

$\Theta_{int} \leftarrow \text{argmin}_{\Theta_{int}} \left(MSE \left(\widetilde{x(t)}, N_{int}(t, x(t); \theta_{int}) \right) + \alpha \text{Lip}(N_{int}) \right)$

Obtain $N_{int}(t, x(t); \theta_{int}) \approx f(t, x(t))$

---

4   Experimental Results and Discussion

In this section we propose numerical examples of our method and comparisons with other methods for system identification. In the examples we use synthetic data with noise amount up to 10%. In this paper we only show one dimensional examples, but we explicitly notice that, since our method is applied component-wise, it can be used for data of any dimension. Three different metrics are used to evaluate the performance of the our method:

1. *Mean Squared Error (MSE)* on test data which measures the distance of the model prediction from the test data. We also report the *generalization gap* (difference between test and training error) obtained with and without Lipschitz regularization in the interpolation network. The smaller the generalization gap the better the network generalizes to unseen data (for a more precise description see [1]).
2. Since we use synthetic data, we have access to the true RHS function $f(t, x)$. This allows to compute the relative MSE between the true $f(t, x)$ and the approximation given by our architecture on arbitrary couples (t, x) in the domain of f. We call this error *recovery error*.
3. Since our method produces a function $N_{int}(t, x)$, it can be used as RHS of a differential equation $\dot{x} = N_{int}(t, x)$. We then solve this differential equation in Python and compute the relative MSE between the solution obtained when using as RHS the network approximation $N_{int}(t, x)$ and when using the true function $f(t, x)$. We call this *error in the solution*.

4.1   Smooth Right-Hand Side

The first example we propose is the recovery of the ODE

$$\dot{x} = xe^t + \sin(x)^2 - x \tag{3}$$

We generate solutions in Python for time steps t in the interval $[0,0.8]$ with $\Delta t = 0.04$ and for 500 initial conditions uniformly sampled in the interval [-3,3]. The hyperparameters for our model are selected in each example by cross validation: the interpolation network N_{int} has $L = 8$ layers, each layer has 20 neurons, while each network N_j of the target data generator has $L_j = 3$ layers with 10 neurons each. The target data generator is made of 20 networks. In Table 1, we report the training MSE, testing MSE, Generalization Gap and estimated Lipschitz constant when 5% and 10% of noise is present in the data. Since our goal here is to compare the performance on test data of the networks with and without regularization, we select the number of epochs during training so as to achieve the same training MSE across all the regularization parameters choices and compare the corresponding Testing errors and Generalization Gaps. We report here only the results obtained for the non-regularized case and for the best regularized one when 5% and 10% of noise is present in the data. We can see from the tables that Lipschitz regularization improves the generalization gap by one order of magnitude for all amounts of noise, that a larger regularization parameter is needed when more noise is present in the data and that, as expected, adding Lipschitz regularization results in a smaller estimated Lipschitz constant. This confirms the findings from our previous paper that Lipschitz regularization improves generalization and avoids overfitting, especially in presence of noise.

Table 1. Test error and Generalization Gap comparison for 5% and 10% noise.

$\dot{x} = xe^t + \sin(x)^2 - x$, 5% Noise				
Regularization Parameter	Training MSE	Testing MSE	Generalization Gap	Estimated Lipschitz Constant
0	0.618%	0.652%	0.034%	7.09
0.004	**0.618%**	**0.619%**	**0.001%**	**6.33**
$\dot{x} = xe^t + \sin(x)^2 - x$, 10% Noise				
Regularization Parameter	Training MSE	Testing MSE	Generalization Gap	Estimated Lipschitz Constant
0	2.01%	2.32%	0.310%	7.72
0.015	**2.01%**	**2.03%**	**0.030%**	**6.38**

In Table 2 we report the error in the recovery for the RHS function $f(t,x) = xe^t + \sin(x)^2 - x$ and the error in the solution of the ODE when using the interpolation network as RHS. We can see that for all amounts of noise in the data, both the reconstruction error and the error in the solution are small, respectively they are less than 0.7% and 0.04%. For larger amounts of noise, the method still works, but provides less accurate approximations. For example with 20% noise the recovery error is approximately 3%. Making the method more robust in higher noise regimes will be object of a future work.

Table 2. Left: Relative MSE in the recovery of the RHS for up to 10% of noise. **Right:** Relative MSE in the solution of the ODE for up to 10% of noise

Relative MSE in the recovery of the RHS of $\dot{x} = xe^t + \sin(x)^2 - x$		Relative MSE in the solution of $\dot{x} = xe^t + \sin(x)^2 - x$	
0% Noise	0.100%	0% Noise	0.016%
5% Noise	0.144%	5% Noise	0.025%
10% Noise	0.663%	10% Noise	0.038%

The left panel of Fig. 1 shows the true and reconstructed RHS and recovery error on the domain on which the original data was sampled for 5% of noise in the data. In the error plot a darker color represents a smaller error. We can see that the largest error is attained at the right boundary of the domain: by design of our architecture the target data generator only generates target data up to the second-last time step. As a consequence the interpolation network has only access to observations up to the second-last time step and so it is forced to predict the value of the RHS function at the last time step by extrapolation. It is then reasonable that the largest recovery error is attained at the right boundary of the domain. In the right panel of Fig. 1 we report the true solution (red line) and the solution predicted when using the interpolation network as RHS (dashed black line) for multiple initial condition and for 5% noise in the data. We notice that the prediction is accurate for all the initial conditions selected, but that it gets worse towards the end of the time interval because of the inaccurate approximation of the RHS at the right boundary of the time interval.

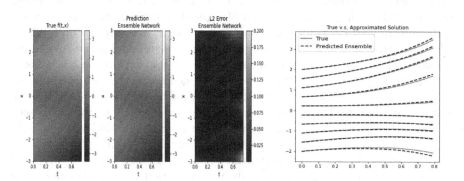

Fig. 1. Left: True RHS, Predicted RHS and recovery error for 5% noise in the data. **Right:** True and Predicted solution for 5% noise in the data

Finally, since the *test error*, the *error in the recovery* and the *error in the solution* are all measured using MSE, it makes sense to compare such homogeneous measurements. The first thing to notice is that the testing errors are larger than the recovery errors. This shows the ability of our network to avoid overfitting and produce reliable approximations of the true RHS even when large amounts of noise are present in the data. In fact, the Test MSE is computed by comparing the value predicted by the network with the value of the corresponding *noisy* observation, while the recovery error is computed by comparing the value predicted by the network with the value of the *true* function f. The disparity between the test error and the recovery error then shows that the interpolation network provides results that successfully avoid fitting the noise in the data. The second thing to notice is the disparity between the recovery error and the error in the solution: the error in the solution is on average smaller than the recovery error. This is due to the data sampling: when recovering the RHS we reconstruct the function on the full domain, while the original data was only sampled on discrete trajectories; for this reason large errors are attained in the parts of the domain where no training data was available. On the other hand the error in the solution is computed on trajectories which were originally part of the training set, so it is reasonable to expect a smaller error in this case.

4.2   Non-smooth Right-Hand Side

We propose is the recovery of

$$\dot{x} = \text{sign}(t - 0.1) \tag{4}$$

and compare the results given by our proposed method and the splines method. Both methods aim at learning a Lipschitz approximation of the right-hand side function. The spline method is based on the notion of classical solution and it is doomed to fail in such a non-smooth setting. In contrast, our proposed method is based on weak notion of solution using integration and is able to accurately reconstruct even non-smooth functions. We generate data for time steps t in the interval [0,0.2] with $\Delta t = 0.02$ and for 500 initial conditions uniformly sampled in the interval $[-0.1, 0.1]$ for noise amounts up to 2%. We only use up to 2% of noise since, the splines model can only provide reliable target data for small noise amounts. The hyperparameters for the models in this example are as follows: each network N_j has $L_j = 3$ layers with 10 neurons each, the interpolation network and the network used in the splines method both have $L = 4$ layers, each layer has 30 neurons. The target data generator is made of 10 networks. As seen in Table 3, because of the low quality target data (approximation of the velocity vector from noisy observations of the positions) obtained by the splines method, this approach fails at reconstructing the non-smooth RHS, while our proposed method is able to produce an accurate reconstruction even in this case. The superior performance of our method over the spline method for this example can also be seen from Fig. 2. From left to right we represent the true, reconstructed RHS and the error in the reconstruction for the spline based method (top row) and for our method (bottom row) when 1% of noise is present in the data. We can see from the figure that the spline method in this case is not even able to find the general form of the RHS function correctly because of the bad quality of the target data. On the contrary, our proposed method, being completely data driven and based on a weak notion of solution, is able to reconstruct RHS functions like $\text{sign}(t - 0.1)$ that are non-smooth in t.

Table 3. Relative MSE in the recovery of the RHS for up to 2% of noise for our method and the splines method.

Relative MSE in the recovery of the RHS of $\dot{x} = \text{sign}(t - 0.1)$		
	Ours	Splines
1% Noise	0.002%	12.5%
2% Noise	0.004%	12.9%

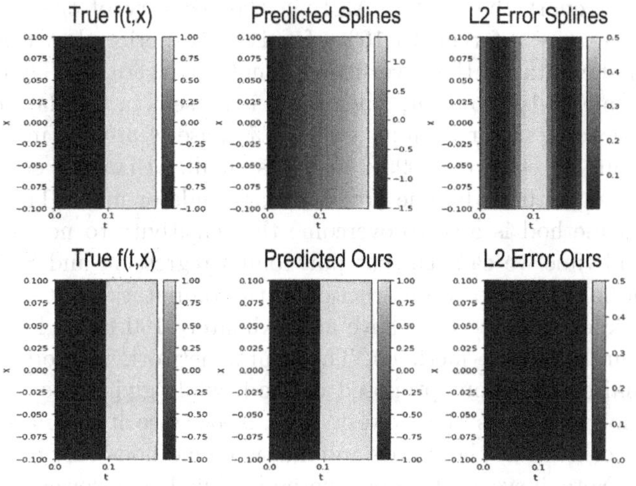

Fig. 2. Top row: Spline method. **Bottom row:** Our proposed method. From left to right: True RHS, Reconstructed RHS and Error in the reconstruction when 1% of noise is present in the data.

4.3 Comparison with Other Methods

We compare our method with the methods proposed in [18] and in [4]. For completeness we also provide a comparison with the splines method [11]. The method proposed in [18], (*multistep method*), is similar to ours: the authors place a neural network prior on the RHS function f, express the differential equation in integral form and use a multistep method to predict the solution at each successive time steps. In contrast with our method they do not use a family of networks and Lispchitz Regularization. The method proposed in [4], (*SINDy*), is based on a sparsity-promoting technique: sparse regression is used to determine, from a dictionary of basis functions, the terms in the dynamic governing equations which most accurately represent the data. Finally, we compare with the splines method described in Sect. 1. We report here the relative error obtained by the different methods in the approximation of the true f as well as the computational time for each method.

We generated the data by computing approximated solutions of

$$\dot{x} = \cos(3x) + x^3 - x \tag{5}$$

for time steps t in the interval [0,1] with $\Delta t = 0.04$ and for 500 initial conditions uniformly sampled in the interval $[-0.7, 0.9]$. The interpolation network has $L = 8$ layers, each layer has 30 neurons, while each network N_j has $L_j = 3$ layers with 20 neurons each. The target data generator is made of 25 networks. We compare the results obtained by our proposed method and the spline, multistep methods, a polynomial regression with degree 20 and SINDy. The dictionary of functions used for SINDy constraints polynomials up to degree 10 as well as other elementary functions: $cos(x), sin(x), sin(3x), cos(3x), e^x, \ln(x), \tan(x)$. In Table 4 we report the relative MSE in the recovery of the RHS function $f = cos(3x) + x^3 - x$ for up to 10% of noise. We notice that when no noise is present in the data, so that overfitting is not a concern, SINDy outperforms all the other methods. However, when noise is present in the data our method gives the best results. For example, when 5% noise is present in the data our method obtains an error of 0.096% which is smaller than the errors obtained by all the other methods by one order of magnitude or more. This shows that our proposed method is able to overcome the sensitivity to noise. In terms of computational time we can see that polynomial regression and SINDy are the fastest at performing the reconstruction with computational time lower than 1 s. This is expected since they have approximately 100 times less parameters than the neural network methods. The neural network methods have higher computational cost and our proposed method, while giving the most accurate results for noisy data, is the slowest. This is because it requires training of multiple networks, while the splines and multistep methods only require training one network. Note, however, that our method, while being slower than the other methods we compare with, provides the most accurate result in under 2 min.

Table 4. Relative MSE and computational time comparison in the recovery of the RHS for up to 10% of noise for our method, the splines and multistep methods, polynomial regression with degree 20, SINDy with custom library.

Relative MSE and computational time comparison for $\dot{x} = cos(3x) + x^3 - x$					
	Ours	*Splines*	*Multistep*	*Polynomial Regression degree 20*	*SINDy custom library*
0% Noise	0.0505%	0.214%	0.116%	6.3e-05%	**5.7e-05%**
5% Noise	**0.0957%**	0.585%	1.20%	3.33%	0.619%
10% Noise	**0.520%**	1.90%	3.51%	17.0%	3.36%
Time (s)	119.5	34.6	26.1	0.60	**0.54**

In Fig. 3 we report the true (red line) and recovered RHS function (blue line) when 5% of noise is present in the data. This figure confirms the findings shown in the previous table: our method is able to reconstruct the true RHS most accurately showing that our method is robust to noise. From the table above we notice that, for noisy data, the worst accuracy was always attained by the polynomial regression. In this case, even if a 20° polynomial has 100 times less parameters than our neural network, increasing the degree of the polynomial increased the error in the recovery. From this figure we can clearly see why that happens: the polynomial regression with degree 20 is already overfitting the noisy data and the largest errors are attained at the boundaries of the domain where the polynomial is highly oscillatory. The other three methods are able to provide approximations that capture the general form of the true RHS function, but only our method is able to provide an accurate approximation even at the boundary of the domain.

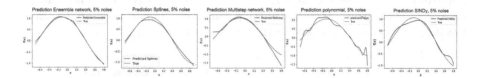

Fig. 3. From left to right, true and recovered RHS for 5% noise in the data obtained by our method, splines method, Multistep Method, Polynomial Regression with degree 20, SINDy with custom library.

4.4 Improving Interpretability Using SINDy

In this section we show how we can improve the interpretability of our method by combining it with SINDy. The strategy is as follows:

1. Given noisy $x(t)$ we use our neural network architecture to find a network N_{int} which approximates the unknown function f.
2. We solve the differential equation $\dot{x}(t) = N_{int}(t, x)$ for multiple initial conditions and obtain new solutions $\bar{x}(t)$. These solutions are a denoised version of the original observations since they were produced using the regularizing neural network architecture.
3. The denoised data $\bar{x}(t)$ is then given to SINDy to produce an interpretable and sparse representation of N_{int}.

We show the results of this strategy for the example proposed in Sect. 4.3. Recall that our goal is to approximate the equation

$$\dot{x} = cos(3x) + x^3 - x \tag{6}$$

In Sect. 4.3 we showed that our neural network architecture is able to reconstruct correctly the RHS and that, when noise it's present in the data, the recovered RHS function is more accurate than the one obtained by SINDy. In this section, we use the network N_{int} found in Sect. 4.3 when 5% of noise is present in the data and use it to produce denoised solutions $\bar{x}(t)$ as explained above. We then use SINDy with the same custom library of functions as in Sect. 4.3 to produce an interpretable and sparse approximation of the original f. When using this technique we obtain the following RHS approximation:

$$\dot{x}(t) \approx 0.898x^3 - 1.055\sin(x) + 0.996\cos(3x) \tag{7}$$

while applying SINDy directly to noisy data gave:

$$\dot{x}(t) \approx -9.431x^3 + 28.141x^5 + 2.730x^6 + -18.665x^7 - 4.902x^9 - 7.477x^{10} + 0.945\cos(3x) \quad (8)$$

We see that Eq. (7) is very close to the true Eq. (6). The main difference is that instead of the term "$-x$" SINDy found "$-1.055\sin(x)$". This is reasonable since for small values of x, like in this example, $x \approx \sin(x)$. On the contrary, Eq. (8) obtained by applying SINDy directly to the noisy data results in an approximated f containing high order terms: this is caused by the noise in the data. As a consequence the MSE in the reconstruction improves from 0.619% to 0.0096%. This can also be seen in the figure below (Fig. 4):

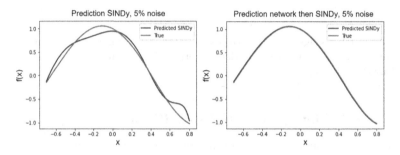

Fig. 4. Left: SINDy reconstruction from 5% noisy data. Right: SINDy reconstruction from denoised network data.

5   Conclusion

In this paper we use a Lipschitz regularized family of neural networks to learn governing equations from data. There are two main differences between our method and other neural network system identification methods in the literature. First, we add a Lipschitz regularization term to our loss function to force the Lipschitz constant of the network to be small. This regularization results in a smoother approximating function and better generalization properties when compared with non-regularized models, especially in presence of noise. Second, we use a family of neural networks instead of a single neural network for the reconstruction. We show that this makes out model robust to noise and able to provide better reconstruction than other state of the art methods for system identification. To our knowledge this is the first time that Lipschitz regularization is added to a family of networks to overcome the sensitivity to noise in a system identification problem. More in detail, our numerical examples, which are representative of a larger testing activity with several different types of right-hand sides $f(x,t)$, show multiple strengths of our method: when noise is present in the data, the Lipschitz regularization improves the generalization gap by one order of magnitude or more. Our architecture is robust to noise and is able to avoid overfitting even when large amounts of noise are present in the data (up to 10%). This robustness to noise is especially an advantage over methods that do not use a family of networks such as [18]. The method is completely data-driven and it is based on weak notion of solution using integration. For this reason, it can be used to reconstruct even non-smooth RHS functions. This is especially an advantage over models that rely on the notion of classical solution like the Splines Method [11]. Since neural networks are universal approximators, we do not need any prior knowledge on the ODE system, in contrast with sparse regression approaches in which a library of candidate functions has to be defined. As shown in Sect. 4.3, direct comparison with polynomial regression and SINDy shows that our model is a better fit when learning from noisy data, even if it comes at the cost of increased computational time. Since our method is applied component-wise, it can be used to identify systems of any dimension, which makes it a valuable approach for high-dimensional real-world problems. As shown in Sect. 4.4, combining our method with SINDy produces a more accurate, intepretable and sparse reconstruction than using SINDy on the original noisy data, thanks to the denoising properties of our architecture.

Future research directions include applying our methods to real world data and extending our methods to the reconstruction of Partial Differential Equations (PDEs). More in detail, first we would like to use our method to reconstruct an equation that approximately describes the evolution in time of COVID-19 infected people. Another interesting application would be to reconstruct the Hodgkin-Huxley model from data. This is a ODE system that describes how action potentials in neurons are initiated and propagated. Second, we would like to generalize our models to the recovery of partial differential equations. Specifically, consider the parabolic PDE $u_t = f(t, x, u, \nabla u, D^2 u)$; given a finite number of observations of $u(t, x)$ the goal is to reconstruct the function f.

Acknowledgements. E. N. is supported by Simons Postdoctoral program at IPAM and DMS 1925919. L. C. is partially supported by NSF DMS 1955992 and Simons Collaboration Grant for Mathematicians 585688.

G. C. is partially supported by the EU Horizon 2020 project GHAIA, MCSA RISE project GA No 777822.

Results in this paper were obtained in part using a high-performance computing system acquired through NSF MRI grant DMS-1337943 to WPI.

References

1. Abu-Mostafa, Y.S., Magdon-Ismail, M., Lin, H.T.: Learning from Data, vol. 4. AMLBook, New York (2012)
2. Berg, J., Nyström, K.: Data-driven discovery of PDEs in complex datasets. J. Comput. Phys. **384**, 239–252 (2019)
3. Billings, S.A.: Nonlinear System Identification: NARMAX Methods in the Time, Frequency, and Spatio-Temporal Domains. Wiley, Hoboken (2013)
4. Brunton, S.L., Proctor, J.L., Kutz, J.N.: Discovering governing equations from data by sparse identification of nonlinear dynamical systems. Proc. Natl. Acad. Sci. **113**(15), 3932–3937 (2016)

5. Budišić, M., Mohr, R., Mezić, I.: Applied Koopmanism. Chaos: Interdisc. J. Nonlinear Sci. **22**(4), 047510 (2012)
6. Champion, K., Lusch, B., Kutz, J.N., Brunton, S.L.: Data-driven discovery of coordinates and governing equations. Proc. Natl. Acad. Sci. **116**(45), 22445–22451 (2019)
7. Chen, R.T., Rubanova, Y., Bettencourt, J., Duvenaud, D.K.: Neural ordinary differential equations. In: Advances in Neural Information Processing Systems, vol. 31 (2018)
8. Jin, P., Lu, L., Tang, Y., Karniadakis, G.E.: Quantifying the generalization error in deep learning in terms of data distribution and neural network smoothness. Neural Netw. **130**, 85–99 (2020)
9. Lusch, B., Kutz, J.N., Brunton, S.L.: Deep learning for universal linear embeddings of nonlinear dynamics. Nat. Commun. **9**(1), 1–10 (2018)
10. Nathan Kutz, J., Proctor, J.L., Brunton, S.L.: Applied Koopman theory for partial differential equations and data-driven modeling of spatio-temporal systems. Complexity **2018** (2018)
11. Negrini, E., Citti, G., Capogna, L.: System identification through Lipschitz regularized deep neural networks. J. Comput. Phys. **444**, 110549 (2021). https://doi.org/10.1016/j.jcp.2021.110549. https://www.sciencedirect.com/science/article/pii/S0021999121004447
12. Oberman, A.M., Calder, J.: Lipschitz regularized deep neural networks converge and generalize. arXiv preprint arXiv:1808.09540 (2018)
13. Ogunmolu, O., Gu, X., Jiang, S., Gans, N.: Nonlinear systems identification using deep dynamic neural networks. arXiv preprint arXiv:1610.01439 (2016)
14. Qin, T., Wu, K., Xiu, D.: Data driven governing equations approximation using deep neural networks. J. Comput. Phys. **395**, 620–635 (2019)
15. Raissi, M.: Deep hidden physics models: deep learning of nonlinear partial differential equations. J. Mach. Learn. Res. **19**(1), 932–955 (2018)
16. Raissi, M., Karniadakis, G.E.: Hidden physics models: machine learning of nonlinear partial differential equations. J. Comput. Phys. **357**, 125–141 (2018)
17. Raissi, M., Perdikaris, P., Karniadakis, G.E.: Machine learning of linear differential equations using Gaussian processes. J. Comput. Phys. **348**, 683–693 (2017)
18. Raissi, M., Perdikaris, P., Karniadakis, G.E.: Multistep neural networks for data-driven discovery of nonlinear dynamical systems. arXiv preprint arXiv:1801.01236 (2018)
19. Raissi, M., Perdikaris, P., Karniadakis, G.E.: Numerical Gaussian processes for time-dependent and nonlinear partial differential equations. SIAM J. Sci. Comput. **40**(1), A172–A198 (2018)
20. Rudy, S.H., Brunton, S.L., Proctor, J.L., Kutz, J.N.: Data-driven discovery of partial differential equations. Sci. Adv. **3**(4), e1602614 (2017)
21. Schaeffer, H.: Learning partial differential equations via data discovery and sparse optimization. Proc. R. Soc. A: Math. Phys. Eng. Sci. **473**(2197), 20160446 (2017)
22. Schaeffer, H., Caflisch, R., Hauck, C.D., Osher, S.: Sparse dynamics for partial differential equations. Proc. Natl. Acad. Sci. **110**(17), 6634–6639 (2013)

GPU for Monte Carlo Search

Lilian Buzer[1]([✉]) and Tristan Cazenave[2]

[1] LIGM, Universite Gustave Eiffel, CNRS, 77454 Marne-la-Vallee, France
`lilian.buzer@esiee.fr`
[2] LAMSADE, Universite Paris Dauphine - PSL, CNRS, Paris, France

Abstract. Monte Carlo Search algorithms can give excellent results for some combinatorial optimization problems and for some games. They can be parallelized efficiently on high-end CPU servers. Nested Monte Carlo Search is an algorithm that parallelizes well. We take advantage of this property to obtain large speedups running it on low cost GPUs. The combinatorial optimization problem we use for the experiments is the Snake-in-the-Box. It is a graph theory problem for which Nested Monte Carlo Search previously improved lower bounds. It has applications in electrical engineering, coding theory, and computer network topologies. Using a low cost GPU, we obtain speedups as high as 420 compared to a single CPU.

Keywords: Monte Carlo Search · GPU · Playouts

1 Introduction

1.1 History of Monte Carlo Search Algorithms

Monte Carlo Tree Search (MCTS) has been successfully applied to many games and problems [4]. It was used to build superhuman game playing programs such as AlphaGo [32], AlphaZero [33] and Katago [35]. It has been recently used to discover new fast matrix multiplication algorithms [21].

Nested Monte Carlo Search (NMCS) [6] is a Monte Carlo Search algorithm that works well for puzzles. It biases its playouts using lower-level playouts. Kinny broke world records at the Snake-in-the-Box applying Nested Monte Carlo Search [22]. He used a heuristic to order moves in the playouts. The heuristic is to favor moves that lead to a state where there is only one possible move. Other applications of NMCS include Single Player General Game Playing [24], Cooperative Pathfinding [1], Software testing [28], Model-Checking [29], the Pancake problem [2], Games [11], Cryptography [14] and the RNA inverse folding problem [27].

Online learning of playout strategies combined with NMCS has given good results on optimization problems [30]. It has been further developed for puzzles and optimization with Nested Rollout Policy Adaptation (NRPA) [31]. NRPA has found new world records at Morpion Solitaire and crossword puzzles. Edelkamp, Cazenave and co-workers have applied the NRPA algorithm to

M. Sellmann and K. Tierney (Eds.): LION 2023, LNCS 14286, pp. 179–193, 2023.
https://doi.org/10.1007/978-3-031-44505-7_13

multiple problems. They have adapted the algorithm for the Traveling Salesman with Time Windows (TSPTW) problem [12,16]. Other applications deal with 3D Packing with Object Orientation [18], the physical traveling salesman problem [19], the Multiple Sequence Alignment problem [20], Logistics [9,17], the Snake-in-the-Box [15], Graph Coloring [10], Molecule Design [8] and Network Traffic Engineering [13]. The principle of NRPA is to adapt the playout policy to learn the best sequence of moves found so far at each level. GNRPA [7] has much improved the result of NRPA for RNA Design [8].

1.2   Generating Playouts

Monte Carlo based algorithms for game AI generate a large number of simulated games called *playouts*. The generation of a playout is presented in Algorithm 1. The main loop of this algorithm is used to play the different moves until the end of the game. At the beginning of each iteration, the game data is analyzed to detect all possible moves. Then, a given policy or a random strategy selects a move among the different possibilities. After that, the chosen move is played and we iterate. When the game ends, the list of played moves is returned with the score. The game AI will then analyze the results of many playouts to generate a new bunch of playouts with better scores.

Algorithm 1: Generation of a playout

Data: *Game*: game object embedding rules and data
Data: *Policy*: function that selects a move

1 **Function** *CreatePlayout(Game, Policy)*:
2     $Playout = [\ ]$ // History of played moves
3   **while** *Game.IsNotTerminated()* **do**
4         $L = Game.GetPossibleMoves()$
5         $ChosenMove = Policy.ChoseMove(L)$
6         $Playout.append(ChosenMove)$
7         $Game.Play(ChosenMove)$
8   **return** *Playout, Game.Score()*

1.3   Why GPUs Have Not Been Considered?

Monte Carlo based approaches for game AI are generally performed on high-end CPUs. Even if these algorithms have demonstrated their performance and quality, the possibility of creating a version on GPU has not been studied. There are many reasons for this:

– The capabilities of a CPU core, in terms of computing power or clock speed, greatly exceed the capabilities of a GPU core.

- Most Monte Carlo based algorithms are written using an iterative approach. GPU processing is based on parallelism and switching from one programming style to the other is difficult.
- On average, a GPU core has only 1 KB of memory cache. It seems difficult to store all the game data inside. Moreover, GPU global memory has a notoriously slow latency.
- Parallel GPU threads must access contiguous data to achieve efficiency. However, depending on the game, it is not always easy to meet this requirement.

These accumulated difficulties do not bode well for the performance. Nevertheless, this paper aims to show that implementing playout simulations on GPU is possible with little difficulty and with performance gain.

1.4 Our Contribution

In this paper, we carefully present the first, to our knowledge, proof of concept showing that it is possible to obtain significant performance for playouts generation on a GPU. We study with precision the performance losses induced by the GPU architecture relative to parallel simulations in Sect. 2, to computing power in Sect. 3 and to memory access in Sect. 4. We finally choose the game 'Snake in the box' to perform a series of benchmarks in Sect. 5. Most of the tests we use are classical tests, however, we use them in the very particular context of parallel simulations doing random memory accesses. The objective of this article is to determine precisely how a GPU behaves in such a specific situation.

2 Parallel Execution

2.1 Warp

We briefly summarize NVIDIA GPUs architecture, the reader can find a more detailed description in [3,26]. A NVIDIA GPU contains thousands of CUDA cores gathered in groups of 32 cores called *warps*. In a warp, all the cores process the same statement at the same time. Thus, the cores belonging to the same warp run the same source code in parallel, only their register values may differ.

When looking at Algorithm 1 used for playouts generation, three steps are required: list the possibles moves, choose one of them and play it. Thus, if a particular game performs the same sequence of instructions to carry out these steps, we can efficiently simulate 32 playouts in parallel in the same warp. But these 32 simulations will obviously differ in their number of rounds. So, when a playout ends before the others, one of the cores becomes idle. The inactive cores are accumulating until the longest playout ends. At this point, all cores wake up and start a new batch of 32 simulations. As opposed to a sequential processing where a new simulation starts as soon as the previous one has finished, some computing resource is wasted by the idle cores waiting for the last simulation to complete. In the following, we theoretically estimate the performance loss inherent to any playouts simulation performed in a warp to verify that this loss is acceptable.

2.2   Theoretical Model

Let us assume that for a given policy, the time spent to simulate a playout can be modeled by a Gaussian distribution. So, the 32 simulations running in a warp can be modeled as different Gaussian random variables denoted $X_i = \mathcal{N}(\mu, \sigma)$ with $i = 1, \ldots, 32$. Now, we want to model the time spent by a warp to complete the generation of a group of 32 simulations. For this, we define a new variable:

$$Z = [\max_i X_i]$$

To simplify the calculation, we use $X_i' = X_i - \mu$ and $Z' = [\max_i X_i']$. By Jensen's inequality, we obtain:

$$e^{t\mathbb{E}[Z']} \leq \mathbb{E}[e^{tZ'}] = \mathbb{E}[\max_i e^{tX_i'}] = \sum_{i=1}^{n} \mathbb{E}[e^{tX_i'}] = ne^{t^2\sigma^2/2}$$

Thus, we can write:

$$\mathbb{E}[Z'] \leq \frac{\log(n)}{t} + \frac{t\sigma^2}{2}$$

With $t = \sqrt{2\log n}/\sigma$ and with $n = 32$, we finally obtain:

$$\mathbb{E}[Z] \leq \mu + \sigma\sqrt{7}$$

In comparison, the average time spent by the 32 simulations follows a Gaussian distribution equal to $Y = (\sum_i Xi)/32 = \mathcal{N}(\mu, \sigma/32)$. Using the empirical rule, we know that 99.7% of the time, the variable Y will satisfy: $Y \geq \mu - 3\sigma/32$. Thus, we can conclude that 99.7% of the time, the ratio Z/Y satisfies :

$$\frac{Z}{Y} = \frac{\max(Xi)}{\text{mean}(X_i)} \leq \frac{\mu + \sigma\sqrt{7}}{\mu - 3\sigma/32} = \frac{\mu + 2.64\sigma}{\mu - 0.09\sigma}$$

The ratio Z/Y bounds the α factor describing the time increase when simulations run on a warp in parallel. As an example, when the mean of the playouts length is equal to 100 moves with a standard deviation of 15, this bound is equal to 1.41. This bound is not tight and may be somewhat overestimated. Thus, we numerically estimate the α factor in the following Section.

2.3   Numerical Estimation

We now try to numerically estimate the α factor corresponding to the ratio between the time spent by a warp to generate 32 playouts and the average time used by the same 32 playouts without considering the idle time. For this, let us assume that the length of each simulation can be modeled as Gaussian variable $X_i = \mathcal{N}(\mu, \sigma)$. In this manner, the α factor is equal to $\mathbb{E}[\max(Xi)/\text{mean}(X_i)]$. Note that the α factor is invariant by scaling the X_i variables. Thus, we present some estimations of the α factor for different Gaussian random variables $X_i =$

$\mathcal{N}(1, \sigma/\mu)$ in Table 1. When the ratio σ/μ increases, the α factor also increases. The known bound from the previous Section for the case $\sigma/\mu = 0.15$ is equal to 1.41, but the numerical estimation is more favorable with an α factor equal to 1.31. This ratio can be considered as an acceptable overhead of 31%.

Table 1. Estimations of the α factor for Gaussian random variables $X_i = \mathcal{N}(\mu, \sigma)$

σ/μ	0.05	0.1	0.15	0.20	0.25	0.30	0.35	0.40
α	1.1	1.2	1.31	1.41	1.51	1.62	1.72	1.82

3   Expected Performance

We compare the performance of a NVIDA GTX 3080 GPU and an AMD Ryzen 9 3900X CPU. Comparisons between CPU and GPU performance can be found in the literature, but usually in a specific context such as linear algebra [23] or neural networks [5]. To our knowledge, no study has been published relative to our specific context of random memory access. Thus, we first benchmark the computing power of these two devices and then, we evaluate their memory latency. All this information allows us to determine what performance gain we may expect, for what problem size and in what way.

3.1   Computing Power

We want to compare the computing power of a GPU core against a CPU core. For this purpose, we set up a function that computes the sums of all possible subsets of $\{k \in \mathbb{N} : k \leq n\}$ without performing any memory access. In this way, we test the performance of two common operations in puzzle games: additions and logical tests. Our two test platforms are a NVIDIA GTX 3080 GPU and an AMD Ryzen 9 3900X CPU. In our test scenario, each CPU or GPU thread performs the same calculations. We perform different tests with 1 or 2 threads per core and with integer or float values. We present the time spent to complete all the threads in Table 2. We choose as reference time the scenario with 1 thread per CPU core and with integer numbers. We notice that using 2 threads per core instead of one seems to be more efficient for both GPUs and CPUs. Finally, we can conclude that GPU execution is about 4 to 5 times longer compared to CPU.

3.2   CPU and GPU Memory Cache Size

A memory cache is used to reduce the average time to access data in memory. The logic behind memory cache is simple: when a core requests data in RAM, it first checks whether a copy exists in the L1 cache and in case of success, this process saves memory access time. For the AMD Ryzen 9, the AMD EPYC and

Table 2. Relative duration of the performance benchmark.

	1 Thread/Core		2 Threads/Core	
	Integer	Float	Integer	Float
CPU vs GPU	1 vs 3.9	1.3 vs 5.6	1.3 vs 6.3	2.0 vs 10.1

the Intel Xeon Platinum family, the L1 cache size is 64 KB per core. For the NVIDIA GTX family, the L1 data cache size is 128 KB but it is shared among 128 CUDA cores. Thus, on average each CUDA core has 1 KB of L1 data cache which is far less than a CPU L1 cache of 64 KB. So if we want a GPU to be able to compete with a CPU, we should process problems with small data size.

3.3 Estimating Memory Latency

After having compared CPU and GPU cache size, we now focus on their response time also called latency. For this, we use the P-Chase method presented in [25,34] which continuously performs the read statement $i = A[i]$ as shown in Algorithm 2. To simulate random memory accesses, we initialize the values in A to perform a random walk of this array as in the example $A[\ ] = \{6, 5, 7, 2, 0, 4, 3, 1\}$. We set up a second test scenario to analyze the latency of Read+Write operations. For this, we still conduct a random walk, but this time each memory read is followed by a memory write at the same location. This behavior simulates a game which is updating the data of its gameboard.

Algorithm 2: Random memory read latency estimation

Data: A: array of Integer, m: number of reads to perform, p: random start

1 **Function P-ChaseReadOnly** (A, m, p):
2 $\quad$ **for** $m/3$ **do**
3 $\quad\quad$ $p = A[p];\quad\quad p = A[p];\quad\quad p = A[p];$ $\qquad\qquad\qquad$ `// 3 reads`

4 **Function P-ChaseReadWrite** (A, m, p):
5 $\quad$ **for** $m/3$ **do**
6 $\quad\quad$ $p2 = A[p];\quad\quad p3 = A[p2];\quad\quad p4 = A[p3];$ $\qquad\quad$ `// 3 reads`
7 $\quad\quad$ $A[p] = p3;\quad\quad A[p2] = p4;\quad\quad A[p3] = p2;$ $\qquad$ `// +3 writes`
8 $\quad\quad$ $p = p4$

3.4 Random Access and CPU L1 Cache Latency

The L1 memory cache on modern CPUs is very efficient. Thus, playout generation can take full advantage of the acceleration provided by the L1 cache. We present the average latency estimated using the P-Chase method in Table 3. We consider different scenarios with 1 or 2 threads per core and with read only or read and write. The estimation we obtain are very stable as long as data resides entirely in the L1 cache. We notice that with 1 or 2 threads per core, performing 1 read or 1 read + 1 write access, the memory latency is very similar. This confirms that Ryzen 9 CPU family is able to handle read and write in parallel, with 2 threads per core and with random access without loss of performance.

Table 3. L1 cache CPU latency for random access.

Threads per core	1	2
Latency in ns - Read Only	1.46	1.53
Latency in ns - Read+Write	1.54	1.60

3.5 Random Access and GPU Latency

The NVIDIA GTX 3080, has 68 Streaming Multiprocessors (SM). Each of these SMs has an internal memory of 128 KB that can be partitioned into L1 cache and shared memory. The SM L1 cache behaves like a CPU L1 cache. Shared memory can be seen as a user-managed memory space that all threads of the same SM have access to. Its size is limited to 100 KB on the 3080 GPU. We know that our parallel playouts simulations will generate mainly random access in memory. In our benchmark, each thread performs its own P-Chase using its own array. In this manner, each thread behaves as if it was performing its own game simulation in a private memory space. So we use the P-Chase algorithm to precisely estimate the memory latency in such a scenario, this information being not documented by NVIDIA. Thus, we have two test scenarios: one where data mainly resides in the L1 cache and another one where data are allocated in shared memory. In the first scenario, we can exceed the size of the L1 cache and use global memory. So, we test arrays up to a size of 2K which requires 16-bit indexing. In the second scenario, data must totally reside in the shared memory space, so we limit arrays to 256 bytes in order to use only 8-bits indexing. We also test the two variants of the P-Chase algorithm: read only or read+write. We present all estimated latency in Table 4.

What are our observations ? When using the L1 cache, latency of read only access is stable until the L1 occupancy remains below 100%. When occupancy is beyond 100%, latency increases rapidly to over 1000ns. When running many threads on the same core, a mechanism called latency hiding is triggered by the GPUs to improve performance. This way, when the active thread is put on hold due to a memory access, a waiting thread can rapidly take its place avoiding a core being idle. Thus with 2 threads per core, we see that the latency reduces by half. Nevertheless, using 2 threads per core divides the memory available for each thread by a factor 2 which increases the strain on the available memory space for each thread. When we perform the Read+Write test, we notice that the performance becomes very bad with a latency nearly 10 times longer. It is not easy to explain this behavior but in any case it seriously harms playouts simulation. When data relies in shared memory, the latency in the read only scenario is better and stable with about 18 ns. But most importantly, the latency during Read + Write tests remained very good with 25ns.

Table 4. GPU latency in nanoseconds for random memory access.

L1 cache + RAM - 16 bits value - X means > 1000									
Buffer size	8	16	32	64	128	256	512	1K	2K
Occupancy	2%	4%	8%	16%	33%	66%	131%	262%	524%
1T/Core Read	26	27	28	30	31	32	524	895	X
1T/Core R+W	79	144	258	296	287	294	X	X	X
Occupancy	4%	8%	16%	33%	66%	131%	262%	524%	1048%
2T/Core Read	13	13	14	17	21	310	868	X	X
2T/Core R+W	46	144	248	297	309	X	X	X	X
SHARED - 8 bits value - X means > 1000									
Buffer size	8	16	32	64	128	256			
Occupancy	1%	2%	4%	8%	16%	33%			
1T/Core Read	17	18	18	19	19	19			
1T/Core R+W	23	24	24	25	25	25			

3.6 Synthesis

We can conclude that the use of shared memory is a wise choice because it provides an optimal latency for random memory accesses, even when reads and writes are performed at the same time. Using shared memory, we must respect the constraint of 100 KB maximum for 128 cores. This will force us to greatly reduce the storage of game data in memory. When looking for performance, we should focus on problems with less than 1 KB of data per simulation.

In terms of computing power, we can conclude that a CPU core is five times faster than a GPU core. When data resides in shared memory, GPU latency (25ns) is 16 times slower compared to CPU latency (1.5ns). Thus memory latency, even when using shared memory, remains the main bottleneck when we speak about performance. We recall that the NVIDIA GTX 3080 has 8704 CUDA cores, thus, when a game performs mainly memory accesses, its computing power will be equivalent to $8704/16/1.3 = 420$ times a single CPU core, the ratio 1.3 being the Warp performance loss factor we present in Sect. 2. This estimation is an approximation, but it gives the level of performance we can expect.

4   Snake in the Box

4.1   Performance Benchmark

We have chosen the game 'Snake in the box' game for several reasons:

- The game rules are intuitive and quickly understandable.
- The source code is easily readable and can be used as a pedagogical example.
- This game generates mainly memory access and finally very few computations in comparison. So, this game allows us to test our scenario where memory latency is the main performance bottleneck.

4.2   Game Rules

A d dimensional hypercube is an analogue of a cube in dimension d with 2^d nodes, each node having d neighbors. The Snake in the box problem consists in searching a longer path among the edges of a hypercube. There are two additional constraints: we cannot turn back and we can not select a new node which is adjacent to a previously visited node (*connectivity constraint*). The score of a playout corresponds to its number of edges in the path.

4.3   Data Structure

We can code each node of a d-dimensional hypercube by an integer value of d bits. The code of two connected corners only differ by one bit. This way, the neighbors of the node 0010b are 1010b, 0110b, 0000b, and 0011b. We associate with each node a 1 bit value named $Usable[i]$ indicating whether that node can be visited. As GPU programming requires optimization of data in the L1 cache/shared memory, we use a bitfield of 2^d bits to store the array $Usable$. In the same way, as there are at most d possible moves at each turn, we can store the sequence of moves using only 4 bits per move when $d < 16$. As the NVIDIA GTX 3080 has a maximum of 100 KB of shared memory, we can test our approach for a value of d ranging from 8 to 11 as shown in Table 5.

Table 5. L1 cache occupancy relative to the dimension.

Dimension of the Snake in the box	8	9	10	11	12
Number of nodes	256	512	1024	2048	4096
Longuest known path	98	190	370	707	1302
Bitfield size in bytes	32	64	128	256	512
Sequence size in bytes	49	95	175	354	651
Data size in KB - 128 playouts	10	20	39	76	145
Shared memory occupancy - 100 KB	10%	20%	40%	78%	149%

5   Nested Monte-Carlo Search

NRPA algorithm uses a lot of memory and it does not suit our constraints. Thus we focus on the NMCS algorithm which is memory efficient and provides very good results for the Snake In the Box problem.

5.1   Algorithm

A Nested Monte-Carlo Search, $NMCS$, returns a sequence of moves used to finish a game. The $NMCS$ algorithm takes two arguments: an integer indicating its recursion level and a game G where n moves have already been played. To complete the game, the NMCS algorithm iteratively plays moves. To choose the next move, the algorithm analyzes all the possible moves. For each move, it creates a copy G' of the current game G, plays the candidate move and performs a recursive call to $NMCS(level-1, G')$. If the sequence returned by the recursive call is associated with a better score, the best known sequence is replaced. After all possible moves have been tested, the algorithm plays the $n+1$-th moves of the current best known sequence and iterates. We point out that at level 0, the $NMCS$ performs only a random playout to build a sequence of moves.

Algorithm 3: NMCS algorithm

```
1  Function NMCS(level, Game G):
2  │  Input: G game in progress (partially started or nearly finished)
3  │  Output: B game completed
4  │  B = G.copy()                          // Current best game
5  │  if level == 0 then
6  │  │  B.playout()               // At level 0, NMCS performs a playout
7  │  else
8  │  │  n = G.Sequence.size              // n turns have been performed
9  │  │  while not G.Terminated() do
10 │  │  │  for move in G.GetPossiblesMoves() do
11 │  │  │  │  G' = G.copy()                    // Create a subgame
12 │  │  │  │  G'.play(move)                     // Test this move
13 │  │  │  │  NMCS(level-1,G')           // Evaluation from level-1
14 │  │  │  │  if G'.score() > B.score() then
15 │  │  │  │  │  B = G'
16 │  │  │  G.play(B.sequence[n])     // n-th move of best known sequence
17 │  │  │  n += 1
18 │  return B
```

5.2   NMCS with Parallel Leaf

To run multiple NMCS algorithms in parallel, we are faced with several difficulties. First, a level-4 NMCS algorithm requires storing 5 games in memory which means the L1 memory will quickly saturate. Second, the NMCS algorithm has been designed as an incremental and also recursive algorithm which makes it almost impossible to migrate to a parallel version. However, we can set up a *parallel leaf* version. For this, instead of building only one playout at level 0, we generate 32 playouts in parallel and select the best one. Other levels of the NMCS algorithm remains in single thread mode. The NMCS with Parallel Leaf remains effective because most of the computation time is spent at level 0.

We recall that the NVIDA GTX 3080 contains 68 Streaming Multiprocessors containing 4 warps of 32 cores. We can run one parallel leaf NMCS per warp to obtain $68 \times 4 = 272$ NMCS running in parallel on this GPU. Each thread in a warp generates a playout. Then when the 32 playouts are over, a single thread, named the *the master thread*, analyzes their results and select the best sequence to be returned to the upper level. As specified in the NVIDIA specification, threads within a warp that wish to communicate via memory must execute the dedicated CUDA function _syncwarp_(). In our case, this function has to be called by the master thread to correctly analyze the playouts.

5.3   NMCS on GPU

While it may seem easy to set up 32 threads running in parallel, there remains a little challenge to address when programming the NMCS algorithm on GPU. In fact, inside a warp, a GPU can easily reduces the number of running threads due to an if statement. But, for the NMCS algorithm, we operate in reverse. Indeed, the higher levels of the NMCS algorithm use only one master thread, and after some recursive calls, playouts generation requests the use of 32 threads in parallel. It is not the usual way a GPU works.

For this, we use a specific trick: when a processing must be performed by the master thread of the algorithm, we precede it with a *filter test* that verifies that the current thread corresponds to the master thread. But, we must keep the 32 threads active until the level 0 of the NMCS algorithm. For this, all threads in the warp must execute recursive calls and their enclosing loops. Threads outside the master thread should do nothing. As the filter test prevents them from performing any processing, they remain active and follow the master thread without performing any processing until level 0.

5.4   Performance Comparison

We compare the performance of a NVIDA GTX 3080 GPU relative to one core of a Ryzen 9 3900X CPU. We validate our GPU implementation by comparing the mean score obtained by the CPU and the GPU versions. Any important deviation is associated to an implementation problem. We set up our GPU version using shared memory in order to obtain better performance.

We show performance gains in Table 6 for the Snake In The Box problem in dimension 8, 9 and 10 using level 1 and 2 of the NMCS algorithm. For level 1, the GPU was able to achieve performance gains by a factor ×390 which is of the same magnitude as the ratio ×420 we estimate in Sect. 3. In a surprising way, we notice that in level 2 of the NMCS algorithm, performance increases reaching ×480. This behavior remains unexplained because the time spent by the higher level is normally negligible. We need to conduct more sophisticated experiments to analyze this phenomenon.

Table 6. Performance gain for the 3080 GPU relative to one CPU core.

Dimension	8	9	10
ÑMCS Level 1	×380	×387	×382
ÑMCS Level 2	×471	×498	×521

5.5 Implementation

To set up our GPU implementation, on the first try, we choose not to optimize data structures for GPU architecture. We thought this task not very useful because the performance bottleneck mainly comes from random memory access. So we import our CPU/C++ source code into our CUDA program. We embed game data and game functions into a C++ class called *SnakeInTheBox* to improve the structure of the code and its readability. We also use a C++ structure called *Info* to gather input and output information of each thread. This first version reaches interesting performance but half the efficiency we show in Table 6.

In a second version, we update our code to use shared memory. We also create an implementation of the list of possible moves specific to GPU. This implementation uses *memory coalescing*, a technique where parallel threads accessing consecutive memory locations combine their requests into only one memory request. Considering all these improvements, we were able to achieve performance ratios shown in Table 6.

The data structures, P-Chase and NMCS algorithms, CUDA source codes and project files for Visual Studio 2022 are available for download at the URL http://anonymousdl.online/LION17/.

6 Conclusion

We have proven that running 32 simulations in parallel on a GPU warp loses an acceptable percentage of performance. Although random memory accesses are known to be extremely costly for a GPU, we were able to show that using shared memory could achieve a memory latencies 16 times slower than CPU memory latency, but with its 8704 cores, the NVIDIA GTX 3080 may achieve a speed of

×420 compared to one Ryzen 9 3900X CPU core. All these observations allowed us to set up the first implementation of the NMCS algorithm on GPU. We test performance gain for the Snake In The Box problem in dimension 8,9 and 10. The performance we obtain corresponds to the order of magnitude that we had previously estimated, which in itself is a great success.

Using shared memory, we must respect the constraint of 100 KB maximum for 128 cores which represents a very important constraint. This forces to greatly reduce game data in memory and to focus on problems with less than 1 KB of data per simulation. But on the other hand, the NVIDIA GTX-4090 card already offers twice as many CUDA cores compared to the 3080 and the next generation with the NVIDIA GTX-5090 will also double performance. We are probably at a technological tipping point where, for some games, it will be more efficient to generate playouts on a GPU than on a CPU. Indeed, the frantic race for performance that GPU founders are waging makes the power/price ratio more and more interesting compared to high-end CPUs.

References

1. Bouzy, B.: Monte-Carlo fork search for cooperative path-finding. In: Computer Games Workshop at IJCAI, pp. 1–15 (2013)
2. Bouzy, B.: Burnt pancake problem: new lower bounds on the diameter and new experimental optimality ratios. In: SOCS, pp. 119–120 (2016)
3. Brodtkorb, A., Hagen, T., Schulz, C., Hasle, G.: GPU computing in discrete optimization. Part I: Introduction to the GPU. EURO J. Transp. Logist. **2** (2013). https://doi.org/10.1007/s13676-013-0025-1
4. Browne, C., et al.: A survey of Monte Carlo tree search methods. IEEE Trans. Comput. Intell. AI Games **4**(1), 1–43 (2012). https://doi.org/10.1109/TCIAIG.2012.2186810
5. Buber, E., Banu, D.: Performance analysis and CPU vs GPU comparison for deep learning. In: 2018 6th International Conference on Control Engineering & Information Technology (CEIT), pp. 1–6. IEEE (2018)
6. Cazenave, T.: Nested Monte-Carlo search. In: Boutilier, C. (ed.) IJCAI, pp. 456–461 (2009)
7. Cazenave, T.: Generalized nested rollout policy adaptation. In: Monte Search at IJCAI (2020)
8. Cazenave, T., Fournier, T.: Monte Carlo inverse folding. In: Monte Search at IJCAI (2020)
9. Cazenave, T., Lucas, J.Y., Kim, H., Triboulet, T.: Monte Carlo vehicle routing. In: ATT at ECAI 2020, Saint Jacques de Compostelle, Spain (2020). https://hal.archives-ouvertes.fr/hal-03117515
10. Cazenave, T., Negrevergne, B., Sikora, F.: Monte Carlo graph coloring. In: Monte Search at IJCAI (2020)
11. Cazenave, T., Saffidine, A., Schofield, M.J., Thielscher, M.: Nested Monte Carlo search for two-player games. In: AAAI, pp. 687–693 (2016)
12. Cazenave, T., Teytaud, F.: Application of the nested rollout policy adaptation algorithm to the traveling salesman problem with time windows. In: Learning and Intelligent Optimization - 6th International Conference, LION, vol. 6, pp. 42–54 (2012)

13. Dang, C., Bazgan, C., Cazenave, T., Chopin, M., Wuillemin, P.-H.: Monte Carlo search algorithms for network traffic engineering. In: Dong, Y., Kourtellis, N., Hammer, B., Lozano, J.A. (eds.) ECML PKDD 2021. LNCS (LNAI), vol. 12978, pp. 486–501. Springer, Cham (2021). https://doi.org/10.1007/978-3-030-86514-6_30

14. Dwivedi, A.D., Morawiecki, P., Wójtowicz, S.: Finding differential paths in ARX ciphers through nested Monte-Carlo search. Int. J. Electron. Telecommun. **64**(2), 147–150 (2018)

15. Edelkamp, S., Cazenave, T.: Improved diversity in nested rollout policy adaptation. In: Friedrich, G., Helmert, M., Wotawa, F. (eds.) KI 2016. LNCS (LNAI), vol. 9904, pp. 43–55. Springer, Cham (2016). https://doi.org/10.1007/978-3-319-46073-4_4

16. Edelkamp, S., Gath, M., Cazenave, T., Teytaud, F.: Algorithm and knowledge engineering for the TSPTW problem. In: 2013 IEEE Symposium on Computational Intelligence in Scheduling (SCIS), pp. 44–51. IEEE (2013)

17. Edelkamp, S., Gath, M., Greulich, C., Humann, M., Herzog, O., Lawo, M.: Monte-Carlo tree search for logistics. In: Clausen, U., Friedrich, H., Thaller, C., Geiger, C. (eds.) Commercial Transport. LNL, pp. 427–440. Springer, Cham (2016). https://doi.org/10.1007/978-3-319-21266-1_28

18. Edelkamp, S., Gath, M., Rohde, M.: Monte-Carlo tree search for 3D packing with object orientation. In: Lutz, C., Thielscher, M. (eds.) KI 2014. LNCS (LNAI), vol. 8736, pp. 285–296. Springer, Cham (2014). https://doi.org/10.1007/978-3-319-11206-0_28

19. Edelkamp, S., Greulich, C.: Solving physical traveling salesman problems with policy adaptation. In: 2014 IEEE Conference on Computational Intelligence and Games (CIG), pp. 1–8. IEEE (2014)

20. Edelkamp, S., Tang, Z.: Monte-Carlo tree search for the multiple sequence alignment problem. In: Proceedings of the Eighth Annual Symposium on Combinatorial Search, SOCS 2015, pp. 9–17. AAAI Press (2015)

21. Fawzi, A., et al.: Discovering faster matrix multiplication algorithms with reinforcement learning. Nature **610**(7930), 47–53 (2022)

22. Kinny, D.: A new approach to the snake-in-the-box problem. In: ECAI 2012. Frontiers in Artificial Intelligence and Applications, vol. 242, pp. 462–467. IOS Press (2012)

23. Li, F., Ye, Y., Tian, Z., Zhang, X.: CPU versus GPU: which can perform matrix computation faster-performance comparison for basic linear algebra subprograms. Neural Comput. Appl. **31**(8), 4353–4365 (2019)

24. Méhat, J., Cazenave, T.: Combining UCT and Nested Monte Carlo Search for single-player general game playing. IEEE Trans. Comput. Intell. AI Games **2**(4), 271–277 (2010)

25. Mei, X., Zhao, K., Liu, C., Chu, X.: Benchmarking the memory hierarchy of modern GPUs. In: Hsu, C.-H., Shi, X., Salapura, V. (eds.) NPC 2014. LNCS, vol. 8707, pp. 144–156. Springer, Heidelberg (2014). https://doi.org/10.1007/978-3-662-44917-2_13

26. NVIDIA: Cuda C++ programming guide (2022). https://docs.NVIDIA.com/cuda/cuda-c-programming-guide, section: arithmetic-instructions

27. Portela, F.: An unexpectedly effective Monte Carlo technique for the RNA inverse folding problem. BioRxiv, p. 345587 (2018)

28. Poulding, S.M., Feldt, R.: Generating structured test data with specific properties using nested Monte-Carlo search. In: GECCO, pp. 1279–1286 (2014)

29. Poulding, S.M., Feldt, R.: Heuristic model checking using a Monte-Carlo tree search algorithm. In: GECCO, pp. 1359–1366 (2015)

30. Rimmel, A., Teytaud, F., Cazenave, T.: Optimization of the nested Monte-Carlo algorithm on the traveling salesman problem with time windows. In: Di Chio, C., et al. (eds.) EvoApplications 2011. LNCS, vol. 6625, pp. 501–510. Springer, Heidelberg (2011). https://doi.org/10.1007/978-3-642-20520-0_51

31. Rosin, C.D.: Nested rollout policy adaptation for Monte Carlo Tree Search. In: IJCAI, pp. 649–654 (2011)

32. Silver, D., et al.: Mastering the game of go with deep neural networks and tree search. Nature **529**, 484–489 (2016)

33. Silver, D., et al.: A general reinforcement learning algorithm that masters chess, shogi, and go through self-play. Science **362**(6419), 1140–1144 (2018)

34. Wong, H., Papadopoulou, M., Sadooghi-Alvandi, M., Moshovos, A.: Demystifying GPU microarchitecture through microbenchmarking. In: 2010 IEEE International Symposium on Performance Analysis of Systems and Software (ISPASS), pp. 235–246. IEEE (2010)

35. Wu, D.J.: Accelerating self-play learning in go. arXiv preprint arXiv:1902.10565 (2019)

Learning the Bias Weights for Generalized Nested Rollout Policy Adaptation

Julien Sentuc[1], Farah Ellouze[1], Jean-Yves Lucas[2], and Tristan Cazenave[1(✉)]

[1] LAMSADE, Université Paris Dauphine - PSL, CNRS, Paris, France
`tristan.cazenave@lamsade.dauphine.fr`
[2] OSIRIS Department, EDF Lab Paris-Saclay, Electricité de France, Palaiseau, France

Abstract. Generalized Nested Rollout Policy Adaptation (GNRPA) is a Monte Carlo search algorithm for single player games and optimization problems. In this paper we propose to modify GNRPA in order to automatically learn the bias weights. The goal is both to obtain better results on sets of dissimilar instances, and also to avoid some hyperparameters settings. Experiments show that it improves the algorithm for two different optimization problems: the Vehicle Routing Problem and 3D Bin Packing.

1 Introduction

Monte Carlo Tree Search (MCTS) [12, 20] has been successfully applied to many games and problems [3]. It originates from the computer game of Go [2] with a method based on simulated annealing [4]. The principle underlying MCTS is learning the best move using statistics on random games.

Nested Monte Carlo Search (NMCS) [5] is a recursive algorithm which uses lower level playouts to bias its playouts, memorizing the best sequence at each level. At each stage of the search, the move with the highest score at the lower level is played by the current level. At each step, a lower-level search is launched for all possible moves and the move with the best score is memorized. At level 0, a Monte Carlo simulation is performed, random decisions are made until a terminal state is reached. At the end, the score for the position is returned. NMCS has given good results on many problems like puzzle solving, single player games [22], cooperative path finding or the inverse folding problem [23].

Based on the latter, the Nested Rollout Policy Adaptation (NRPA) algorithm was introduced [26]. NRPA combines nested search, memorizing the best sequence of moves found, and the online learning of a playout policy using this sequence. NRPA achieved world records in Morpion Solitaire and crossword puzzles and has been applied to many problems such as object wrapping [17], traveling salesman with time window [10, 15], vehicle routing problems [8, 16] or network traffic engineering [13].

GNRPA (Generalized Nested Rollout Policy Adaptation) [6] generalizes the way the probability is calculated using a temperature and a bias. It has been applied to some problems like Inverse Folding [7] and Vehicle Routing Problem (VRP) [27].

This work presents an extension of GNRPA using bias learning. The idea is to learn the parameters of the bias along with the policy. We demonstrate that learning the bias

M. Sellmann and K. Tierney (Eds.): LION 2023, LNCS 14286, pp. 194–207, 2023.
https://doi.org/10.1007/978-3-031-44505-7_14

parameters improves the results of GNRPA for Solomon instances of the VRP and for 3D Bin Packing.

This paper is organized as follows. Section 2 describes the NRPA and GNRPA algorithms, as well as its extension. Section 3 presents the experimental results for the two problems studied: VRP and 3D Bin Packing. Finally, the last section concludes.

2 Monte Carlo Search

This section presents the NRPA algorithm as well as its generalization GNRPA. The formula for learning the bias weights is introduced. A new optimization for GNPRA, based on conventional ones, is then presented.

2.1 NRPA and GNRPA

The Nested Rollout Policy Adaptation (NRPA) [26] algorithm is an effective combination of NMCS and the online learning of a playout policy. NRPA holds world records for Morpion Solitaire and crosswords puzzles.

In NRPA/GNRPA each move is associated to a weight stored in an array called the policy. The goal of these two algorithms is to learn these weights thanks to the solutions found during the search, thus producing a playout policy that generates good sequences of moves.

NRPA/GNRPA use nested search. In NRPA/GNRPA, each level takes a policy as input and returns a sequence and its associated score. At any level > 0, the algorithm makes numerous recursive calls to lower levels, adapting the policy each time with the best solution to date. It should be noted that the changes made to the policy do not affect the policy in higher levels (line 7–8 of Algorithm 1). At level 0, NRPA/GNRPA return the sequence obtained by playout function as well as its associated score.

The playout function sequentially constructs a random solution biased by the weight of the moves until it reaches a terminal state. At each step, the function performs Gibbs sampling, choosing the actions with a probability given by the softmax function.

Let w_{ic} be the weight associated with move c in step i of the sequence. In NRPA, the probability of choosing move c at the index i is defined by:

$$p_{ic} = \frac{e^{w_{ic}}}{\sum_k e^{w_{ik}}}$$

GNRPA [6] generalizes the way the probability is calculated using a temperature tau and a bias $beta_{ic}$. The temperature makes it possible to vary the exploration/exploitation trade-off. The probability of choosing the move c at the index i then becomes:

$$p_{ic} = \frac{e^{\frac{w_{ic}}{\tau} + \beta_{ic}}}{\sum_k e^{\frac{w_{ik}}{\tau} + \beta_{ik}}}$$

By taking $\tau = 1$ and $\beta_{ik} = 0$, we find the formula for NRPA.

In NRPA, policy weights can be initialized in order to accelerate convergence towards good solutions. The original weights in the policy array are then not uniformly set to 0, but to an appropriate value according to a heuristic relevant to the problem to solve. In GNRPA, the policy initialization is replaced by the bias. Furthermore, it is sometimes more practical to use β_{ij} biases than to initialize the weights as we will see later on.

When a new solution is found (line 8 of Algorithm 1), the policy is then adapted to the best solution found at the current level (line 13 of Algorithm 1). The policy is passed by reference to the Adapt function.

The current policy is first saved into a temporary policy array named polp before modifying it. The policy copied into polp is then modified in the Adapt function, while the current policy will be used to calculate the probabilities of possible moves. After modification of the policy, the current policy is replaced by polp. The principle of the Adapt function is to increase the weight of the chosen moves and to decrease the weight of the other possible moves by an amount proportional to their probabilities of being played.

The NRPA algorithm therefore strikes a balance between exploration and exploitation. It exploits by shifting the policy weights to the best current solution and explores by picking moves using Gibbs sampling at the lower level. NRPA is a general algorithm that has been shown to be effective for many optimization problems. The idea of adapting a simulation policy has been applied successfully for many games such as Go [18].

It should be noted that in the case of optimization problems such as the VRP, we aim at minimizing the score (consisting of a set of penalties). $bestScore$ is therefore initialized to $+\infty$ (line 5 of Algorithm 1) and we update it each time we find a new $result$ such that $result \leq bestScore$ (line 9 of Algorithm 1).

Algorithm 1. The GNRPA algorithm.

 1: GNRPA ($level, policy$)
 2:    **if** level == 0 **then**
 3:        **return** playout (root, $policy$)
 4:    **else**
 5:        $bestScore \leftarrow -\infty$
 6:        **for** N iterations **do**
 7:            $polp \leftarrow policy$
 8:            (result,new_seq) $\leftarrow$ GNRPA($level - 1, polp$)
 9:            **if** result $\geq$ bestScore **then**
10:                bestScore $\leftarrow$ result
11:                best_seq $\leftarrow$ new_seq
12:            **end if**
13:            Adapt (policy, best_seq)
14:        **end for**
15:        **return** (bestScore, seq)
16:    **end if**

2.2   Learning the Bias

The advantage of the bias over weights initialization relies on its dynamic aspect. It can therefore take into account factors related to the current state. The goal of the extension proposed in this paper is to learn the parameters of the bias. For example, if we consider a bias formula made up of several criteria, such as in [27], we obtain in the case of 2 criteria β_1 and β_2: $\beta_{ic} = w_1 * \beta_1 + w_2 * \beta_2$, where β_1 and β_2 describe two different characteristics of a move. For VRP, it can be the time wasted while waiting to service a customer, the distance traveled, etc.

For some instances, a criterion is a sufficient feature, while others emphasize on another. It is therefore difficult or even impossible to find a single formula that would be appropriate for all instances. To tackle this problem, we propose a simple, yet effective modification of the GNRPA Algorithm, which we name Bias Learning GNRPA (BLGNRPA). We aim at learning the parameters of the bias in order to improve the results on different instances. The idea of learning the bias parameters w_1 and w_2 lies in adapting the importance of the different criteria along with the policy to the specific instance that we are trying to solve.

The probability of choosing the move c at the index i with this bias is:

$$p_{ic} = \frac{e^{\frac{w_{ic}}{\tau} + (w_1 \times \beta_{1ic} + w_2 \times \beta_{2ic})}}{\sum_k e^{\frac{w_{ik}}{\tau} + (w_1 \times \beta_{1ik} + w_2 \times \beta_{2ik})}}$$

Let $A_{ik} = e^{\frac{w_{ik}}{\tau} + (w_1 \times \beta_{1ik} + w_2 \times \beta_{2ik})}$.
The formula for the derivative of $f(x) = \frac{g(x)}{h(x)}$ is :

$$f'(x) = \frac{g'(x) \times h(x) - g(x) \times h'(x)}{h(x)^2}$$

So the derivative of p_{ic} relative to w_1 is:

$$\frac{\delta p_{ic}}{\delta w_1} = \frac{\beta_{1ic} A_{ic} \times \sum_k A_{ik} - A_{ic} \times \sum_k \beta_{1ik} A_{ik}}{(\sum_k A_{ik})^2}$$

$$\frac{\delta p_{ic}}{\delta w_1} = \frac{A_{ic}}{\sum_k A_{ik}} \times (\beta_{1ic} - \frac{\sum_k \beta_{1ik} A_{ik}}{\sum_k A_{ik}})$$

$$\frac{\delta p_{ic}}{\delta w_1} = p_{ic} \times (\beta_{1ic} - \frac{\sum_k \beta_{1ik} A_{ik}}{\sum_k A_{ik}})$$

The cross-entropy loss for learning to play a move is $C_i = -log(p_{ic})$. In order to apply the gradient, we calculate the partial derivative of the loss: $\frac{\delta C_i}{\delta p_{ic}} = -\frac{1}{p_{ic}}$. We then calculate the partial derivative of the softmax with respect to the weight:

$$\nabla w_1 = \frac{\delta C_i}{\delta p_{ic}} \frac{\delta p_{ic}}{\delta w_1} = -\frac{1}{p_{ic}} \times p_{ic}(\beta_{1ic} - \frac{\sum_k \beta_{1ik} A_{ik}}{\sum_k A_{ik}}) =$$

$$\frac{\sum_k \beta_{1ik} A_{ik}}{\sum_k A_{ik}} - \beta_{1ic}$$

If we use α_1 and α_2 as learning rates, we update the weight with (line 16 of Algorithm 2):

$$w_1 \leftarrow w_1 + \alpha_1 \left(\beta_{1ic} - \frac{\sum_k \beta_{1ik} A_{ik}}{\sum_k A_{ik}} \right)$$

Similarly, the formula for w_2 is (line 17 of Algorithm 2):

$$w_2 \leftarrow w_2 + \alpha_2 \left(\beta_{2ic} - \frac{\sum_k \beta_{2ik} A_{ik}}{\sum_k A_{ik}} \right)$$

Optimizations for GNRPA exist and are presented in [6]. A new optimization inspired by the previous ones is presented below.

Avoid Recomputing the Biases. In some cases, the computation of the bias for all possible moves can be costly. In the same way we avoid recomputing all the possible moves, we store the values of the bias in a β matrix.

Algorithm 2. The new generalized adapt algorithm

1: Adapt $(policy, sequence)$
2:    $polp \leftarrow policy$
3:    $w_{1temp} \leftarrow w_1$
4:    $w_{2temp} \leftarrow w_2$
5:    $state \leftarrow root$
6:    **for** $move \in sequence$ **do**
7:        $polp[code(move)] \leftarrow polp[code(move)] + \frac{\alpha}{\tau}$
8:        $w_{1temp} \leftarrow w_{1temp} + \beta_1(move)$
9:        $w_{2temp} \leftarrow w_{2temp} + \beta_2(move)$
10:       $z \leftarrow 0$
11:       **for** $m \in$ possible moves for $state$ **do**
12:           $z \leftarrow z + e^{\frac{policy[code(m)]}{\tau} + w1\beta_1(m) + w2\beta_2(m)}$
13:       **end for**
14:       **for** $m \in$ possible moves for $state$ **do**
15:           $polp[code(m)] \leftarrow polp[code(m)] - \frac{\alpha}{\tau} \times \frac{e^{\frac{policy[code(m)]}{\tau} + w1\beta_1(m) + w2\beta_2(m)}}{z}$
16:           $w_{1temp} \leftarrow w_{1temp} - \alpha_1 \beta_1(m) \frac{e^{\frac{policy[code(m)]}{\tau} + w1\beta_1(m) + w2\beta_2(m)}}{z}$
17:           $w_{2temp} \leftarrow w_{2temp} - \alpha_2 \beta_2(m) \frac{e^{\frac{policy[code(m)]}{\tau} + w1\beta_1(m) + w2\beta_2(m)}}{z}$
18:       **end for**
19:       $state \leftarrow play(state, b)$
20:   **end for**
21:   $policy \leftarrow polp$

---

3 Experimental Results

We now present experiments with bias weights learning for 3D Bin Packing and Vehicle Routing.

Table 1. Results of LSAH, HM, NRPA,GNRPA and BLGNRPA on the 3D bin packing problem

Method/Set	w_1	w_2	Set1		Set2		Set3		Set4		Set5		Set6	
			Uti	N	Uti	N	Uti	N	Uti	N	Uti	N	Uti	N
LSAH			0.502	39	0.527	15	0.623	27	0.675	24	0.431	15	0.641	30
Heightmap			0.502	39	0.463	14	0.623	27	0.738	27	0.836	31	0.565	27
NRPA			0.743	46	0.836	27	0.843	38	0.852	31	**0.868**	**33**	0.807	37
GNRPA	1.00	1.00	0.796	48	0.836	27	**0.887**	**41**	0.852	31	**0.868**	**33**	0.807	37
BLGNRPA	1.00	1.00	**0.808**	**50**	**0.916**	**28**	**0.887**	**41**	**0.913**	**33**	**0.868**	**33**	0.807	37
GNRPA	2.68	10.84	**0.808**	**50**	0.836	27	**0.887**	**41**	0.852	31	**0.868**	**33**	0.807	37
BLGNRPA	2.68	10.84	**0.808**	**50**	**0.916**	**28**	**0.887**	**41**	**0.913**	**33**	**0.868**	**33**	**0.892**	**39**

3.1 3D Bin Packing

The 3D bin Packing Problem is a combinatorial optimization problem in which we have to store a set of boxes into one or several containers. The goal is to minimize the unused space in the containers and put the greatest possible number of items into each of them, or, alternatively, to minimize the number of container used to store all the boxes.

We based our experiments on the problem modeled in the paper [30].
We kept the same capacity for the unique container and the same intervals for the items dimensions. However, as opposed to the paper, we worked on the offline variation of 3D Bin Packing, where the set of boxes are known a priori and taken into account in a given order. Also, the boxes dimensions are considered to be integers.

Heuristics. We used two heuristics proposed in the paper cited above. The first heuristic is the Least Surface Area Heuristic (LSAH) that aims to minimize the surface area of the bin that could hold all the items that we need to pack. The candidates are selected in the structured coordinates (Empty Maximal Space-EMS). It is described by a linear program detailed in the following article [19].

The second one is the Heightmap Minimization (HM) heuristic which is described in the following article [29]. It aims at minimizing the volume increase of the object pile as observed from the loading direction.

The Bias. The BLGNRPA uses these two heuristics to compute the bias using the following formula: $1/Score-of-the-heuristic$ and updating the weights of each move.

Its purpose is to adapt itself to the current situation of the problem and make it easier to choose the next legal move through learning with the bias. It enables having a prior on moves given the current state.

Modeling the Problem. To represent each possible move, we use the coordinates (x, y) where the bottom-left corner of the lower side of the object will be placed. To encode the rotation, we use the dimensions of the object along the three axes (x, y and z) for every possible rotation.

Results. The results are shown in the Table 1. The first column of each set refers to the utilization ratio of the container and the second one to the number of boxes that were put in it. NRPA, GNRPA and BLGNRPA outperform LSAH and Heightmap heuristics across all instances. GNRPA obtained better scores than NRPA on 2 instances (Set 1 and 3) and the same score on the other 4 when using $w_1 = 1$ and $w_2 = 1$. With this initialization of the bias weights, BLGNRPA improves the results of GNRPA on 3 instances (Set 1,2 and 4) and obtains the same score on the 3 others. The final average bias weights found by BLGNRPA over all instances ($w_1 = 2.68$ and $w_2 = 10.84$) were then used to initialize GNRPA and BLGNRPA (line 6 and 7 in Table 1). First, we can see that the use of the average of the weights found by BLGNRPA improved the GNRPA score on one of the sets (Set 1). As for the initialization of the weights of the bias to 1, BLGNRPA (with $w_1 = 2.68$ and $w_2 = 10.84$) performs better than GNRPA (with $w_1 = 2.68$ and $w_2 = 10.84$) on 3 sets (Set 2,4 and 6) and obtains the same score on the 3 others. Finally, using the average bias weights for BLGNRPA improves the results for the last set (Set 6). This suggests that using better starting weights improves the results of the algorithm.

3.2 The Vehicle Routing Problem

The Vehicle Routing Problem is one of the most studied optimization problems. It was first introduced in 1959 by G.B. Dantzig and J.H. Ramser in "The Truck Dispatching Problem" [14]. The goal is to find a set of optimal paths given a certain number of delivery vehicles, a list of customers or places of intervention as well as a set of constraints. We can therefore see this problem as an extension of the traveling salesman problem. In its simplest version, all vehicles leave from the same depot. The goal is then to minimize an objective function, generally defined by these 3 criteria given in order of importance: the number of customers that were not serviced, the number of vehicles used, and finally the total distance traveled by the whole set of vehicles. These 3 criteria may be assigned specific weights in the objective function, or a lexicographic order can be taken into account. The vehicle routing problem is NP-hard, so there is no known algorithm able to solve any instance of this problem in polynomial time. Although exact methods such as Branch and Price exist, approximate methods like Monte-Carlo Search are nonetheless useful for solving difficult instances. Many companies with a fleet of vehicles find themselves faced with the vehicle routing problem [9]. Many variations of the vehicle routing problem have therefore appeared through the years. This paper focuses on the CVRPTW which adds a demand to each customer (e.g., the number of parcels they have purchased) and a limited carrying capacity for all vehicles. Each customer also have a time window in which he must be served. The depot also has a time window, thus limiting the duration of the tour.

Solomon Instances. This work uses the 1987 Solomon instances [28] for the CVRPTW problem. Solomon instances are the main benchmark for CVRPTW to evaluate the different algorithms. The benchmark is composed of 56 instances, each of them consisting of a depot and 100 customers with coordinates included in the interval [0,100]. Vehicles start their tours with the same capacity defined in the instance. A time window is defined for each client as well as for the depot. The distances and the durations correspond to the Euclidean distances between the geometric points.

The Solomon problems are divided into six classes, each having between 8 and 12 instances. For classes C1 and C2 the coordinates are cluster based while classes R1 and R2 coordinates are generated randomly. A mixture of cluster and random structures is used for the problems of classes RC1 and RC2. The R1, C1, and RC1 problem sets have a short time horizon and only allow a few clients per tour (typically up to 10). On the other hand, the sets R2, C2 and RC2 have a long time horizon, allowing many customers (more than 30) to be serviced by the same vehicle.

Use of the Bias. In this paper, we used the dynamic bias introduced in [27]. It is made up of 3 parts. First, the distance, like previous works [1]. Second, the waiting time on arrival. Third, the "lateness". This consists in penalizing an arrival too early in a time window. The formula used for the bias is thus:

$$\beta_{ic} = w_1 * \beta_{distance} + w_2 * \beta_{waiting} + w_3 * \beta_{lateness},$$

with $w_1, w_2, w_3 > 0$.

$$\beta_{distance} = \frac{-d_{ij}}{max_{kl}(d_{kl})}$$

$$\beta_{lateness} = \frac{-(dd_j - max(d_ij + vt, bt_j))}{biggest\ time\ window}$$

$$\beta_{waiting} = 0\ if\ vt + d_{ij} > bt_j$$

$$\beta_{waiting} = \frac{-(bt_j - (d_{ij} + vt))}{biggest\ time\ window}\ if\ i \neq depot$$

$$\beta_{waiting} = \frac{-(bt_j - max(ftw, d_{ij} + vt))}{biggest\ time\ window}\ if\ i = depot$$

where d_{ij} is the distance between customer i and j, bt_j is the beginning of customer j time window, dd_j the end of customer j time window, vt is the departure instant and ftw is the beginning of the earliest time window. In the previous formulas, using $-value$ instead of $\frac{1}{value}$ enables zero values for the waiting time or the lateness. To avoid too much influence from $\beta_{waiting}$ at the start of the tour, where the waiting time can be big, we used $max(ftw, d_{ij} + vt)$. The idea is to only take into account the "useful time" lost. The underlying principle behind learning the bias weights for VRP is to increase the importance of the different criteria depending on the instance. For some instances, distance is the major factor (if for example the time windows are very large). For others, the emphasis will be put on wasted time in order to reduce the number of cars needed. The idea is therefore to adapt the bias formula to make it more relevant for the corresponding instance.

It should be noted that since the bias is dynamic, it is necessary to calculate it many times. As a result, the bias must be updated quickly to reduce the impact on the running time.

Results. In this section, the parameters used for testing NRPA, GNRPA and BLGN-RPA are 3 levels, $\alpha = 1$ and 100 iterations per level. For BLGNRPA, α_1, α_2 and α_3 (for $\beta_{distance}$, $\beta_{waiting}$ and $\beta_{lateness}$) are all initially set to 1 and the bias weights are learned at all level. We compare BLGNRPA with all the bias weights initially set to 0 (that we denote BLGNRPA(0)) with NRPA (GNRPA with all the weights set to 0). We also compare BLGNRPA with weights initialized to the vector of weights W (that we denote BLGNRPA(w)) and GNRPA. For both algorithm, the weights are either initialized or set to the values used in [27]. The weights used are therefore $w_1 = 15$, $w_2 = 75$ and $w_3 = 10$. The results given in Table 1 are the best runs out of 10 with different seeds. The running times for NRPA, GNRPA, BLGNRPA are close to each other and smaller than 1800 s.

We also compare our results with the OR-Tools library. OR-Tools is a Google library for solving optimization problems. It can solve many types of VRP problems, including CVRPTW. OR-Tools offers different choices to build the *first solution*. In our experiments, we used "PATH_CHEAPEST_ARC" parameter. Starting from a start node, the algorithm connects it to the node which produces the cheapest route segment, and iterates the same process from the last node added. Then, OR-Tools uses local search in order to improve the solution. Several options are also possible here. We used the "GUIDED_LOCAL_SEARCH", which guides local search to escape local minima. OR-Tools is run for 1800 s on each problem.

The results are compared with the lexicographical approach, first taking into account the number of vehicles used NV and then the total distance traveled Km. The best score among the different algorithms is put in bold and when the best known score is obtained an asterisk is added at the end of the vehicle number.

BLGNRPA(0) performed better than NRPA on all instances. NRPA only obtained the same result for the easiest instance c101, for which NRPA and BLGNRPA(0) got the best known solution. NRPA obtained the best known solution only on instance c101 while BLGNRPA(0) got the best known solution for 10 instances and the best results among all algorithms for 17 of the 56 instances. BLGNRPA(0) obtained similar or even better results than GNRPA and BLGNRPA(W) on some instances such as C-type instances, r101, r111, etc. However, it also gets much worse results for other instances (r107, rc201,...). For some instances, the initialization of the bias weights with $w_1 = 15$, $w_2 = 75$ and $w_3 = 10$, that we found manually thanks to many tries, is clearly not the most suitable. For example, on instance r101, previous experiments showed that a good set of weights for the instance is $w_1 = 20$, $w_2 = 20$ and $w_3 = 0$. Interestingly, for BLGNRPA(0) the weights at the end of the best run are $w_1 = 21.07$, $w_2 = 19.44$ and $w_3 = 1.42$. However, we could not guarantee the same favorable behavior on all instances, due to the random aspect of the algorithm, and also due to the amount of time required by the learning process in order to reach appropriate weights values.

We observe that BLGNRPA has a better score than GNRPA on 36 instances, the same score on 12 instances and a worse one on 8 instances. Therefore, learning the weights of the bias seems to improve the results of GNRPA. OR-Tools obtains a better result on the majority of the instances. However, GNRPA got a better score than OR-Tools for 12 instances and an equivalent score for 9 of them, BLGNRPA(W) got a better score than OR-Tools for 18 instances and also an equivalent score for 9 of them.

We can see that for both GNRPA and BLGNRPA the results are better on instances of class R1 and RC1 than on instances of class R2 and RC2. Instances of classes R2 and RC2 have their time window constraints weakened (larger time windows). Whenever dealing with weak constraints, local search performs better than Monte Carlo search with or without the learning of the bias weights. This observation was also made in [11] [27] (Table 2).

Table 2. The different algorithms tested on the 56 standard instances

Instances	NRPA		BLGNRPA(0)		GNRPA		BLGNRPA(w)		OR-Tools		Best Known	
	NV	Km	NV	Km	NV	Km	NV	Km	NV	Km	NV	Km
c101	10*	**828.94**	10*	**828.94**	10*	**828.94**	10*	**828.94**	10*	**828.94**	10	828.94
c102	10	1,011.40	10	843.57	10	843.57	10	843.57	10*	**828.94**	10	828.94
c103	10	1,105.10	10	844.86	10	843.02	10	828.94	10*	**828.06**	10	828.06
c104	10	1,112.66	10	831.88	10	839.96	10	**828.94**	10	846.83	10	824.78
c105	10	896.93	10*	**828.94**	10*	**828.94**	10*	**828.94**	10*	**828.94**	10	828.94
c106	10	853.76	10*	**828.94**	10*	**828.94**	10*	**828.94**	10*	**828.94**	10	828.94
c107	10	891.22	10*	**828.94**	10*	**828.94**	10*	**828.94**	10*	**828.94**	10	828.94
c108	10	1006.69	10*	**828.94**	10*	**828.94**	10*	**828.94**	10*	**828.94**	10	828.94
c109	10	962.35	10	**834.85**	10	**834.85**	10	836.60	10	857.34	10	828.94
c201	4	709.75	3*	**591.56**	3*	**591.56**	3*	**591.56**	3*	**591.56**	3	591.56
c202	4	929.93	3	609.23	3	611.08	3	611.08	3*	**591.56**	3	591.56
c203	4	976.00	3	599.33	3	611.79	3	605.58	3	**594.23**	3	591.17
c204	4	995.19	3	595.65	3	614.50	3	597.74	3	**593.82**	3	590.60
c205	3	702.05	3*	**588.88**	3*	**588.88**	3*	**588.88**	3*	**588.88**	3	588.88
c206	4	773.28	3*	**588.49**	3*	**588.49**	3*	**588.49**	3*	**588.49**	3	588.49
c207	4	762.73	3	592.50	3	592.50	3	592.50	3*	**588.29**	3	588.29
c208	3	741.98	3*	**588.32**	3*	**588.32**	3*	**588.32**	3*	**588.32**	3	588.32
r101	19	1,660.01	19*	**1,650.80**	19*	**1,650.80**	19	1,654.67	19	1,653.15	19	1,650.80
r102	17	1,593.73	17	1,499.20	17	1,508.83	17	1,501.11	17	**1489.51**	17	1,486.12
r103	14	1,281.89	14	1,235.31	13	1,336.86	13	1,321.17	13	**1,317.87**	13	1,292.68
r104	11	1,098.30	10	1,000.52	10	1,013.62	10	**996.61**	10	1,013.23	9	1,007.31
r105	15	1,436.75	14	1,386.07	14	**1,378.36**	14	1385.76	14	1,393.14	14	1,377.11
r106	12	1,364.09	12	1,269.82	12	1,274.47	12	**1,265.97**	13	1,243.0	12	1,252.03
r107	11	1,241.15	11	1,079.96	10	1,131.19	10	1,132.95	10	**1,130.97**	10	1,104.66
r108	11	1,106.14	10	953.15	10	990.18	10	**941.74**	10	963.4	9	960.88
r109	12	1,271.13	12	1,173.57	12	1,180.09	12	**1,171.70**	12	1,175.48	11	1,194.73
r110	12	1,232.03	11	1,116.64	11	1,140.22	11	**1,094.84**	11	1,125.13	10	1,118.84
r111	12	1,200.37	11	**1,071.84**	11	1,104.42	11	1,073.74	11	1,088.01	10	1,096.72
r112	10	1,162.47	10	**965.43**	10	1,013.50	10	974.56	10	974.65	9	982.14
r201	5	1,449.95	5	1,250.16	4	1,316.27	4	1,293.38	4	**1,260.67**	4	1,252.37
r202	4	1,335.96	4	1,124.91	4	1,129.89	4	1,122.80	4	**1,091.66**	3	1,191.70
r203	4	1,255.78	4	930.58	3	1,004.49	3	970.45	3	**953.85**	3	939.50
r204	3	1,074.37	3	765.47	3	787.69	3	772.22	3	**755.01**	2	852.52
r205	4	1,299.84	3	1,047.53	3	1,043.81	3	1,052.15	3	**1,028.6**	3	994.43

(continued)

Table 2. (*continued*)

Instances	NRPA		BLGNRPA(0)		GNRPA		BLGNRPA(w)		OR-Tools		Best Known	
	NV	Km	NV	Km	NV	Km	NV	Km	NV	Km	NV	Km
r206	3	1,270.89	3	982.50	3	990.88	3	959.89	3	**923.1**	3	906.14
r207	3	1,215.47	3	871.66	3	900.17	3	878.91	3	**832.82**	2	890.61
r208	3	1,027.12	3	726.34	2	779.25	2	737.50	2	**734.08**	2	726.82
r209	4	1,226.67	3	954.02	3	981.82	3	960.40	3	**924.07**	3	909.16
r210	4	1,278.61	3	970.30	3	995.50	3	991.87	3	**963.4**	3	939.37
r211	3	1,068.35	3	821.79	3	850.33	3	798.84	3	**786.28**	2	885.71
rc101	15	1,745.99	15	1,636.50	14	**1,702.68**	15	1,636.50	15	1,639.54	14	1,696.95
rc102	14	1,571.50	13	1,497.11	13	1,509.86	**13**	**1,496.16**	13	1,522.89	12	1,554.75
rc103	12	1,400.54	**11**	**1,265.80**	11	1,287.33	11	1,273.28	12	1,322.84	11	1,261.67
rc104	11	1,264.53	10	1,147.69	10	1,160.55	**10**	**1,146.36**	10	1,155.33	10	1,135.48
rc105	15	1,620.43	14	**1,553.43**	14	1,587.41	14	1,563.18	14	1,614.98	13	1,629.44
rc106	13	1,486.81	**12**	**1,385.21**	12	1,397.55	12	1,388.80	13	1,401.73	11	1,424.73
rc107	12	1,338.18	11	1,238.04	11	1,247.80	**11**	**1,233.76**	11	1,255.62	11	1,230.48
rc108	11	1,286.88	**10**	**1,150.68**	10	1,213.00	10	1152.61	11	1,148.16	10	1,139.82
rc201	5	1,638.08	5	1,354.84	4	1,469.50	4	1,469.16	**4**	**1,424.01**	4	1,406.94
rc202	4	1,593.54	4	1,260.11	4	1,262.91	4	1,203.10	**4**	**1,161.82**	3	1,365.65
rc203	4	1,431.32	4	1,010.99	3	1,123.45	3	1,141.27	**3**	**1,095.56**	3	1,049.62
rc204	3	1,260.05	3	841.48	3	864.24	3	822.39	**3**	**803.06**	3	789.46
rc205	5	1,578.73	4	1,359.74	4	1,347.86	4	1,333.95	**4**	**1,315.72**	4	1,297.65
rc206	4	1,412.26	3	1,294.77	3	1,208.52	3	1,246.48	**3**	**1,157.2**	3	1,146.32
rc207	4	1,395.02	4	1,066.06	3	1,164.99	3	1,124.15	**3**	**1,098.61**	3	1,061.14
rc208	3	1,182.55	3	911.34	3	948.82	3	906.01	**3**	**843.02**	3	828.14

4   Discussion

The use of a bias in the softmax has some similarities with the formula used in Ant Colony Optimization (ACO) [21, 24, 25, 31] since a priori knowledge of a fixed prior bias associated to actions are also used in ACO with a kind of softmax. The originality of our approach is that we learn the parameter that multiplies the prior bias associated to actions, dynamically and on each instance. We also provide a theoretical and mathematical derivation of the way the parameters of the bias are updated.

Different kind of algorithms are used for different variations on the VRP. For example in the recent DIMACS challenge on VRP, the number of vehicles was not taken into account to evaluate solutions which makes fair comparison with our algorithm difficult (the total distance of the tours might be reduced with one more vehicle).

5   Conclusion

In this paper, we introduced a new method to learn the bias weights for the GNRPA algorithm with BLGNRPA. This new method partially removes the need to choose hand-picked weights for GNRPA. However, GNRPA and BLGNRPA have several limitations. First they are less efficient on weakly constrained problems as we presented in the results section. In addition, GNRPA/BLGNRPA are designed for complete information problems. Finally, the bias must be simple to compute. Indeed, for a GNRPA/BLGNRPA search, the bias must be calculated $100^{level} \times \bar{c}$ times, where $\bar{c}$ is the average number of moves considered in the playout function. In order to have a fast and efficient search, the computation of the bias must therefore be fast.

The results we obtained show that the learning of the bias improves the solutions for GNRPA. For 3D Bin Packing, BLGNRPA got a better score on 3 out of the 6 sets and the same score on the 3 others. For VRP, BLGNRPA provided better solutions than GNRPA on 36 out of the 56 instances, the same score on 12 instances and a worse one on 8 instances. For both problems, it seems that having the bias parameters already initialized with good values for BLGNRPA improves the results compared to initializing the values to 0 or 1.

These preliminary results look promising, so in future work we plan to test some enhancements of the GNRPA on BLGNRPA such as the Stabilized GNRPA (SGNRPA). Similarly to the Stabilized NRPA, in SGNRPA the Adapt function is not systematically run after each level 1 playout, but with an appropriate periodicity. Finally, we plan to work on finding better values for the bias weights initialization, possibly by running a preliminary phase consisting of solely learning the bias weights but not the BLGNRPA policy.

Acknowledgment. Thanks to Clément Royer for advising us to use a gradient when possible. This work was supported in part by the French government under the management of Agence Nationale de la Recherche as part of the "Investissements d'avenir" program, reference ANR19-P3IA-0001 (PRAIRIE 3IA Institute).

References

1. Abdo, A., Edelkamp, S., Lawo, M.: Nested rollout policy adaptation for optimizing vehicle selection in complex VRPs, pp. 213–221 (2016)
2. Bouzy, B., Cazenave, T.: Computer go: an AI oriented survey. Artif. Intell. **132**(1), 39–103 (2001)
3. Browne, C., et al.: A survey of Monte Carlo tree search methods. IEEE Trans. Comput. Intell. AI Games **4**(1), 1–43 (2012)
4. Brügmann, B.: Monte Carlo Go. Max-Planke-Inst. Phys., Munich, Technical report (1993)
5. Cazenave, T.: Nested Monte-Carlo search. In: Boutilier, C. (ed.) IJCAI, pp. 456–461 (2009)
6. Cazenave, T.: Generalized nested rollout policy adaptation. In: Monte Carlo Search at IJCAI (2020)
7. Cazenave, T., Fournier, T.: Monte Carlo inverse folding. In: Monte Carlo Search at IJCAI (2020)
8. Cazenave, T., Lucas, J.Y., Kim, H., Triboulet, T.: Monte Carlo vehicle routing. In: ATT at ECAI (2020)
9. Cazenave, T., Lucas, J.Y., Triboulet, T., Kim, H.: Policy adaptation for vehicle routing. AI Commun. **34**, 21–35 (2021)
10. Cazenave, T., Teytaud, F.: Application of the nested rollout policy adaptation algorithm to the traveling salesman problem with time windows. In: Hamadi, Y., Schoenauer, M. (eds.) LION 2012. LNCS, pp. 42–54. Springer, Heidelberg (2012). https://doi.org/10.1007/978-3-642-34413-8_4

11. Cornu, M.L: Local search, data structures and Monte Carlo search for multi-objective combinatorial optimization problems. (recherche locale, structures de données et recherche Montecarlo pour les problèmes d'optimisation combinatoire multi-objectif) (2017)
12. Coulom, R.: Efficient selectivity and backup operators in Monte-Carlo tree search. In: van den Herik, H.J., Ciancarini, P., Donkers, H.H.L.M.J. (eds.) CG 2006. LNCS, vol. 4630, pp. 72–83. Springer, Heidelberg (2007). https://doi.org/10.1007/978-3-540-75538-8_7
13. Dang, C., Bazgan, C., Cazenave, T., Chopin, M., Wuillemin, P.-H.: Monte Carlo search algorithms for network traffic engineering. In: Dong, Y., Kourtellis, N., Hammer, B., Lozano, J.A. (eds.) ECML PKDD 2021. LNCS (LNAI), vol. 12978, pp. 486–501. Springer, Cham (2021). https://doi.org/10.1007/978-3-030-86514-6_30
14. Dantzig, G.B., Ramser, J.H.: The truck dispatching problem. Manage. Sci. 6(1), 80–91 (1959)
15. Edelkamp, S., Gath, M., Cazenave, T., Teytaud, F.: Algorithm and knowledge engineering for the TSPTW problem. In: Computational Intelligence in Scheduling (SCIS), 2013 IEEE Symposium on, pp. 44–51. IEEE (2013)
16. Edelkamp, S., Gath, M., Greulich, C., Humann, M., Herzog, O., Lawo, M.: Monte-Carlo tree search for logistics. In: Clausen, U., Friedrich, H., Thaller, C., Geiger, C. (eds.) Commercial Transport. LNL, pp. 427–440. Springer, Cham (2016). https://doi.org/10.1007/978-3-319-21266-1_28
17. Edelkamp, S., Gath, M., Rohde, M.: Monte-Carlo tree search for 3D packing with object orientation. In: Lutz, C., Thielscher, M. (eds.) KI 2014. LNCS (LNAI), vol. 8736, pp. 285–296. Springer, Cham (2014). https://doi.org/10.1007/978-3-319-11206-0_28
18. Graf, T., Platzner, M.: Adaptive playouts in Monte-Carlo tree search with policy-gradient reinforcement learning. In: Plaat, A., van den Herik, J., Kosters, W. (eds.) ACG 2015. LNCS, vol. 9525, pp. 1–11. Springer, Cham (2015). https://doi.org/10.1007/978-3-319-27992-3_1
19. Hu, H., Zhang, X., Yan, X., Wang, L., Xu, Y.: Solving a new 3D bin packing problem with deep reinforcement learning method. arXiv:1708.05930 (2017)
20. Kocsis, L., Szepesvári, C.: Bandit based Monte-Carlo planning. In: Fürnkranz, J., Scheffer, T., Spiliopoulou, M. (eds.) ECML 2006. LNCS (LNAI), vol. 4212, pp. 282–293. Springer, Heidelberg (2006). https://doi.org/10.1007/11871842_29
21. Maniezzo, V., Gambardella, L.M., de Luigi, F.: Ant colony optimization. In: New Optimization Techniques in Engineering. Studies in Fuzziness and Soft Computing, vol. 141, pp. 101–121. Springer, Heidelberg (2004). https://doi.org/10.1007/978-3-540-39930-8_5
22. Méhat, J., Cazenave, T.: Combining UCT and nested Monte Carlo Search for single-player general game playing. IEEE Trans. Comput. Intell. AI Games 2(4), 271–277 (2010)
23. Portela, F.: An unexpectedly effective Monte Carlo technique for the RNA inverse folding problem. BioRxiv, p. 345587 (2018)
24. Qi, C., Sun, Y.: An improved ant colony algorithm for VRPTW. In: 2008 International Conference on Computer Science and Software Engineering, vol. 1, pp. 455–458. IEEE (2008)
25. Rizzoli, A.E., Montemanni, R., Lucibello, E., Gambardella, L.M.. Ant colony optimization for real-world vehicle routing problems. Swarm Intell. 1(2), 135–151 (2007)
26. Rosin, C.D.: Nested rollout policy adaptation for Monte Carlo Tree Search. In: IJCAI 2011, Proceedings of the 22nd International Joint Conference on Artificial Intelligence, pp. 649–654 (2011)
27. Sentuc, J., Cazenave, T., Lucas, J.Y.: Generalized nested rollout policy adaptation with dynamic bias for vehicle routing. In: AI for Transportation at AAAI (2022)
28. Solomon, M.M.: Algorithms for the vehicle routing and scheduling problems with time window constraints. In: Operations Research (1985)
29. Wang, F., Hauser, K.: Stable bin packing of non-convex 3D objects with a robot manipulator. arXiv:1812.04093v1 (2018)

30. Zhao, H., Yu, Y., Xu, K.: Learning efficient online 3D bin packing on packing configuration trees. In: International Conference on Learning Representations (2022)
31. Zhen, T., Zhang, Q., Zhang, W., Ma, Z.: Hybrid ant colony algorithm for the vehicle routing with time windows. In: 2008 ISECS International Colloquium on Computing, Communication, Control, and Management, vol. 1, pp. 8–12. IEEE (2008)

Heuristics Selection with ML in CP Optimizer

Hugues Juillé$^{(\boxtimes)}$, Renaud Dumeur, and Paul Shaw

IBM France Lab, Orsay, France
{hugues.juille,renaud.dumeur,paul.shaw}@fr.ibm.com

Abstract. IBM® ILOG® CP Optimizer (CPO) is a constraints solver that integrates multiple heuristics with the goal of handling a large diversity of combinatorial and scheduling problems while exposing a simple interface to users. CPO's intent is to enable users to focus on problem modelling while automating the configuration of its optimization engine to solve the problem. For that purpose, CPO proposes an *Auto* search mode which implements a hard-coded logic to configure its search engine based on the runtime environment and some metrics computed on the input problem. This logic is the outcome of a mix of carefully designed rules and fine-tuning using experimental benchmarks. This paper explores the use of Machine Learning (ML) for the off-line configuration of CPO solver based on an analysis of problem instances. This data-driven effort has been motivated by the availability of a proprietary database of diverse benchmark problems that is used to evaluate and document CPO performance before each release. This work also addresses two methodological challenges: the ability of the trained predictive models to robustly generalize to the diverse set of problems that may be encountered in practice, and the integration of this new ML stage in the development workflow of the CPO product. Overall, this work resulted in a speedup improvement of about 14% (resp. 31%) on Combinatorial problems and about 5% (resp. 6%) on Scheduling problems when solving with 4 workers (resp. 8 workers), compared to the regular CPO solver.

Keywords: Combinatorial Optimization · Machine Learning · Lifecycle management

1 Introduction

The search landscape of any combinatorial optimization problem exhibits specific structures. Hence, algorithms that can exploit these structures more efficiently are likely to outperform those that don't. Over the past decades, there has been significant research to determine how to select or adapt search algorithms to improve performance on a specific problem class or problem instance [13]. The algorithm selection problem entails many sub-questions and sub-problems that have been heavily studied. For instance, should a single or a pool of algorithms be

M. Sellmann and K. Tierney (Eds.): LION 2023, LNCS 14286, pp. 208–222, 2023.
https://doi.org/10.1007/978-3-031-44505-7_15

selected, how to distribute compute-time among the selected algorithms, should the algorithm selection be performed before-hand (off-line) or adjusted during search space exploration. Similar questions are raised for computing the features that will characterize a problem class or a problem instance.

While approaches based on hand-crafted heuristics have shown promising results, the design of such heuristics assumes that some internal structures reflecting the intrinsic complexity of problems have been identified and an algorithm can efficiently exploit them. However, because of the ill-defined nature of the search space for hard problems, many of these heuristics are based on rules of thumb rather than computable rigorous mathematical formulations. Therefore, data-driven machine learning methods are natural candidates to address this heuristic design difficulty. The underlying motivation is that ML methods may capture properties of a problem class or a problem instance in a decision model.

Here again, many approaches have been explored for introducing ML for solving combinatorial optimization problems [2,8]. In the more extreme approaches, ML handles problem solving end-to-end by designing new algorithms that integrate trainable models to drive the exploration of the search space. These approaches are usually specific to some problem classes and a frequent goal is to analyze the generalization capability of the trained algorithm to larger problem instances [3,7]. On the other hand, one may choose to capitalize on the large effort that has been put in the design of competitive combinatorial optimization algorithms. In that domain, the two main approaches consist either in using ML before invoking solvers (for instance, by learning how to configure algorithms [9] or by learning some portfolio-based algorithm selection strategy [15]), or in integrating ML in the inner logic of the algorithms to learn rules controlling decision making at runtime [6]. Our approach is the former and, using the taxonomy proposed by [4], can be defined as *ML-in-MH* used to improve the *Algorithm selection* stage.

Objectives and Motivations

The goal of the work presented in this paper is to use Machine Learning to improve the performance of the CPO solver over a large range of application domains and real-world problems (that is, not specific to a problem class). Multiple challenges must be addressed to achieve this goal:

1. **Algorithm selection**: this problem consists in determining how to allocate search heuristics to the available workers (CPU threads) involved in a CPO solve. Predictive models are trained to make this decision, based on an off-line computation of metrics on the input problem.
2. **Training methodology robustness**: Predictive models must not specialize on benchmarks used for training so that CPO solver extended with ML will improve performance over a large class of unseen application domains. Therefore, the training methodology will be designed carefully to limit overfitting issues.

3. **Trained models lifecycle management**: CPO solver is continuously evolving. Existing heuristics efficiency is improved and new heuristics are designed. Therefore, our ML approach must take into account these changes and be integrated in the product development workflow. This means that the training process must be automated as much as possible (e.g., to be ultimately executed from a continuous delivery environment) and reproducible.

The following items summarize the main tasks involved in the implementation of our ML approach:

1. define features to be computed on input problems
2. train a predictive model from these features to drive algorithm selection
3. embed the trained model in CPO executable
4. benchmark this *CPO with ML* version against *regular CPO* version

Section 2 introduces how features are computed based on the formulation of input problems in CPO modelling language. Then, our repository of benchmark problems and our performance assessment process are presented. Section 3 details our formulation of algorithm selection as a ML problem along with the steps involved in training a predictive model and embedding this model in CPO. Robustness and lifecycle management challenges are also addressed in that section. Finally, experimental results for training and benchmarking are presented in Sect. 4.

2   CPO Modelling Language and Features Definition

2.1   CPO Modelling Language

CPO [10] proposes a rich set of constructs to model combinatorial optimization problems. The core of combinatorial problems consists in assigning values to a number of integer decision variables, subject to a number of constraints which enforce conditions on valid domains of these variables. Exploring the problem search space consists in finding valid assignments to its decision variables. A solution to the problem corresponds to a situation where all decision variables are assigned a value while satisfying all constraints. In addition, a measure of quality for solutions (an objective) may be formulated, and the goal is to find a solution that maximize (or minimize) this objective.

CPO supports all regular constraints on integer decision variables: equality, difference, ordering... (low level constraints) along with: alldiff, count, element, distribute, pack... (global constraints)

In addition, CPO introduces the concept of interval variables for the formulation of scheduling problems. An interval is characterized by a start value, an end value, a size and an intensity. Also, interval variables can be optional; that is, one can decide not to consider them in the solution schedule. A number of constructs exploits interval variables to enforce constraints like spanning, precedence, presence, alternative..., etc. Interval variables are also used to define higher level abstractions like sequences of intervals, over which specific constraints may also be enforced.

2.2   Features Definition

A combinatorial problem model can be represented as a directed graph. There is one vertex for each decision variable, one for each constant, one for each constraint and one for each expression. For each constraint (resp. expression), edges connect the associated node with all nodes corresponding to the parameters of the constraint (resp. expression) definition. As a result, vertices with only outgoing edges correspond to decision variables and constants, while vertices with at least one incoming edge correspond to constraints, expressions, or objective definition.

Before actually solving a model, CPO performs a pre-solve stage which consists in a reformulation of the input model. The purpose of this stage is to improve solve performance by removing useless entities or rewriting expressions for faster computation and search space reduction. As a result, this "pre-solved" model can be quite different from the input model and metrics computed on this second model may also vary significantly from those computed on the original model. One will explore whether training the predictive model on one of these constraints models or the other impacts performance.

Our approach to an off-line analysis of input problems consists in engineering a collection of features that captures information about structural properties of the problem. In order to compute features on the graph representing the input problem, the following attributes are defined for each node:

- **Type**: an identifier of the detailed nature of the node (e.g., *integer variable, alldiff constraint...*),
- **Flavor**: an identifier that defines a coarser grouping based on the nature of nodes (e.g., *constraint, integer expression...*),
- **Constraint**: a flag indicating if the node corresponds to a constraint,
- **Leaf**: a flag indicating if the node has no child (e.g., decision variables, constants...),
- **Root**: a flag indicating if the node has no parent (e.g., objective function, constraints that are not argument of another constraint or expression...),

Using this terminology, Table 1 describes the features that have been explored in this work. This table identifies a group of *mandatory* features that will always be used, while different combinations of features related to search space size will be investigated. In this table, *search space size* refers to the cartesian product of all variables domain. CPO propagation engine reduces the domain of decision variables. As will be seen in the experimental section, considering search space size *before* or *after* this initial propagation (hence resulting in twice as many features) impacts the performance of predictive models.

2.3   Benchmark Problems and Performance Assessment

CPO involves non-deterministic algorithms that are continually tuned to enhance their performances. This tuning needs input from performance tests to be accurate. To enable such testing, the CPO Product Improvement Platform (CPOPIP)

Table 1. Features definition

		Features
"Mandatory" features	**Density-based features** • For each **Type**, density of all nodes of this Type with respect to all other nodes with same value of **Constraint** attribute • For each **Flavor**, density of all nodes of this Flavor with respect to all other nodes with same value of **Constraint** attribute	158 41
	Misc ratios based on number of vars, constraints, types Ratio of leaf nodes or constraint nodes over all nodes in graph, ratio of Integer and Interval variables among leaf nodes...	11
	Distribution-based statistics features $log1p()$ of mean, standard deviation and skewness for distributions: • Number of children for *Constraint* nodes • Number of parents for *Leaf* nodes • Number of parents for non-*Leaf* and non-*Constraint* nodes • Number of non-*Leaf* and non-*Constraint* child nodes	12
Log Search Space size features	**Search Space size-based features** First, compute BEFORE and AFTER initial propagation: • For all Integer variables: log of size of domain • For all Interval variables: log of size of interval start domain Then, compute following features: • Sum of logs for Integer variables, Interval variables and overall • For all "density-based" and "misc ratios" features, product of original feature by: - total log search space size over all Integer variables only - total log search space size over all variables	3×2 210×2 210×2

has been developed and deployed on a dedicated cluster. CPOPIP has been performing benchmarks to monitor enhancements of the CPO product over time. These benchmark problems are based on a repository of a few thousands tests that have been collected from miscellaneous public benchmarks or from real world projects. It is continuously enriched with new tests. All these tests are tagged with attributes to organize them by problem type (integer optimization, scheduling, feasibility) and by family. Problems in a family share some common structure (usually, a family is built from multiple instances of a same problem with varying sizes for decision variables along with customized constraints). At the time of writing of this paper, our repository of tests is composed of 6142 CPO models (2140 combinatorial optimization problems, 483 combinatorial feasibility problems, and 3519 scheduling problems), that are grouped in 327 families. The families size varies from a single instance to 440 for the largest family. CPOPIP embeds multiple tools to support performance analysis at different granularity levels.

A common setup for a benchmarking *campaign* is to execute 10 runs (CPO solve) for each test, with a time limit of 1000 s per run. Latest CPO released version can solve to optimality about 45% of the test for all runs, before the time limit. For less than 2% of these tests, no run can found a first solution. These corresponds to the hardest problems of the repository of tests. For the remaining tests (about 53%), some runs didn't prove optimality but at least one run found a first solution. This corresponds to difficult optimization problems which focus our effort for improving CPO heuristics.

Improvements between CPO versions are assessed by comparing their corresponding benchmarking campaigns. One of the metrics that is computed when comparing two campaigns is the average test speedup. For each model, this indicator evaluates the average ratio of runtimes for the two campaigns to achieve a same performance with respect to the objective value. The speedup value is above 1.0 for tests where the first campaign is faster on average than the second campaign to achieve a same value of the objective (or to find a solution for satisfiability problems), over the 10 runs. The range of values for the test speedup is limited to the interval $[0.01, 100.0]$.

3   General Approach

3.1   Algorithm Selection Problem Formulation

The CP Optimizer search is based on *constructive search*, which is a search technique that attempts to build a solution by fixing decision variables to values. While the built-in CP Optimizer search is based on this technique, the optimizer also uses other heuristics to improve search. These heuristics (or *SearchType* in CPO terminology) are named: *Restart*, *MultiPoint*, *DepthFirst* and *IterativeDiving*.

- *DepthFirst* search is a tree search algorithm such that each fixing, or instantiation, of a decision variable can be thought of as a branch in a search tree. The optimizer works on the subtree of one branch until it has found a solution or has proven that there is no solution in that subtree.
- *Restart* is a depth-first search that is restarted after a certain number of failures that increases after each restart.
- *IterativeDiving* is a search method that attempts to quickly build feasible solutions and improve them using consecutive iterations of backtrack-free search. This heuristics is specialized for Scheduling problems.
- *MultiPoint* search creates a set of solutions and combines the solutions in the set in order to produce better solutions. Multi-point search is more diversified than depth-first or restart search, but it does not necessarily prove optimality or the inexistence of a solution.

Each heuristic can also be manually fine-tuned with specific parameters. *DepthFirst* and *IterativeDiving* are restricted variations of *Restart* heuristics. On the other hand, *Restart* and *MultiPoint* heuristics implement very different approaches to search. Our experience has shown that, depending on the problem instance, one heuristic can be much faster than the other to solve it. For this reason, we decided to focus on the problem of predicting a score that correlates with the probability that the *MultiPoint* heuristic outperforms the *Restart* heuristic, given a problem instance.

Moreover, when the CPO's *Workers* parameter exceeds 1, multiple worker threads are started when search begins. When the *SearchType* parameter is set to *Auto*, these threads will employ a variety of search heuristics, such as *MultiPoint* and *Restart*. Through the exchange of information, such as intermediate solutions and nogoods, these worker threads can collaborate and cooperate in the search process. Hence, identifying the right mix of heuristics to assign to workers has a strong impact on search performance. In our approach, the predicted score determines the allocation of each heuristic to the different workers, by using a proportionality rule.

This section has formulated the algorithm selection problem as a binary classification problem. Before detailing the ML workflow that will be used to solve this problem in Sect. 3.4, the next two sections introduce how the robustness and life-cycle management challenges have been addressed.

3.2   Training Methodology Robustness

From a Machine Learning methodological point of view, a number of challenges must be overcome. First, even if a significant number of benchmark problems compose the dataset available for training, a few thousand data points very quickly expose us to overfitting and variance issues. Second, the dataset is structured into families that group similar problems together. This raises the issue of diversity in the training data, along with the risk of overfitting to the actual data used for benchmarking and generalizing poorly on unseen problems.

These issues have been mitigated as follows. First, an aggressive splitting strategy has been implemented, using: 30% of problems for training, 30% for validation and 40% for testing. Second, splitting has been performed using stratification based on pairs *(family id, target)*. By using a small fraction of data for training the predictive model and about the same amount of data for validation, overfitting is limited. Indeed, a large validation set enables a more reliable evaluation of generalization so that training is stopped before overfitting occurs. Keeping a large chunk (40%) of all problems for the final evaluation of the predictive model also makes this final measure more reliable. Moreover, stratified splitting introduces more diversity in the training data and improves generalization capability for the trained predictive model.

The LightGBM algorithm [5,11] has been selected for training predictive models. Several reasons motivated our choice of this Gradient Boosted Trees (GBT) framework. First, training models with LightGBM is very fast, which makes heavy Hyper Parameters Optimization (HPO) more manageable to investigate configuration options. Second, this framework exhibits interesting properties for a smooth integration in CPO: serializing decision trees and implementing their evaluation logic is simple (a decision tree is a list of tuples (subtree id, split feature, threshold value, left and right subtrees id) for internal nodes, and tuples (subtree id, value) for leaf nodes), also the memory footprint of serialized trained models is small.

Also, the logic implemented by decision trees is easy to interpret. Being able to get insights about the main features used to compute class probabilities helps to build confidence in the model and to engineer additional features. Finally, training a LightGBM model is reproducible (given a seed value), which is desirable for our ML workflow.

3.3   Trained Models Lifecycle Management

Heuristics embedded in CPO are continuously improving over time. Therefore, a predictive model trained for a specific version has to be adjusted to reflect the new relative performance of the different heuristics in following versions. The two main challenges to address this particular issue concern: the end-to-end automation of the workflow, and reproducibility.

CPO development and release follow the Continuous Delivery (CD) approach which aims at automating the build, test and release steps of a software. In order to embed a trained predictive model in CPO, all the steps needed to produce this model must be automated in a reliable pipeline. This pipeline covers all the regular ML steps (data preparation, model training, HPO, feature selection...) along with the generation of the actual resources to be packaged in the delivered product. Details about this pipeline are discussed in Sect. 3.4. This pipeline can be executed without any human interaction and can be added to the Continuous Delivery workflow.

Reproducibility is important because the workflow may be executed multiple times for a same version. Therefore, it is important that the outcome of the workflow be identical at each execution. In particular, this means that CPO performance evaluation does not depend on non-deterministic behaviors in the ML workflow. This has been achieved by controlling random seeds of all algorithms involved in the workflow that make non-deterministic decisions: LightGBM, stratified K-fold algorithm, BayesianOptimization [12], features importance assessment algorithm.

3.4   Machine Learning Workflow

This section introduces the different steps involved in the Machine Learning workflow:

- Training dataset preparation:
  - Target definition: For each test problem, compute the speedup between the *MultiPoint* and *Restart* heuristics to assess their relative performance. These speedups result from the comparison of two campaigns performed on our CPOPIP platform. Each campaign executes a single worker that is configured either with *MultiPoint* or *Restart* as the search heuristic. The target for the classification problem corresponds to the winning heuristics

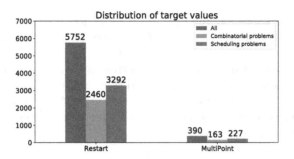

Fig. 1. Number of problems for which each heuristic (*Restart* or *MultiPoint*) outperforms the other, detailed by problem type.

(0 for *Restart*, 1 for *MultiPoint*). The distribution of target values plotted in Fig. 1 illustrates that the two classes are imbalanced. The actual speedup value will be used for weighting training samples.

- Compute all features listed in Table 1 for all problems in our dataset.
- Perform the stratified split, using 60% as training/validation dataset and keeping the other 40% aside for testing (as described in Sect. 3.2).
- Model training using LightGBM algorithm:
  - As detailed in Sect. 3.1, algorithm selection has been formulated as a binary classification problem. In that case, cross-entropy is the usual loss function for training.
  - Two-folds cross-validation is used for training. Each fold takes 50% of the input dataset for training (that is 30% of all problems) and the other half for validation. LightGBM training is stopped when no progress is observed for the cross-entropy loss computed on validation data for 50 iterations. The final score of trained models is this out-of-fold (OOF) cross-entropy score.
- Hyper-parameter optimization (HPO): HPO aims at exploring the space of values for the training algorithm parameters in order to identify a configuration for these parameters that optimizes a selected performance metric. At each HPO iteration, 10 runs (changing seed at each run) are executed to account for the randomness incurred by the `feature_fraction` and `bagging_fraction` LightGBM parameters (which control sampling of features and data points). Each run performs a complete two-fold cross-validation training. The OOF cross-entropy loss averaged over all these runs is the actual performance metric that is optimized by HPO. The final output is an assignment of values for selected parameters. In our experiments, HPO is using the BayesianOptimization library [12] and is invoked twice in the Machine Learning workflow:
  1. To identify a good initial configuration for the following list of LightGBM parameters: `max_depth`, `max_bin`, `feature_fraction`, `bagging_fraction`, `bagging_freq`, `min_data_in_leaf`, `lambda_l1` and `lambda_l2`. A detailed documentation of these parameters can be found

in the on-line LightGBM documentation [11]. Features importance will be evaluated based on the configuration of parameters identified after this first HPO round.

2. To select the number of features to keep for training, along with an updated configuration for the above list of LightGBM parameters. Selecting a good subset of features reduces the risk of overfitting and helps generalization.

– Features importance assessment: The purpose of evaluating features importance is to rank features based on their relevance for the training task. We used the *permutation importance* method [1] for ranking features. In the second round of HPO, the number of features to keep in this sorted list is one of the hyper parameter to optimize.

– Selection and serialization of the model to embed in CPO: The LightGBM model with the best OOF score is selected as the final predictive model to be embedded in CPO. This model is serialized as C++ data structures in a header source file that is added to CPO source code.

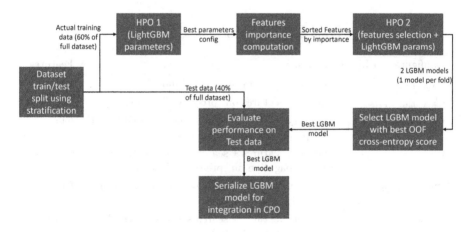

Fig. 2. Training workflow

The output of this Machine Learning workflow is the serialized LightGBM trained model. Figure 2 illustrates how these different steps are sequenced.

3.5   Integration in CPO and Final Performance Evaluation

The CPO solver integrates the features computation code along with the trained model. At runtime, features are computed for the current input problem so that the predictive model can return a score. This score is used as a ratio and drives the strategy to assign heuristics (or *SearchType*) to the different workers involved in search (the number of workers depends on the runtime environment and can be set by the user).

When evaluating the performance of CPO using the predictive model, the overhead introduced by computing features and evaluating prediction for heuristics selection is accounted for in the overall solve time.

The final performance of this problem-specific heuristics selection strategy is assessed by executing two benchmarking campaigns configured with 4 and 8 workers respectively. It is then compared to the benchmarks performed for the regular CPO that performs heuristics selection based on a hard-coded strategy.

4 Experimental Results

4.1 Experimental Setup and Features Sets

In order to explore the relevance of the different categories of features, the configurations detailed in Table 2 will be used in our experiments. All these configurations have a common subset of *mandatory features* that group all features except those related to search space size. As discussed in Sect. 2.2, these features may be computed either on the original model, or on the "pre-solved" model.

Table 2. Configurations for features set

Configuration Id	Description	Features
$C_{\text{Mandatory,Orig}}$	Mandatory features on original model	222
$C_{\text{Mandatory,Presol}}$	Mandatory features on "pre-solved" model	222
$C_{\text{BeforeProp,Orig}}$	Mandatory features + search-space size BEFORE initial propagation on original model	645
$C_{\text{BeforeProp,Presol}}$	Mandatory features + search-space size BEFORE initial propagation on "pre-solved" model	645
$C_{\text{AfterProp,Orig}}$	Mandatory features + search-space size AFTER initial propagation on original model	645
$C_{\text{AfterProp,Presol}}$	Mandatory features + search-space size AFTER initial propagation on "pre-solved" model	645
$C_{\text{Bef\&AftProp,Orig}}$	Mandatory features + search-space size BEFORE and AFTER initial prop. on original model	1068
$C_{\text{Bef\&AftProp,Presol}}$	Mandatory features + search-space size BEFORE and AFTER initial prop. on "pre-solved" model	1068

The next two sections present the experimental results for the training workflow (to identify features and parameters value that result in best performance), and for benchmarking CPO extended with ML.

We will also look at the details of feature importance assessment to identify some of the most relevant features.

4.2 Training Workflow Results

Table 3 summarizes the outcome of the LightGBM training experiments. For each configuration of features set, the number of features optimized by the second round of HPO along with the corresponding averaged cross-entropy loss is reported (at each iteration, HPO performs 10 training runs). Then, using parameters values returned by HPO, 50 additional runs are performed to monitor the evolution of selected metrics during training. In addition to cross-entropy loss, ROC AUC (the area under the receiver operating characteristics curve) is reported for all configurations of features sets. ROC AUC is a metric that provides insights about the sensitivity of binary classifiers to threshold selection for separating positive and negative examples. This information is useful for imbalanced datasets, like in our case.

Table 3. Number of selected features and Out-Of-Fold metrics associated with best parameters value found by HPO, for each configuration of features set.

Configuration Id	Total number of features	Number of selected features	HPO best OOF cross-entropy loss (10 runs)	OOF cross-entropy loss (50 runs)	OOF ROC AUC ROC AUC
$C_{\text{Mandatory,Orig}}$	222	20	0.05641 [8]	0.05641	0.86491
$C_{\text{Mandatory,Presol}}$	222	55	0.05464 [5]	0.05464	0.87390
$C_{\text{BeforeProp,Orig}}$	645	120	0.05387 [2]	0.05387	0.87398
$C_{\text{BeforeProp,Presol}}$	645	296	**0.05369 [1]**	**0.05381**	**0.87847**
$C_{\text{AfterProp,Orig}}$	645	118	0.05415 [3]	0.05415	0.86594
$C_{\text{AfterProp,Presol}}$	645	344	0.05495 [7]	0.05511	0.8803
$C_{\text{Bef\&AftProp,Orig}}$	1068	314	0.05420 [4]	0.05420	0.87384
$C_{\text{Bef\&AftProp,Presol}}$	1068	536	0.05467 [6]	0.05468	0.8754

Table 3 indicates that the best performance is achieved by selecting the top 296 features from the $C_{\text{BeforeProp,Presol}}$ configuration (after sorting by importance). The corresponding trained predictive model is then serialized and embedded in CPO to select heuristics to be used by each worker. Experimental results for this extended CPO version are presented in the next section.

4.3 Benchmarking Results for CPO with ML

A CPO executable embedding the trained predictive model is built and evaluated on our CPOPIP test platform. The baseline for all experiments in this section corresponds to a regular CPO runtime. These regular and *extended with ML* CPO versions differ only in the ML specific logic. All figures reported in this section correspond to solve speedup compared to the baseline.

In order to evaluate performance, a Virtual Best Solver has also been designed by evaluating CPO performance for all possible configurations for assigning

Table 4. Performance of CPO with ML vs Virtual Best Solver **over all problems in benchmark**, by problem type, for 4 and 8 workers

		4 Workers		8 Workers	
		Family geometric av.	All tests geometric av.	Family geometric av.	All tests geometric av.
Combinatorial problems	Virtual Best	1.20	1.23	1.45	1.44
	CPO with ML	**1.08**	**1.11**	**1.28**	**1.29**
Scheduling problems	Virtual Best	1.34	1.27	1.47	1.40
	CPO with ML	**1.09**	**1.06**	**1.12**	**1.07**
Overall	Virtual Best	1.27	1.26	1.46	1.41
	CPO with ML	**1.09**	**1.08**	**1.19**	**1.13**

Restart and *MultiPoint* heuristics to the pool of workers (of size 4 or 8). Then, for each problem, the best speedup among all configurations is kept. The performance of this *Oracle* is then compared to the baseline, providing an upper bound on the best performance that our ML approach can achieve.

Table 5. Performance of CPO with ML vs Virtual Best Solver **for TEST problems only**, by problem type, for 4 and 8 workers

		4 Workers		8 Workers	
		Family geometric av.	All tests geometric av.	Family geometric av.	All tests geometric av.
Combinatorial problems	Virtual Best	1.19	1.24	1.42	1.40
	CPO with ML	**1.13**	**1.14**	**1.30**	**1.31**
Scheduling problems	Virtual Best	1.37	1.27	1.50	1.40
	CPO with ML	**1.08**	**1.05**	**1.09**	**1.06**
Overall	Virtual Best	1.28	1.26	1.46	1.40
	CPO with ML	**1.10**	**1.07**	**1.18**	**1.12**

Tables 4 and 5 present the results, separating Combinatorial and Scheduling problems. The first table presents performance results on the full set of benchmark problems, while only problems from the *Test* dataset (that have not been involved in the training process at all) are considered in the second table. One can notice that the drop in performance between the two setups is minor, which makes us confident in the robustness of the approach (in particular since some families in the test data are not seen at all during training because of their small size).

4.4 Features Importance Analysis

A side question of this work is related to the analysis of the most relevant features that are exploited by trained models. Investigating which metrics computed on an input problem correlate better with the learning target can help engineering new features and provide some insights to design new heuristics.

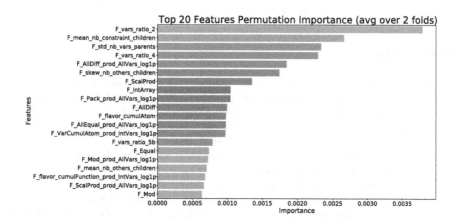

Fig. 3. Top 20 features ranked by permutation importance.

Figure 3 plots the top 20 features ranked by permutation importance for the best predictive model. Interestingly, 5 of the top 6 features in this chart correspond to global structural properties of the directed graph associated with the problem that are independent of constraints types. For instance, the top feature is the proportion of nodes in the graph that are variables nodes, the second one is the mean number of children for constraint nodes...

5   Concluding Remarks

The aim of the work presented in this paper is to extend the CP Optimizer product with some automated heuristics selection capabilities driven by a few hundreds of metrics computed on input problems. A predictive model trained from an extensive repository of benchmark problems supports the heuristics selection strategy. Beyond the challenge of improving the performance of the solver using a Machine Learning approach, embedding a predictive model in a product entails several technical constraints. In particular, the lifecycle of embedded models must be managed so that they are updated when needed. This implies that all decisions involved in the workflow for creating the final predictive model are clearly identified and automated. This involves in particular preparing the training data by benchmarking individual heuristics, assessing and selecting the best configuration of features by performing multiple HPO sessions, or including the serialized final model in the product code.

This work confirms the benefit and the feasibility of using ML methods for improving a productized solver. Our next step is to extend this effort to automatically configure more parameters controlling CPO internal heuristics. Our goal was a real challenge because of the difficulty to assemble a vast dataset of diverse problems and not limit the domain of application to a few classes of problem. For this reason, adversarial strategies like in [14] are promising approaches that we also intend to investigate in our work.

References

1. Altmann, A., Tolosi, L., Sander, O., Lengauer, T.: Permutation importance: a corrected feature importance measure. Bioinformatics (Oxford, England) **26**, 1340–1347 (2010). https://doi.org/10.1093/bioinformatics/btq134
2. Bengio, Y., Lodi, A., Prouvost, A.: Machine learning for combinatorial optimization: a methodological tour d'horizon. Eur. J. Oper. Res. **290**, 405–421 (2021)
3. Hottung, A., Tierney, K.: Neural large neighborhood search for the capacitated vehicle routing problem. arXiv abs/1911.09539 (2020)
4. Karimi-Mamaghan, M., Mohammadi, M., Meyer, P., Karimi-Mamaghan, A.M., Talbi, E.G.: Machine learning at the service of meta-heuristics for solving combinatorial optimization problems: a state-of-the-art. Eur. J. Oper. Res. **296**(2), 393–422 (2022)
5. Ke, G., et al.: LightGBM: a highly efficient gradient boosting decision tree. In: Advances in Neural Information Processing Systems, vol. 30, pp. 3146–3154 (2017)
6. Khalil, E.B., Morris, C., Lodi, A.: MIP-GNN: a data-driven framework for guiding combinatorial solvers. In: Proceedings of the AAAI Conference on Artificial Intelligence, vol. 36, no. 9, pp. 10219–10227 (2022). https://doi.org/10.1609/aaai.v36i9.21262. https://ojs.aaai.org/index.php/AAAI/article/view/21262
7. Kool, W., van Hoof, H., Welling, M.: Attention, learn to solve routing problems! In: ICLR (2019)
8. Kotary, J., Fioretto, F., Hentenryck, P.V., Wilder, B.: End-to-end constrained optimization learning: a survey. arXiv abs/2103.16378 (2021)
9. Kruber, M., Lübbecke, M.E., Parmentier, A.: Learning when to use a decomposition. In: CPAIOR (2017)
10. Laborie, P., Rogerie, J., Shaw, P., Vilím, P.: IBM ILOG CP optimizer for scheduling. Constraints **23**(2), 210–250 (2018). https://doi.org/10.1007/s10601-018-9281-x
11. Microsoft: LightGBM documentation. https://lightgbm.readthedocs.io (2021)
12. Nogueira, F.: Bayesian Optimization: open source constrained global optimization tool for Python (2014). https://github.com/fmfn/BayesianOptimization
13. Smith-Miles, K., Lopes, L.: Measuring instance difficulty for combinatorial optimization problems. Comput. Oper. Res. **39**, 875–889 (2012)
14. Tang, K., Liu, S., Yang, P., Yao, X.: Few-shots parallel algorithm portfolio construction via co-evolution. IEEE Trans. Evol. Comput. **25**(3), 595–607 (2021)
15. Xu, L., Hutter, F., Hoos, H.H., Leyton-Brown, K.: SATzilla: portfolio-based algorithm selection for SAT. J. Artif. Intell. Res. **32**, 565–606 (2008)

Model-Based Feature Selection for Neural Networks: A Mixed-Integer Programming Approach

Shudian Zhao[1], Calvin Tsay[2], and Jan Kronqvist[1]([⊠])

[1] Optimization and Systems Theory, Department of Mathematics,
KTH Royal Institute of Technology, Stockholm, Sweden
`jankr@kth.se`
[2] Department of Computing, Imperial College London, London, UK

Abstract. In this work, we develop a novel input feature selection framework for ReLU-based deep neural networks (DNNs), which builds upon a mixed-integer optimization approach. While the method is generally applicable to various classification tasks, we focus on finding input features for image classification for clarity of presentation. The idea is to use a trained DNN, or an ensemble of trained DNNs, to identify the salient input features. The input feature selection is formulated as a sequence of mixed-integer linear programming (MILP) problems that find sets of sparse inputs that maximize the classification confidence of each category. These "inverse" problems are regularized by the number of inputs selected for each category and by distribution constraints. Numerical results on the well-known MNIST and FashionMNIST datasets show that the proposed input feature selection allows us to drastically reduce the size of the input to ∼15% while maintaining a good classification accuracy. This allows us to design DNNs with significantly fewer connections, reducing computational effort and producing DNNs that are more robust towards adversarial attacks.

Keywords: Mixed-integer programming · Deep neural networks · Feature selection · Sparse DNNs · Model reduction

1 Introduction

Over the years, there has been an active interest in algorithms for training sparse deep neural networks (DNNs) or sparsifying trained DNNs. By sparsifying a DNN we mean removing some connections (parameters) in the network, which can be done by setting the corresponding weights to zero. Examples of algorithms for sparsifying or training sparse networks include, dropout methods [13,15,31], optimal/combinatorial brain surgeon [12,36], optimal brain damage [19], and regularization based methods [21,23,33]. Carefully sparsifying the network, *i.e.,* reducing the number of parameters wisely, has shown to reduce over-fitting and improve overall generalizability [13,18]. This paper focuses on feature selection

© The Author(s), under exclusive license to Springer Nature Switzerland AG 2023
M. Sellmann and K. Tierney (Eds.): LION 2023, LNCS 14286, pp. 223–238, 2023.
https://doi.org/10.1007/978-3-031-44505-7_16

for DNNs, which can also be interpreted as "sparsifying" the first/input layer, and we show that we can significantly reduce the number of parameters, *i.e.*, non-zero weights, while keeping a good accuracy. Throughout the paper we focus on image classification, but the framework is general.

This work focuses on finding the salient input features for classification using a DNN. We hypothesize that the number of inputs to DNNs for classification can often be greatly reduced by a "smart" choice of input features while keeping a good accuracy (*i.e.*, feature selection). We build the hypothesis on the assumption that not all inputs, or pixels, will be equally important. Reducing the number of inputs/parameters has the potential to: i) reduce over-fitting, ii) give more robust DNNs that are less sensitive for adversarial attacks (fewer degrees of freedom for the attacker), and iii) reduce computational complexity both in training and evaluating the resulting classifier (fewer weights to determine and fewer computational operations to evaluate the outputs). The first two are classical focus areas within artificial intelligence (AI), and the third is becoming more important with an increasing interest in so-called green AI [26]. Most strategies for feature selection can be grouped as either *filter* methods, which examine the data, *e.g.*, for correlation, and *wrapper* methods, which amount to a guided search over candidate models [4,20]. Feature selection can also be incorporated directly into model training using *embedded* methods, *e.g.*, regularization. This paper and the numerical results are intended as a proof of concept to demonstrate mixed-integer linear programming (MILP) as an alternative technology for extracting the importance of input features from DNNs.

Input feature selection is an active research area, *e.g.*, see the review papers [8,37], and a detailed comparison to state-of-the-art methods is not within the scope of this paper.

Our proposed method leverages trained models that achieve desirable performance, and attempts to select a feature set that replicates the performance using mixed-integer programming. We build on the idea that, given a relatively well-trained DNN, we can analyze the DNN in an inverse fashion to derive information about the inputs. Specifically, to determine the most important inputs, or pixels in the case of image classification, for a given label, we solve an optimization problem that maximizes the classification confidence of the label with a cardinality constraint on the number of non-zero inputs to the DNN. We consider this as an "inverse problem", as the goal is to determine the DNN inputs from the output. We additionally propose some input distribution constraints to make the input feature selection less sensitive to errors in the input-output mapping of the DNN. We only consider DNNs with the rectified linear unit (ReLU) activation function, as it enables the input feature selection problem to be formulated as a MILP problem [7,22]. However, the framework can be easily generalized to CNN architectures and other MIP representable activation functions, e.g., max pooling and leaky ReLU.

Optimizing over trained ReLU-based DNNs has been an active research topic in recent years [14], and has a wide variety of applications including verification [2,22,28], lossless compression [27], and surrogate model optimization [10,35]. There even exists software, such as OMLT [3], for directly incorporating ReLU

DNNs into general optimization models. Optimizing over a trained ReLU-based DNN through the MILP encoding is not a trivial task, but significant progress has been made in terms of strong formulations [1, 16, 29], solution methods [5, 25], and techniques for deriving strong valid inequalities [1, 2]. In combination with the remarkable performance of state-of-the-art MILP solvers, optimization over DNNs appears computationally tractable (at least for moderate size DNNs). This work builds upon recent optimization advancements, as reliably optimizing over ReLU-based DNNs is a key component in the proposed method.

The paper is structured as follows. Section 2 first describes the MILP problem to determine which inputs maximize the classification confidence. Some enhancements for the input selection are presented, and the complete input selection algorithm is presented in Sect. 2.4. Numerical results are presented in Sect. 3, where we show that we can obtain a good accuracy when downsizing the input to 15% by the proposed algorithm, and that the resulting DNNs are more robust towards adversarial attacks in the ℓ_∞ sense. Section 4 provides some conclusions.

2 Input Feature Selection Algorithm

Our feature selection strategy is based on the idea of determining a small optimal subset of inputs, or pixels for the case of image classification, that are allowed to take non-zero values to maximize the classification confidence for each label using a pre-trained DNN. By combining the optimal subsets for each label, we can determine a set of salient input features. These input features can be considered as the most important for the given DNN, but we note that the DNN might not offer a perfect input-output mapping. To mitigate the impact of model errors, we propose a technique of using input distribution constraints to ensure that the selected input features are to some extent distributed over the input space. This framework could easily be extended to use optimization over an ensemble DNN model [32] for input selection, where the inputs would be selected such that the ensemble classification confidence is maximized. While using DNN ensembles can further mitigate the effect of errors in individual DNNs, our initial tests did not indicate clear advantages of using DNN ensembles for this purpose.

Here we focus on fully connected DNNs that classify grayscale images into 10 categories. While this setting is limited, the proposed method is applicable to classification of RGB images, and other classification problems in general. The input features are scaled between 0 and 1, with 255 (white) corresponding to 1 and 0 remaining black. We start by briefly reviewing MILP encoding of DNNs in the next subsection, and continue with more details on the input feature selection in the following subsections.

2.1   Encoding DNNs as MILPs

In a fully connected ReLU-based neural network, the l-th layer with input x^l and output x^{l+1} is described as

$$x^{l+1} = \max\{0, W^l x^l + b^l\},$$

where $W^l \in \mathbb{R}^{n_{l+1} \times n_l}$ is the weight matrix and $b^l \in \mathbb{R}^{n_{l+1}}$ is the bias vector.

The input-output mapping of the ReLU activation function is given by a piece-wise linear function, and is mixed-integer representable [30]. There are different formulations for encoding the ReLU activation function using MILP, where the big-M formulation [7,22] was the first presented MILP encoding and remains a common approach. For the i-th ReLU node at a fully connected layer with input x^l, the big-M formulation for input-output relation is given by

$$
\begin{aligned}
(w_i^l)^\top x^l + b_i^l &\leq x_i^{l+1}, \\
(w_i^l)^\top x^l + b_i^l - (1-\sigma)LB_i^{l+1} &\geq x_i^{l+1}, \\
x_i^{l+1} &\leq \sigma UB_i^{l+1}, \\
\sigma \in \{0,1\}, \ x_i^{l+1} &\geq 0,
\end{aligned}
\tag{1}
$$

where w_i^l is the i-th row vector of W^l, b_i^l is the i-th entry of b^l, LB_i^{l+1} and UB_i^{l+1} are upper and lower bounds on the pre-activation function over x_i^{l+1}, such that $LB_i^{l+1} \leq (w_i^l)^\top x^l + b_i^l \leq UB_i^{l+1}$.

The big-M formulation is elegant in its simplicity, but it is known to have a weak continuous relaxation which may require the exploration of a huge number of branch-and-bound nodes in order to solve the problem [1]. Anderson et al. [1] presented a so-called extended convex hull formulation, which gives the strongest valid convex relaxation of each individual node, and a non-extended convex hull formulation. Even though the convex hull is the strongest formulation for each individual node, it does not in general give the convex hull of the full input-output mapping of the DNN. Furthermore, the convex hull formulation results in a large problem formulation that can be computationally difficult to work with. The class of partition-based, or P-split, formulations, was proposed as an alternative formulation with a stronger continuous relaxation than big-M and computationally cheaper than the convex hull [16,29]. Computational results in [17,29] show that the partition-based formulation often gives significant speed-ups compared to the big-M or convex hull formulations. Here, we do not focus on the computational efficiency of optimizing over ReLU-based DNNs, and, for the sake of clarity, we use the simpler big-M formulation (1). In fact, for the problems considered in this work, the computational time to solve the MILP problems did not represent a limiting factor (big-M tends to actually perform relatively well for simple optimization problems). But, alternative/stronger formulations could directly be used within our framework.

2.2   The Optimal Sparse Input Features (OSIF) Problem

With the MILP encoding of the DNN, we can rigorously analyze the DNN and find extreme points in the input-output mapping. Recently Kronqvist et al. [17] illustrated a similar optimal sparse input features (OSIF) problem. This problem aims to maximize the probability of at most $\aleph$ non-zero input features being classified with a certain label i for a given trained DNN. The problem is formed by encoding the ReLU activation function for each hidden node by

MILP. Instead of the softmax function, the objective function is x_i^L, where x^L is the output vector, thus maximizing the classification confidence of label i.

Using the big-M formulation, the OSIF problem can be stated as

$$\max x_i^L \tag{2a}$$
$$\text{s.t. } W^l x^l + b^l \le x^{l+1}, \ \forall l \in [L-1], \tag{2b}$$
$$W^l x^l + b^l - \text{diag}(LB^{l+1})(1 - \sigma^{l+1}) \ge x^{l+1}, \ \forall l \in [L-1], \tag{2c}$$
$$x^l \le \text{diag}(UB^l)\sigma^l, \ \sigma^l \in \{0,1\}^{n_l}, \forall l \in \{2,\dots,L\}, \tag{2d}$$
$$x^L = W^{L-1} x^{L-1} + b^{L-1}, x^L \in \mathbb{R}^{10} \tag{2e}$$
$$x^l \in \mathbb{R}_+^{n_l}, \forall l \in [L-1], \tag{2f}$$
$$y \ge x^1, y \in \{0,1\}^{n_1}, \tag{2g}$$
$$1^\top y \le \aleph_0, \tag{2h}$$

where n_1 is the size of the input data, $x^L \in \mathbb{R}^{10}$ is the output, L is the number of layers, LB^l and UB^l are the bounds on x^l, 1 denotes the all-ones vector, $\text{diag}(\cdot)$ denote the matrix with $\cdot$ on the diagonal and 0 on other entries. Equation (2g) and (2h) describe the cardinality constraint $\|x^1\|_0 \le \aleph_0$, which limits the number of selected inputs. Figure 1 shows some example results of solving problem (2) with $\aleph \in \{10, 20\}$ and class $i \in \{0, 1\}$. Given a larger cardinality number $\aleph$, the latter is more visually recognizable from the selected pixels.

2.3   Input Distribution Constraints

To extract information across the whole input image, we propose to add constraints to force the selected pixels to be distributed evenly across some predefined partitioning of the input space. Forcing the selected pixels to be more spread-out may also mitigate the effect of inaccuracy of the DNN used in the OSIF problem, *e.g.,* by preventing a small area to be given overly high priority. There are various ways to partition the input variables. In this paper, we focus on image classification problems with square images as input. Furthermore, we assume that the images are roughly centered. Therefore, we denote each input variable as matrix $X \in \mathbb{R}^{n \times n}$ and define the partition as k^2 submatrices of equal size, *i.e.,* $X^{ij} \in \mathbb{R}^{\frac{n}{k} \times \frac{n}{k}}$ for $i, j \in [k]$. For instance, given n is even and $k = 2$, a natural partition of the matrix is

$$X = \begin{pmatrix} X^{11} & X^{12} \\ X^{21} & X^{22} \end{pmatrix}.$$

In this way, we denote $x^1 := \text{vec}(X)$ the input data and $n_1 := n^2$ the size of the input data, then I_{ij} is the index set for entries mapped from X_{ij}

$$I_{ij} = \{(i_1 - 1)n + i_2 \mid i_1 \in \{(i-1)\frac{n}{k} + 1, \dots, i\frac{n}{k} - 1\}, i_2 \in \{(j-1)\frac{n}{k} + 1, \dots, j\frac{n}{k} - 1\}\}.$$

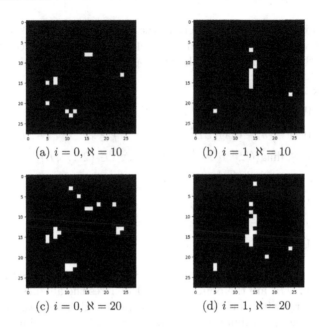

(a) $i = 0$, $\aleph = 10$ (b) $i = 1$, $\aleph = 10$

(c) $i = 0$, $\aleph = 20$ (d) $i = 1$, $\aleph = 20$

Fig. 1. The results of OSIF for class 0 and 1 on MNIST ($\aleph_0 \in \{10, 20\}$). Note that the selected pixels are white.

We denote the collection of index sets for the partition as $\mathcal{I} := \{I_{i,j}\}_{\forall i,j \in [k]}$. To limit the number of pixels selected from each box for each category we add the following constraints

$$\lfloor \frac{\aleph_0}{k^2} \rfloor \leq \sum_{i \in I_t} y_i \leq \lceil \frac{\aleph_0}{k^2} \rceil, \forall I_t \in \mathcal{I}, \tag{3}$$

The constraint (3) forces the pixels to spread evenly between all partitions, while allowing some to contain one more selected pixel for each category.

To illustrate on the impact on the distribution constraints, Fig. 2 compares the selected pixels for MNIST with $k \in \{1, 2\}$. Compared to the result without distribution constraints (equivalent to $k = 1$), pixels selected with $k = 2$ are more scattered over the whole image and are more likely to identify generalizable distinguishing features of the full input data, assuming the dataset has been pre-processed for unused areas of the images matrices.

2.4   Controlling the Number of Selected Features

Repeatedly solving the OSIF problem (2) for each label, *i.e.*, $i \in \{0, \ldots, 9\}$, and taking the union of all selected pixels does not give us full control of the total number of selected pixels. Specifically, some of the pixels can be selected by the OSIF problem for multiple classes, resulting in fewer combined pixels (the union of selected subsets) than an initial target.

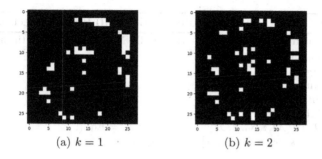

(a) $k = 1$ (b) $k = 2$

Fig. 2. Optimal input features for MNIST ($\aleph = 50$)

Therefore, we present an approach to control the number of selected inputs, which we use in the proposed MILP-based feature selection algorithm. The main idea is to allow freedom over features already selected by previous classes in the current OSIF problem and adjust the cardinality constraint (2h) to

$$\sum_{i \in [n_1] \setminus J} y_i \leq \aleph_0, \tag{4}$$

where J is the index set for input features selected by previous models. Similarly, constraints (3) are adjusted as

$$\lfloor \frac{\aleph_0}{k^2} \rfloor \leq \sum_{i \in I_t \setminus J} y_i \leq \lceil \frac{\aleph_0}{k^2} \rceil, \forall I_t \in \mathcal{I}. \tag{5}$$

Finally, we formulate the OSIF problem with input distribution and total number control as

$$OSIF(\mathcal{M}, i, \aleph_0, \mathcal{I}, J) = \operatorname{argmax} \{x_i^L \mid (2\mathrm{b})\text{-}(2\mathrm{g}), (4), (5)\}. \tag{6}$$

Based on the described techniques, we introduce the input feature selection algorithm, which is presented as pseudo code in Algorithm 1.

3 Computational Results

In this paper, we focus on image classification problems for the MNIST [6] and FashionMNIST [34] datasets. Both datasets consist of a training set of 60,000 examples and a test set of 10,000 examples. Each sample image in both datasets is a 28×28 grayscale image associated with labels from 10 classes. MNIST is the dataset of handwritten single digits between 0 and 9. FashionMNIST is a dataset of Zalando's article images with 10 categories of clothing. There is one fundamental difference between the two data sets, besides FashionMNIST being a somewhat more challenging data set for classification. In MNIST there are significantly more pixels that do not change in any of the training images, or

---

Algorithm 1: MILP-based feature selection (MILP-based selection)

Data: the number of features $\aleph$, a trained DNN $\mathcal{M}$, a matrix partition set $\mathcal{I}$,
   class set $\mathcal{C}$;
Input: $J \leftarrow \emptyset$;
Output: Index set J;
$\aleph_0 \leftarrow \aleph/10$;
for $i \in \mathcal{C}$ **do**
   $\quad x \leftarrow OSIF(\mathcal{M}, i, \aleph_0, \mathcal{I}, J)$;   # Eq. (6)
   $\quad J \leftarrow J \cup \{s \mid x_s^1 \neq 0, s \in [n_1]\}$;
end

---

only change in a few images, compared to FashionMNIST. The presence of such "dead" pixels is an important consideration for input feature selection.

Image preprocessing and training DNNs are implemented in PyTorch [24], and the MILP problems are modeled and solved by Gurobi through the Python API [11]. We trained each DNN with 2 hidden layers of the same size.

3.1   Accuracy of DNNs with Sparse Input Features

The goal is to illustrate that Algorithm 1 can successfully identify low-dimensional salient input features. We chose to focus on small DNNs, as DNN 2×20 can already achieve an accuracy of 95.7% (resp. 86.3%) for MNIST (resp. FashionMNIST) and larger DNNs did not give clear improvements for the input selection. For such models, the computational cost of solving the MILPs is low[1].

Table 1 present the accuracies of DNNs with sparse input features on MNIST and FashionMNIST. It is possible to obtain a much higher accuracy by considering a more moderate input reduction (about 0.5% accuracy drop with 200–300 inputs), but this defeats the idea of finding low dimensional salient features. For grayscale input image of 28×28, we select at most 15% input features and present the results with $\aleph \in \{50, 100\}$.

MILP-Based Feature Selection. Table 1 compares the accuracy of classification models with different architectures, *i.e.*, with 2×20 vs. with 2×40. We select sparse input features by Algorithm 1 with OSIF models 2×10 and 2×20. Since the distribution constraints are supposed to force at least one pixel selected in each submatrix, we select partition number $k \in \{1, 2\}$ and $k \in \{1, 2, 3\}$ for instances with $\aleph = 50$ and $\aleph = 100$ respectively.

First, we investigate the effect of the distribution constraints. Table 1 show that the accuracy increases when adding the distribution constraints (*i.e.*, from $k = 1$ to $k = 2$) for $\aleph \in \{50, 100\}$. However, the distribution constraints become

---

[1] On a laptop with a 10-core CPU, Gurobi can solve instances with $\aleph = 100$ and a DNN of 2×20 under 15 s. However, previous research [1,29] has shown that significant speed-ups can be obtained by using a more advanced MILP approach.

Table 1. Accuracy of DNNs of different architectures with sparse input features selected by Algorithm 1 on MNIST and FashionMNIST

DNNs of 2×20 hidden layers for MNIST dataset

$\aleph$	OSIF Model	k	Acc.	OSIF Model	k	Acc.
50	2×10	1	80.6%	2×20	1	80.5%
	2×10	2	**85.3%**	2×20	2	**86.6%**
100	2×10	1	88.8%	2×20	1	89.0%
	2×10	2	**91.2%**	2×20	2	**90.6%**
	2×10	3	89.3%	2×20	3	89.3%

DNNs of 2×40 hidden layers for MNIST dataset

$\aleph$	OSIF Model	k	Acc.	OSIF Model	k	Acc.
50	2×10	1	83.1%	2×20	1	82.7%
	2×10	2	**87.6%**	2×20	2	**89.2%**
100	2×10	1	91.4%	2×20	1	91.4%
	2×10	2	**93.4%**	2×20	2	**92.9%**
	2×10	3	92.3%	2×20	3	91.8%

DNNs of 2×20 hidden layers for FashionMNIST dataset

$\aleph$	OSIF Model	k	Acc.	OSIF Model	k	Acc.
50	2×10	1	76.6%	2×20	1	77.2%
	2×10	2	**77.0%**	2×20	2	**77.9%**
100	2×10	1	81.1%	2×20	1	81.3%
	2×10	2	**82.3%**	2×20	2	81.8%
	2×10	3	82.2%	2×20	3	**82.1%**

DNNs of 2×40 hidden layers for FashionMNIST dataset

$\aleph$	OSIF Model	k	Acc.	OSIF Model	k	Acc.
50	2×10	1	78.3%	2×20	1	78.6%
	2×10	2	**79.4%**	2×20	2	**79.6%**
100	2×10	1	82.4%	2×20	1	82.6%
	2×10	2	83.1%	2×20	2	83.2%
	2×10	3	**83.7%**	2×20	3	**84.0%**

less important as the number of selected features $\aleph$ increases; the best k for MNIST and FashionMNIST varies for $\aleph = 100$ (noting that the choice of k also affects accuracy less). For MNIST with $\aleph = 100$, the accuracy of instances drops slightly as the k increases from 2 to 3, while using input features selected with $k = 3$ leads to slightly higher accuracy for FashionMNIST with $\aleph = 100$. One reason behind this difference could be that input pixels of MNIST and FashionMNIST are activated in different patterns. In MNIST, there are more active pixels in the center, and the peripheral pixels stay inactive over the training data. Given the distribution constraints with $k = 2$ (see Fig. 3b), the selected pixels stay away from the peripheral area. However, when $k = 3$ (see Fig. 3c), more pixels are forced to be chosen from the upper right-hand and bottom left-hand corners. In contrast, the active pixels are more evenly spread across the full input in FashionMNIST. Hence, as k increases (see Fig. 3e and Fig. 3f), the active pixels remain well-covered by the evenly scattered selected pixels.

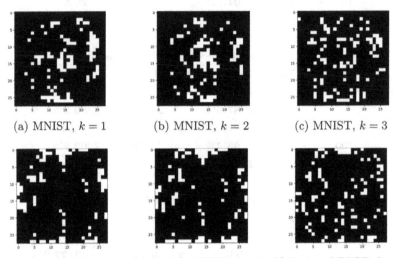

(a) MNIST, $k = 1$ (b) MNIST, $k = 2$ (c) MNIST, $k = 3$

(d) FashionMNIST, $k = 1$ (e) FashionMNIST, $k = 2$ (f) FashionMNIST, $k = 3$

Fig. 3. MNIST and FashionMNIST input features selected by Algorithm 1 with $\aleph = 100$ and $k \in \{1, 2, 3\}$

Table 1 also compares OSIF using different DNN architectures, *i.e.*, 2×10 and 2×20. The accuracy is 94% (resp. 84%) for the former and 96% (resp. 86%) for the latter for MNIST (resp. FashionMNIST). The results show that even using a simple DNN for feature selection using OSIF can produce feature sets that achieve good accuracy, when appropriately large classification models are trained on the selected features. For MNIST, the former model has accuracy at most 2 points worse than the latter model when $\aleph = 50$. When $\aleph = 100$, both models achieve similar levels of performance. As for Fashion, the difference is at most 1 point for $\aleph \in \{50, 100\}$. Hence, we cannot observe a clear difference between the two OSIF DNN models in terms of feature selection quality.

Finally, we would like to make a brief remark on the improvement in accuracy by increasing the size of the DNNs for classification, *i.e.*, from 2×20 to 2×40, for both MNIST and FashionMNIST. Unsurprisingly, using larger DNNs results in overall higher accuracy. More importantly, the performance of the proposed input feature selection seems to be robust toward the final architecture. For both architectures, we observe a similarly slight reduction in classification accuracy related to the reduction in number of input features (pixels).

Comparisons Between Feature Selection Approaches. In this section, we compare the performance of the MILP-based features-selection approach (*i.e.*, Algorithm 1) to some other simple feature-selection approaches. The other feature selection techniques considered in the comparison are random feature selection, data-based feature selection, and DNN weights-based. In the following paragraphs, we briefly describe the feature selection algorithms that are used as reference points for the comparison.

The random feature selection uniformly samples a subset of $\aleph$ input features, and we present the average accuracy with the standard deviation over 5 DNNs trained with input features independently selected by this approach. This approach is included, as it is the simplest approach to down-sample the full input.

The data-based feature selection is conducted in the following way: i) we first calculate the mean value of each input feature over the whole train dataset; ii) the features with the largest $\aleph$ mean are selected. The motivation behind this simple heuristic is that it selects the pixels that are most strongly colored over the training data, *i.e.*, the strongest signals. For example, selecting a pixel uncolored in all images of the training data does not make sense as that input does not contain any information for the training data.

In the DNN weights-based approach, we use the same DNN models as we use in the MILP-based selection, but the inputs are now selected based on the weights of the inputs. For each input, we sum up the absolute values of all the weights from the input to the nodes in the consecutive layer and select the ones with the largest sum. This can be seen as a form of pruning of inputs, and the motivation is that inputs with small, or almost zero, weights should be less important as these inputs have less impact in the DNN.

In Table 2, we compare MILP-based feature selection (*i.e.*, Algorithm 1) with random selection, data-based selection, and weights-based selection. From the table, it can be observed Algorithm 1 achieves the best result with the accuracy of DNNs with sparse input features only 5 points (with $\aleph = 100$) less than the accuracy with full input features. It can be observed that our method has the best overall performance. For MNIST, the random selection has the worst performance, but the data-based selection and weights-based selection achieve a slightly worse performance than our method.

Table 4 compares the performance of features selections on FashionMNIST. The results show a different pattern to MNIST. Our methods still have the best overall performance over different settings by maintaining the accuracy to 84% (resp. 88%) with $\aleph = 100$ for DNNs 2×20 (resp. 2×40), while the accuracy of DNNs with full input features are 86% (resp. 88%). While delivering the worst performance on MNIST, random selection has a very close performance to our method on FashionMNIST. The weights-based selection still lies in third overall, while the data-based selection is much worse than the other three methods (*e.g.*, 58% for the DNN 2×20 with $\aleph = 100$ and 62% for the DNN 2×40 with $\aleph = 100$). Moreover, we run experiments on MNIST with a random 2 pixel wide frame, see Table 3, forming a dataset with redundant inputs. The data-based method can achieve a maximum accuracy of 57% while the other methods can obtain similar performance as on the original dataset. In Table 5, we can observer very similar results on experiments with a random frame on FashionMNIST.

Table 2. Accuracy of DNNs with sparse input features selected by different methods on MNIST

DNN	Feature Selection Approaches					$\aleph = 784$
	$\aleph$	MILP-based	Random	Data-based	Weights-based	
2×20	50	86.6%	$77.4 \pm 2.2\%$	81.3%	80.7%	95.7%
	100	91.2%	$86.0 \pm 2.5\%$	89.2%	89.0%	
2×40	50	89.2%	$76.0 \pm 3.9\%$	85.1%	83.3%	97.1%
	100	93.4%	$90.6 \pm 1.1\%$	92.4%	91.2%	

Table 3. Accuracy of DNNs with sparse input features selected by different methods on MNIST with a two pixel wide random frame

DNN	Feature Selection Approaches					$\aleph = 1024$
	$\aleph$	MILP-based	Random	Data-based	Weights-based	
2×20	50	83.5%	$72.7 \pm 3.4\%$	53.7%	80.8%	95.5%
	100	90.1%	$81.5 \pm 1.5\%$	53.5%	89.8%	
2×40	50	86.3%	$76.0 \pm 6.1\%$	56.7%	82.6%	96.8%
	100	93.2%	$88.1 \pm 1.6\%$	56.4%	91.6%	

Table 4. Accuracy of DNNs with sparse input features selected by different methods on FashionMNIST

DNN	Feature Selection Approaches					$\aleph = 784$
	$\aleph$	MILP-based	Random	Data-based	Weights-based	
2×20	50	77.9%	$77.2 \pm 0.8\%$	49.6%	73.8%	86.3%
	100	82.3%	$80.6 \pm 1.1\%$	58.4%	80.3%	
2×40	50	79.6%	$78.8 \pm 0.4\%$	51.6%	74.6%	87.5%
	100	84.0%	$82.8 \pm 0.4\%$	62.3%	81.7%	

Table 5. Accuracy of DNNs with sparse input features selected by different methods on FashionMNIST with a two pixel wide frame

DNN	Feature Selection Approaches					$\aleph = 1024$
	$\aleph$	MILP-based	Random	Data-based	Weights-based	
2×20	50	79.2%	$75.5 \pm 1.8\%$	49.4%	73.3%	85.9%
	100	82.2%	$79.9 \pm 0.5\%$	59.8%	80.2%	
2×40	50	80.6%	$77.0 \pm 1.8\%$	52.3%	75.2%	87.2%
	100	83.9%	$80.7 \pm 0.9\%$	62.2%	81.9%	

The weights-based selection performs decently on both data sets compared to random and data-based selection. However, based on the results it is clear that MILP-based selection (*i.e.*, Algorithm 1) can extract more knowledge from the DNN model regarding the importance of inputs compared to simply analyzing the weights. The overall performance of MILP-based selection is more stable than other feature selection methods on both datasets.

3.2 Robustness to Adversarial Inputs

The robustness of a trained DNN classifier can also be analyzed using MILP, *e.g.*, in verification or finding adversarial input. We use the minimal distorted adversary as a measure of model robustness x under l_∞ norm [9]. For a given image x_{image}, the minimal adversary problem [29] can be formulated as

$$\min \epsilon$$
$$\text{s.t. (2b)–(2f)}, \tag{7}$$
$$x_i^L \le x_j^L, \|x^1 - x_{\text{image}}\|_\infty \le \epsilon,$$

where i is the true label of image x_{image} and j is an adversarial label. Simply put, problem (7) finds the smallest perturbation, defined by the ℓ_∞ norm, such that the trained DNN erroneously classifies image x_{image} as the adversarial label j. We hypothesize that DNNs trained with fewer (well-selected) features are more robust to such attacks, as there are fewer inputs as degrees of freedom. Furthermore, we note that the robustness of smaller DNNs can be analyzed with significantly less computational effort. Table 6 shows the minimal adversarial distance (mean and standard deviation over 100 instances), defined by (7) for DNNs trained on MNIST and FashionMNIST with MILP-based feature selection. The adversaries are generated for the first 100 instances of the respective test datasets, with adversarial labels selected randomly. Furthermore, we report the mean percentage increase, Δ, in minimal adversarial distance over the 100 instances for the reduced input DNNs compared to the full input. In all cases, reducing the the number of inputs ℵ results in a more robust classifier. For the 2×40 DNN trained on FashionMNIST, reducing the number of inputs from 784 to 50 increases the mean minimal adversarial distance by almost 90%, with a loss in accuracy of $<10\%$.

Table 6. Minimal adversarial distance ϵ for trained DNNs.

DNN	MNIST			FashionMNIST		
	ℵ	ϵ ($\times 10^{-2}$)	Average Δ	ℵ	ϵ ($\times 10^{-2}$)	Average Δ
2×20	50	16.1±7.9	69.9%	50	12.1±7.2	65.4%
	100	14.0±6.1	33.6%	100	11.7±5.8	44.3%
	784	10.1±4.7	–	784	8.9±3.7	–
2×40	50	15.2±7.2	57.2%	50	12.9±7.6	89.1%
	100	12.4±4.6	42.6%	100	11.5±5.5	52.7%
	784	10.4±5.1	–	784	8.4±3.4	–

4   Conclusion

In the paper, we have presented an MILP-based framework using trained DNNs to extract information about salient input features. The proposed algorithm is able to drastically reduce the size of the input by using the input features that are most important for each category according to the DNN, given a regularization on the input size and spread of selected features. The numerical results show that the proposed algorithm is able to efficiently select a small set of features for which a good prediction accuracy can be obtained. The results also show that the proposed input feature selection can improve the robustness toward adversarial attacks.

Acknowledgment. This research is supported by C3.ai Digital Transformation Institute, and Digital Futures.

References

1. Anderson, R., Huchette, J., Ma, W., Tjandraatmadja, C., Vielma, J.P.: Strong mixed-integer programming formulations for trained neural networks. Math. Program. **183**(1), 3–39 (2020)
2. Botoeva, E., Kouvaros, P., Kronqvist, J., Lomuscio, A., Misener, R.: Efficient verification of Relu-based neural networks via dependency analysis. In: Proceedings of the Conference on AAAI Artificial Intelligent, vol. 34, pp. 3291–3299 (2020)
3. Ceccon, F.: OMLT: optimization & machine learning toolkit. J. Mach. Learn. Res. **23**(349), 1–8 (2022)
4. Chandrashekar, G., Sahin, F.: A survey on feature selection methods. Comput. Electr. Eng. **40**(1), 16–28 (2014)
5. De Palma, A., Behl, H.S., Bunel, R., Torr, P.H., Kumar, M.P.: Scaling the convex barrier with sparse dual algorithms (2021). arXiv:2101.05844
6. Deng, L.: The mnist database of handwritten digit images for machine learning research. IEEE Signal Process. Mag. **29**(6), 141–142 (2012)
7. Fischetti, M., Jo, J.: Deep neural networks and mixed integer linear optimization. Constraints **23**(3), 296–309 (2018)
8. Ghojogh, B., et al.: Feature selection and feature extraction in pattern analysis: a literature review (2019). arXiv preprint
9. Goodfellow, I.J., Shlens, J., Szegedy, C.: Explaining and harnessing adversarial examples (2014). arXiv:1412.6572
10. Grimstad, B., Andersson, H.: Relu networks as surrogate models in mixed-integer linear programs. Comput. Chem. Eng. **131**, 106580 (2019)
11. Gurobi Optimization, LLC: Gurobi Optimizer Reference Manual (2022). https://www.gurobi.com
12. Hassibi, B., Stork, D.: Second order derivatives for network pruning: optimal brain surgeon. In: Hanson, S., Cowan, J., Giles, C. (eds.) Proceedings of NIPS 1992, vol. 5 (1992)
13. Hinton, G.E., Srivastava, N., Krizhevsky, A., Sutskever, I., Salakhutdinov, R.R.: Improving neural networks by preventing co-adaptation of feature detectors (2012). arXiv preprint

14. Huchette, J., Muñoz, G., Serra, T., Tsay, C.: When deep learning meets polyhedral theory: A survey (2023). arXiv:2305.00241
15. Kingma, D.P., Salimans, T., Welling, M.: Variational dropout and the local reparameterization trick, vol. 28 (2015)
16. Kronqvist, J., Misener, R., Tsay, C.: Between steps: intermediate relaxations between big-M and convex hull formulations. In: Stuckey, P.J. (ed.) CPAIOR 2021. LNCS, vol. 12735, pp. 299–314. Springer, Cham (2021). https://doi.org/10.1007/978-3-030-78230-6_19
17. Kronqvist, J., Misener, R., Tsay, C.: P-split formulations: A class of intermediate formulations between big-M and convex hull for disjunctive constraints (2022). arXiv:2202.05198
18. Labach, A., Salehinejad, H., Valaee, S.: Survey of dropout methods for deep neural networks (2019). arXiv:1904.13310
19. LeCun, Y., Denker, J., Solla, S.: Optimal brain damage. In: Touretzky, D. (ed.) Proceedings of NIPS 1989, vol. 2 (1989)
20. Li, J., et al.: Feature selection: a data perspective. ACM Comput. Surv. 50(6), 1–45 (2017)
21. Liu, B., Wang, M., Foroosh, H., Tappen, M., Pensky, M.: Sparse convolutional neural networks. In: Proceedings of the IEEE Conference on Computer Vision and Pattern Recognition, pp. 806–814 (2015)
22. Lomuscio, A., Maganti, L.: An approach to reachability analysis for feed-forward ReLU neural networks (2017). arXiv:1706.07351
23. Manngård, M., Kronqvist, J., Böling, J.M.: Structural learning in artificial neural networks using sparse optimization. Neurocomputing 272, 660–667 (2018)
24. Paszke, A., et al.: Pytorch: an imperative style, high-performance deep learning library. In: Proceedings of NeurIPS 2019, pp. 8024–8035. Curran Associates, Inc. (2019)
25. Perakis, G., Tsiourvas, A.: Optimizing objective functions from trained ReLU neural networks via sampling (2022). arXiv:2205.14189
26. Schwartz, R., Dodge, J., Smith, N.A., Etzioni, O.: Green AI. Commun. ACM 63(12), 54–63 (2020)
27. Serra, T., Kumar, A., Ramalingam, S.: Lossless compression of deep neural networks. In: Hebrard, E., Musliu, N. (eds.) CPAIOR 2020. LNCS, vol. 12296, pp. 417–430. Springer, Cham (2020). https://doi.org/10.1007/978-3-030-58942-4_27
28. Tjeng, V., Xiao, K., Tedrake, R.: Evaluating robustness of neural networks with mixed integer programming (2017). arXiv:1711.07356
29. Tsay, C., Kronqvist, J., Thebelt, A., Misener, R.: Partition-based formulations for mixed-integer optimization of trained ReLU neural networks. In: Ranzato, M., Beygelzimer, A., Dauphin, Y., Liang, P., Vaughan, J.W. (eds.) Proceedings of NeurIPS 2021, vol. 34, pp. 3068–3080. Curran Associates, Inc. (2021)
30. Vielma, J.P.: Mixed integer linear programming formulation techniques. SIAM Rev. 57(1), 3–57 (2015)
31. Wan, L., Zeiler, M., Zhang, S., Le Cun, Y., Fergus, R.: Regularization of neural networks using DropConnect. In: Proceedings of the 30th ICML, pp. 1058–1066. PMLR (2013)
32. Wang, K., Lozano, L., Cardonha, C., Bergman, D.: Acceleration techniques for optimization over trained neural network ensembles (2021). arXiv:2112.07007
33. Wen, W., Wu, C., Wang, Y., Chen, Y., Li, H.: Learning structured sparsity in deep neural networks. In: Lee, D., Sugiyama, M., Luxburg, U., Guyon, I., Garnett, R. (eds.) Proceedings of NeurIPS 2016, vol. 29 (2016)

34. Xiao, H., Rasul, K., Vollgraf, R.: Fashion-MNIST: a novel image dataset for bench-marking machine learning algorithms (2017). arXiv:1708.07747
35. Yang, D., Balaprakash, P., Leyffer, S.: Modeling design and control problems involving neural network surrogates. Comput. Optim. Appl. 1–42 (2022)
36. Yu, X., Serra, T., Ramalingam, S., Zhe, S.: The combinatorial brain surgeon: pruning weights that cancel one another in neural networks. In: Chaudhuri, K., Jegelka, S., Song, L., Szepesvari, C., Niu, G., Sabato, S. (eds.) Proceedings of the 39th ICML, vol. 162, pp. 25668–25683. PMLR (2022)
37. Zebari, R., Abdulazeez, A., Zeebaree, D., Zebari, D., Saeed, J.: A comprehensive review of dimensionality reduction techniques for feature selection and feature extraction. J. Appl. Sci. Technol. Trends. 1(2), 56–70 (2020)

An Error-Based Measure for Concept Drift Detection and Characterization

Antoine Bugnicourt[1,2]([✉]), Riad Mokadem[1], Franck Morvan[1], and Nadia Bebeshina[2]

[1] IRIT Laboratory, Université Paul Sabatier - Toulouse III, Toulouse, France
antoine.bugnicourt@irit.fr
[2] MeetDeal, Rivesaltes, France
https://meetdeal.fr/

Abstract. Continual learning is an increasingly studied field, aiming at regulating catastrophic forgetting for online machine learning tasks. In this article, we propose a prediction error measure for continual learning, to detect concept drift induced from learned data input before the learning step. In addition, we check this measure's ability for characterization of the drift. For these purposes, we propose an algorithm to compute the proposed measure on a data stream while also estimating concept drift. Then, we calculate the correlation coefficients between this estimate and our measurement, using time series analysis. To validate our proposal, we base our experiments on simulated streams of metadata collected from an industrial dataset corresponding to real conversation data. The results show that the proposed measure constitutes a reliable criterion for concept drift detection. They also show that a characterization of the drift relative to components of the stream is possible thanks to the proposed measure.

Keywords: Online learning · Continual learning · Concept drift · Change detection

1 Introduction

In the last years, successive advancements in machine learning allowed its use in multiple fields. The most common approach is batch learning, with multiple learning steps over a fixed dataset. Some tasks may require to learn from a continuous stream of data. Among those, there are applications requiring *plasticity*, i.e. the learning model being able to take into account concept drift: changes in the distribution of data [1,11]. For batch learning, this creates an over-cost on both storage (the dataset must be updated with every new input) and computation time (the model must be re-trained regularly on the dataset) [17].

In this context, online learning comes as an alternative to batch learning; the model learns iteratively on a data stream, without the need to store data. This new approach allows for change detection and integration in real time. However, the learning model will tend to favoritize newly learned knowledge, and to forget past knowledge. In the literature, this phenomenon is named catastrophic forgetting [20,24]. Catastrophic forgetting infringes on model *stability*, the model's ability to retain knowledge.

To counter catastrophic forgetting, a wide array of works consider a kind of online learning named continual learning [26]. It aims to improve online learning models to be more resistant to outliers, and to provide them with a better stability of acquired knowledge.

As an example, let's suppose we want to analyze user interactions on a business website, to build a question answering (QA) system. This site undergoes regular changes to catalogue items, their availability, prices, etc. Changes can also occur in users' interactions on the website. Every one of these changes must be integrated into the QA process as quickly as possible, in order to always produce relevant answers for the users. A continual learning approach trained on a stream of interactions on the website would allow the model to acquire new information while staying accurate for regular questions. Let's then consider that one key product undergoes a sale, preparing for the release of a new version of the same product. The sale needs to be taken into account in the QA system, but we also wish to notice when the sale ends, in order to "forget" its effect on the system, i.e. to come back to the state of the model prior to the sale integration.

In the literature, few works have considered solutions where a model could need to come back to a previous state of knowledge. Continual learning as a field consists of various methods to counter catastrophic forgetting, which can be classified in two main categories derived from Biesialska et al. [6]: **model-based methods**, that use dynamic adaptations over learning parameters [12,15,30] or the model structure [16,19] to consolidate knowledge and avoid forgetting; and **memory-based methods** [2,14,25,27] that use an external memory and a rehearsal or replay mechanism to have the model remember previously learned knowledge.

Model-based methods are mostly used in multi-task systems. In those systems, returning back to a previous state doesn't make sense, as the goal is to manage an ever-increasing set of skills. On the other hand, memory-based methods, used in this work, have a distinct advantage: using the rehearsal mechanism specific to these methods and the notion of catastrophic forgetting, it is indeed possible to have the model come back to a previous state.

Considering we induce forgetting through the aforementioned rehearsal, one remaining issue is determining when to trigger that rehearsal mechanism; we need to preserve the balance between stability and plasticity. For this reason, we wish to be able to detect concept drift and to decide on a sequence of actions based on some characteristic traits of that drift. To do so, a characterization of the current *context* (i.e. the set of all concepts and the associated probabilities of occurrence) available at all times is required. The decision process itself is two-fold:

1. Detecting concept drift: determining whether the new input integrated to the model creates a drift.

2. Determining the direction of the trend shift: a model under concept drift could be either evolving towards a new model (leading to an update of the sampled replay memory to memorize the older state of knowledge) or devolving towards an older state (leading to the use of the sampled replay memory for induced forgetting over the previous registered changes).

In this paper, we propose a new measure of the concept drift, which is used to determine the significance of the change in the data, and as such, to perform concept drift detection. Furthermore, this measure pave the way towards characterization of the drift. The measure is based on the predictive errors by the model. To measure this detection process, we use statistical measures (e.g. distance between labels) that allow to notice a significant correlation between forecast error and model shift. In order to validate the detection of concept drift, an algorithm is proposed to compute this correlation.

The next step would be to explore the characterization potential. This would then allow us to decide which approach to follow: classical continual learning if the model is evolving or returning back if the model is devolving. We defer this last step to a future work.

To validate our proposal, we use a proprietary dataset based on metadata from online chats provided by a corporation. Our experiments show that using the proposed algorithm, we compute correlation values that allow us to consider a characterization of concept drift. In summary, the main contribution of this paper is a correlation-based measure allowing both concept drift detection and characterization decision.

The paper follows as described: in Sect. 2, we establish a state of the art for continual learning approaches, specifically on memory-based methods, and for concept drift detection. In Sect. 3, we propose an approach allowing concept drift detection and the criteria to consider for concept drift characterization using this approach. In Sect. 4, we present our experimental results. Finally, we conclude this paper with a summary of our work and some future considerations.

2   Related Work

The core idea of catastrophic forgetting is a loss of knowledge stability in a machine learning model due to integrating a degree of plasticity towards new inputs. This phenomenon has long been an issue of concern [15,20], and a significant goal for both incremental and continual learning is to counter it [21].

Continual learning encompasses a broad range of approaches [6], all aiming to control catastrophic forgetting and to preserve the balance between stability and plasticity. We have compiled a classification of these approaches.

Model-Based Methods

Various approaches deal with adaptation to concept drift by enacting structural change of the machine learning model. Those model-based methods can be split into two main categories:

Regularization methods use weight manipulation and forgetting factors to influence the learning process and preserve stability. As examples: Kirkpatrick et al. [15] propose an algorithm called *Elastic Weight Consolidation* (EWC) operating on neural networks, singling out essential neurons for memorization of specific knowledge and slowing the learning on those neurons. Gupta et al., on their STAFF tool [12] propose a generally weaker forgetting thanks to a stabilization coefficient. Finally, Yu and Webb [30] link their forgetting factor's value to the analysis of concept drift on the data stream over a period of time.

Architectural methods consist of a set of dynamic adaptation processes on the neuronal architecture (i.e. addition of layers or parameters) in order to integrate new behaviors. Examples include the work of Li et al. [16], in which a model learns to adapt weight values or to create new neurons in parallel with the learning process; or Masana et al. [19], who store in an external structure various masks and normalization parameters related to specific tasks. These methods are particularly suitable for multitask applications, but are otherwise sub-optimal for repetitive tasks over long periods of time. They also don't include any flexibility over the plasticity mechanics: the applied transformations to the model are not revertible by default.

Memory-Based Methods

Memory-based methods use incremental learning plasticity itself to prevent catastrophic forgetting. These methods set up a replay or rehearsal of various informations (learning data) in the model input, mixed with the input stream, to create stability. There are multiple ways to proceed:

Rehearsal or **replay methods** rely on maintaining a separate dataset of examples parallel to the learning process, that is re-integrated to the learning process on a regular basis. Works to mention include iCarl from Rebuffi et al. [25]. Their method, used for multi-class classification, relies on task-specific examples dataset, updated with each new class. Aljundi et al. [2] define a "Maximal Interference" criterion, applied post-learning over model predictions, in order to identify data entries most susceptible to suffer from catastrophic forgetting, and to use them as a replay dataset.

Pseudo-rehearsal methods don't use an entry dataset, but store generalizations built from rehearsal data entries. Biesialska et al. mention two examples: DCR [27] and FearNet [14], using generative adversarial networks (GAN) and autoencoders respectively, to make this generalization.

For all of these methods, knowledge stability is consolidated by a memory module distinct from the model. This kind of approach is more suited to the notion of bringing back a previous context, since rehearsal allows the recreate a previous state of the model. The idea of reusing past concepts already exists in other works related to multi-task learning, such as Wang et al.'s SDR architecture [29]. SDR relies on checking for similarities in new data entries compared to previously learned tasks, in order to determine if a new entry is an instance of one of these tasks, or instead illustrates a new task to be integrated to the model.

Some works on concept drift [1,11] define two categories of drift, depending on the considered source of change: **virtual concept drift**, a measurable change on distribution of the values of specific data fields; and **real concept drift**, a change in the relation between features and labels. Sometimes a third kind of drift is considered: *label* or *prior-probability shift*, related to a change in the distribution of the prediction labels.

There are various existing tests to detect concept drift based on model performance. A number of them are mentioned in the Bayram et al. review of concept drift methods [5]:

- A first set of detectors called "Statistical Process Control" detectors, watch the evolution of the error rate of the learner as an indicator of concept drift. Drift Detection Method (DDM) [10] tracks the rate of prediction errors over the data stream, compared to specific thresholds for warning (announcing a potential drift) and drift itself, following the hypothesis that a rising rate means a drift is either coming or happening. Early Drift Detection Method (EDDM) EDDM [4] uses the same principle, but looks at the distance between consecutive errors rather than their rate, which improves detection in gradual drift.
- Another set of detectors compare statistical measures over sliding windows to detect change. Those include ADWIN (short for Adaptive Windows) [7], that tracks the change in the distribution of a variable in two sliding windows of varying sizes, dynamically adapting the size to optimize the detection. Raab et al. propose another detector called KSWIN [23] (KS being short for Kolmogorov-Smirnov), using the Kolmogorov-Smirnov test [18] over two sliding windows to check for concept drift.

The base measure we propose in this article has elements from both categories. It is error-based, as it relies on prediction error for new data inputs—although we're not specifically interested in the rate of those errors, but rather on their amplitude. Furthermore, the drift detection decision is based on aggregates of those errors in sliding windows.

Our general approach differs from previously mentioned works by the attention given to the chronology of consecutive concepts, which requires an accurate characterization of those concepts and even more so, of the concept drift. We not only want to detect concept drift, but also to be able measure its amplitude and *direction*—in the sense of whether the drift is towards a new concept, or a previously occurring one. This measure must come before the learning step (like in Wang et al.'s SDR), in order to use it in further work for decision over the kind of rehearsal to enact.

In summary, we propose a new measure for concept drift detection and characterization in new learning data inputs, based on prediction error, and available before the learning step.

3   Proposal

Our process involves analyzing data from a single data stream: we first aim to detect a variation in this stream regarding the relation between a vector their corresponding labels, and for a second step, we will consider if characterization of this relation is possible. The variation of the relation between vector and label is measurable through the variation of the predictions performed by a learning model over a test dataset of labeled vectors.

In this paper, let's consider that a continual learning model M is set up over two distinct data streams: $S_{predict}$, made of data vectors X for which we want to make a forecast; and S_{learn}, made of labeled vectors (X, y) to be integrated to the model through learning.

In the example from Sect. 1, the prediction stream $S_{predict}$ is made on questions asked by the users while browsing on the website. The learning stream S_{learn} contains questions with relevant answers given by an outside source— for example a human expert. The model then learns on the questions/answers couples from $S_{predict}$ so that it may improve answers given for the questions in S_{learn}.

Our goal here is to confirm the existence of a correlation between the changes induced on the model by the learning of data entries from S_{learn}, and the prediction error measured on those same data entries before learning them. When this is confirmed, a significant change in the model—i.e. a concept drift—might be anticipated using the prediction made by the model itself. We describe in the following segments some tools and a method aiming to measure this correlation between predictions and model change.

3.1   General Notations

To represent the learning stream over which our proposal operates, we consider a discrete time period $T \subseteq \mathbb{N}$. The labeled vectors of data (X_t, y_t) of this stream are indexed by time periods $t \in T$. It is then necessary to explicitly define the essential operations of our continual learning model M:

The operation of learning a labeled vector (X, y) to update a model M into a new model M' has the following notation:

$$M' = learn(M, X, y) \tag{1}$$

The operation of predicting a label $\hat{y}$ for a vector X, with a trust value $p \in [0, 1]$ associated to the prediction, has the following notation:

$$(\hat{y}, p) = predict(M, X) \tag{2}$$

To give a formal notation for the prediction error on a measure, we need a comparison criterion—a *distance* measure between predicted and actual labels. The distance between two labels y, y' is noted as follows:

$$d = dist(y, y') = \|y - y'\| \in \mathbb{R} \tag{3}$$

The meaning of the calculation for a norm $\|y - y'\|$ varies depending of the use case, since it represents a relation between each pair of labels. If those labels are already represented by numerical values in $\mathbb{R}$, we can simply use $|y - y'|$. In our example, the distance between two answers may simply be a binary value (good or bad answer), with the following:

$$dist(y, y') = \begin{cases} 0 \text{ if } y = y' \\ 1 \text{ otherwise} \end{cases} \tag{4}$$

If more than a pair of categories of answers are considered, the notion of gap can become more complex, and as such, may justify to use label values from $\mathbb{R}$, or even vectors with values in $\mathbb{R}$.

The formalism for time series is required to invoke the measures computed in our algorithm. Time series are sequences of values with a time index. They are useful to study the interdependency of these values through time. **Cross correlation** is a measure of this dependency; it compares two series shifted by a lag τ [9,22]. The cross correlation between variables A and B on a discrete period of time T (with values in $\mathbb{C}$) is a function of τ defined as:

$$(B \star A)(\tau) = \sum_{t \in T} A(t + \tau)\overline{B(t)} \tag{5}$$

where $\overline{B(t)}$ is the complex conjugate of $B(t)$: $\forall z \in \mathbb{C}, z = a + ib \implies \bar{z} = a - ib$.

We finally define a specific notation for the cross correlation measured at $\tau = 0$ (a significant value for our proposal, considering the relation we want to measure would happen without lag): $corr(A, B) = (B \star A)(0)$.

3.2 Algorithm

The proposed algorithm aims to determine the relation between the model prediction over a given data vector, and the evolution of the model induced by the integrated labeled vector. This Algorithm 1 applied to a model M, operates on a stream of labeled vectors $(X_t, y_t)_{t \in T}$.

For each time period t of T, we retrieve the corresponding labeled vector (X_t, y_t), and first measure the prediction error over this vector (lines 3 and 4 of the algorithm). This error is defined as the distance between the prediction 2 and the true label. This error is weighted by the trust value given to the prediction, also obtained through the prediction operation. The error function for predictions from current model M_t is:

$$err(X, y, t) = p \cdot dist(y, \hat{y}) \text{ where } (\hat{y}, p) = predict(M_t, X) \tag{6}$$

This error value is then used in a computation over error values from previous data entries (in a sliding window). We call $Err(S, W)$ the linear combination of the error values in the sliding window $S = (e_t)_{t \in [0;n]}$, weighted by $W = (w_t)_{t \in [0;n]}$ such that:

$$Err(S, W) = \sum_{w \in W, e \in S} w \cdot e \tag{7}$$

The linear combination value is saved in an array $\widehat{D}$. Next step (line 6) is M_t learning the labeled vector, and following Eq. 1, returning a new model $M_{t+1} = learn(M_t, X_t, y_t)$. Both models M_t and M_{t+1} are compared to measure the model shift induced by learning the new entry.

To compare the models, we can use Eq. 6 to make a variation function, defined for a pair of models $M = M_t$ and $M' = M_{t+k}$, and for a test dataset $\langle \mathcal{X}, Y \rangle = (X'_i, y'_i)_{i \in [1,N]}$, as:

$$\Delta_M^{M'}(\mathcal{X}, Y) = \sum_{i=1}^{N} \Big(err(X_i, y_i, t + k) - err(X_i, y_i, t) \Big) \tag{8}$$

The variation value $\Delta_M^{M'}(\mathcal{X}, Y)$ is stored in an array D.

When the last value of T is reached (end of the stream), cross correlation values are computed for any value of lag $\tau \in [-N \dots N]$. The remaining lines of the algorithm (from line 9 to the end) correspond to the correlation checking. We consider the value $corr(\widehat{D}, D)$, i.e. the cross correlation value for lag 0 (cf Eq. 5) for values in arrays $\widehat{D}$ and D (line 9).

If the value of $corr(\widehat{D}, D)$ measured through the algorithm is not 0, values stored in $\widehat{D}$ can be used—to a certain extent—to predict shift in value of D. This would account for the measure's ability to detect change, but would not immediately solve the characterization issue, which will require further work.

Algorithm 1: Concept drift estimation and computation of correlation coefficients

---

Inputs : time period $T \subseteq \mathbb{N}$, classification model M, test dataset $\langle \mathcal{X}, Y \rangle$
Data : labeled vectors $(X_t, y_t)_{t \in T}$
Output : correlation coefficient C

1 initialization of $\widehat{D}[\ ]$ and $D[\ ]$

2 **For all** $t \in T$ **do**

3     $X, y \leftarrow X_t, y_t$

4     $S \leftarrow S.update(err(X, y, t))$ /* measure of prediction error */

5     $\widehat{D}[t] \leftarrow Err(t, S, W)$ /* computation with values from S */

6     $M' \leftarrow learn(M, X, y)$

7     $D[t] \leftarrow \Delta_M^{M'}(\mathcal{X}, Y)$ /* measure of model shift over (X, y) */

8     $M \leftarrow M'$

9 $C \leftarrow corr\left(\widehat{D}, D\right)$ /* cross correlation with lag $\tau = 0$ */

10 **If** $C \neq 0$ **then**

11   **return** C /* correlation is identified */
                                                        /* (possible characterization) */

12 **else**

13   **return** $\perp$ /* no correlation revealed */

---

4   Evaluation

4.1   Protocol

To evaluate our proposal, we use a dataset from a proprietary source. The dataset is made of various metadata collected over a volume of real online conversations provided by a corporation, each associated through the result of the conversation, to a positive or negative label. Five categories of metadata are considered for just over 10,000 conversations, with all or a subset of these categories being used at once. Online learning classification is implemented using the River library[1]; the model used is Hoeffding tree classifiers [8,13], as implemented in the same library. The classification model is used with default parameters.

Through our experiments, we first want to check if the prediction error measure can be used for model shift detection. To show this, we apply our Algorithm 1 on a simulated stream of data, for the full set of features available, and measure the cross correlation around lag 0. Streaming simulation is also performed through the River library.

In order to qualify the relevancy of our prediction measure err_M as a concept drift detection tool, we apply our Algorithm 1 over the KSWIN test [23] in parallel to the measure, with the same window size for both. KSWIN can only detect concept drift over a single feature stream at once. To compensate for that fact, we use an aggregate (specifically a logical disjunction or OR operation) of detection values from all possible streams: one for each feature, and one for the label. Cross correlation is also measured between this aggregate and the model shift.

We first perform the experiment on a window of size 100, which is the default size for KSWIN. We then perform multiple runs with various window sizes to find an optimal size for cross correlation for our dataset.

We then consider the issue of characterization of the measured model shift. In order to use an already established detection measure as a characterization criterion, one way could be to look for a function mapping from the measure to the actual model shift. To evaluate this possibility, we compute the ratio between both values of the prediction measure and the model shift and look for an identifiable trend between the two quantities.

4.2   Results

Concept Drift Detection. During the first experiment, we measured values for the measure, identified the vectors for which KSWIN was detecting a model shift, and generated two figures: Fig. 1 shows compared values of concept drift computed on the model, and of the normalized measure, over time; Fig. 2 shows cross correlation curves for the measure and KSWIN, both in relation to concept drift.

---

[1] https://riverml.xyz/0.14.0/.

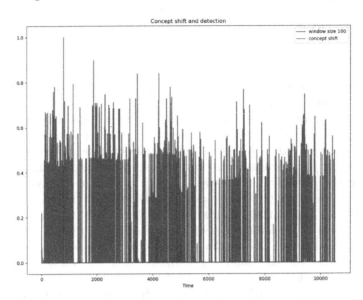

Fig. 1. Computed real concept drift (in red) against detected shift using the error-based measure, with a window size of 100 (in blue) (Color figure online)

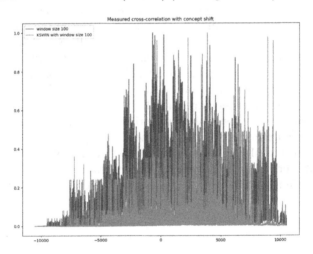

Fig. 2. Cross-correlation with computed model concept drift from the error-based measure err_M *(in blue)* and the KSWIN aggregate *(in orange)* (Color figure online)

Even though Fig. 1 presents a very noisy curve for the measure, Fig. 2 reveals that cross correlation for our method offers a similar curve shape to KSWIN's. Furthermore, when looking at the cross correlation values at lag 0 and under (i.e. when past values of the criteria align with future values of concept drift, giving a *forecast* of sorts), our method outperforms KSWIN.

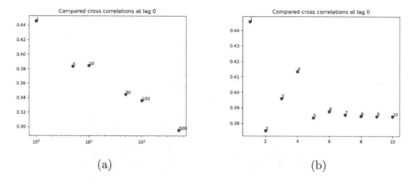

Fig. 3. Cross-correlation values (Y-axis) at lag 0, depending on the window size (X-axis and indices of points): (a) in low grain on a logarithmic scale; (b) in finer grain, from 1 to 10.

By comparing the measure and KSWIN, we can say that the former seems efficient as a concept drift detection mechanism, or even as a concept drift predictor. However, the aggregate of KSWIN detection values we compared the measure to isn't a real concept drift detector, but merely an improved virtual drift detector, and the performance seen here is relative to various experiment criteria, among which the aggregation method itself.

Setting the Measure Parameters. We started with a low grain setting, in order to select a range of window sizes optimizing correlation at lag 0, shown in Fig. 3a. This first setting shows that lower window sizes give better cross correlation values. We then restrict our setting to a finer grain, low sizes range (Fig. 3b). Again, the lower sizes, specifically between 1 and 4 (2 being an outlier, probably due to the weight computation method) perform better on cross correlation.

Comparing various window sizes allows us to conclude that a window of size 1 is the optimal choice for both this data source and this weighting of the window. This conclusion is not an absolute, and both the data and the weighting approach influence the optimal outcome.

Characterization. A detection criterion can be used as a characterization of concept drift criterion if there is a function mapping criterion values to the amplitude of the concept drift. In order to check for that potential characterization, we measured various ratios between the measure values and computed drift. The following measures are made with a window of size 1, following the conclusion from Subsect. 4.2. We consider various subsets of features $\{A, B, C, D\}$ to generate Figs. 4 (A, B, C), 5a (A, B) and 5b (B, C, D).

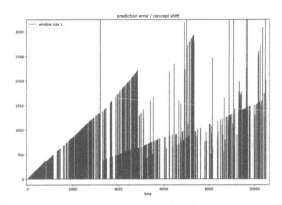

Fig. 4. Ratio of error-based measure over concept drift for a subset of features $\{A, B, C\}$

In the resulting figures, we can see that multiple *trends* appear in various intervals of the data stream, since in those intervals, the ratio follows linear trajectories. This can be seen in Figs. 5a (with one such interval visible between time indices 0 and 3000, and another, sparser one with a steeper slope, between 3000 and 9000) and 5b (two noticeable intervals respectively between 0 and 5000, and between before 5000 and after 7000). Looking back at model shift values on Fig. 1, we notice that those trends happen over intervals of low model shift, or between high model shift periods. We also note that those trends don't necessarily happen in succession to each other, but can overlap (e.g. between 3000 and 5000 on Fig. 4).

Even though none of these trends can describe a single, constant relation between the measure and real concept drift, they each take part in a set of relations between both of them. The aforementioned overlap between trends suggest that this variety of relations is not only due to model shift, but also (and most importantly) to specific values taken by some of the features in those intervals.

Mutiple remarks can be made about characterization. The experiments we performed here show that the measure highlights a number of trends during the learning process, seemingly related to the values taken by specific features. This allows us to hypothesize that a more particular analysis of these feature values, further developed in the conclusion of this article, would allow an even more thorough characterization than expected, even for other works.

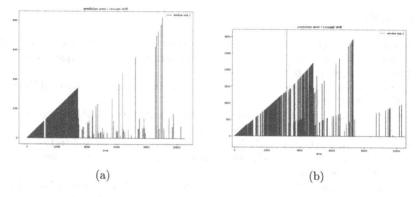

Fig. 5. Ratio of error-based measure over concept drift for subsets of features: (a) $\{A, B\}$ and (b) $\{B, C, D\}$

5   Conclusion

In this paper, we proposed a concept drift detection and characterization measure related to a continual learning model. Our measure uses model prediction to determine if and to what degree a new data entry creates a concept drift. Our experiments show that real concept drift detection is possible using the measure, and that various parameters allow an adaptation to multiple kinds of data or models.

The experiments also show a set of distinct trends that can be observed in the relation between the measure and real drift, thus opening new perspectives for characterization. No simple aggregate of those trends can be used as a general characterization of concept drift; however, those trends and their overlapping suggest that this behavior is related directly to distinct trends in the learning features themselves.

As a future work, explainability approaches, as studied in the Explainable AI (XAI) field of research [3,28], could be combined to our measure to produce a more accurate characterization. Furthermore, if we were to store the parameters related to each trend as representations of the state of the model, this would be a step towards building a continual learning system sensitive to concept temporality and recurrence. We could also take into account real use conditions of continual learning in terms of memory resource capacity and computation time. Finally, although the experiments were already lead on a real dataset, we project to experiment over a real datastream instead of a simulation.

References

1. Agrahari, S., Singh, A.K.: Concept drift detection in data stream mining: a literature review. J. King Saud Univ. Comput. Inf. Sci. (2021). https://doi.org/10.1016/j.jksuci.2021.11.006
2. Aljundi, R., et al.: Online Continual Learning with Maximally Interfered Retrieval. arXiv:1908.04742 (2019)
3. Arrieta, A.B., et al.: Explainable Artificial Intelligence (XAI): Concepts, Taxonomies, Opportunities and Challenges toward Responsible AI (2019)
4. Baena-Garcıa, M., Gavalda, R., Morales-Bueno, R.: Early drift detection method. In: Fourth International Workshop on Knowledge Discovery from Data Streams, vol. 6, pp. 77–86 (2006)
5. Bayram, F., Ahmed, B.S., Kassler, A.: From Concept Drift to Model Degradation: An Overview on Performance-Aware Drift Detectors (2022). https://doi.org/10.48550/arXiv.2203.11070
6. Biesialska, M., Biesialska, K., Costa-jussà, M.R.: Continual lifelong learning in natural language processing: a survey. In: Proceedings of the 28th International Conference on Computational Linguistics, pp. 6523–6541 (2020). https://doi.org/10.18653/v1/2020.coling-main.574
7. Bifet, A., Gavaldà, R.: Learning from time-changing data with adaptive windowing. In: Proceedings of the 2007 SIAM International Conference on Data Mining (SDM), pp. 443–448. Society for Industrial and Applied Mathematics (2007). https://doi.org/10.1137/1.9781611972771.42
8. Bifet, A., et al.: MOA: massive online analysis, a framework for stream classification and clustering. In: Proceedings of the First Workshop on Applications of Pattern Analysis, pp. 44–50 (2010)
9. Bracewell, R.: Pentagram notation for cross correlation. In: The Fourier Transform and Its Applications, vol. 46, p. 243. McGraw-Hill, New York (1965)
10. Gama, J., Medas, P., Castillo, G., Rodrigues, P.: Learning with drift detection. In: Bazzan, A.L.C., Labidi, S. (eds.) SBIA 2004. LNCS (LNAI), vol. 3171, pp. 286–295. Springer, Heidelberg (2004). https://doi.org/10.1007/978-3-540-28645-5_29
11. Gama, J., Žliobaitė, I., Bifet, A., Pechenizkiy, M., Bouchachia, A.: A survey on concept drift adaptation. ACM Comput. Surv. **46**(4), 44:1–44:37 (2014). https://doi.org/10.1145/2523813
12. Gupta, U., Babu, M., Ayoub, R., Kishinevsky, M., Paterna, F., Ogras, U.Y.: STAFF: online learning with stabilized adaptive forgetting factor and feature selection algorithm. In: Proceedings of the 55th Annual Design Automation Conference, San Francisco, California, pp. 1–6. ACM (2018). https://doi.org/10.1145/3195970.3196122
13. Hulten, G., Spencer, L., Domingos, P.: Mining time-changing data streams. In: Proceedings of the Seventh ACM SIGKDD International Conference on Knowledge Discovery and Data Mining, San Francisco, California, pp. 97–106. ACM (2001). https://doi.org/10.1145/502512.502529
14. Kemker, R., Kanan, C.: FearNet: brain-inspired model for incremental learning. In: International Conference on Learning Representations (2022)
15. Kirkpatrick, J., et al.: Overcoming catastrophic forgetting in neural networks. Proc. Natl. Acad. Sci. **114**(13), 3521–3526 (2017). https://doi.org/10.1073/pnas.1611835114
16. Li, X., Zhou, Y., Wu, T., Socher, R., Xiong, C.: Learn to grow: a continual structure learning framework for overcoming catastrophic forgetting. In: International Conference in Machine Learning, p. 10 (2019)

17. Lin, J.: The lambda and the kappa. IEEE Internet Comput. **21**(5), 60–66 (2017)
18. Lopes, R.H.C.: Kolmogorov-Smirnov test. In: Lovric, M. (ed.) International Encyclopedia of Statistical Science, pp. 718–720. Springer, Heidelberg (2011). https://doi.org/10.1007/978-3-642-04898-2_326
19. Masana, M., Tuytelaars, T., van de Weijer, J.: Ternary Feature Masks: Zero-forgetting for task-incremental learning. arXiv:2001.08714 (2021)
20. McCloskey, M., Cohen, N.J.: Catastrophic interference in connectionist networks: the sequential learning problem. In: Psychology of Learning and Motivation, vol. 24, pp. 109–165. Elsevier (1989). https://doi.org/10.1016/S0079-7421(08)60536-8
21. Nguyen, C.V., Achille, A., Lam, M., Hassner, T., Mahadevan, V., Soatto, S.: Toward Understanding Catastrophic Forgetting in Continual Learning. arXiv:1908.01091 (2019)
22. Papoulis, A.: The fourier integral and its applications. Polytechnic Institute of Brooklyn, McCraw-Hill Book Company Inc., USA (1962). ISBN 67-048447-3
23. Raab, C., Heusinger, M., Schleif, F.M.: Reactive soft prototype computing for concept drift streams. Neurocomputing **416**, 340–351 (2020). https://doi.org/10.1016/j.neucom.2019.11.111
24. Ramasesh, V.V., Dyer, E., Raghu, M.: Anatomy of Catastrophic Forgetting: Hidden Representations and Task Semantics. arXiv:2007.07400 (2020)
25. Rebuffi, S.A., Kolesnikov, A., Sperl, G., Lampert, C.H.: iCaRL: incremental classifier and representation learning. In: 2017 IEEE Conference on Computer Vision and Pattern Recognition (CVPR), Honolulu, HI, pp. 5533–5542. IEEE (2017). https://doi.org/10.1109/CVPR.2017.587
26. Ring, M.B.: Continual Learning in Reinforcement Environments. GMD-Bericht (1994)
27. Shin, H., Lee, J.K., Kim, J., Kim, J.: Continual Learning with Deep Generative Replay. arXiv:1705.08690 (2017)
28. Tjoa, E., Guan, C.: A survey on explainable artificial intelligence (XAI): toward medical XAI. IEEE Trans. Neural Netw. Learn. Syst. **32**(11), 4793–4813 (2021). https://doi.org/10.1109/TNNLS.2020.3027314
29. Wang, S., Choi, Y., Chen, J., El-Khamy, M., Henao, R.: Toward Sustainable Continual Learning: Detection and Knowledge Repurposing of Similar Tasks (2022)
30. Yu, H., Webb, G.I.: Adaptive online extreme learning machine by regulating forgetting factor by concept drift map. Neurocomputing **343**, 141–153 (2019). https://doi.org/10.1016/j.neucom.2018.11.098

Predict, Tune and Optimize for Data-Driven Shift Scheduling with Uncertain Demands

Michael Römer[(✉)] [ORCID], Felix Hagemann, and Till Frederik Porrmann

Department of Management Science and Business Analytics, Bielefeld University, Bielefeld, Germany
{michael.roemer,felix.hagemann,till.porrmann}@bielefeld.de

Abstract. When it comes to data-driven optimization under uncertainty, it is well known that a naïve predict-then-optimize pipeline in which point forecasts are plugged into a deterministic optimization model typically leads to a poor expected decision quality. In stochastic programming, one aims at obtaining better decisions by explicitly representing the joint probability distribution in the optimization model, e.g. in form of a sample approximation. A downside of that approach is that it gives rise to large-scale model instances that are hard to solve. An alternative approach that recently attracted considerable interest aims to train prediction models in a way that the expected decision quality obtained with the (prediction-informed) deterministic model is maximized, this approach is referred to as decision-focused learning or predict and optimize in the literature. In this paper, we propose to generalize this idea by optimizing not only parameters affecting the prediction but also additional parameters influencing other (non-stochastic) parts of the optimization model. Specifically, we propose to simultaneously optimize both types of parameters with the goal of maximizing expected decision quality and refer to this approach as predict, tune and optimize. We demonstrate the usefulness of the approach for a multi-activity shift scheduling problem under demand uncertainty. Specifically, we show that while decision-oriented tuning of point forecasts usually yields better results than a simple predict-then-optimize approach, adding the possibility to modify additional parameters considerably improves the expected performance which becomes competitive with a stochastic programming approach.

Keywords: Optimization under uncertainty · Decision-focused learning · Parameter tuning

1 Introduction

Combinatorial optimization problems (COPs) under uncertainty are notoriously hard to solve. While it is well-known that naively predicting uncertain problem parameters and solving a deterministic COP using these predictions (a so-called *predict-then-optimize* approach) typically yields solutions with poor

M. Sellmann and K. Tierney (Eds.): LION 2023, LNCS 14286, pp. 254–269, 2023.
https://doi.org/10.1007/978-3-031-44505-7_18

expected performance, more appropriate approaches such as stochastic programming tend to be very challenging from computational perspective. Recently, an emerging stream of research referred to as *decision-focused learning* or *predict-and-optimize*, see [8] for a survey, aims at aligning the prediction with the downstream optimization problem. This is typically performed by adapting the loss function of the prediction model in a way that it optimizes the expected performance of the downstream decisions instead of focusing on prediction accuracy.

In this paper, we propose to generalize the idea of decision-focused learning by optimizing not only "prediction parameters" affecting the prediction of the stochastic parameters of a COP but also other "tuning parameters" affecting other (deterministic) parameters of the optimization model. We propose to simultaneously optimize both sets of parameters with the goal of maximizing the expected performance of the resulting decisions and refer to this approach as predict, tune and optimize.

We demonstrate the usefulness of the approach for a multi-activity shift scheduling problem under uncertainty in which almost 1000 demand parameters are affected by uncertainty. In a set of experiments, we show that while decision-oriented tuning of point forecasts usually yields better results than a simple predict-then-optimize approach, adding the possibility to optimize additional tuning parameters considerably improves the expected performance which becomes competitive with a sample-average-approximation-based stochastic programming approach, albeit at a much smaller online computational cost.

The remainder of this paper is structured as follows: In the next section, we describe the problem setting addressed in this paper as well as the key idea of predict, tune and optimize (PTO). In Sect. 3 we describe the multi-activity shift scheduling problem under demand uncertainty that we use for evaluating the PTO approach along with the MILP formulation used in the experiments. Section 4 presents and discusses the computational results, and Sect. 5 discusses related work.

2   Predict, Tune and Optimize

In this section, we provide a description of the type of two-stage optimization problem considered in this paper as well as a description of the PTO and related approaches that can be used to solve this type of problem.

Deterministic Two-Stage Optimization Problem. We start by considering a two-stage combinatorial optimization problem P of the form

$$P = \min f(x) + Q(x) \text{ s.t. } x \in \mathcal{X} \tag{1}$$

where x is a vector of (continuous or discrete) first-stage decision variables, $f(x)$ is an objective function and $\mathcal{X}$ forms the feasible set of the variables x. As an example, if P is a mixed-integer linear programming (MILP) problem, $\mathcal{X}$ is characterized by a set of linear inequalities and a set of integrality constraints, and $f(x)$ is a linear function. The first-stage decisions affect the second-stage problem $Q(x)$ given by:

$$Q(x) = \min g(x,y) \text{ s.t. } y \in \mathcal{Y}(x) \tag{2}$$

Here, y is a vector set of (continuous or discrete) second-stage decision variables, $g(x,y)$ is the second-stage objective function and $\mathcal{Y}$ is the feasible set of y; both depend on the first-stage decisions x. Examples for problems that can be expressed using this two-stage framework are problems in logistics where first stage may involve location decisions and the second stage deals with distribution decisions. In this paper, we consider another example: A personnel scheduling problem in which the start and end time of daily shifts are decided a priori (e.g. on a monthly basis) and the concrete assignment of work activities and the timing of breaks is decided in the second stage (e.g. on the day of operation).

If both the first and the second-stage problems are deterministic, that is, if all parameters affecting the objective functions and feasible sets are known at the time the optimization problem is solved, P can be solved as a single deterministic optimization problem, here referred to as the nominal problem P^N:

$$P^N = \min f(x) + g(x,y) \text{ s.t. } x \in \mathcal{X}, y \in \mathcal{Y}(x) \tag{3}$$

Two-Stage Optimization Problem Under Uncertainty. In many practical settings, the first-stage decisions have to be taken way ahead of the second-stage decisions. As a result, it is unrealistic to assume that all information affecting the second-stage optimization problem Q is known at the time when the first-stage decisions have to be taken. As an example, the exact demand to be satisfied in a personnel scheduling problem may only become available at the day of operation which means that the second-stage decisions are affected by demand information not yet available in the first stage. In such a setting, the second-stage problem Q is affected by uncertainty, and this is the setting we focus on in this paper. Specifically, we assume that some of the parameters of Q are uncertain at the time when we have to take the first-stage decisions, and that they become known before we have to take the second-stage decisions. We write the stochastic parameters as ξ, and we note that ξ may affect both the feasible set Y and the objective function g of the second-stage problem. For a realization $\hat{\xi}$ of ξ, we can write the second-stage problem as:

$$Q(x,\hat{\xi}) = \min g(x,y,\hat{\xi}) \text{ s.t. } y \in \mathcal{Y}(x,\hat{\xi}) \tag{4}$$

In the full two-stage problem under uncertainty (referred to as P^U) we aim at finding a solution that minimizes the sum of the first-stage objective and the expected value of the second-stage problem $Q(x,\xi)$:

$$P^U = \min f(x) + E_\xi Q(x, \xi) \text{ s.t. } x \in \mathcal{X} \tag{5}$$

P^U forms a two-stage stochastic optimization problem, and the variables y are also referred to as recourse decision variables, see e.g. [2]. In the context of this paper, we assume complete recourse, that is, we assume that Q is feasible for any choice of $x \in \mathcal{X}$ (and for any realization of ξ). Furthermore, note that this setting also covers the case of a single-stage optimization problem in which only the objective function is affected by uncertainty. In that case, the set of second-stage variables is empty, and the stochastic objective function g only involves the first-stage variables x.

Predict, Then Optimize (P). A naive approach to solve P^U is to replace the uncertain parameters ξ by their expected values $E(\xi)$ and to solve the corresponding deterministic problem P^P:

$$P^P = \min f(x) + Q(x, E(\xi)) \text{ s.t. } x \in \mathcal{X} \tag{6}$$

This approach is sometimes referred to as predict, then optimize, in particular if $E(\xi)$ is obtained by a prediction model taking into account contextual information. The problem with this approach is that while it yields a relatively easy-to-solve optimization problem, the expected performance of the first-stage decisions resulting from this problem is typically not very good.

Formally, for a given set of first-stage decisions x', the expected performance under uncertainty is $f(x') + E_\xi(Q(x', \xi))$. One can approximate $E_\xi(Q(x', \xi))$ by a so-called sample average approximation (SAA). Denoting a set of samples from ξ with S, and the sample realization in sample s with $\hat{\xi}_s$, we can approximate $E_\xi Q(x, \xi)$ by:

$$Q^{SAA}(x, S) = \frac{1}{|S|} \sum_{s \in S} Q(x, \hat{\xi}_s) \tag{7}$$

Sample Average Approximation. While the SAA can be used for evaluating a given set of first-stage decisions, it can also be used to form a so-called two-stage stochastic optimization model that simultaneously optimizes x and the (sample-dependent) recourse decisions maximizing the (approximate) expected performance. The resulting optimization problem can be written as follows:

$$P^{SAA}(S) = \min_x f(x) + Q^{SAA}(x, S) \text{ s.t. } x \in \mathcal{X} \tag{8}$$

If the nominal problem P^N forms a MILP, then $P^{SAA}(S)$ can either be solved directly using standard software, or using specialized decomposition approaches such as the so-called (integer) L-shaped method, see e.g. the monograph [2] for an in-depth overview of stochastic programming models and techniques.

In any case, depending on the complexity of the nominal problem P^N and on the number of samples/scenarios, solving $P^{SAA}(S)$ (or more general stochastic programming formulations) is typically very challenging and requires much more effort than solving a deterministic nominal problem P^N (as is happening in the predict-then-optimize approach).

Predict and Optimize (PO). An alternative idea to improve upon a naive predict-then-optimize approach relies on the observation that in such an approach, the prediction is completely decoupled from the decision. Specifically, when fitting parametric prediction models to data, one aims at maximizing predictive accuracy. This estimation does not account for the effect of the predictions on the downstream optimization problem. As described in the introduction, these shortcomings gave rise to a the stream of research called decision-focused learning (DFL) or predict and optimize. In DFL, one aims at choosing parameters that yield the best decisions when plugged into the subsequent optimization model. In this paper, we consider a somewhat simplified setting compared to most approaches dealing with DFL: While in DFL, one considers predictions based on features representing contextual information, we assume a stationary setting for which a set of sample data is available (or can be generated) that forms an approximation of the true joint distribution of the stochastic parameters ξ.

For our purposes, we formalize such a (simplified) predict-and-optimize approach PO to solve P^U as follows. We assume that we have a parametric prediction model m which, given a vector of parameters θ^p and a set of samples S, returns predictions $\hat{\xi} = m(S, \theta^p)$ for the stochastic model parameters ξ. These predictions can then be used to solve the following variant of the nominal problem P^N in which all uncertain parameters are replaced by their predictions:

$$P^{PO}(\theta^p, S) = \operatorname*{argmin}_{x} f(x) + Q(x, m(\theta^p, S)) \text{ s.t. } x \in \mathcal{X} \qquad (9)$$

Now, in the spirit of decision-focused learning, we aim at determining the prediction parameters θ^p that, when being fed into the problem P^{PO} yield the first-stage decisions with best expected performance with respect to the optimization problem P^U. The resulting tuning problem (which can be solved offline) can be stated as follows:

$$\operatorname*{argmin}_{\theta^p} Q^{SAA}\big(P^{PO}(\theta^p, S), S^{eval}\big) \qquad (10)$$

Note that in (10), we account for the fact that the set of samples S used for the prediction may be different from the set S^{eval} used in the SAA-based evaluation using Q^{SAA}.

Predict, Tune and Optimize (PTO). One of the arguments for the often-observed superior performance of a predict-and-optimize approach over a naive predict-then-optimize approach is that a decision-focused prediction manages to provide accurate predictions in those parts of the distribution ξ where it matters for the decision-making. Another argument, however, is that different predictions may modify the optimization model in a way that the decisions obtained with the nominal model are "pushed" into a direction that result in a better expected performance. If we consider the latter argument, it is a natural idea to also modify other (deterministic) model parameters in order to achieve similar effects. Actually, this observation is a main motivation for introducing a set of *tuning parameters* θ^t that are used to modify the nominal deterministic optimization problem in a way that the resulting first-stage solution yields a better expected performance under uncertainty. Please note that in general, the tuning parameters θ^t are not identical to the parameters of the optimization model, indeed, ideally, a tuning parameter affects multiple model parameters (e.g. multiple objective function coefficients, or multiple right hand side values) at once. In general, as can be seen below, the tuning parameters may affect every part of the nominal model:

$$P^{PTO}(\theta^p, \theta^t, S) = \operatorname*{argmin}_{x} f(x, \theta^t) + Q(x, \theta^p, \theta^t, S) \text{ s.t. } x \in \mathcal{X}(\theta^t) \qquad (11)$$

Analogously to the PO case, the tuning problem for the PTO approach can be stated as follows:

$$\operatorname*{argmin}_{\theta^p, \theta^t} Q^{SAA}\big(P^{PTO}(\theta^p, \theta^t, S), S^{eval}\big) \qquad (12)$$

Training/Parameter Tuning. The approaches PO and PTO involve tuning a set of parameters affecting a nominal (deterministic) model to be solved for obtaining the first-stage decisions. We propose to carry out this tuning using a standard black box optimization solver that does not make any structural assumptions regarding the prediction model or with respect to the response of the optimization model to a set of parameter values. A straightforward approach to choose the initial values for the prediction parameters is to use the expected values of the uncertain parameters, the tuning parameters can be chosen in a way that the parameters of the nominal model are not adjusted.

During the tuning, the black box optimizer proposes new parameter values based on the feedback (response) of previous iterations. In each iteration (trial), two steps are performed: First, given the parameters θ^p (in case of PO) or θ^p, θ^t (in case of PTO), the parameterized nominal model $P^{PO}(\theta^p)$ or $P^{PTO}(\theta^p, \theta^t)$ is solved to obtain a set of first-stage decisions x'. In the second step, these decisions are used in an out-of-sample evaluation by solving $Q^{SAA}(x', S)$ for a set S of samples of the uncertain parameters ξ to approximate the expected performance $f(x') + Q^{SAA}(x', S)$. Observe that in this evaluation, the original (non-adjusted) cost parameters are used. The expected performance is then provided as feedback to the black box optimizer. Observe that, for a fixed x', $Q^{SAA}(x', S)$ is separable by sample (scenario) $s \in S$, and typically, it is faster to solve $Q(x', \hat{\xi}_s)$ for each sample s and compute the mean afterwards than directly solving $Q^{SAA}(x', S)$. After a fixed number of iterations, the tuning process is stopped and the set of best-performing parameters is returned.

Speeding Up the Training. In case of combinatorial optimization problems, it can often be observed that small parameter changes do not change the optimal solution. This effect is amplified in our case where we are not interested in the full solution of each parameterized optimization model but only in the values of the first-stage decision variables x. During training, we exploit the behavior by storing all first-stage solutions and their performance values in a hash table, allowing to skip the evaluation of previously seen solutions. We can further speed up the solution of the parameterized nominal problems by passing all existing first-stage solutions to the solver as (partial) warm starts. In particular when used with a MILP solver, these warm start solutions often fall within the optimality tolerance of the solver, which means that the optimization ends in the root node of the branch-and-bound search of the MILP solver.

Offline Tuning and Online Solution Time. While the parameter tuning described above can be time-consuming, it can be carried out offline, that is, before the actual decisions have to be taken. Once a good set of parameters was found, these parameters can be used whenever an (online) problem has to be solved. In fact, one of the key motivations for our approach is to shift some effort to an offline phase, and get high-quality solutions quickly in the online phase when the actual first-stage decisions have to be taken.

3   Multi-activity Shift Scheduling Under Uncertainty

To illustrate the predict, tune and optimize approach and to compare it to a sample average approximation approach, we use a Multi-Activity Shift Scheduling Problem (MASSP) with demand uncertainty. Our study is based on the instances for the deterministic MASSP introduced in [5], and we augment these instances by introducing demand uncertainty.

The MASSP Considered Here. The MASSP variant introduced in [5] consists in designing a set of work shifts covering demands $d_{a,t}$ given per 15-min-period $t \in T$ (T is full day with 96 periods) and per activity $a \in A$; in the largest instances, the set of activities A contains 10 activities. The number n of (anonymous) employees (and thus shifts to be scheduled) is given, and it is assumed that all employees can perform each activity and are subject to the same set of shift legality rules (that is, they have the same working contract). There are two types of feasible shifts: Short shifts consist of two blocks of consecutive work periods (work blocks) separated by a one-period break; the minimal (maximal) number of work periods in a short shift is 12 (23) periods. Long shifts consist of four work blocks separated by two one-period breaks and one 4-period breaks; the minimal (maximal) number of work periods in a long shift is 24 (32) periods. Irrespective of the type of work shift, each work block needs to comprise at least 4 periods (1 h) and within such a work block, no activity changes are permitted. The objective of the MASSP is to minimize the sum of the costs for the total work time and the costs incurred by under- and overcovering demand.

The deterministic version of the MASSP has been considered in various papers and has for example been tackled using grammar-based MIP formulations [3], implicit MIP formulations [4] and matheuristics [7]. In this paper, we employ the MILP formulation proposed in [11] that will be sketched next.

Deterministic MILP Formulation. A key idea of the formulation proposed in [11] is to encode all rules governing shift feasibility in so-called block-based state-expanded networks in which edges correspond to assignments of work blocks and break blocks and nodes are associated with state attributes containing rule-related information such as the number of periods worked so far or the number and types of breaks taken so far in a shift. In addition to those state nodes, the state-expanded network $G = (N, E)$ also contains a source and a sink node as well as a circulation edge from the sink to the source for "counting" the number of shifts to be created. By construction, every source-sink path in the network forms a feasible shift. As an example, a short shift comprises 5 edges: One edge from the source to a node representing the start of a shift followed by a work block edge, a break edge, another work block edge and then an edge pointing to the sink node.

In the mathematical model, the state-expanded network is embedded as a network flow component. In particular, each edge $e \in E$ is associated with an (integer) flow variable x_e, allowing for a flow comprising multiple units (anonymous employees) through a single network. Observe that in the model, the work block edges are "activity"-agnostic, that is, they only represent a possible work block $b \in B$ where B is the set of all possible (and feasible) work blocks in a given instance. The assignment of concrete activities for a given work block is delegated to the integer variables $z_{b,a}$ representing the number of employees whose shifts comprise work block b that are assigned to perform activity a in that block b. Finally, the variables $y_{a,t}^o$ and $y_{a,t}^u$ represent over- and undercovering the demand for activity a in period t. Using those variables, the full mathematical model reads as follows:

$$\min \sum_{e \in E} c_e x_e + \sum_{a \in A} \sum_{t \in T} (c^o y_{a,t}^o + c^u y_{a,t}^u) \tag{13}$$

$$\sum_{e \in v^{\mathrm{in}}} x_e = \sum_{e \in v^{\mathrm{out}}} x_e \qquad \forall v \in N \qquad (14)$$

$$x_{e^{\mathrm{circ}}} = n \qquad\qquad\qquad (15)$$

$$\sum_{e \in E_b^{\mathrm{block}}} x_e = \sum_{a \in A^b} z_{b,a} \qquad \forall b \in B \qquad (16)$$

$$\sum_{b \in B_{a,t}^{\mathrm{cov}}} z_{b,a} + y_{a,t}^{\mathrm{u}} = d_{a,t} + y_{a,t}^{\mathrm{o}} \qquad \forall a \in A, t \in T \qquad (17)$$

$$x_e \in \mathbb{Z}_0^+ \qquad\qquad \forall e \in E \qquad (18)$$

$$z^{b,a} \in \mathbb{Z}_0^+ \qquad\qquad \forall a \in A, b \in B_a \qquad (19)$$

$$y_{a,t}^{\mathrm{o}} \geq 0, \quad y_{a,t}^{\mathrm{u}} \geq 0 \qquad\qquad \forall a \in A, p \in P \qquad (20)$$

The objective function minimizes the sum of the costs incurred by work block assignments (c_e is the cost coefficient of arc e) and the costs associated with over- and undercovering (the respective cost coefficients are c^o and c^u). Constraints (14) are the flow balance constraints, and constraint (15) fixes the total flow in the network to the number of employees available. The constraints (16) links the activity-agnostic blocks b resulting from the flow in the network on edges E_b^{block} representing these blocks to the assignment variables $z_{b,a}$, and the constraints (17) compute the amount of under- and overcovering for each activity and each period. The remaining constraints determine the domains of the decision variables. For a more detailed exposition of this model, we refer to [11]. Observe that in the context of the formalization introduced in the previous section, the model presented here forms a formulation for the nominal problem P^N.

Two-Stage Problem with Demand Uncertainty. In this paper, we consider a stochastic variant of the MASSP problem described above. Specifically, we assume that the demand parameters $d_{a,t}$ are subject to uncertainty, that is $\xi = (d)_{a \in A, t \in T}$. Furthermore, we assume that the first-stage decisions consist in determining the start and end times for the shifts. We assume that demand information becomes available before we have to decide about the detailed work and break assignments within each shift (which form the decisions in the second-stage problem $Q(x)$). The described extension gives rise to a two-stage problem under uncertainty of the form P^U as described in the previous section. With regard to the mathematical model for the MASSP presented above, this means that all flow variables associated with edges starting and ending at the source and the sink node constitute first-stage decision variables. And all other flow variables as well as the y- and z-variables are second-stage variables. Correspondingly, all constraints only involving first-stage variables (in this case, the flow balance constraints associated with the source and with the sink node) are first-stage constraints, while all other constraints are second-stage constraints (and thus part of the problem Q). If we model this problem as a two-stage-stochastic program (e.g. using an SAA approach to form a problem of type P^{SAA}), all second-stage variables and constraints occur once per sample (scenario).

This paper is not the first one in which the MASSP instances were extended to include demand uncertainty: In [10], the authors use these instance in a so-called tour scheduling problem under uncertainty. In particular, we use a similar approach as [10] to generate randomized demand samples based on the demand data of the original MASSP instances from [5].

Prediction and Tuning Parameters. Given a set of random samples from the joint distribution of the uncertain demands, the prediction task consist in determining point estimates $\hat{d}_{a,t}$ to be used in the parameterized nominal model P^{PO}. In a predict-then-optimize setting, one would simply use expected values of the respective parameters, but in a predict-and-optimize setting, we would like to be able to control the prediction in a way that we achieve a good expected performance of the first-stage decisions obtained by solving model P^{PO}. We thus decided to use quantiles of the sample values of each uncertain parameter $d_{a,t}$. More precisely, we introduce a single prediction parameter θ^p. Then, for a given value $0 \leq \theta^p \leq 1$, we consider each demand parameter $d_{a,t}$ separately, and we use the empirical θ^p-quantile based on the sample values $d^s_{a,t}, s \in S$ as the point prediction to be used in the model P^{PO} for each uncertain demand parameter.

In addition, we introduce three tuning parameters:

θ^t_1 aims at penalizing shifts that start late. For each edge e representing the first work block of a shift, the cost coefficient c_e is multiplied by a factor of $1 + \theta^t_1 t^{\text{rel}}$, where t^{rel} is the relative start time of work block (that is, the start time normalized in an interval between 0 and 1).

θ^t_2 aims at penalizing shifts with very high workload. If an edge e represents a work block that ends with a workload which is greater than the average workload of an employee given the demand data considered in the parameterized nominal instance, the a factor of $1 + \theta^t_2 w^{\text{rel}}$, where w^{rel} is the normalized difference between the periods worked after the work block and the average per-employee workload.

θ^t_3 is used as a factor the over-covering penalty costs c^o, that is, c^o in the original model is replaced by $\theta^t_3 c^o$.

As can be seen from the description, the tuning parameters θ^t affect certain objective function coefficients, namely the coefficients of the edge flow variables and the coefficients representing the penalty for overcovering demand. To clarify this, we can write the parameterized objective function as follows:

$$\min \sum_{e \in E} c_e(\theta^t) X_e + \sum_{a \in A} \sum_{p \in P} \left(c^o(\theta^t) Y^o_{a,p} + c^u Y^u_{a,p} \right) \tag{21}$$

Note that the tuning parameters were mainly chosen based on intuition gained from examining different solutions and their performance under uncertainty. Probably, better parameters can be found, but investigating these is beyond the scope of this paper. In general, it can be assumed that domain knowledge, experience and intuition is helpful for determining a small but useful set of tuning parameters.

4   Computational Experiments

The evaluation was performed with a subset of the MASSP instances from [5]. The original instance set contains 10 groups of 10 instances each. Each group is characterized by its number of activities (1 to 10 activities). From these instances, we picked the the instance groups with an even number of activities, and for each group, we took the first five instances, yielding 25 instances in total. Note that within each group, the instances vary considerably with respect to the number of available employees and with respect to the workload per employee.

In our experiments, we compare the following approaches:

P a naive predict-then-optimize approach using the expected values as prediction for the uncertain parameters

PO predict and optimize. For this variant, we consider two parameter tuning variants: instance-specific tuning where parameters are tuned for a specific instance and global tuning where parameters are tuned to exhibit the best average performance on all considered instances

PTO predict, tune and optimize. Again, we consider both instance-specific and global parameter tuning

SAA a classical two-stage stochastic program in which the scenarios correspond to equally weighted random samples

Observe that only the approaches PO and PT involve an (offline) training phase in which parameters are tuned; the other approaches (the naive P approach as well as the more sophisticated SAA approach) involve no training but rely on directly solving a model using online available information. Observe, however, that, as we will see later, the SAA approach requires substantially more computational effort during the online phase than the P and PO approaches.

All experiments were implemented in Python, and all MILP models solved with Gurobi 9.5.1 with standard parameters, except that we set the *mipgap* tolerance to 0.005. The training/parameter tuning for the PO and PTO approaches was carried out with the hyperparameter tuning framework Optuna [1]. The computer used for the experiments was a standard laptop with an Intel Core i7 10750H processor clocked at 2.66 GHz with 6 cores and 32 GB RAM.

Training/Parameter Tuning. The approaches P and PO require tuning the parameters θ^p and (in case of PTO) θ^t. The start values of the parameters to tune are the 50 %-quantiles for the demand parameters and the "nominal" values, that is, the original non-adjusted values in case of the tuning parameters. As briefly mentioned above, we perform two different types of training for both approaches: An instance-specific training in which we determine a specific parameter vector for each of the 25 instances, and a global training in which we tune a single parameter vector that is optimized for a good average performance on all instances. Irrespective of the type of training, we use 1000 samples per instance for approximating the θ^p quantile values of the uncertain demands per activity and period, and 100 samples for the sample-average approximation

Q^{SAA} used to evaluate the training performance of a given parameter vector. Both sets of samples are used in each trial when dealing with a certain instance, but they are different from the samples that are used later for the out-of-sample evaluation when comparing the different approaches.

For the training, we performed 100 tuning iterations per instance and per approach (PO and PTO). Some statistics regarding the (instance-specific) training times for both approaches are reported in Table 1. Specifically, for both approaches, it shows the training time when using a "plain" training approach and compares them to the reduced training times that result from applying the speedup techniques explained in Sect. 2. For the PO (PTO) approach, the training time was reduced by more than 50% (40%) on average. The reason for the reduction is that for the PO (PTO) training, in on average 70% (56%) of all trials, the parameter settings yielded first stage solutions that had already been found before – this means that the solution of the parameterized model P^{PTO} was faster to solve since the warm-starting solutions were most likely already within the desired optimality range, and that no performance evaluation using Q^{SAA} was needed. Also, it can be observed that even for the most complex instances, using the speedup techniques, the training time for the PTO approach never exceeded 3.5 h.

Table 1. Average training times (in s) with and without the speedup techniques explained in Sect. 2 (instance-specific tuning)

inst. group	PO			PTO		
(# act)	time(plain)	skip(%)	time(speedup)	time(plain)	skip(%)	time(speedup)
2	3403	83	977	2782	75	1003
4	6936	77	2526	5449	64	2575
6	12072	68	5627	8521	52	4955
8	20607	58	11651	11426	41	7745
10	33681	65	16282	19159	48	11764
total avg	15340	70	7413	9467	56	5608

Observe that Table 1 only reports the times required for the instance-specific training. We do not report the times of the generic training here because the average training time spent per instance is almost identical to the instance-specific training, and thus, the total training time is very similar to the total time spent for the instance-specific training.

Testing/Out-of-Sample Evaluation. To evaluate the performance of the PTO approach and to compare it to alternative approaches for making first-stage decisions, we use a so-called out-of-sample evaluation. We draw two new sets of samples (which are different to those used for training) for each of the 25 instances: One sample set comprising 1000 samples is used to compute the prediction parameter θ^p, and the other set comprising 200 samples is used for computing the approximate performance of the first-stage decisions obtained with each approach. Table 2 reports the out-of-sample performance of the approaches and their variants explained above. Specifically, the results are expressed in relative expected cost (in %) compared to the naive predict-then-optimize approach P. Each cell represents the average of these percentages across the five instances in each group, and the last row displays the total average across all considered instances. It turns out that all evaluated approaches yield significantly better costs than the approach P. Overall, the PTO approach performs significantly better than the PO approach. Interestingly, even if using a single "globally tuned" parameter vector for all instances, PTO consistently performs better than the PO approach with parameters tuned specifically for each instance. When comparing the PTO approach to the SAA approach, it turns out that on average, the instance-specific PTO approach performs better than the SAA approach with 3 samples (scenarios), while the SAA with 5 scenarios performs slightly better than the PTO approach.

Table 2. Average expected cost in % of expected cost from predict-then-optimize approach P (out-of-sample evaluation)

| inst. group | P | PO | | PTO | | SAA | |
(# act)	mean	specific	global	specific	global	$|S| = 3$	$|S| = 5$
2	100.0	91.8	94.2	90.5	94.1	94.3	92.0
4	100.0	92.1	92.2	91.3	91.7	90.4	89.8
6	100.0	91.2	91.8	88.4	90.0	88.4	87.6
8	100.0	92.3	93.0	89.0	89.8	90.6	88.2
10	100.0	95.3	95.1	92.4	93.7	93.2	91.2
total avg	100.0	92.5	93.3	90.3	91.9	91.4	89.7

Let us now consider the solution time needed for to solve the models for determining first-stage decisions for each of the approaches. Table 3 reports the average solution times needed for each instance group (5 instances per group) and for all 25 instances. As can be expected, it turns out that the approaches relying on solving a parameterized deterministic model can be solved faster than the SAA-approaches in which the model size is in the order of the number of samples times of the size of the deterministic model. As an example, while for the 10-activity instances, the deterministic model exhibits about 80,000 (integer) decision variables, the SAA model with a sample size $|S| = 5$ has almost 390,000 decision variables. This is reflected in the solution times of the approaches. As an example, while the average solution time for the models from the PTO approach is under one minute, the SAA model with $|S| = 5$ takes almost 20 min to solve on average. Even the SAA model with $|S| = 3$ requires an order of magnitude more time on average to solve than the parameterized deterministic models in the PTO approach. Interestingly, while the models to be solved in the P, PO and PTO approaches have exactly the same size, the PTO models are solved significantly faster on average than the models from the other two approaches. We suspect that the parameterization of many objective function coefficients in PTO has a positive effect on the solution time since it basically forms a sort of objective function perturbation.

Table 3. Average solution time in seconds per instance to obtain first-stage decisions

inst. group (# act)	P mean	PO specific	PO global	PTO specific	PTO global	SAA $|S| = 3$	SAA $|S| = 5$
2	15	25	24	23	27	144	461
4	50	67	80	67	66	247	753
6	65	76	85	50	48	322	1287
8	90	115	121	61	43	426	1650
10	136	174	175	73	92	524	1429
total avg	71	91	97	55	55	333	1116

5   Related Work

As mentioned in the introduction, our work combines ideas from the emerging field of decision-focused learning with tuning deterministic model parameters for decision-making under uncertainty. Compared to decision-focused learning (see [8] for a recent survey), where the training of a prediction model is aligned with the downstream optimization problem, we consider a simplified setting since we do not use contextual feature information for prediction, but assume the availability of a sample of (or a sampling mechanism for) the uncertain problem parameters. Note, however, that our approach can be combined with a probabilistic prediction model that can provide a way to sample from a distribution that is conditional on contextual information. Also note that while most approaches for decision-focused learning are restricted to single-stage problems and to uncertain parameters in the objective function, our approach is capable of handling two-stage problems with recourse decisions and uncertainty in the constraints.

Modifying parameters of deterministic optimization models for better performance under uncertainty is not a new idea. In fact, it is often used in practice, e.g. by adding artificial slack in scheduling problems to increase robustness of schedules against small disruptions or delays. In multi-stage stochastic optimization settings, the idea of tuning parameters of deterministic optimization models that are used in a type of lookahead policy has been recently considered in different works. In [9], this idea is termed *parametric cost function approximation*, and in [12], tuning so-called "virtual parameters" affecting a deterministic optimization model is part of the generic UNIFY framework for multi-stage optimization under uncertainty. Moreover, following [6], such a parameter tuning can be an offline component of an integrated offline/online approach to solving multi-stage optimization problems under uncertainty. Note, however, that we are not aware of any approach applying this idea to an inherently two-stage stochastic optimization problem.

6   Conclusions

In this paper, we propose the approach predict, tune and optimize for solving two-stage optimization problems under uncertainty. The approach combines the idea of making decision-focused predictions affecting stochastic model parameters with tuning non-stochastic model parameters. We illustrate the approach using a multi-activity shift scheduling problem with demand uncertainty, showing that the new approach performs better than a pure (simplified) predict and optimize approach. In future work, we aim at extending our work to settings in which the prediction part involves contextual data, that is, in a setting that can be framed as contextual stochastic optimization.

References

1. Akiba, T., Sano, S., Yanase, T., Ohta, T., Koyama, M.: Optuna: a next-generation hyperparameter optimization framework. In: Proceedings of the 25th ACM SIGKDD International Conference, KDD 2019, pp. 2623–2631. Association for Computing Machinery, New York (2019)
2. Birge, J.R., Louveaux, F.: Introduction to Stochastic Programming. Springer, Berline (2011)
3. Côté, M.C., Gendron, B., Rousseau, L.M.: Grammar-based integer programming models for multiactivity shift scheduling. Manag. Sci. **57**(1), 151–163 (2010)
4. Dahmen, S., Rekik, M., Soumis, F.: An implicit model for multi-activity shift scheduling problems. J. Sched. **21**(3), 285–304 (2018)
5. Demassey, S., Pesant, G., Rousseau, L.-M.: Constraint programming based column generation for employee timetabling. In: Barták, R., Milano, M. (eds.) CPAIOR 2005. LNCS, vol. 3524, pp. 140–154. Springer, Heidelberg (2005). https://doi.org/10.1007/11493853_12
6. Filippo, A.D., Lombardi, M., Milano, M.: The blind men and the elephant: integrated offline/online optimization under uncertainty. In: Twenty-Ninth International Joint Conference on Artificial Intelligence, vol. 5, pp. 4840–4846 (2020)
7. Hernández-Leandro, N.A., Boyer, V., Salazar-Aguilar, M.A., Rousseau, L.M.: A matheuristic based on Lagrangian relaxation for the multi-activity shift scheduling problem. Eur. J. Oper. Res. **272**(3), 859–867 (2019)
8. Kotary, J., Fioretto, F., Van Hentenryck, P., Wilder, B.: End-to-End Constrained Optimization Learning: A Survey (2021)
9. Powell, W.B., Ghadimi, S.: The Parametric Cost Function Approximation: a new approach for multistage stochastic programming (2022)

10. Restrepo, M.I., Gendron, B., Rousseau, L.M.: A two-stage stochastic programming approach for multi-activity tour scheduling. Eur. J. Oper. Res. **262**(2), 620–635 (2017)
11. Roemer, M.: Block-Based State-Expanded Network Models for Multi-Activity Shift Scheduling (2022)
12. Silvestri, M., De Filippo, A., Lombardi, M., Milano, M.: UNIFY: A Unified Policy Designing Framework for Solving Constrained Optimization Problems with Machine Learning (2022)

On Learning When to Decompose Graphical Models

Aleksandra Petrova$^{(\boxtimes)}$ (iD) and Javier Larrosa (iD)

UPC Barcelona Tech, Barcelona, Spain
{apetrova,larrosa}@cs.upc.edu

Abstract. Decomposition is a well-known algorithmic technique for Graphical Models. It is commonly believed that such a technique is cost-effective for instances with low width. In this paper, we show on a large data set of real-life inspired instances that this is not the case. To better understand this result, we narrow our study and consider *k*-tree instances where the width is well controlled and get similar results. Finally, we show that by adding a few simple features and using simple Machine Learning models we can predict the convenience to decompose with an accuracy of more than 85%, which produces time reductions in standard benchmarks of nearly 90%.

Keywords: Machine Learning · Graphical Models · Tree Decomposition · Discrete Optimization

1 Introduction

Graphical Models [15] is an umbrella term that covers a broad number of modeling languages for combinatorial problems such as *Bayesian Networks*, *Markov Networks* [12], *(Weighted) CSPs*, *etc.*, that have attracted intense research for roughly four decades [10, 15]. What they all have in common is that problems are modeled as a set of *variables* and a set of *functions*, the scope of which is a small subset of the variables. Therefore, the problem is given in a factorized form, as a set of *local* pieces of information[1]

Most state-of-the-art algorithms are based on search. The simplest search space is the so-called OR tree. Each path from the root to a leaf corresponds to a complete assignment of the variables where each step of the path is the assignment of one more variable. In order to make the search efficient many sophisticated techniques have been developed over the years. For instance, algorithms for CSPs and Weighted CSPs enforce local consistencies at each node. Enforcing a local consistency propagates the effect of the node's associated partial assignment and simplifies the node's associated subproblem [33].

---

[1] A line of work under the name of *soft global constraints* extends this definition by allowing large scope cost functions as long as the functions are tractable.

Supported by grant PID2021-122830OB-C43, funded by MCIN/AEI/10.13039/501100011033 and by "ERDF: A way of making Europe".

M. Sellmann and K. Tierney (Eds.): LION 2023, LNCS 14286, pp. 270–285, 2023.
https://doi.org/10.1007/978-3-031-44505-7_19

The structure of a graphical model is captured by its *interaction graph* where vertices correspond to variables and edges correspond to explicit interactions between them (from the functions). One of the many uses of the interaction graph is to identify *conditional independencies*, which means that during the search, algorithms can *decompose* the current sub-problem into independent components that can be solved separately. The incorporation of this idea corresponds to transforming the OR tree into a more compact search space called AND/OR tree. If each time a sub-problem is solved its solution is cached in order to avoid redundant solving, then the search space becomes the AND/OR graph [16]. The size of the OR tree is exponential on the number of variables, while the size of the AND/OR graph is exponential on a structural parameter called *width*, always smaller than the number of variables.

Since the AND/OR graph is always smaller than the OR tree one could assume that decomposing is always a good idea. However, decomposition-based algorithms (i.e. those traversing AND/OR search spaces) have some disadvantages. For instance, they require more sophisticated data structures, which cause a non-negligible overhead. Over more, they limit, make more complicated, or less effective the use of some techniques (e.g. dynamic variable ordering, pruning, ...). Because of that, for some problem instances, it is more convenient to decompose and for others, it is better not to decompose. This is the problem that we address in this paper: given a problem instance and a search strategy, predict whether we should use this search strategy on OR or AND/OR spaces.

Considering the size of the search spaces, the answer seems straightforward: one should use a decomposition-based algorithm for problems whose width is small and a non-decomposition-based algorithm when the width is large. The only thing that would remain is to investigate where the transition occurs, and what happens in the whereabouts. Surprisingly enough, there is a huge bibliography about decomposition-based algorithms but, to the best of our knowledge, nobody has systematically analyzed how well the width predicts the choice of the best algorithm.

The first contribution of the paper is an unexpected observation that contradicts conventional wisdom. On a very large data set of real-life inspired WCSPs and using depth-first search, low width is completely uncorrelated with the instances for which decomposition is best. To better understand such findings, we repeat the experiment with k-trees which are synthesized instances where we can control the width while other features are kept fixed or made random. We see that, again, the width does not tell when to decompose. We then approach the problem as a Machine Learning problem and augment the set of features. A simple SVM model provides reasonable results with an accuracy between 80% and 90% on different scenarios.

For the sake of simplicity and lack of space, in this paper we restrict ourselves to WCSPs (i.e. minimizing an additive objective function made of local tabular cost functions) solved with the toolbar2 solver. WCSPs have many practical applications in many different domains. Because of that, there are many benchmarks for testing purposes and several available solvers. Toolbar2 is an award winning solver that offers three different algorithms (DFS, RDS, HBFS) with their decomposition-based version. In the three cases decompositions are implemented following BTD [26].

2 Preliminaries

2.1 Graphical Models

A *Graphical Model* is a tuple $P = (X, D, F)$ where $X = \{x_1, \ldots, x_n\}$ is the set of *variables*, $D = \{d_1, \ldots, d_n\}$ is the (finite) set of *domains* (d_i is the domain of x_i), F is a set of *functions*. Each function $f_S \in F$ has associated a subset of variables $S \subseteq X$, called *scope*, and the function assigns a value to each possible assignment of these variables.

A Graphical Model implicitly specifies a global function $F(X) = \oplus_{f_S \in F} f_S(S)$ where $\oplus$ is a well-defined operator for the corresponding values. Depending on the domain there may be different queries of interest associated with the objective function $F(X)$. In general, answering a query corresponds to computing $\otimes_X F(X) = \oplus_{f_i \in F} f_S(S)$ where, again, $\otimes$ is a well-defined operator for the outcomes of the global function.

In this paper, we consider the arguably most common and best-studied case where functions return natural numbers, $\oplus = \sum$ and $\otimes = \min$. In probabilistic problems (i.e. the objective function has a probabilistic interpretation) this task is called *most probable explanation*. In non-probabilistic problems it is usually referred to as *Weighted CSP*.

Fig. 1. Interaction graph of a graphical model with 7 variables (one per node) and arity-2 cost functions (one per edge) (left) and one of its possible tree-decompositions (right).

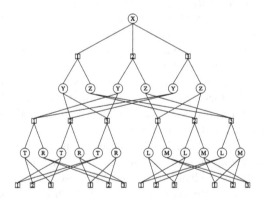

Fig. 2. Minimal AND/OR search graph for the problem in Fig. 1. All variables have domain size 2, except X having domain size 3. Circle and square nodes correspond to OR and AND nodes, respectively. Children of AND nodes represent sub-problem decompositions.

The term *Graphical Model* comes from the existence of an underlying graph that captures important structural properties. Given a Graphical Model $P = (X, D, F)$, its *Interaction graph* $G_P = (V, E)$ is an undirected graph with vertices V and edges E. There is one vertex $i \in V$ associated with each variable $x_i \in X$, and there is an edge $(i, j) \in E$ if and only if there some cost function $f_S \in F$ with $\{i, j\} \subseteq S$. Thus, the interaction graph tells pairs of variables that are linked (or connected) via cost functions.

Figure 1 (left) shows the interaction graph of a graphical model that has 7 variables (one associated with each graph vertex). The graph having edge (X, Y) indicates that there is one cost function having variables X and Y in their scope.

A **tree-decomposition** of a graphical model $P = (X, D, F)$ is a tree $T = (V, A)$. For every vertex $e \in V$ there is a cluster $C_e \subseteq X$. The set of clusters must cover all the variables (i.e., $\cup_{e \in V} C_e = X$) and all the cost functions (for every $f_S \in F$ there is some cluster C_e such that $S \subseteq C_e$). Furthermore, if a variable x_i appears in two clusters C_e and C_k, it must also appear in all the clusters on the unique path from e to k (this is called the running intersection property).

The **tree-width** (or just *width*) of a tree decomposition, noted w, is $\max_{e \in V} \{|C_e|\} - 1$. The width of G is the minimum tree width of all tree decompositions of G. Figure 1 (right) shows one tree decomposition associated with the interaction graph on the left. Its width is 1, which corresponds to the extreme case of the interaction graph being acyclic.

2.2 Decomposition-Based Backtracking Algorithms

The simplest way to systematically generate all possible assignments in a graphical model is by extending partial assignments to one more variable at a time. From a partial assignment, select one unassigned variable and make one child for each value in its domain. Starting from the empty assignment, the rule is applied recursively until nodes correspond to complete assignments. This is OR search tree and it is easy to see that its size is exponential on the number of variables $O(\exp(n))$.

Consider the WCSP-producing graphs in Fig. 1. The assignment of variables X produces a sub-problem made of two independent components whose variables are $\{Y, T, R\}$ and $\{Z, L, M\}$. The search space associated with solving independent components separately is called AND/OR tree. Note that each tree node corresponds to a sub-problem and different nodes may represent the same sub-problem. When it happens, traversing the AND/OR tree means solving the same problem more than once. To overcome such inefficiency the algorithm may record each sub-problem it solves. Then the search space becomes the so-called AND/OR graph [16][2]. Tree decompositions can be used to direct the generation of an AND/OR graph. For instance, the tree decomposition in Fig. 1 (right) using cluster (X, Y) as root determines the AND/OR graph of Fig. 2. It is known that the size of the AND/OR graph determined by a tree decomposition of width w is $O(\exp(w))$.

Decomposing graphical models is an important technique, but the efficiency of current state-of-the-art solvers mostly depends on other complementary ideas. One line

---

[2] Note that this corresponds to solving the problem with dynamic programming implemented with memoization.

of work studies the use of different search strategies. While the most common one is depth-first search (DFS), some authors have identified benefits for best-first search (BFS) [34], hybrids (i.e., HBFS [2]) or nested (i.e., RDS [40]), to name a few. In our description of search spaces we did not mention in which order variables are found along paths and values are found among children. Such orderings may have a dramatic effect on performance. Another very relevant line of work is on finding cheap yet useful simplifications to be applied dynamically during the search. Arguably, the most common approach to this is through *local consistencies* which apply equivalence-preserving transformation (EPTs) at each sub-problem that identify unfeasible values and anticipate backtracking. The conjunction of all these techniques and some other constitutes state-of-the-art solvers such as toulbar2. However, the combination of all these techniques with decomposition-based algorithms is possible, but very often limited or problematic [2, 35, 38].

3   A Preliminary Experiment

Conventional wisdom supported by theoretical results about the size of search spaces says that if we have a problem for which we know a tree decomposition with a small width, we should use a decomposition-based algorithm (i.e. an algorithm that traverses the AND/OR search space). Similarly, if we do not have a tree decomposition with a small width, we should use a non-decomposition version of the algorithm because it may cause overhead or may compromise other algorithmic techniques.

In this section, we want to test this hypothesis in practice. For the sake of simplicity, we focus on weighted CSPs (WCSPs) because there is a rich collection of instances that have been used for testing over the years, and because there are many solvers available. In particular, we will be using toulbar2 because of its proven efficiency across a variety of domains [25] and because it provides implementations of several algorithms with and without decomposition. Besides, the decomposition version of toulbar2 implementations always follows the principles of BTD [39].

For the experiment, we collected a large sample of real-world inspired instances from the 2 largest WCSP repositories: *EvalGM*[3] and *Cost Function Library*[4]. Instances come from many different applications such as *Gene sequencing, Satellite allocation, Warehouse allocation, Protein design, Object detection*, etc. Each instance without global constraints (which are out of the scope of this work) was VAC-preprocessed and subsequently solved with toulbar2[5] with depth-first-search with and without decomposition. We will refer to these algorithms as DFS-BTD and DFS, respectively.

From the original set of 8109 instances, we removed those that were too difficult (neither DFS-BTD nor DFS could solve them with less than 10^8 backtracks) or too easy (both algorithms required less than 10 s). This reduced the dataset to a total number of 2550 instances. Note that we use the number of backtrackings as a proxy of CPU time as it is known that both measures are highly correlated and backtrackings have

---

[3] http://genoweb.toulouse.inra.fr/~degivry/evalgm/.

[4] https://forgemia.inra.fr/thomas.schiex/cost-function-library.

[5] https://toulbar2.github.io/toulbar2/index.html.

the advantage of being machine independent, which makes experiments much easier to conduct.

Let b_i^d and b_i denote the number of backtrackings for instance i with and without decomposition, respectively. Figure 3 exhibits b_i^d versus b_i. It clearly shows that there are many instances for which one algorithm is orders of magnitude faster than the other. In this dataset DFS-BTD outperforms DFS 57.5% of the time. The aggregated running of DFS and DFS-BTD is $\sum_i b_i = 3.4 \times 10^{10}$ number of backtrackings and $\sum_i b_i^d = 3.2 \times 10^{10}$ number of backtrackings, respectively. Therefore, in aggregated form both algorithms are roughly equally efficient. However, if we could predict and solve with the best option the total number of backtrackings would be $\sum_i \min\{b_i, b_i^d\} = 1.3 \times 10^9$ which is more than a 95% reduction. On average choosing the best algorithm with respect to the worst produces a backtracks reduction of 98% from the worst (80% reduction in CPU time).

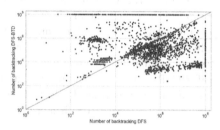

Fig. 3. Performance of DFS vs DFS-BTD on EvalGM and Cost Function Library instances. Note the logarithmic scale.

To analyze the relation between width and algorithmic advantage we model a linear regression of $\{(w_i, y_i)\}$ where y_i is the order of magnitude of one algorithm being better than the other (negative for DF-BTD) defined as,

$$y_i = \begin{cases} \log \frac{b_i^d}{b_i} & \text{if } b_i \leq b_i^d \\ -\log \frac{b_i}{b_i^d} & \text{otherwise} \end{cases} \quad (1)$$

This data set is plotted in Fig. 4 on the left, which shows that the advantage of one algorithm over the other is unrelated to the width. The Pearson correlation is 0.07. Next, we study the potential of the width as a predictor (i.e. a classification problem). For this purpose we consider dataset $\{(w_i, c_i)\}$ where c_i represents the algorithm of choice, defined as,

$$c_i = \begin{cases} 0 & \text{if } b_i \leq b_i^d \\ 1 & \text{otherwise} \end{cases} \quad (2)$$

This dataset is plotted in Fig. 4 on the right. It also shows that predicting the winning algorithm cannot be done with a logistic regression based on the width in isolation. The R^2 value of a Logistic Regression is 0.018.

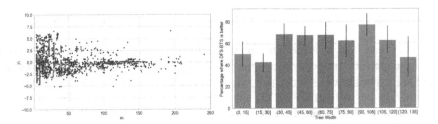

Fig. 4. Comparison of DFS vs DFS-BTD with respect to width on EvalGM and Cost Function Library instances. On the left orders of magnitude of advantage (advantage of DFS-BTD is negative). On the right percentage of instances where DFS-BTD outperforms DFS

There are two relevant observations to be made from this experiment: *i*) being able to predict when to decompose may produce significant gains and therefore has practical implications, and *ii*) *width* (at least in isolation) is not a good feature to make the prediction (at least for general instances). Beyond the interest of the surprising results, the first observation justifies the relevance of the topic of our research, and the second observation motivates further analysis. One possible explanation of our empirical finding is that the space of WCSPs is too large and/or our data is not general enough. Another explanation is that other features may need to be added to increase the accuracy in the prediction. The following section tries to confirm or discard the explanations.

4 Machine Learning for Decomposition

4.1 Random k-Trees

Since the space of WCSPs is huge and the dataset from the WCSP libraries is not too diverse, in the following we will generate synthetic data that allows us a much more controlled experiment. In particular, we want to generate instances where we can control the width. For that purpose, we implemented a generator of *random k-trees* [37] of n nodes as follows: starting with a $k + 1$-clique, add nodes one by one. Each time a new node is added, it is connected to all the nodes of a randomly selected k-clique embedded in the current graph. It is well-known that given a k-tree there is a tree decomposition of width k (it can be easily obtained using the min-fill order). Additionally, there is no tree decomposition with a width less than k.

Recall that the width of a tree decomposition is the maximum size of its clusters. It often happens that such maximal size is rare among the many clusters. Then, a high width may be a misleading measure of bad decomposability, since it will also be exceptional during the execution of decomposition-based algorithms. One advantage of k-trees is that they are more robust in terms of width because all clusters in the tree decomposition have size k. Therefore, their width is a much more precise measure of the amount of decomposability that can be achieved.

4.2 Instances with Random Cost Functions

In our first experiment, we generated random k-tree structured WCSPs of domain size m with *random cost functions* of density d. A *random cost function of density d* is generated by associating cost 1 to d out of the m^2 cost function entries. Thus, random instances are generated subject to four parameters (k, m, d, n), where $k = w$ is the instance width.

We generated an *independent and identical distributed* (IID) data set of this type of problem with parameters bounded by $3 \leq w \leq 20$, $2 \leq m \leq 10$, $m^2 \times 0.1 \leq d \leq m^2 \times 0.5$ and $21 \leq n \leq (1000/w + (20 - n)/17 \times 250) \times (-0.5 \times d + 2)$. The upper bound for n which depends on the other parameters was decided based on previous experiments to obtain instances of reasonable difficulty. We generated 26000 instances that were solved with both DFS and DFS-BTD (after VAC pre-processing) with a time out of 10^8 backtrackings. 6822 instances were discarded because neither algorithm could finish within the timeout leaving a total of 19178 instances. In this dataset, DFS outperforms DFS-BTD 67.1% of the time. The aggregated number of backtrackings for DFS and DFS-BTD is $\sum_i b_i = 6 \times 10^{10}$ and $\sum_i b_i^d = 3.8 \times 10^{10}$, respectively, which means that also in this much more specific benchmark, both algorithms are roughly equally efficient in the long run. If we could predict and execute each instance with the best-performing algorithm, the aggregated number of backtrackings would be $\sum_i \min\{b_i, b_i^d\} = 6 \times 10^9$, which is more than 80% reduction. The average reduction of choosing the best-performing algorithm with respect to the worst-performing is 93.46% percent. The plot of b_i vs b_i^d (Fig. 5 left) shows a much stronger correlation between the difficulty of instances for both algorithms (b_i vs b_i^d). Also, it shows that DFS is better for easy instances (low number of backtrackings) and DFS-BTD is better for the hardest instances.

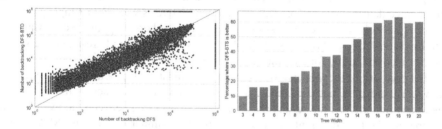

Fig. 5. Performance of DFS vs DFS-BTD on random k-tree instances. Note the logarithmic scale (left). Percentage of instances where DFS-BTD outperforms DFS vs width on random k-tree instances (right).

We again consider the classification problem of predicting c_i. Figure 5 shows, for each width, the percentage of times that DFS-BTD is better than DFS. In the preliminary experiment reported in Sect. 3 we observed that width could not explain the class (Fig. 3 right). Now the result is even more surprising: the width is related to the class, but not as expected. The higher the width, the most likely DFS-BTD will be the winning algorithm.

Because plot 5 shows an unexpected but clear trend, we conducted a simple Logistic Regression of $\{(w_i, c_i)\}$. We split the data into training, testing and validation. Upon training the model with just the width as a feature, we then tested the model and obtained a 71% accuracy. However, the confusion matrix (Fig. 6 (left)) shows that although the accuracy is high it is quite unbalanced. In particular class DFS is predicted quite accurately, but predictions for class DFS-BTD are very inaccurate.

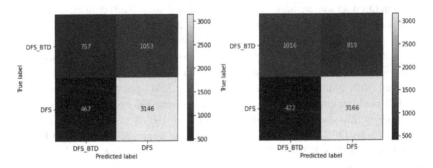

Fig. 6. Confusion matrix of the Logistic Regression with only width as a feature (left), and confusion matrix of the Logistic Regression with features (w, n, m, d, d') (right).

The lesson learned from the previous experiment is that width in isolation cannot be used to advise on whether to decompose or not, even in nicely structured k-trees. Thus, we need to incorporate other features and other ML models. Accordingly, we incorporate as features the other parameters that characterize the creation of instances: the number of variables (n), domain size (m), and density (d). Since instances were pre-processed using VAC, which can alter the density of cost functions, we included the average density after pre-processing (d') as a feature as well. Four simple Machine Learning models (Logistic Regression, k Nearest Neighbors, Support Vector Machine and Decision Trees) and two Ensemble Method models (Random Forest and XGBoost) were trained and tested for classification.

For each one of the models, we followed the standard ML pipeline. Figure 6 (right) shows the confusion matrix of the Logistic Regression. The accuracy increases to 78% and, most importantly, the results are slightly more balanced between the two classes. Comparing the four different methods, they produce similar results with SVM being slightly better. Its accuracy is 82%, precision and recall for DFS is 0.82 and 0.93, respectively and for DFS-BTD it is 0.80 and 0.60. Ensemble models produce additional gains in balancing the quality of results between both classes, with XGBoost being the best. Its accuracy is 82%. The precision and recall of the DFS are 0.83 and 0.92 respectively. The precision and recall for DFS-BTD are up to 0.79 and 0.62. The aggregated running time of DFS-BTD and DFS on the test set is 10^{10} and 1.6×10^{10}. The aggregated running time of executing the SVM predicted best algorithm is 5.5×10^9 which means that using this model we can reduce the time more than 50%. Table 1 reports the results of selected ML models.

Next, we want to evaluate if these results can be transferred to other algorithms. So we repeated the experiment with *Hybrid Best First Search* (HBFS) [2]. HBFS is a hybrid of Depth and Best First Search. It maintains a set of open nodes à la BFS, but each time a node is selected for expansion, it is done according to the DFS strategy for a fixed amount of time. The advantage of HBFS over DFS is two-fold: it makes a more flexible traversal of the search space not being committed to early decisions and, it produces both any-time lower and upper bounds during the solving process. Toulbar2 implements HBFS with and without tree decomposition.

We took the same set of instances and solved them with HBFS and HBFS-BTD at the same time out of 10^8 backtrackings. We removed unsolved and trivial instances (as we did with DFS) which left us with 19209. With this data set we performed Transfer Learning with the SVM model, because it is robust and as a simple model had similar performance with more complex models. To be able to do the transfer learning we kept the same hyperparameters of the model and trained the model with 70% of the data. No tuning of the model was performed. The model was then tested and achieved the following results. Accuracy of 77%, precision and recall for HBFS of 0.83 and 0.84, and precision and recall for HBFS-BTD of 0.60 and 0.58. Although the results are not as high as in the DFS case we can say that we can successfully use the same model with the same hyperparameters on another algorithm.

Table 1. Summary of the results on random k-trees with random cost functions for different Machine Learning models

Dataset	Random	Random	Random	Random
Features	w	w, n, m, d, d'	w, n, m, d, d'	w, n, m, d, d'
Model	Log. Reg.	Log. Reg.	SVM	XGBoost
Accuracy	72%	78%	82%	82%
Precision decomp.	0.72	0.71	0.80	0.79
Recall decomp.	0.42	0.55	0.60	0.62
Precision n-decomp.	0.75	0.80	0.82	0.83
Recall n-decomp.	0.87	0.89	0.93	0.92

4.3 Instances with Deterministic Cost Functions

In the previous subsection instances had random cost functions. So we had two sources of randomness: the k-tree structure and cost functions. In this subsection we want to see if more accuracy of ML models can be obtained by restricting the randomness of the instances. With this goal, we repeated the experiment with random k-trees but deterministic cost functions. We decided to use cost functions reminiscent of *Frequency Assignment Problems* (FAP) [7] where cost functions give higher costs to closer values,

$$f(x_i, x_j) = m - 1 - |x_i - x_j|$$

We generated IID random instances within the following intervals: $3 \le w \le 21, 2 \le m \le 10$ and $21 \le n \le 1000/w \times 2$. Note that instances with $m = 2$ correspond to a weighted version of the 2 Graph Coloring problem.

We generated 15000 instances that were VAC pre-processed. All instances were solved with DFS and DFS-BTD with the usual time out of 10^8 backtracks. Unsolved and trivial instances were removed leaving 11513 instances. In this dataset, DFS-BTD outperforms DFS 65.5% of the time. The aggregated number of backtrackings for DFS is 2.5×10^{11}, whereas for DFS-BTD it is 2×10^{10}. This time we can see a more significant difference between the efficiency of both algorithms in the long run. Taking the best-performing algorithm for each instance would result in an aggregation of 3.3×10^9. Which is a 83.5% time reduction. The average advantage of the best-performing algorithm compared to the worst gives a reduction of 98.77%. In plot 7 we can see the relationship between DFS and DFS-BTD. Here the scatter is much more uniform, with DFS performing better for instances that require less number of backtrackings, and DFS-BTD taking a small lead in the opposite scenario.

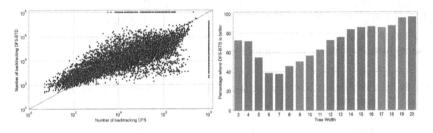

Fig. 7. Scatter plot of the Frequency Assignment dataset with backtracking time, on a logarithmic scale (left) and Winning DFS-BTD percentage per width (right).

Figure 7 on the right shows the percentage of times DFS-BTD is better than DFS for each width. Compared to the previous section, here we see that the majority of times DFS-BTD outperforms DFS. Interestingly, this also happens in the small widths, which wasn't the trend with random cost functions. We see the special case of the Frequency Assignment problems once again as a classification problem. Here we use the same features except for the density because all of the instances keep before preprocessing the maximal amount of tuples per function given the formula.

We turn directly to SVM as our selected model given the success in the previous section. The accuracy is 78%. Precision and recall for DFS are 0.70 and 0.65, whereas for DFS-BTD are 0.82 and 0.85. The aggregated running time of DFS-BTD and DFS on the test set is 5.6×10^9 and 7.6×10^{10}. As seen the model doesn't perform very differently from the previous section above. To ensure consistency we also train and test the other Machine Learning models we previously used. All of them report similar results. From here we hypothesize that cost function information is not a factor that affects the decision of whether the version of the algorithm should be with be decomposition based on non-decomposition based.

Since the results are similar we also test the idea that we can use the same SVM model set up to predict any specific subtype of problem that exists. We use transfer learning, where we keep the same hyperparameters of the model and train the model on 70% of the dataset and later test it on a never before seen testing dataset, which is 30%

of the data. The result achieved is an accuracy of 77%. As it can be seen, the results are quite similar, and transfer learning can happen from a wide range of problems to a specific subtype of problem.

The same procedure of transfer learning is applied also to HBFS. An accuracy of 78% is achieved, with precision and recall for HBFS being 0.72 and 0.83, compared to HBFS-BTD which are 0.85 and 0.75. All these results are summarized in Table 2.

4.4 Benchmark Instances

Finally, we repeated the procedure of SVM transfer learning to the benchmark instances considered in Sect. 3. An accuracy of 87% is achieved, with precision and recall for DFS being 0.80 and 0.94, compared to DFS-BTD which are 0.95 and 0.81.

The aggregated running time of DFS-BTD and DFS on the test set is 10^{10} and 8.8×10^9. The aggregated running time of executing the SVM predicted best algorithm is 10^9 which means that using this model we can reduce the time by nearly 89%.

Table 2. The results of the Machine Learning models on the testing data for the Frequency Assignment problem (FAP) and results from the Transfer Learning. The results from the original SVM model that was used for Transfer learning, are given in the last column for comparison.

Data	FAP DFS	FAP DFS	FAP DFS	FAP HBFS	Bench. DFS	Random DFS
Features	w, n, m, d'	w, n, m, d'	w, n, m, d'	w, n, m, d'	w, n, m, d, d'	w, n, m, d, d'
Model	Log. Reg	SVM	SVM	SVM	SVM	SVM
Set-up	Traditional	Traditional	Transfer	Transfer	Transfer	Traditional
Accuracy	66%	78%	77%	78%	87%	81%
Precision decomp	0.66	0.82	0.82	0.85	0.95	0.78
Recall decomp	1.00	0.85	0.84	0.75	0.81	0.60
Precision n-decomp	0.97	0.70	0.68	0.72	0.80	0.82
Recall n-decomp	0.02	0.65	0.64	0.83	0.94	0.91

5 Related Work

5.1 Decomposition-Based Algorithms

The exploitation of conditional independencies in graphical models has been a long and well-established line of research with a consistent presence in the literature. For a detailed review the reader is urged to check Chapters 8 and 9 in [12] and Chapters 9 and 10 in [14]. There are two works from 1973, when the term Graphical Model was not even established, that deserve to be mentioned. On the one hand, Bertele and Briochi [5] studied the use of dynamic programming to optimization problems given as sets of cost functions and called it non-serial dynamic programming. Although not explicitly said, their algorithms identify and exploit conditional independencies and record solved subproblems. In parallel, and addressing systems of linear equations, the work in [21] also proposed to decompose independent sub-problems.

In the field of *Bayesian Networks* a line of work has looked to exploit a very specific type of decomposition obtained by conditioning on the so-called loop-cutset which is a set of variables that, when instantiated, decompose the problem into acyclic independent sub-problems. The idea of decomposing recursively during the search was proposed in [9] and refined in [1, 11].

In the related field of *Constraint Processing* the first proposal to recursively decompose conditionally independent sub-problems is [18] using the concept of *pseudo-tree search*. In [30] it was shown that the time complexity of the algorithm is bounded by $O(wlogw)$. Recording solved sub-problems to avoid redundancies and therefore decreasing the time complexity to $O(w)$, was also proposed and engineered in [26, 28, 29]. The notion of loop-cutset was taken to this field in [13]. Since in Constraint Satisfaction Problems, cost functions are boolean, better decompositions can be achieved [22].

Weighted Constraint Satisfaction Problems (WCSPs) generalize classical CSPs from boolean to cost functions. In this context, search algorithms use sophisticated propagation techniques (called soft local consistencies) and the difficulty of introducing decomposition techniques was to make them work in practice. Pseudo-tree search was first tested in [32] and sub-problem recording was incorporated in [27]. Some practical improvements appeared in [31] and popular algorithms such as Russian Doll Search and Hybrid Best First Search had the decomposition-based version in [2, 38]

5.2   Machine Learning in Graphical Models

The use of Machine Learning in the context of Graphical Models has been a recent intense topic of research [24, 36]. One approach has been model elicitation, where the goal is to alleviate the user from modeling instances by relying on examples and ML techniques. For example, decision Trees based models are used in [6, 20, 23]

Another approach, closer to our work, is the use of ML to improve the performance of algorithms. In [19] a Deep Learning network is created to guide the search process, and in [4] the use of Reinforcement Learning is seen for the same purpose but aiming at the Traveling Salesman Problem. Some authors focus on learning efficient variable and value ordering heuristics [3, 8, 17, 41]. The closest work to us, to the best of our knowledge is the work of Guerri et al. [23] that applies decision trees to decide if the instance should be solved using IP or CP (the model is applied to Bid Evaluation problems).

6   Conclusions and Future Work

In this paper we address the problem of predicting when it is convenient to use decomposition while solving Graphical Models. We consider Weighted CSPs and classical search strategies. We challenge the usual hypothesis that instance width is the main feature driving such prediction. The first contribution of the paper is the observation that, on a large data set of realistic problems, such a hypothesis does not hold. Furthermore, in a much more width-controlled experiment on k-trees where the space of possible interaction graphs is smaller, the hypothesis does not hold either.

Motivated by this surprising result, the second contribution of the paper is to analyze the potential of Machine Learning models to automatize this classification problem. We restrict ourselves to simple ML models and simple features and observed that all models produced similar results with a simple ensemble model XGBoost being the best. Overmore, models seem to be successfully transferable to different search strategies (such as HBFS) and to both more specific and more general problems.

Our work leaves many interesting open lines of future work. One of them is to corroborate the irrelevance of width in other computational tasks for Graphical Models (i.e. counting or summation problems). Another one is to augment the set of features. In our work, we only used static features that can be obtained by inspecting the instance, but that says nothing about how different cost functions with common variables in their domain interact or how local consistencies may propagate costs. We conjecture that these, more sophisticated features, are the key to further increasing the accuracy of the ML models.

References

1. Allen, D., Darwiche, A.: New advances in inference by recursive conditioning. In: Meek, C., Kjærulff, U. (eds.) UAI 2003, Acapulco, Mexico, 7–10 August 2003, pp. 2–10. Morgan Kaufmann (2003)
2. Allouche, D., de Givry, S., Katsirelos, G., Schiex, T., Zytnicki, M.: Anytime hybrid best-first search with tree decomposition for weighted CSP. In: Pesant, G. (ed.) CP 2015. LNCS, vol. 9255, pp. 12–29. Springer, Cham (2015). https://doi.org/10.1007/978-3-319-23219-5_2
3. Arbelaez, A., Hamadi, Y., Sebag, M.: Continuous search in constraint programming. In: Hamadi, Y., Monfroy, E., Saubion, F. (eds.) Autonomous Search, pp. 219–243. Springer, Heidelberg (2011). https://doi.org/10.1007/978-3-642-21434-9_9
4. Bello, I., Pham, H., Le, Q.V., Norouzi, M., Bengio, S.: Neural combinatorial optimization with reinforcement learning. In: ICLR 2017, Toulon, France, 24–26 April 2017, Workshop Track Proceedings (2017)
5. Bertelè, U., Brioschi, F.: On non-serial dynamic programming. J. Comb. Theory Ser. A **14**(2), 137–148 (1973)
6. Bonfietti, A., Lombardi, M., Milano, M.: Embedding decision trees and random forests in constraint programming. In: Michel, L. (ed.) CPAIOR 2015. LNCS, vol. 9075, pp. 74–90. Springer, Cham (2015). https://doi.org/10.1007/978-3-319-18008-3_6
7. Cabon, B., de Givry, S., Lobjois, L., Schiex, T., Warners, J.P.: Radio link frequency assignment. Constraints Int. J. **4**(1), 79–89 (1999). https://doi.org/10.1023/A:1009812409930
8. Cappart, Q., Moisan, T., Rousseau, L.M., Prémont-Schwarz, I., Cire, A.A.: Combining reinforcement learning and constraint programming for combinatorial optimization. In: Proceedings of the AAAI Conference on Artificial Intelligence, vol. 35, pp. 3677–3687 (2021)
9. Cooper, G.F.: Bayesian belief-network inference using recursive decomposition. Technical report, Knowledge Systems Laboratory, Stanford, CA (1990)
10. Cooper, M.C., de Givry, S., Sánchez-Fibla, M., Schiex, T., Zytnicki, M., Werner, T.: Soft arc consistency revisited. Artif. Intell. **174**(7–8), 449–478 (2010)
11. Darwiche, A.: Recursive conditioning. Artif. Intell. **126**(1–2), 5–41 (2001)
12. Darwiche, A.: Modeling and Reasoning with Bayesian Networks. Cambridge University Press, Cambridge (2009)

13. Dechter, R.: Bucket elimination: a unifying framework for reasoning. Artif. Intell. **113**(1–2), 41–85 (1999)
14. Dechter, R.: Constraint Processing. Elsevier/Morgan Kaufmann (2003)
15. Dechter, R.: Reasoning with Probabilistic and Deterministic Graphical Models: Exact Algorithms. Synthesis Lectures on Artificial Intelligence and Machine Learning, 2nd edn. Morgan & Claypool Publishers (2019)
16. Dechter, R., Mateescu, R.: AND/OR search spaces for graphical models. Artif. Intell. **171**(2–3), 73–106 (2007)
17. Erdeniz, S.P., Felfernig, A.: Cluster and learn: cluster-specific heuristics for graph coloring. In: PATAT 2018, pp. 401–404 (2018)
18. Freuder, E.C., Quinn, M.J.: Taking advantage of stable sets of variables in constraint satisfaction problems. In: IJCAI 1985, Los Angeles, CA, USA, August 1985, pp. 1076–1078 (1985)
19. Galassi, A., Lombardi, M., Mello, P., Milano, M.: Model agnostic solution of CSPs via deep learning: a preliminary study. In: van Hoeve, W.-J. (ed.) CPAIOR 2018. LNCS, vol. 10848, pp. 254–262. Springer, Cham (2018). https://doi.org/10.1007/978-3-319-93031-2_18
20. Gent, I.P., et al.: Learning when to use lazy learning in constraint solving. In: ECAI 2010, Lisbon, Portugal, 16–20 August 2010, Proceedings, vol. 215, pp. 873–878. IOS Press (2010)
21. George, A.: Nested dissection of a regular finite element mesh. SIAM J. Numer. Anal. **10**(2), 345–363 (1973)
22. Gottlob, G., Greco, G., Leone, N., Scarcello, F.: Hypertree decompositions: questions and answers. In: Proceedings of the 35th ACM SIGMOD-SIGACT-SIGAI, PODS 2016, San Francisco, CA, USA, 26 June–01 July 2016, pp. 57–74. ACM (2016)
23. Guerri, A., Milano, M.: Learning techniques for automatic algorithm portfolio selection. In: ECAI 2004, Valencia, Spain, 22–27 August 2004, pp. 475–479. IOS Press (2004)
24. Huang, L., et al.: Branch and bound in mixed integer linear programming problems: a survey of techniques and trends. CoRR abs/2111.06257 (2021). https://arxiv.org/abs/2111.06257
25. Hurley, B., et al.: Multi-language evaluation of exact solvers in graphical model discrete optimization. Constraints Int. J. **21**(3), 413–434 (2016). https://doi.org/10.1007/s10601-016-9245-y
26. Jégou, P., Terrioux, C.: Hybrid backtracking bounded by tree-decomposition of constraint networks. Artif. Intell. **146**(1), 43–75 (2003)
27. Jégou, P., Terrioux, C.: Decomposition and good recording for solving Max-CSPs. In: ECAI 2004, Spain, 22–27 August 2004, pp. 196–200 (2004)
28. Jégou, P., Terrioux, C.: Combining restarts, nogoods and decompositions for solving CSPs. In: ECAI 2014, Czech Republic, 18–22 August 2014, vol. 263, pp. 465–470 (2014)
29. Jégou, P., Terrioux, C.: Combining restarts, nogoods and bag-connected decompositions for solving CSPs. Constraints Int. J. **22**(2), 191–229 (2017). https://doi.org/10.1007/s10601-016-9248-8
30. Bayardo Jr., R.J., Miranker, D.P.: On the space-time trade-off in solving constraint satisfaction problems. In: IJCAI 1995, Québec, Canada, 20–25 August 1995, vol. 2, pp. 558–562 (1995)
31. Kitching, M., Bacchus, F.: Exploiting decomposition in constraint optimization problems. In: Stuckey, P.J. (ed.) CP 2008. LNCS, vol. 5202, pp. 478–492. Springer, Heidelberg (2008). https://doi.org/10.1007/978-3-540-85958-1_32
32. Larrosa, J., Meseguer, P., Sánchez-Fibla, M.: Pseudo-tree search with soft constraints. In: ECAI 2002, Lyon, France, July 2002, pp. 131–135. IOS Press (2002)
33. Larrosa, J., Schiex, T.: Solving weighted CSP by maintaining arc consistency. Artif. Intell. **159**(1–2), 1–26 (2004)
34. Marinescu, R., Dechter, R.: AND/OR branch-and-bound for graphical models. In: IJCAI-2005, Edinburgh, Scotland, UK, 30 July–5 August 2005, pp. 224–229 (2005)

35. Otten, L., Dechter, R.: Anytime AND/OR depth-first search for combinatorial optimization. AI Commun. **25**(3), 211–227 (2012)
36. Popescu, A., et al.: An overview of machine learning techniques in constraint solving. J. Intell. Inf. Syst. **58**(1), 91–118 (2022). https://doi.org/10.1007/s10844-021-00666-5
37. Robertson, N., Seymour, P.D.: Graph minors. III. Planar tree-width. J. Comb. Theory Ser. B **36**(1), 49–64 (1984)
38. Sánchez-Fibla, M., Allouche, D., de Givry, S., Schiex, T.: Russian doll search with tree decomposition. In: IJCAI 2009, California, USA, 11–17 July 2009, pp. 603–608 (2009)
39. Terrioux, C., Jégou, P.: Bounded backtracking for the valued constraint satisfaction problems. In: Rossi, F. (ed.) CP 2003. LNCS, vol. 2833, pp. 709–723. Springer, Heidelberg (2003). https://doi.org/10.1007/978-3-540-45193-8_48
40. Verfaillie, G., Lemaître, M., Schiex, T.: Russian doll search for solving constraint optimization problems. In: IAAI 1996, Portland, Oregon, USA, 4–8 August 1996, pp. 181–187 (1996)
41. Xu, H., Koenig, S., Kumar, T.K.S.: Towards effective deep learning for constraint satisfaction problems. In: Hooker, J. (ed.) CP 2018. LNCS, vol. 11008, pp. 588–597. Springer, Cham (2018). https://doi.org/10.1007/978-3-319-98334-9_38

Inverse Lighting with Differentiable Physically-Based Model

Kazem Meidani[1,2(✉)], Igor Borovikov[1], Amir Barati Farimani[2,3],
and Harold Chaput[1]

[1] Electronic Arts, Redwood City, CA, USA
[2] Department of Mechanical Engineering, Carnegie Mellon University, Pittsburgh,
PA, USA
mmeidani@andrew.cmu.edu
[3] Machine Learning Department, Carnegie Mellon University, Pittsburgh, PA, USA

Abstract. The design of scene lighting in video games and computer graphics can be a challenging and time-consuming task for lighting artists. Automating the lighting in problems such as stadium lighting design in sports games would help the artists by making this tedious process more efficient. In this work, we explore several practical solutions to this problem via optimization and data-driven models. First, we evaluate evolutionary and swarm intelligence gradient-free algorithms with black-box Physically-Based Rendering (PBR) models. Next, by implementing a differentiable PBR model, we leverage gradients to apply gradient-descent optimization to find an optimal solution. We exploit this differentiable model to develop a data-driven framework to learn the mapping from the illumination field to the lighting parameters via minimizing the loss between the illumination and its reconstructed field using a differentiable PBR decoder. Having the learned model, we directly predict the lighting configuration given a user-defined target illumination. In general, we show that all the mentioned methods can reach acceptable solutions, however, based on the conditions, one method can be preferred among others.

Keywords: Inverse Lighting · Deep Learning · Physically-Based Rendering · Optimization

1 Introduction

Lighting design including placing lights and tuning their parameters plays a prominent role in Computer Graphics (CG) and video games. This process, however, can be quite challenging. The challenges might come from the necessity to meet certain criteria (e.g., to have uniform lighting with predefined illuminance levels in certain areas) and/or constraints (e.g., to use less than a predefined number of lights of a particular type and to place them in the vicinity of the models of the light fixtures). Therefore, manual placement and tuning of lights can be tedious and time-consuming. Also, it is rarely a creative task for the lighters working on such problems. Such challenges make the lighting of the CG

M. Sellmann and K. Tierney (Eds.): LION 2023, LNCS 14286, pp. 286–300, 2023.
https://doi.org/10.1007/978-3-031-44505-7_20

or game scenes more expensive in terms of labor and time than it has to be. This process becomes more difficult as the scenes get more complex, with more features to consider, and it becomes more inefficient when there are several similar scenes that each have to be manually addressed (e.g., a wide variety of stadiums in sports games).

Automating the lighting design can help the lighting artists by making this process fast, reliable, and less error-prone. There are several studies to develop such streamlined frameworks by introducing lighting estimation models [2] and applying optimization methods [5, 14] to various types of scenes [8, 26]. Depending on the scene type and the application, the design goals can be energy efficiency, illuminance uniformity, or detail highlighting [6, 15]. In this work, we focus on user-defined illumination goals for the scene. Hence, we target to find a set of lighting parameters to achieve a known ideal illumination result as well as possible.

To find the luminary parameters of light sources from a given image or illumination field, inverse rendering [22] and lighting estimation methods [4] have been studied in recent years. Learning-based inverse models utilize neural networks to estimate the illumination from 2D images. These models are mainly used for applications such as virtual object insertion and multi-view scene re-rendering [13]. To optimize for the best lighting design solution, gradient-free methods propose a relatively simple approach that fits the complex nature of lighting models [14]. It has been shown that techniques like hierarchical coverage optimization [5] or intensity distance fields [30] can be used to find a suitable number of light sources and intensities for isotropic point light sources in simple scenes, but these methods might not be easily generalizable to other applications.

In this work, we explore practical and general ways to solve the problem of lighting design. First, we try gradient-free and gradient-based optimization algorithms to iteratively converge to a feasible and acceptable solution for our experiments on stadium lighting. We compare these two methods and discuss their advantages over each other. Next, we introduce a data-driven framework to predict the lighting parameters from the target illumination field. To train this model, we have implemented a vectorized differentiable model of the PBR module to exploit its gradients in an end-to-end pipeline. We show that this Deep Inverse Lighting (DIL) model can directly estimate acceptable solutions for a given target illumination field. In this paper, we formulate a particular task of lighting stadiums for a video game. This task remains our focus though-out the manuscript; however, the problem formulation and the proposed methods can naturally generalize to other types of lighting problems.

2   Related Works

Inverse Rendering and Lighting Estimation. Inverse rendering is a well-known task in computer graphics with the aim of recovering scene physical properties including geometries, reflectance, and lighting from image(s) [22]. Recently, learning-based models are introduced to jointly predict these scene properties from a single image even in complex indoor scenes [13, 22, 28]. Lighting estimation can be viewed as a sub-task of inverse rendering where the focus is to predict the illumination from a single image [26, 29]. Estimating an environment map from a single low dynamic range image that captures illumination incidents at different locations in the image can be used for applications such as relighting for virtual object insertion [3, 23–25]. However, a single environment map might not represent the illumination, especially in cases like indoor scenes where light sources are close to the objects. This has led to the use of spatially-varying lighting estimation [4, 26] and parametric models [2].

Parametric models aim to globally estimate the 3D scene illumination via estimating the parameters of a discrete number of light sources. These models have been used for both indoor and outdoor scenes. Learning-based parametric models can predict the geometric and photometric properties of light sources from a single 2D image by leveraging the power of neural networks. Hold-Geoffroy et al. [9] implement a 2D CNN model to directly learn the mapping from pixels of an outdoor image to its corresponding lighting parameters. Similarly, Gardner et al. [2] train a CNN model with various output sizes to predict parameters from indoor images. The use of reconstruction loss in auto-encoder like architectures have shown to be effective in parametric models [8, 31]. As a part of our work (DIL), we follow a similar approach to optimize the parameters of light sources to achieve the desired illumination.

Lighting Design Optimization. While lighting estimation aims to find the parameters for a given illumination field, design problems seek optimal lighting configuration to achieve a goal. The goals include reaching a desired illumination level [5, 6] or energy considerations [15]. The majority of these techniques assume point light sources and aim to estimate or optimize the parameters of these point sources. Zhang et al. [30] make use of intensity distance fields to localize the isotropic light sources and find their parameters including the number of lights and their intensity levels in a 2D domain. To optimize for the lighting parameters, most of the previous works leverage gradient-free algorithms due to the nonlinear and complex nature of the lighting. For instance, Genetic Algorithm (GA) has been used in different applications, e.g., stadium lighting, to design the lighting configuration [14, 15, 20, 27]. Optimizing the light parameters altogether is shown to be a challenging task due to the large search space and nonlinearity of its nature. Therefore, many of the previous works solve a simplified or limited version of the general problem. In a close work, Gkaravelis and Papaioannou [5] construct a hierarchical tree of light sources and perform light clustering steps along with nonlinear optimization to find the optimal number of lights and their positions for interior lighting design. In their later work [6], a voting mechanism is used to select light sources for the task of highlighting the details of complex objects.

Differentiable Rendering. Rendering techniques such as PBR are complex and thus not readily differentiable. Differentiable rendering methods aim to provide and make use of the gradients in this process to enable end-to-end optimization and inverse rendering [10, 32]. Physics-based and performance-oriented differentiable renderers have been used to predict lighting and other scene properties by taking advantage of the rendering gradients combined with the power of neural networks [1]. For the purpose of lighting estimation, the approximation methods are more suitable to provide gradients for optimization using gradient descent [18]. Differentiable PBR plays a key role in our proposed gradient-based optimization and deep inverse lighting model.

3 Problem Formulation

A concrete example we use for illustrating the main ideas of the paper is the lighting of stadiums in sports games. This problem is easier to formalize than more general lighting problems in CG and video games. Figure 1 depicts the idea by showing an example of manually authored stadium lighting where the illuminance levels are shown across the stadium. In the example that we consider, the target is for the pitch to be uniformly lit, the sidelines to receive less light, and the stands to be gradually darker away from the pitch. The pitch has to have a consistent look from multiple angles during gameplay or cinematics. Uniform lighting of the surface of the pitch is not the only criterion. The level of illuminance received by it has to be PBR compliant. Also, the players on the pitch have to look "good", meaning, they receive light from multiple directions, shadows are not too dark, and the players do not look flat. To summarize: we need uniform lighting across the pitch at a predetermined level, controlled spill, and lights must support the realistic appearance of the players. The combination of lights including their location, direction, intensity, cone and falloff angles, as well as some other parameters describe the *light rig*.

Manual authoring of a light rig to meet the described requirements is challenging. A lighter has to go over a long and tedious iterative process. It starts with constructing a light rig or using an existing template. Then there is a lengthy loop of adjusting the rig, lighting parameters, visualization of the stadium, and tuning. It takes experience and patience to make this process converge into the desired result. Also, when constructing a rig, there is a trade-off on the number of lights to maintain acceptable computational performance. Therefore, we cannot simply fix imperfections by adding more lights to the rig.

Problem and Objective Function. To capture the outlined setup more formally, we denote the entire set of parameters of a light rig as a vector $\mathcal{W} = \{w_1, \ldots, w_N\}$, where w_i denotes a vector slice of the parameters describing the i-th light and N is the number of lights. All lights in this discussion are of the same type, and all w_i, $i = 1, \ldots, N$, have the same dimensionality d. However, the assumption of having the same types of lights in the rig is not critical. We explain components of w later where we explain the PBR light model. In our application, N usually takes values under 20, with fewer lights being beneficial for real-time performance. Next, we define pitch as P, sidelines as A, and stands

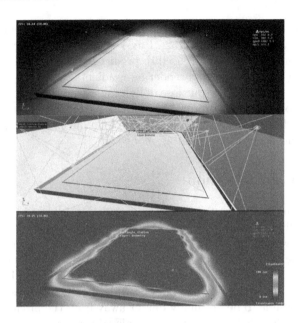

Fig. 1. An example of a manually lit simplified stadium with acceptable lighting design with the visualization of illuminance levels inside the stadium. The illumination is close to uniform across the pitch.

as B. The union $S = P \cup A \cup B$ covers the entire area S of the stadium. Points $s \in S$ receive light according to the PBR lighting model which we denote as a function $\mathcal{I}(s, \mathcal{W})$.

Now, we introduce the objective function T term by term. Uniform illumination is desired on the pitch. Considering a fixed target lux (illuminance) level $\widetilde{\mathcal{I}}$ on the pitch, the most straightforward way is to define a loss function between the resulting illumination $\mathcal{I}(s, \mathcal{W})$ and $\widetilde{\mathcal{I}}$ as

$$T_{P,q}(\mathcal{W}) = \int_P |\mathcal{I}(s, \mathcal{W}) - \widetilde{\mathcal{I}}|_q \, ds.$$

Here, q is the norm, where $q = 1$ (absolute loss) and $q = 2$ (quadratic loss) are the first choices to explore. Similar to the pitch, we define the loss terms for the sidelines and stands as L_q norms between the resulting illumination and the target illumination. For these areas, we have more freedom in defining the objective function and its choice can include aesthetic preferences as well as computational efficiency considerations. However, we stick to the usual loss in the scope of this study. Consequently, the main loss function consists of the following terms:

$$T_q(\mathcal{W}) = T_{P,q}(\mathcal{W}) + T_{A,q}(\mathcal{W}) + T_{B,q}(\mathcal{W}).$$

The placement of lights in the scene should loosely correspond to the location of the visible geometry of the light fixtures. This places a constraint on the possible values of the light positions. Including these constraints into the optimization problem can happen explicitly or via a penalty term that we add to the objective function. One way to define the penalty is to make it depend on the distance of a particular light from the corresponding bounding box for its possible locations. We denote that penalty as $\Theta(\mathcal{W})$ in such a way that it is additive across lights and the corresponding term is zero when the lights reside inside the desired volumes. All in all, our target function gets the form:

$$\min_{\mathcal{W}} \mathcal{L} = T_q(\mathcal{W}) + \Theta(\mathcal{W}) \tag{1}$$

Physically-Based Lighting Model. We use the model of PBR light as defined in [12] where we only consider rigs including punctual lights. However, our techniques are not sensitive to the particular structure of the PBR light model and can work as long as the illuminance function $\mathcal{I}(s, \mathcal{W})$ does not deviate from the physics in any significant manner (i.e., it is differentiable and monotonous in distances and angles). In our model, each light source comprises several parameters where some of them represent its geometrical features and the others determine its photometric properties. The first parameters include the light position in 3D space and its direction. In this work, the latter is defined by the coordinates of a point in the pitch surface where the light is directed toward. The other parameters are the light intensity, its inner (radius) and outer (falloff) angles, and its attenuation radius and offset. The original PBR model computes the illuminance function $\mathcal{I}(s, \mathcal{W})$ for each output grid $s \in S$ by summing the effect of every light on the rig.

Considering that stadiums in many sports games are mainly symmetrical, we enforce some symmetrical conditions for the light sources to reduce the number of unknown parameters in $\mathcal{W}$. In this work, we only define the position of light sources in a set quarter of the pitch in $x - z$ surface such that $x \geq 0, z \geq 0$ where the origin is defined at the center of the pitch surface. Subsequently, we replicate the light source in other quarter regions symmetrically. Hence, we will have only one slice of vector w_i for four lights on the rig. Another important remark is on defining plausible ranges for the parameters. First, some parameters have to be bounded by their definition or scene constraints. The examples are the angles, the position of point sources, and their direction point on the pitch surface. Second, properties like intensity can take continuous values in different scales of magnitude. Defining a reasonable bound based on the common values of such parameters for our application is significantly effective to reduce the size of search space and facilitate the optimization process. Without such constraints, the optimization space might get so large that the model cannot find a feasible solution.

4   Methods

Gradient-Free Optimization. As our first method to explore, we have considered the use of gradient-free, black-box, optimization algorithms to find a sufficiently good optimal solution for our problem. These groups of methods have been broadly used due to their simplicity, flexibility, and their ability to avoid local optima. To this end, we can consider various metaheuristic algorithms such as evolutionary methods or swarm intelligence (SI) algorithms to approach our problem defined by Eq. (1). Evolution Strategies (ES) and its variants have shown successful applications in different fields such as reinforcement learning [7,21]. In this algorithm, we try to estimate the gradient by sampling neighbors of the search agent in the search space. Then, we can move toward the opposite direction of the estimated gradient to approach the optimal solution in a minimization problem. On the other hand, swarm intelligence algorithms are mostly sample-based, and try to sample the search space using a population of agents, and to converge to a solution with simple rules following the elite samples. To showcase, we have used one of the recent SI algorithms named Grey Wolf Optimizer (GWO) [17] which has shown outstanding performances in the literature [16].

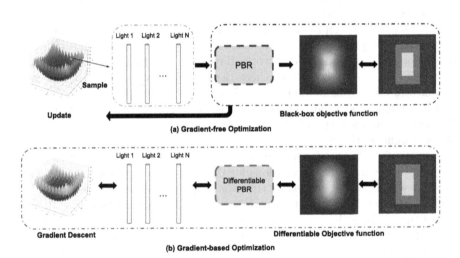

Fig. 2. (a) Framework for stadium lighting design with gradient free optimization algorithms. Lighting parameters are sampled from a bounded search space. The PBR and the objective function are viewed as black-box functions to evaluate the parameters. The evaluation is then used to update the search agents of the optimizer. (b) Framework for stadium lighting design with gradient-based optimization. The whole process is and end-to-end pipeline. The forward pass evaluates a set of parameters with the objective function. The gradients of this process are used in backpropagation to update the parameters in the search space via gradient descent.

Figure 2(a) shows the general scheme of using gradient-free optimization algorithms for inverse lighting problem. Since the number of light sources should not exceed some limited number in practical situations, we try optimizing parameters for different scenarios of light numbers and their relevant positioning. We enforce the symmetrical positioning of light in both axes of the surface, resulting in four light sources defined by a single set of parameters.

Differentiable Gradient-Based Optimization. The original PBR computations calculate the illumination in each grid of the space and then sum them up from different sources of lights. We first vectorize these calculations to compute the illumination in the whole field instead of specific output grids. In the case of 2D stadium illumination, we observe up to more than *200 times speed up* in the PBR computations. This accelerated computation enables us to make use of differentiable models for the PBR. To this end, we utilize PyTorch [19] to implement all steps of the simplified PBR in a differentiable fashion so we can compute the gradients via backpropagation. Starting from an initial parameter set, we can leverage built-in PyTorch optimizers such as Adam optimizer [11]. An important remark here is that similar to gradient-free optimization algorithms, there is no guarantee of reaching the global optima for gradient-based methods. However, it is expected that they perform better for problems with large search spaces. We show the results and discuss this in more details in the experiments section.

Similar to the gradient-free case, we can enforce symmetrical conditions where suitable. Also, we normalize the parameters into the range of $[-1, 1]$ in order for the gradients to be in a comparable range with each other to facilitate the gradient descent steps [33]. The end-to-end differentiable pipeline is illustrated in Fig. 2(b). We would like to emphasize that the essential requirement for using gradient-based optimization is that every step of the model, including the loss and the constraints, be differentiable. Therefore, for this model, we use a quadratic loss ($q = 2$ in (1)). To incorporate constraints on the location of light sources in the loss function, we use soft conditions implemented with differentiable torch operations. In this study, we penalize the placement of light sources outside the bounding box in each direction linearly using the ReLU function.

Deep Inverse Lighting Model. In this section, we aim to develop a data-driven framework to predict the lighting configuration including locations and other parameters to achieve target illumination. By training a model that can map the illumination field to its corresponding lighting parameters, we will bypass the iterative optimization process, and directly estimate the pseudo-optimal parameters. Therefore, lighting parameters can be predicted with a forward pass of network for any target illumination. The assumption here is that the model is trained on samples that have the same statistical distribution as the target illumination distribution.

The illumination field $\mathcal{I}(S)$, defined on the region S, is mapped to the vector of lighting parameters $\mathcal{W}$ via function $\Phi(\theta) : \mathcal{I}(S) \to \mathcal{W}$ where θ are the parameters of the model. On the other hand, we have the PBR model as a function mapping the lighting parameters to the illumination field, i.e., $\text{PBR} : \mathcal{W} \to \mathcal{I}(S)$. Hence, $\Phi(\theta)$ can be viewed as the inverse of PBR model: $\Phi(\theta) = \text{PBR}^{-1}$. Autoencoders are thus conceivable candidates where we learn these two inverse functions as encoder and decoder. To this end, the differentiable PBR model is implemented as explained in the previous section to be used as the decoder. Figure 3 depicts a scheme of the framework. By using the differentiable PBR decoder, we can reconstruct the illumination field $\mathcal{I}(S)$ from the predicted parameters $\mathcal{W}$ and exploit the reconstruction loss to learn the parameters of encoder Φ.

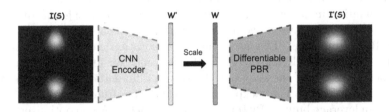

Fig. 3. A scheme of deep inverse model for stadium lighting design. CNN encoder maps illumination fields to normalized vector of parameters. After scaling, these parameters pass through a differentiable PBR model to reconstruct the illumination field. The model can be trained by this reconstruction loss, and used to predict lighting parameters from the target illumination in the test stage.

An important remark here is on the limitations of this framework. While this model is advantageous in terms of fast and direct prediction and flexibility to various targets, we should note that to get acceptable results from the network, the target illumination should be from the same statistical distribution as the training data. Also, in the current form of the model, we have a fixed bottleneck vector size which requires the model to be trained on a fixed number of light sources.

5   Experiments

Experimental Setup. To fairly evaluate the performance of different methods that we presented in the methods section, we have to first define the metrics for evaluation. We assume that there is no ground truth parameter set to compare our optimal solution with. Therefore, we pass the solution through a PBR model and evaluate its resulting illumination field upon the user-defined target illumination. The metric used to report the final result is the normalized root mean squared error (NRMSE) defined as:

$$\mathcal{L} = \frac{\sqrt{\frac{\sum_{s \in S}(\mathcal{I}(\mathcal{W},s) - \widetilde{\mathcal{I}}(s))^2}{|S|}}}{\max_s \widetilde{\mathcal{I}}(s)},$$

where $|S|$ is the total number of grids in a 64×64 output size of S, $\widetilde{\mathcal{I}}(s)$ is the target illumination at grid s, and $\mathcal{I}(\mathcal{W}, s)$ is the computed illumination from the optimal solution.

Gradient-Free and Gradient-Based Optimization. We initialize the lighting parameter by randomly sampling a population of points in the normalized bounded search space. After updating the parameters at each iteration, we clip them inside the bounds to avoid wasting computation time. The final results and the total optimization time of using ES and GWO for this experiment are reported in Table 1. Since these methods are stochastic, numbers are reported after 10 independent trials. The results indicate that sampling-based algorithms like GWO are more effective in the lighting design task than gradient estimation. The reason could be that plausible solutions are sparse in the large search space. Hence, without a very good initialization, gradient estimation has a lower chance to fall into a very good optimum.

Table 1. Comparison of two gradient-free optimizers applied on the stadium lighting problem. The computation time heavily depends on the number of lights, number of iterations, and the population size. We generally observe a better performance (reported as average and median) from Grey Wolf Optimizer compared to Evolutionary Search. N denotes population size.

Optimizer		ES		GWO	
(N, Iterations)		(50, 300)	(100, 500)	(50, 300)	(100, 500)
4 lights	Ave. Loss	0.13246	0.10351	0.09572	**0.09448**
	Med. Loss	0.12001	0.10442	0.09581	**0.09429**
	Ave. Time (s)	53	162	50	158
16 lights	Ave. Loss	0.10449	0.11107	0.09013	**0.08573**
	Med Loss	0.10029	0.10490	0.09157	**0.08497**
	Ave. Time (s)	164	484	155	491

The objective function for gradient-free optimizers is defined as Eq. (1) and includes the penalty for the constraints of the light positions. Also, to strictly enforce the constraints for these optimizers, we clip the parameters inside their bounds at every iteration. For the gradient-based differentiable optimization, we implement a differentiable PyTorch-based model of these constraints to be included in the differentiable loss function. However, the final solution might violate some of the constraints depending on the extent to which we enforce the penalties. Table 2 shows the results of using Adam optimizer in two cases whether constraint penalties are applied or not. The results indicate that applying this penalty can lower the constraints violation but it may slightly sacrifice the NMSE loss between the result and the target illumination. Another important observation from Table 2 is that gradient-based algorithms show their superior performance only for more lights (16 and 64) which means larger search spaces. The reason lies behind the fact that in larger dimensions, the solutions are too sparse to be found by sampling methods while gradient can be a trusted guide toward the optimal solution.

Deep Inverse Lighting Model. To train the model, we generate 100K/10K train/test samples of PBR-compliant illumination fields where the samples are randomly selected from the bounded parameter space. Similar to previous experiments, we train models for the cases of 4 and 16 lights (with symmetrical positioning).

Table 2. Performance of differentiable optimization with gradient descent (Adam) optimizer for different number of lights and compared with gradient-free optimizer (GWO) when applied on the stadium lighting problem. Target result indicates the normalized loss between the optimal solution and the target illumination. Const. denotes the loss corresponding to the violation of positional constraints, and Total is their summation.

| Optimizer | | Gradient-free | Differentiable Gradient-based | | | | | |
| | | | w/o const. | | | w const. | | |
Loss		Total	Target	Const.	Total	Target	Const.	Total
4 lights	Ave.	**0.09572**	0.10354	0.00096	0.10451	0.11243	0.00068	0.11311
	Med.	**0.09581**	0.09814	0.00108	0.09923	0.09837	0.00058	0.09865
	Time (s)	50	53 (15000 iterations)					
16 lights	Ave.	**0.09013**	0.08817	0.01475	0.10292	0.09756	0.00066	0.09822
	Med.	**0.09157**	0.08390	0.01424	0.09813	0.09778	0.00040	0.09818
	Time (s)	155	74 (5000 iterations)					
64 lights	Ave.	0.09306	0.07841	0.01522	0.09363	0.08876	0.00079	**0.08797**
	Med.	0.09335	0.07761	0.01590	0.09351	0.08423	0.00090	**0.08377**
	Time (s)	536	343 (5000 iterations)					

Table 3. Performance of deep inverse lighting model for two types of learning. Using differentiable decoder and reconstruction loss significantly improves the final solution.

Model	Reconstruction		Parameters	
# of Lights	4 Lights	16 Lights	4 Lights	16 Lights
Loss	**0.11012**	**0.10670**	0.13609	0.16220

To show the importance of the differentiable PBR module, we compare the results with a CNN model that maps the illumination fields to the parameter space (trained by parameter-space loss). The results reported in Table 3 show that using reconstruction loss on the illumination image-space ($\mathcal{I}(S)$) is far more effective than using the loss in the parameter-space ($\mathcal{W}$). The final target loss achieved by DIL model is not as good as optimization-based models in Table 2,

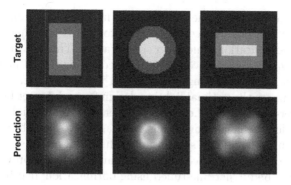

Fig. 4. Illumination predictions of the trained deep inverse lighting model with 16 lights for different test targets.

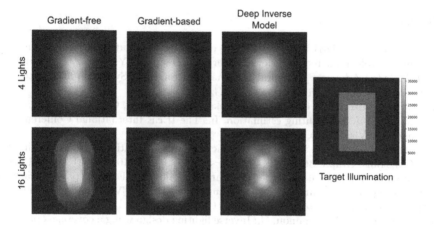

Fig. 5. Sample results of illumination field achieved by different inverse models studied in this work compared to the target illumination. We observe that almost all of the models can achieve acceptable results. Gradient-free models are better for less number of lights while gradient-based models can cover more details with 16 lights. DIL model achieves a reasonable result with a one-time prediction.

however, it enables us to predict for different targets quickly without an iterative process. Figure 4 shows the model's predictions at test time for different choices of target.

6 Discussion and Future Work

In this work, we explored several approaches to the inverse lighting problem and showcased the results for stadium lighting design in sports games. All in all, we can state that all the methods that we explored in this paper can provide acceptable solutions to some extent. This is also shown qualitatively in Fig. 5 where final solutions of different methods are compared with the target illumination.

However, we can still select the best method in different conditions. Gradient-free methods are suitable choices when the problem is low-dimensional and we can afford the computation time for a single target. This is while differentiable gradient-based methods work very well in unconstrained and high dimensional problems. Deep Inverse Lighting (DIL) model is a data-driven framework that can directly estimate the optimal lighting design for each target by learning with an auto-encoder-like model where the decoder is a differentiable PBR model. While this model has some limitations such as a fixed number of lights, constraint definition problem, and sub-optimal solution, it opens several opportunities for future work. For instance, the model can be used for providing an initial solution that can be further fine-tuned using optimization algorithms. Also, the other limitations such as fixed bottleneck size can be addressed by using more flexible models such as transformers.

References

1. Chen, W., et al.: DIB-R++: learning to predict lighting and material with a hybrid differentiable renderer. In: Beygelzimer, A., Dauphin, Y., Liang, P., Vaughan, J.W. (eds.) Advances in Neural Information Processing Systems (2021). https://openreview.net/forum?id=gRqHB07GGz3
2. Gardner, M.A., Hold-Geoffroy, Y., Sunkavalli, K., Gagne, C., Lalonde, J.F.: Deep parametric indoor lighting estimation. In: The IEEE International Conference on Computer Vision (ICCV) (2019)
3. Gardner, M.A., et al.: Learning to predict indoor illumination from a single image. ACM Trans. Graph. **36**(6), 1–14 (2017). https://doi.org/10.1145/3130800.3130891
4. Garon, M., Sunkavalli, K., Hadap, S., Carr, N., Lalonde, J.F.: Fast spatially-varying indoor lighting estimation. In: Proceedings of the IEEE/CVF Conference on Computer Vision and Pattern Recognition (CVPR) (2019)
5. Gkaravelis, A., Papaioannou, G.: Inverse lighting design using a coverage optimization strategy. Vis. Comput. **32**, 771–780 (2016). https://doi.org/10.1007/s00371-016-1237-9
6. Gkaravelis, A., Papaioannou, G.: Light optimization for detail highlighting. In: Computer Graphics Forum, vol. 37 (2018)
7. Hansen, N.: The CMA evolution strategy: a tutorial (2016). https://doi.org/10.48550/ARXIV.1604.00772. https://arxiv.org/abs/1604.00772
8. Hold-Geoffroy, Y., Athawale, A., Lalonde, J.F.: Deep sky modeling for single image outdoor lighting estimation. In: 2019 IEEE/CVF Conference on Computer Vision and Pattern Recognition (CVPR), pp. 6920–6928 (2019)
9. Hold-Geoffroy, Y., Sunkavalli, K., Hadap, S., Gambaretto, E., Lalonde, J.F.: Deep outdoor illumination estimation. In: 2017 IEEE Conference on Computer Vision and Pattern Recognition (CVPR), pp. 2373–2382 (2017). https://doi.org/10.1109/CVPR.2017.255
10. Kato, H., et al.: Differentiable rendering: a survey (2020). https://doi.org/10.48550/ARXIV.2006.12057. https://arxiv.org/abs/2006.12057
11. Kingma, D.P., Ba, J.: Adam: a method for stochastic optimization. CoRR abs/1412.6980 (2015)
12. Lagarde, S.: SIGGRAPH 2014: moving frostbite to physically based rendering V3 (2014). https://seblagarde.wordpress.com/2015/07/14/siggraph-2014-moving-frostbite-to-physically-based-rendering/. Accessed 09 Mar 2022

13. Li, Z., Shafiei, M., Ramamoorthi, R., Sunkavalli, K., Chandraker, M.: Inverse rendering for complex indoor scenes: shape, spatially-varying lighting and SVBRDF from a single image. In: Proceedings of the IEEE/CVF Conference on Computer Vision and Pattern Recognition, pp. 2475–2484 (2020)
14. Lima, G.F.M., Tavares, J., Peretta, I.S., Yamanaka, K., Cardoso, A., Lamounier, E.: Optimization of lighting design usign genetic algorithms. In: 2010 9th IEEE/IAS International Conference on Industry Applications - INDUSCON 2010, pp. 1–6 (2010). https://doi.org/10.1109/INDUSCON.2010.5740021
15. Madias, E.N.D., Kontaxis, P.A., Topalis, F.V.: Application of multi-objective genetic algorithms to interior lighting optimization. Energy Build. **125**, 66–74 (2016). https://doi.org/10.1016/j.enbuild.2016.04.078. https://www.sciencedirect.com/science/article/pii/S0378778816303553
16. Meidani, K., Hemmasian, A., Mirjalili, S., Barati Farimani, A.: Adaptive grey wolf optimizer. Neural Comput. Appl. **34**(10), 7711–7731 (2022). https://doi.org/10.1007/s00521-021-06885-9
17. Mirjalili, S., Mirjalili, S.M., Lewis, A.: Grey wolf optimizer. Adv. Eng. Softw. **69**, 46–61 (2014). https://doi.org/10.1016/j.advengsoft.2013.12.007. https://www.sciencedirect.com/science/article/pii/S0965997813001853
18. Nieto, G., Jiddi, S., Robert, P.: Robust point light source estimation using differentiable rendering (2018). https://doi.org/10.48550/ARXIV.1812.04857. https://arxiv.org/abs/1812.04857
19. Paszke, A., et al.: PyTorch: an imperative style, high-performance deep learning library. In: Wallach, H., Larochelle, H., Beygelzimer, A., d'Alché-Buc, F., Fox, E., Garnett, R. (eds.) Advances in Neural Information Processing Systems, vol. 32, pp. 8024–8035. Curran Associates, Inc. (2019). http://papers.neurips.cc/paper/9015-pytorch-an-imperative-style-high-performance-deep-learning-library.pdf
20. Petranović, D.: Stadium reflector aiming using genetic algorithms. In: 2012 Proceedings of the 35th International Convention MIPRO, pp. 1070–1075 (2012)
21. Salimans, T., Ho, J., Chen, X., Sutskever, I.: Evolution strategies as a scalable alternative to reinforcement learning. ArXiv ArXiv:1703.03864 (2017)
22. Sengupta, S., Gu, J., Kim, K., Liu, G., Jacobs, D., Kautz, J.: Neural inverse rendering of an indoor scene from a single image. In: 2019 IEEE/CVF International Conference on Computer Vision (ICCV), pp. 8597–8606 (2019). https://doi.org/10.1109/ICCV.2019.00869
23. Song, S., Funkhouser, T.: Neural illumination: lighting prediction for indoor environments. In: Proceedings of 33th IEEE Conference on Computer Vision and Pattern Recognition (2019)
24. Srinivasan, P.P., Mildenhall, B., Tancik, M., Barron, J.T., Tucker, R., Snavely, N.: Lighthouse: predicting lighting volumes for spatially-coherent illumination. In: Proceedings of the IEEE/CVF Conference on Computer Vision and Pattern Recognition (CVPR) (2020)
25. Wang, L.-W., Siu, W.-C., Liu, Z.-S., Li, C.-T., Lun, D.P.K.: Deep relighting networks for image light source manipulation. In: Bartoli, A., Fusiello, A. (eds.) ECCV 2020, Part III. LNCS, vol. 12537, pp. 550–567. Springer, Cham (2020). https://doi.org/10.1007/978-3-030-67070-2_33
26. Wang, Z., Philion, J., Fidler, S., Kautz, J.: Learning indoor inverse rendering with 3D spatially-varying lighting. In: Proceedings of International Conference on Computer Vision (ICCV) (2021)
27. Xiao, H., Fang, J., Zhu, P., Yin, W., Kang, Q.: Energy-saving optimization of football field lighting via genetic algorithm. Sens. Lett. **12**, 264–269 (2014). https://doi.org/10.1166/sl.2014.3265

28. Yu, Y., Smith, W.: InverseRenderNet: learning single image inverse rendering. In: 2019 IEEE/CVF Conference on Computer Vision and Pattern Recognition (CVPR), pp. 3150–3159 (2019)
29. Zhan, F., et al.: GMLight: lighting estimation via geometric distribution approximation. IEEE Trans. Image Process. **31**, 2268–2278 (2022). https://doi.org/10.1109/TIP.2022.3151997
30. Zhang, E., Cohen, M.F., Curless, B.: Discovering point lights with intensity distance fields. In: The IEEE Conference on Computer Vision and Pattern Recognition (CVPR) (2018)
31. Zhang, J., Sunkavalli, K., Hold-Geoffroy, Y., Hadap, S., Eisenmann, J., Lalonde, J.F.: All-weather deep outdoor lighting estimation. In: IEEE International Conference on Computer Vision and Pattern Recognition (2019)
32. Zhang, K., Luan, F., Wang, Q., Bala, K., Snavely, N.: PhySG: inverse rendering with spherical gaussians for physics-based material editing and relighting. In: The IEEE/CVF Conference on Computer Vision and Pattern Recognition (CVPR) (2021)
33. Zhao, S.Y., Xie, Y.P., Li, W.J.: On the convergence and improvement of stochastic normalized gradient descent. Sci. China Inf. Sci. **64**(3), 132103 (2021). https://doi.org/10.1007/s11432-020-3023-7

Repositioning Fleet Vehicles: A Learning Pipeline

Augustin Parjadis[1]([✉]), Quentin Cappart[1], Quentin Massoteau[2],
and Louis-Martin Rousseau[1]

[1] Polytechnique Montréal, Montreal, Canada
{augustin.parjadis-de-lariviere,quentin.cappart,
louis-martin.rousseau}@polymtl.ca
[2] Fastercom, Montreal, Canada
quentin.massoteau@fastercom.ca

Abstract. Managing a fleet of vehicles under uncertainty requires careful planning and adaptability. We consider a *ride-hailing problem* where the operator manages vehicle repositioning to maximize responsiveness. This paper introduces a supervised learning pipeline that uses past trip data to reposition vehicles while adapting to fleet activity, a geographical zone, and seasonal or daily request variation. The pipeline incorporates trip features, such as medical motives of transportation for ambulances and the time and location of the trips. This provides a better estimate of the probability that a given vehicle will be required in a particular sector and provides insights into which events and trip features should be incorporated into the decision-making process for better fleet management and improved reactivity. This tool has been developed for, and used by, operators of an ambulance company in Belgium. Using predictors for ambulance repositioning reduces at least 10% of the overall fleet reaction distance.

Keywords: Dynamic ride-hailing · Supervised learning · Fleet management

1 Introduction

Fleet management in a ride-hailing context is a challenging problem due to the stochastic nature of the demand and the dynamic decisions that must be made. Medical transportation for unplanned requests faces the same challenges and aims at providing a wide-spread and fast coverage to ensure medical services availability; they have been a regular focus of routing problems in general, as cities need efficient and cost-effective medical transportation [12,14].

This work focuses on learning from limited historical data for ride-hailing in a medical transportation setting and providing easy tools to set up, interpret and apply to small fleets of ambulances. The contribution of this paper is an effective learning pipeline that can provide an adaptable repositioning agent capable of

M. Sellmann and K. Tierney (Eds.): LION 2023, LNCS 14286, pp. 301–317, 2023.
https://doi.org/10.1007/978-3-031-44505-7_21

recognizing daily and hourly patterns and adapting to evolving demand distribution. The current strategy used by the company taking care of the repositioning yields a low total distance driven by the ambulance fleet but provides poor reactivity to requests; our learning pipeline improves this without a significant increase in the total distance driven on real-world data. The paper is organized as follows. Section 2 reviews the relevant literature. Then, Sect. 3 introduces and formalizes the ride-hailing problem. Section 4 presents the pipeline structure. Section 5 presents a case study where we apply our repositioning agent to a fleet of ambulances in Belgium for unplanned medical trips.

2   Literature Review

Vehicle routing problems (VRPs), focusing on finding optimal routes for vehicles between multiple locations and under various constraints, are ubiquitous in the flow of people and goods in our increasingly interconnected cities, networks, etc. As a traveling salesman problem at its core, these problems are hard [15,39]. They are often subject to additional constraints when modeling real problems, often linked to time windows, traffic rules, environmental considerations, or many other potential obstacles [9,23,26].

Ride-hailing problems are vehicle routing problems, more precisely dial-a-ride problems where multiple users request trips with given time windows and are transforming logistics and sustainability in the context of rising awareness towards efficiency and environmental issues. This transformation of urban mobility can stimulate the adoption of new technologies and positively impact energy consumption and quality of life [4,19]. Nevertheless, the success of these systems depends on their effective efficiency and ease of use, which is a challenge given the complex nature of vehicle routing problems in dynamic environments. Extensive literature can be found for dynamic VRPs and dial-a-ride problems in recent surveys [16,32]. The dynamic nature of ride-hailing problems, with new requests arriving at unknown times and locations, creates different challenges and requires novel solution approaches than do not simply assume perfect information.

The first challenge of ride-hailing problems is to assign vehicles to incoming requests. This has been tackled by traffic flow study [25], oversupply analysis [6], and negotiation techniques between vehicles [36]. Several studies approach fleet management from the opposite direction, focusing on vehicle repositioning to anticipate future requests. Miao et al. [29] use a receding horizon for distribution estimation, Zhang and Pavone [42] focus on a queuing-theoretical model for supply rebalancing, while Braverman et al. [7] introduce fluid-based optimization methods to control car flow and maximize the expected value of served requests. Other works study assignment and repositioning for a more complete approach to the ride-hailing problem with either heuristic strategies at global scales for autonomous vehicles [11] or large-scale optimization-based methods [1,5,10]. Furthermore, increasing interconnectivity and data availability allow enriched models and more precise control of the vehicles. For example, real-time information such as traffic congestion or driving times opens up the possibility of adapting programmed tasks and trip trajectories [13,35].

Recent machine learning developments allow the creation of new tools to tackle complex problems and integrate historical data. From the plethora of regressors and classifiers, [31], multi-layer perceptrons and deep learning emerged as powerful tools to learn from unstructured data [24]; deep reinforcement learning (DRL) [2,38] allowed the training of agents to interact with their environment. These approaches are incredibly prolific in image classification [37], natural language processing [33], super-human performance in games [20], and have also been considered by the operations research community for mathematical optimization [3]. Optimization coefficients can, for example, be learned inside a more traditional learning algorithm [40], multi-agent DRL allows the training of a fleet of agents to maximize shared profit [17,30], and can integrate traditionally complex variants like electric vehicle charging [22]. The ride-hailing problem can thus be tackled with a hybrid approach of learning, and optimization [34], or learning from optimization models [41]. Handling a whole fleet in a dynamic environment is a complex task, often solved by reinforcement learning in which fleets are represented by an agent [18], or multiple agents [27,28], that explore and refine fleet management policies. These learning approaches are efficient but require a large amount of data and are not easy to use and interpret for dispatchers of a small fleet. Based on this context, our contribution focuses on a smaller granularity, both time-wise and vehicle-wise, aiming at developing a pipeline that works with a limited amount of data and is easy to interpret for the operators.

3   The Ride-Hailing Problem

This section presents the ride-hailing problem and the relevant notation used throughout this work.

3.1   Problem Definition

Let V be a fleet of vehicles and R the set of unplanned requests arriving throughout the day. A vehicle $v \in V$ is characterized by its geographical position and its status. The possible status are as follows: (1) driving to a client, (2) serving a client, (3) repositioning, (4) idling at a location, and (5) waiting for repositioning instructions. A request $r \in R$ is emitted at a time r_t and is characterized by its origin and destination coordinates.

Each time a request arises, it is assigned to the closest vehicle. Vehicles idling or repositioning immediately go serving the request they have been assigned. However, vehicles currently serving a request might also be chosen to serve the new request after their current ones have been fulfilled, as there is no maximum waiting time. The requests can arise and lead anywhere, usually hospitals and patient houses; they will all get served by the fleet.

A repositioning decision is made when a vehicle finishes serving a request and does not have another to serve next. Let L be the set of locations containing predefined geographical points where vehicles are sent to idle between requests.

The requests can arise anywhere, whereas vehicles must wait at the predefined locations L. The goal is to reposition vehicles to the idling locations to minimize the average reaction distance, defined as the distance driven by an ambulance from where it is when assigned to a request to the pick-up point of the request. The reaction distance includes the distance driven by ambulances that have to finish current trips before answering the request.

Visual Example. An illustration of possible decisions is proposed in Fig. 1: (1) A request for a patient is first received and assigned to a nearby vehicle chosen during the assignment process. (2) After the ride, the vehicle is repositioned to a location chosen during the repositioning process. (3) Let us point out that the location does not have to be the closest one, nor does it have to be its original idling location. (4) Finally, another vehicle can be repositioned to balance the supply.

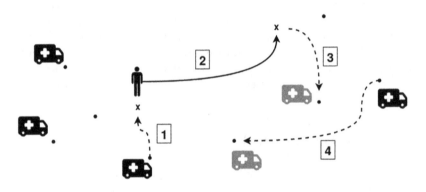

Fig. 1. Ride-hailing situation - example.

3.2   Modelling the Repositioning Task

A *decision* must be carried out each time a new request is received and each time a vehicle finishes serving a trip request. An *episode* refers to a full day of operations or the fleet and contains as many decisions as requests received and trips finished. When a decision must be made, the system is given a *state*. The formal definitions of a state and a decision are proposed below.

States. A state $s = (s_R, s_r, s_t, s_V)$ is composed of a prior requests vector (s_R), a request vector (s_r), a time vector (s_t) and a vehicle vector (s_V). The prior requests vector is composed of tuples indicating the location of origin, arrival, and the medical motive of the past requests within a given horizon H: $(r_i^O, r_i^A, r_i^m)_{i \in H}$. The request vector s_r provides the geographical coordinates of the origin and destination of the request received at the start of the decision epoch. It is empty if the epoch was triggered by a vehicle becoming idle. The time vector $s_t = (d, t)$ indicates the day of the week d and hour of the day t as one-hot encoded vectors. The vehicle vector s_V indicates the status and positions of the fleet's vehicles.

Decisions. A repositioning decision $a = (a_1, a_2, \ldots, a_V)$ is composed of vehicle repositioning instructions. For all vehicles, $v \in V$, the instruction $a_v \in L \cup \{\emptyset\}$ indicates the location $l \in L$ at which a vehicle must reposition or an absence of instruction. Idle vehicles that finished serving a trip request must be repositioned to idle at a valid location.

4   Learning Pipeline for Vehicle Repositioning

This section presents a learning pipeline used in the specific context of ambulance management, which is general enough to be adapted to other situations. The pipeline consists of a learning algorithm and a prediction agent. It is illustrated in Fig. 2 and is divided into 4 steps: (1) Past data is collected, used to build context vectors, and split to build the training and test sets, (2) an agent is trained on this dataset and is used operationally to predict the next location most likely to receive a request, and (3) this information is fed to a repositioning algorithm adapted to the fleet. Each step is detailed in the next paragraphs.

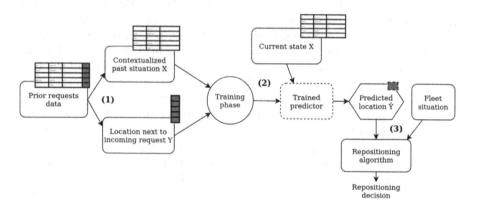

Fig. 2. Pipeline structure.

Step 1: Data Preparation. The data comprises the set of feature vectors X, encoding the state and context in which a trip request arrived, and the associated set of target values vector Y, containing the locations of the origin of the requests. Several features are used to make a prediction, such as time, location, and motives of previous requests. Geographical distributions will, for example, be affected by the population moving during work days or days off, resulting in hourly and weekly variations in requests. Previous requests' origin, destination, and motives are also likely to provide relevant information about whether a patient will be emitting another request and when; we encode their frequency and recency with discounted factors in a context vector that aggregates the information from H requests as follows.

Prior requests origins (PRO), prior requests arrivals (PRA), and medical motives are features computed from the components of the prior requests vector. For each location l, we define the prior requests origins feature $s_l^O = \frac{1}{H} \sum_{i \in [1,H]} \delta(l, r_i) \beta^i$, where $\beta \in]0,1]$ is a constant, and

$$\delta(l, r_i) = \begin{cases} 1 & \text{if } l = r_i^O \\ 0 & \text{otherwise} \end{cases} \tag{1}$$

indicates which location a prior request emerged next to, with r_i^O being the closest location from r_i's origin. Here, r_1 is the latest request at time t, r_2 the one before it, etc., with r_H being the oldest request considered on our horizon. For the requests arrivals, $(s_l^A)_{l \in L}$ is computed similarly with arrival locations. For each medical motive $m \in M$, we define $s_m = \frac{1}{H} \sum_{i \in [1,H]} \mu(m, r_i) \beta^i$, with

$$\mu(m, r_i) = \begin{cases} 1 & \text{if } m = r_i^m \\ 0 & \text{otherwise} \end{cases} \tag{2}$$

where r_i^m is the motive of request i, indicating the number of requests with motive $m \in M$ and their recency. $(s_l^O)_{l \in L}$, $(s_l^A)_{l \in L}$, and $(s_m)_{m \in M}$ thus indicates which hospitals requests emerged from and arrived at, and which medical motives were observed. The set of medical motives M can include simple consultations, transfers, specialized interventions, etc. Table 1 provides a summary of the state features for our case study related to healthcare. All feature values are in $[0,1]$. The relevance of such features is analyzed in Appendix 1.

The training set is built by taking whole days of data and extracting the arising requests as well as the state in which they arrived to build the feature vector set $X = \{(s^O, s^A, s_m, s_t)\}$ and the targets $Y = \{r_i^O\}$. A request r_i arises at time r_{it} with r_i^O the closest location from its origin. We then train a classifier to try and infer the next request location, given the current state.

Table 1. State features

Feature	Number of values
Time s_t	24+7
PRO s_l^O	$\|L\|$
PRA s_l^A	$\|L\|$
Motives s_m	$\|M\|$

Step 2: Location Classifier Training. A learning model F is defined and trained to predict which locations Y will most likely receive the following request based on the features X describing the problem's state and the fleet's previous activity during the day. We consider random forests, gradient boosting methods, and multilayer perceptrons for our classifiers. Details about the training are provided for the case study in Sect. 5.2. Once trained, the models give as an output a vector quantifying the probability of each location being the next receiving a request, $F(x) = \hat{y}$. It is then used to issue a repositioning decision to vehicles finishing a trip and idle for too long at an inactive location.

Step 3: Algorithm for Vehicle Repositioning. The repositioning procedure is illustrated in Algorithm 1. It is called each time a decision must be made, i.e. when a request is issued or when an ambulance has finished its trip. The algorithm generates a prediction for the different locations with the trained model (line 9) and chooses locations to reposition the vehicles that have been idle for too long and the vehicle that finished its trip, if applicable (lines 11 to 13). The repositioning location is obtained by chooseLocation$(\hat{y}, v, \mathcal{L})$ procedure (line 12), which is formalized in Algorithm 2. It is done for a vehicle $v \in V$, based on the score obtained for each position $(\hat{y})$ and on the current idling location of each vehicle $(\mathcal{L})$. The repositioning instructions are stored in $\mathcal{L}$ (line 12). The idling time for each vehicle is also updated (lines 13 and 15).

Algorithm 1: Fleet Repositioning

---

1 ▷ **Pre:** x is the current state, as defined in Step 1 (data preparation).
2   F is the learned model (gives a score for each location from L).
3   V is the set of all vehicles.
4   ρ is the vehicle that has finished its trip and is to be repositioned.
5       $\mathcal{L}[v]$ is the idling location of vehicle v.
6       $I[v]$ is the number of iterations during which v has been idle.
7   Θ is a threshold triggering a repositioning, if exceeded.
8
9 $\hat{y} := F(x)$ ▷ *Calling the trained model*
10 **for each** $v \in V$ **do**
11 | **if** $v = \rho \vee I[v] > \Theta$ **then**
12 | | $\mathcal{L}[v] := $ chooseLocation$(\hat{y}, v, \mathcal{L})$ ▷ *Procedure defined in Algorithm 2*
13 | | $I[v] := 0$
14 | **else**
15 | $I[v] := I[v] + 1$

The goal of Algorithm 2 is to find the closest empty location likely to receive a request. First, the number of vehicles present at each location is computed (line 11). Let $\hat{y} = \langle \hat{y}_l \rangle_{l \in L}$ be the score vector returned by the classifier for each location. The higher $\hat{y}_l$, the more confident the classifier is in having the subsequent request arising next to location l. Candidate locations Λ for repositioning are selected by taking all probable unfilled locations with a tolerance factor defined by $\gamma \in \,]0,1[$ (line 15). A location is *unfilled* if it contains fewer vehicles than a given parameter $\tau \in \mathbb{N}^+$. For example, setting this parameter to 1 only allows picking an empty location until all locations have one idling vehicle. From those candidates, the vehicle is sent to the closest one (line 17). If all candidate locations are filled, we consider decreasingly probable locations until an unfilled one is found, which is done by ignoring the most probable location and setting their score to 0 for the next iterations (line 19) as it did not yield a valid repositioning location. If all the locations are filled, the closest location is selected regardless of the number of vehicles idling there (line 21). This algorithm allows spreading the vehicles from the highest to lowest-valued locations, with the shortest distance being the decisive criterion for equally probable locations. We evaluate the practical use of the pipeline in a case study in the next section.

Algorithm 2: Selecting a Location for a Vehicle

---

1 ▷ **Pre:** $\hat{y} = \langle \hat{y}_1, \ldots, \hat{y}_L \rangle$ is the predictor output for each location.
2   v is the vehicle to reposition.
3   L is the set of all locations.
4      $\mathcal{L}[v]$ is the idling location of vehicle v.
5      $d[v,l]$ is the distance function from v to the location l.
6      $\gamma \in \,]0,1[$ is a prediction tolerance parameter.
7      $\tau \in \mathbb{N}^+$ is a filled location parameter.
8
9 chooseLocation($\hat{y}, v, \mathcal{L}$) :
10   **for each** $l \in L$ **do**
11     $\quad$ $C[l] := \left| \{ v \in V \mid \mathcal{L}[v] = l \} \right|$ ▷ *Number of vehicles at each location*
12     $l^* := \perp$ ▷ *Store the best location to reposition*
13   **while** $l^* = \perp \wedge \max(\hat{y}) > 0$ **do**
14     $\quad$ $k^* := \arg\max(\hat{y})$ ▷ *Taking the index of location with highest score*
15     $\quad$ $\Lambda := \{ l \in L \mid \hat{y}_l \geq (1-\gamma)\hat{y}_{k^*} \wedge C[l] < \tau \}$ ▷ *Filtering the locations*
16     $\quad$ **if** $\Lambda \neq \emptyset$ **then**
17     $\quad\quad$ $l^* := \arg\min_{l \in \Lambda} \left(d[v,l] \right)$
18     $\quad$ **else**
19     $\quad\quad$ $y_{k^*} := 0$
20   **if** $l^* = \perp$ **then**
21     $\quad$ $l^* := \arg\min_{l \in L} \left(d[v,l] \right)$
22   **return** l^*

---

5   Case Study: Ambulance Fleet in Belgium

This project is conducted in collaboration with a company to develop automated assistance for managing a fleet of ambulances. We consider a centralized dispatcher for the fleet that assigns identical ambulances to requests for medical transport and repositions idle ambulances to nearby hospitals. Even though requests arise randomly, patterns can be exploited as the requests tend to emerge more frequently in specific zones at certain times. Furthermore, transport between hospitals might depend on why patients are being brought to the hospitals. Some interventions are short and patients will quickly be sent back home, some are long and require patients to spend the night at the hospital, and others might require an inter-hospital transfer. We use past trip data with geographical coordinates, time, and medical motives in our pipeline to study what impact a trained predictor has on overall fleet efficiency during real-world operating days. The pipeline is used to obtain a more reactive ambulance fleet by idling closer to oncoming requests; in particular, the company is interested in reducing the average reaction time, or the average reaction distance as a proxy. A study of the retraining of the predictor to improve over time and adapt to new data is presented in Appendix 2.

5.1 Use of the Pipeline

We use collected data from 500 operating days to train our predictor. The data from past requests are transformed and rescaled for anonymity and better learning. The geographical coordinates of a request are mapped to the closest location (there are seven predefined hospitals used as waiting locations for ambulances), and the transport motive is encoded as a one-hot vector with predefined categories $M = \{\text{consultation}, \text{radiotherapy}, \dots\}$. As presented in Sect. 4, these encoded requests are used to build a state representation $x(t) = (s^O, s^A, s_m, s_t)$ at time t. The training set is composed of all previous requests, the state they arrived in, and the closest location to that request $\{x(r_{it}), r_i^O\}$.

5.2 Training Phase

Roughly 8000 real trips are available for training. We split this dataset into a training set and a validation set (85%-15%). We also keep 248 real trips (10 days of operations) for the test set. Three models have been implemented for the predictor: a *neural network*, *random forest*, and *gradient boosting*. The neural network is implemented with *PyTorch* and consists of a self-attention layer coupled to a two hidden-layer feed-forward neural networks with 250 and 150 neurons in each hidden layer, respectively. Training is carried out with the *Adam* optimizer [21], and we use an *early stopping* method to avoid overfitting. A binary cross-entropy loss is used for the cost function ($F_\theta(x) = \hat{y}$ trained to minimize $C(\theta) = -\frac{1}{N} \sum_{i=1}^{N} y_i \log(F_\theta(x_i)) + (1 - y_i) \log(1 - F_\theta(x_i))$ on batches of size N). Random forest and gradient boosting are implemented with *scikit-learn*. Each model is trained on a personal computer (i7-8750H 2.20GHz CPU) in less than 1 min. The best values for the context vector parameters β and H are obtained by grid-search ($\beta = 0.75$ and $H = 10$). The evolution of the loss during training is reported in Fig. 3(b) for the neural network. As comparison baselines we also integrate a random selection, and a constant heuristic *Most Frequent* always assigning the location closest on average to the requests most frequently observed in the historical data. The accuracy (percentage of correctly predicted location) of the neural network is reported in Fig. 3(a) and includes comparisons with gradient boosting, random, and the best constant heuristic *Most Frequent*.

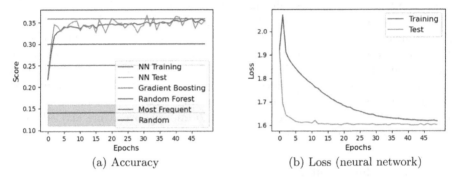

(a) Accuracy   (b) Loss (neural network)

Fig. 3. Statistics for the training phase.

5.3   Results: Location Prediction

Table 2 summarizes the main results of the different predictors, obtained with a 5-fold cross-validation method for predicting the next request, without any integration with the repositioning algorithm. The accuracy and the average distance between the predicted locations and the requested positions are reported. We can observe that the three predictors outperform the heuristic, both in terms of accuracy and in terms of average distance, which is the result intended. Among the three predictors, *GradBoost* provides the best performance. We evaluate the impact of the improvement of the location prediction on the fleet performance in the next section.

Table 2. Results of the different predictors for the next request. The percentage of improvement compared to the most frequent heuristic is also proposed.

Predictor	Accuracy	Average distance to request (km)
Random	14%	13.1 ± 5.1
Most frequent	25%	8.58 ± 4.1
Random Forest	30%	8.53 ± 4.3 (−0%)
NN	35%	8.05 ± 4.3 (−6%)
GradBoost	**36%**	**6.71** $\pm$ 4.1 (−21%)

5.4   Results: Ambulances Repositioning

Thanks to an *application programming interface* (API) provided by the partner company, we were able to obtain the travel time between each pair of ambulance depots and request position. With this information, we replayed *a posteriori* the requests of 10 historical days and re-positioned the ambulances with different policies: (1) positioning ambulances to the closest hospital after a trip (*Closest*), (2) sending ambulances to an empty hospital that was historically closest to the most requests (*Most frequent*, which is the previously seen heuristic restricted to empty locations), (3) the relocation procedure proposed in Algorithm 2 ($\gamma = 0.05$ and $\tau = 1$) with the three predictors (*NN*, *Random Forest*, and *GradBoost*). Results are illustrated in Fig. 4, and Table 3 presents the numerical results. The *reaction distance* corresponds to the accumulated travel times of all the ambulances between their current position to the location of accepted requests, and the *total distance travelled* is the accumulated travel times of all the ambulances, including the repositioning operation.

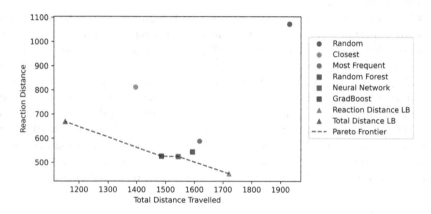

Fig. 4. Distance to drive to requests in 10 simulated days (data in Table 3)

The results show that the learned predictors help ensure more precise placement of vehicles for better reactivity, improving the *Most frequent* strategy by 10% and the *closest* strategy by 35%. By design, *Most frequent* ensures reactivity but results in greater overall distance travelled by the fleet. This can be corrected by using the learned predictors, especially the gradient boosting method that brings the total distance travelled back to *Closest* levels. *Closest* yields a low total distance because most ambulance movements are reactive. The ambulances reposition to the closest hospital after fulfilling a trip and then go directly to the next arising request. This ensures that few detours are made, but the reaction distance might be large because of a lack of anticipation. As we replayed historical days, it is possible to compute the optimal repositioning decisions with perfect information, minimizing either the reaction distance or the total distance. Although not feasible in practice, it provides lower bounds

Table 3. Distance (km) to drive to requests in 10 simulated days.

Predictor	Reaction distance (locations to requests)	Total distance traveled
Random	1073	1930.8
Closest	812.5	1412.1
Most frequent	587.6	1617.6
Random Forest	544.2	1593.4
NN	525.9	1552.6
GradBoost	526.3	1470.1
Reaction distance LB	453.8	1721.3
Total distance LB	669.2	1052.1

on the distance that can be achieved. The results show that the learned predictors approach these theoretical lower bounds while keeping the total distance travelled relatively low.

We deployed our repositioning strategy on the partner company server and ran a live experiment on 20 new days. The prediction pipeline is used as described in the previous section, and we evaluate the impact it would have had compared to the real-life instructions given by the operators (partly *Closest* and partly *Most Frequent* strategies most of the time) by fetching data in the database and providing real-time predictions for repositioning. The results are summarized in Table 4 for 20 operational days and show that the reaction distance could have been diminished, by between 15% to 21%, while the total distance would have increased, by around 10%. Thus, the predictors succeed in improving fleet reactivity with minimal deterioration of the total distance; an intermediate strategy can be used if the fleet is averse to small increases in total distance.

Table 4. Distance (km) to drive to requests, averaged over 20 days.

Predictor	Reaction distance (locations to requests)	Total distance traveled
Dispatcher instructions	1002.7	1712.3
NN	787.6 (−21%)	1898.5 (+11%)
GradBoost	846.8 (−15%)	1882.0 (+10%)

6   Conclusion

We developed an adaptive learning pipeline to reposition fleet vehicles in a ride-hailing problem with centralized control. Prior data on trips and requests allowed us to train classifiers to determine the zone most likely to receive requests, and vehicles that need to be repositioned are dispatched according to the prediction and state of the fleet. Our simulated application to a real ambulance fleet in

Belgium shows a theoretical improvement of more than 10% in reactivity for arising requests, and this result was confirmed by the implementation of our predictor in an industrial vehicle routing and fleet management program for ambulances.

Appendix 1.   Analysis: Features Importance

In addition to the learning-based repositioning strategy, the company is also interested to know which features (time, location, motives, etc.) are the most important to explain a specific decision. To do so, we use *permutation feature importance* which measures the drop in performance when the dataset is corrupted by randomly shuffling a feature [8]. Figure 5 shows the relative importance of different features for the neural network and the gradient boosting method. The higher the number the more important; the features are grouped by time, PRO, PRA, and prior requests motives. We observe that *arrivals* (PRA) provide the most information for predictors, as they provide information about the locations of users that might require an additional trip in the future. We confirm these results by training with only parts of the features as shown in Table 5. As suggested by the previous analysis, motives have a small impact on performance, raising the accuracy only by one percent or two, whereas prior requests information is essential to training accurate predictors.

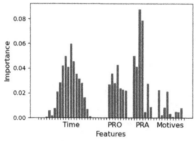

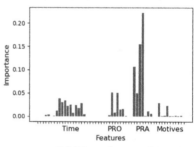

(a) Gradient boosting.   (b) Neural network.

Fig. 5. Features importance.

Table 5. Accuracy of the different predictors for the next request.

State features	NN accuracy	GradBoost accuracy
Time	28%	31%
Prior requests	32%	34%
Motives	29%	27%
Time, prior requests	33%	35%
Time, motives	30%	32%
Prior requests, motives	34%	35%
Time, prior requests, motives	**35%**	**36%**

Appendix 2.   Analysis: Online Learning

The environment in which our predictor works might vary through changes in request distribution or the way hospitals operates, so it is desirable to be able to adapt the predictor and retrain it on newer and more relevant data. This is commonly referred to as *online learning*. This can be carried out in several ways, for example, by expanding the training set with new data and retraining the model, or by freezing layers of the model and fitting on the new data. We test the efficiency of retraining approaches.

Our procedure is as follows. The dataset is split into quintiles (groups of 20% of the dataset), and at iteration $i \in \{1, \ldots, 5\}$, the i first quintiles are used to train and the quintile $i+1$ serves as the test set. This simulates new days of operation as tests, then used to retrain the predictors. This is important for further refining the accuracy of the model, and allows it to adapt to progressive environment changes; Fig. 6 shows how accuracy evolves in time by retraining on all the data available and testing on the subsequent days of operation, averaged over 100 iterations. The quintiles are generated either by shuffling the dataset, or by keeping the data sorted time-wise to replicate retraining in practice. The temporal coherence in small data samples allows the model to capture recent temporal variations in requests but hinders access to a wider selection of examples. Overall, this appears to have a small beneficial impact on the neural network but a negative one for gradient boosting. It is clear that retraining is important for this application when there is relatively little trip data available for initial training.

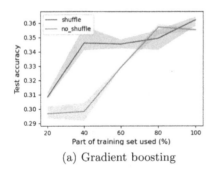
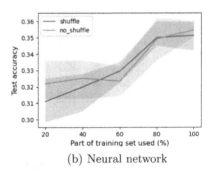

(a) Gradient boosting   (b) Neural network

Fig. 6. Retraining accuracy

References

1. Alonso-Mora, J., Samaranayake, S., Wallar, A., Frazzoli, E., Rus, D.: On-demand high-capacity ride-sharing via dynamic trip-vehicle assignment. Proc. Natl. Acad. Sci. **114**(3), 462–467 (2017). https://doi.org/10.1073/pnas.1611675114
2. Arulkumaran, K., Deisenroth, M.P., Brundage, M., Bharath, A.A.: A brief survey of deep reinforcement learning (2017). http://arxiv.org/abs/1708.05866

3. Bengio, Y., Lodi, A., Prouvost, A.: Machine learning for combinatorial optimization: a methodological tour d'horizon. Eur. J. Oper. Res. **290**(2), 405–421 (2021)
4. Bertsimas, D., Jaillet, P., Martin, S.: Flexbus: improving public transit with ride-hailing technology. Dow Sustainability Fellowship (2017). http://sustainability. umich.edu/media/files/dow/Dow-Masters-Report-FlexBus.pdf
5. Bertsimas, D., Jaillet, P., Martin, S.: Online vehicle routing: the edge of optimization in large-scale applications. Oper. Res. **67**(1), 143–162 (2019). https://doi.org/ 10.1287/opre.2018.1763
6. Bischoff, J., Maciejewski, M.: Simulation of city-wide replacement of private cars with autonomous taxis in berlin. In: The 7th International Conference on Ambient Systems, Networks and Technologies, vol. 83, pp. 237–244 (2016)
7. Braverman, A., Dai, J.G., Liu, X., Ying, L.: Empty-car routing in ridesharing systems. Oper. Res. **67**(5), 1437–1452 (2019). https://doi.org/10.1287/opre.2018. 1822
8. Breiman, L.: Random forests. Mach. Learn. **45** (2001)
9. Caceres-Cruz, J., Arias, P., Guimarans, D., Riera, D., Juan, A.A.: Rich vehicle routing problem: survey. ACM Comput. Surv. **47**(2) (2014). https://doi.org/10. 1145/2666003
10. Dandl, F., Hyland, M., Bogenberger, K., Mahmassani, H.S.: Evaluating the impact of spatio-temporal demand forecast aggregation on the operational performance of shared autonomous mobility fleets. Transportation **46**(6), 1975–1996 (2019). https://doi.org/10.1007/s11116-019-10007-9
11. Fagnant, D.J., Kockelman, K.M.: The travel and environmental implications of shared autonomous vehicles, using agent-based model scenarios. Transp. Res. Part C Emerg. Technol. **40**, 1–13 (2014). https://doi.org/10.1016/j.trc.2013.12.001
12. Gendreau, M., Laporte, G., Semet, F.: Solving an ambulance location model by tabu search. Locat. Sci. **5**(2), 75–88 (1997). https://doi.org/10.1016/S0966-8349(97)00015-6
13. Gmira, M., Gendreau, M., Lodi, A., Potvin, J.Y.: Managing in real-time a vehicle routing plan with time-dependent travel times on a road network. Transp. Res. Part C Emerg. Technol. **132**, 103379 (2021)
14. Goldberg, J.B.: Operations research models for the deployment of emergency services vehicles. EMS Manag. J. **1**(1), 20–39 (2004)
15. Held, M., Karp, R.M.: A dynamic programming approach to sequencing problems. J. Soc. Ind. Appl. Math. **10**(1), 196–210 (1962). https://doi.org/10.1137/0110015
16. Ho, S.C., Szeto, W., Kuo, Y.H., Leung, J.M., Petering, M., Tou, T.W.: A survey of dial-a-ride problems: literature review and recent developments. Transp. Res. Part B Methodol. **111**, 395–421 (2018)
17. Holler, J., et al.: Deep reinforcement learning for multi-driver vehicle dispatching and repositioning problem. In: 2019 IEEE International Conference on Data Mining (ICDM), pp. 1090–1095. IEEE Computer Society (2019). https://doi.org/10.1109/ ICDM.2019.00129
18. Jiao, Y., et al.: Real-world ride-hailing vehicle repositioning using deep reinforcement learning. Transp. Res. Part C Emerg. Technol. **130**, 103289 (2021). https:// doi.org/10.1016/j.trc.2021.103289
19. Jones, E.C., Leibowicz, B.D.: Contributions of shared autonomous vehicles to climate change mitigation. Transp. Res. Part D Transp. Environ. **72**, 279–298 (2019). https://doi.org/10.1016/j.trd.2019.05.005
20. Kaiser, L., et al.: Model-based reinforcement learning for atari. In: International Conference on Learning Representations, ICLR (2020)

21. Kingma, D.P., Ba, J.: Adam: a method for stochastic optimization. arXiv preprint arXiv:1412.6980 (2014)
22. Kullman, N.D., Cousineau, M., Goodson, J.C., Mendoza, J.E.: Dynamic ride-hailing with electric vehicles. Transp. Sci. **56**(3), 775–794 (2022)
23. Kumar, S., Panneerselvam, R.: A survey on the vehicle routing problem and its variants. Intell. Inf. Manag. **4**(3), 66–74 (2012)
24. LeCun, Y., Bengio, Y., Hinton, G.: Deep learning. Nature **521**, 436–44 (2015). https://doi.org/10.1038/nature14539
25. Lee, D.H., Wang, H., Cheu, R., Teo, S.H.: Taxi dispatch system based on current demands and real-time traffic conditions. Transp. Res. Rec. **1882**, 193–200 (2004). https://doi.org/10.3141/1882-23
26. Lin, C., Choy, K., Ho, G., Chung, S., Lam, H.: Survey of green vehicle routing problem: past and future trends. Expert Syst. Appl. **41**(4, Part 1), 1118–1138 (2014)
27. Liu, K., Li, X., Zou, C.C., Huang, H., Fu, Y.: Ambulance dispatch via deep reinforcement learning. In: SIGSPATIAL 2020, pp. 123–126. Association for Computing Machinery (2020). https://doi.org/10.1145/3397536.3422204
28. Luo, Q., Huang, X.: Multi-agent reinforcement learning for empty container repositioning. In: 2018 IEEE 9th International Conference on Software Engineering and Service Science (ICSESS), pp. 337–341 (2018)
29. Miao, F., et al.: Taxi dispatch with real-time sensing data in metropolitan areas: a receding horizon control approach. IEEE Trans. Autom. Sci. Eng. **13**(2), 463–478 (2016). https://doi.org/10.1109/TASE.2016.2529580
30. Oda, T., Joe-Wong, C.: Movi: a model-free approach to dynamic fleet management. In: IEEE INFOCOM 2018 - IEEE Conference on Computer Communications, pp. 2708–2716. IEEE Press (2018). https://doi.org/10.1109/INFOCOM.2018.8485988
31. Pedregosa, F., et al.: Scikit-learn: machine learning in Python. J. Mach. Learn. Res. **12**, 2825–2830 (2011)
32. Psaraftis, H.N., Wen, M., Kontovas, C.A.: Dynamic vehicle routing problems: three decades and counting. Networks **67**(1), 3–31 (2016)
33. Qiu, X.P., Sun, T.X., Xu, Y.G., Shao, Y.F., Dai, N., Huang, X.J.: Pre-trained models for natural language processing: a survey. Science China Technol. Sci. **63**(10), 1872–1897 (2020). https://doi.org/10.1007/s11431-020-1647-3
34. Riley, C., van Hentenryck, P., Yuan, E.: Real-time dispatching of large-scale ride-sharing systems: integrating optimization, machine learning, and model predictive control. In: IJCAI-20. International Joint Conferences on Artificial Intelligence Organization, pp. 4417–4423 (2020). https://doi.org/10.24963/ijcai.2020/609. Special track on AI for CompSust and Human well-being
35. Rossi, F., Zhang, R., Hindy, Y., Pavone, M.: Routing autonomous vehicles in congested transportation networks: structural properties and coordination algorithms. Auton. Robot. **42**(7), 1427–1442 (2018). https://doi.org/10.1007/s10514-018-9750-5
36. Seow, K.T., Dang, N.H., Lee, D.H.: A collaborative multiagent taxi-dispatch system. IEEE Trans. Autom. Sci. Eng. **7**(3), 607–616 (2010). https://doi.org/10.1109/TASE.2009.2028577
37. Sermanet, P., Eigen, D., Zhang, X., Mathieu, M., Fergus, R., LeCun, Y.: Overfeat: integrated recognition, localization and detection using convolutional networks (2014)
38. Sutton, R.S., Barto, A.G.: Reinforcement Learning: An Introduction. MIT Press, Cambridge (2018)

39. Woeginger, G.J.: Exact algorithms for NP-hard problems: a survey. In: Jünger, M., Reinelt, G., Rinaldi, G. (eds.) Combinatorial Optimization — Eureka, You Shrink! LNCS, vol. 2570, pp. 185–207. Springer, Heidelberg (2003). https://doi.org/10.1007/3-540-36478-1_17
40. Xu, Z., et al.: Large-scale order dispatch in on-demand ride-hailing platforms: a learning and planning approach. In: Proceedings of the 24th ACM SIGKDD International Conference on Knowledge Discovery & Data Mining, pp. 905–913 (2018)
41. Yuan, E., Chen, W., Van Hentenryck, P.: Reinforcement learning from optimization proxy for ride-hailing vehicle relocation. J. Artif. Intell. Res. (JAIR) **75**, 985–1002 (2022). https://doi.org/10.1613/jair.1.13794
42. Zhang, R., Pavone, M.: Control of robotic mobility-on-demand systems: a queueing-theoretical perspective. Int. J. Robot. Res. **35**(1–3), 186–203 (2016)

Bayesian Decision Trees Inspired from Evolutionary Algorithms

Efthyvoulos Drousiotis[1]([✉]), Alexander M. Phillips[1], Paul G. Spirakis[2],
and Simon Maskell[1]

[1] Department of Electrical Engineering and Electronics, University of Liverpool,
Liverpool L69 3GJ, UK
{E.Drousiotis,A.M.Philips,S.Maskell}@liverpool.ac.uk
[2] Department of Computer Science, University of Liverpool, Liverpool L69 3BX, UK
P.Spirakis@liverpool.ac.uk

Abstract. Bayesian Decision Trees (DTs) are generally considered a
more advanced and accurate model than a regular Decision Tree (DT) as
they can handle complex and uncertain data. Existing work on Bayesian
DTs uses Markov Chain Monte Carlo (MCMC) with an accept-reject
mechanism and sample using naive proposals to proceed to the next
iteration. This method can be slow because of the burn-in time needed.
We can reduce the burn-in period by proposing a more sophisticated way
of sampling or by designing a different numerical Bayesian approach. In
this paper, we propose a replacement of the MCMC with an inherently
parallel algorithm, the Sequential Monte Carlo (SMC), and a more effective sampling strategy inspired by the Evolutionary Algorithms (EA).
Experiments show that SMC combined with the EA can produce more
accurate results compared to MCMC in 100 times fewer iterations.

Keywords: Swarm Particle Optimisation · Bayesian Decision Trees ·
Machine Learning

1 Introduction and Relevant Work

Obtaining and calculating random samples from a probability distribution is
challenging in Bayesian statistics. Markov Chain Monte Carlo (MCMC) is a
widely used method to tackle this issue. MCMC can characterise a distribution
without knowing its analytic form by using a series of samples and it has been
used to solve problems in various domains, including psychology [9], forensics
[24], education [10,11], and chemistry [15], among other areas. Monte Carlo
applications are generally considered embarrassingly parallel since each chain can
run independently on two or more independent machines or cores. However, this
method is not very effective. Even though you increase the number of chains and
decrease the number of samples produced from each chain, the burn-in process
for each chain would remain unchanged. Nevertheless, the principal problem
is that processing within each chain is not embarrassingly parallel. When the

M. Sellmann and K. Tierney (Eds.): LION 2023, LNCS 14286, pp. 318–331, 2023.
https://doi.org/10.1007/978-3-031-44505-7_22

feature space and the proposal are computationally expensive, we can only do a little to improve the running time.

Much research has been done to improve the efficiency of MCMC methods. The improvements can be grouped into several categories [20]. One of these categories is based on understanding the geometry of the target density function. An example of this is Hamiltonian Monte Carlo [14] (HMC), which uses an auxiliary variable called momentum and updates it using the gradient of the density function. Different methods, such as symplectic integrators of various precision levels, have been developed to approximate the Hamiltonian equation [3]. HMC tends to generate less correlated samples than the Metropolis-Hastings algorithm, but it requires the gradient of the density function to be available and computationally feasible.

Another approach involves dividing complex problems into smaller and more manageable components. For example, as discussed earlier, using multiple parallel MCMC chains to explore the parameter space and then combining the samples obtained from these chains [19]. However, this approach does not achieve faster convergence of the chains to the stationary distribution. This is because all chains have to converge independently of each other in MCMC. It is also possible to partition the data space, which is already implemented in the context of Bayesian DTs [13], or partition the parameter space [1] into simpler pieces that can process independently, which are proven not to be much effective.

MCMC DTs simulations can take a long to converge on a good model when the state space is large and complex. This is due to both the number of iterations needed and the complexity of the calculations performed in each iteration, such as finding the best features and structure of a DT. MCMC, in general, and as will be explained detailed in Sect. 2, generates samples from a specific distribution by proposing new samples and deciding whether to accept them based on the evaluation of the posterior distribution. Because the current sample determines the next step of the MCMC, it cannot be easy to process a single MCMC chain simultaneously using multiple processing elements. A method [12] is proposed to parallelise a single chain on MCMC Decision trees but the speedup is not always guaranteed.

Another method of speeding up MCMC focuses on enhancing the proposal function, which is the approach we pursue in this paper. This can be achieved through techniques such as simulated tempering [18], adaptive MCMC [8], or multi-proposal MCMC [17]. Having a good proposal function in Markov Chain Monte Carlo (MCMC) and, in general, Monte Carlo methods is crucial for the efficiency and accuracy of the algorithm. Poor proposals can lead to slow convergence, poor mixing, and biased estimates. Good proposal functions can efficiently explore the target distribution and reduce the correlation between successive samples.

Several papers describe novelties specifically on Bayesian DTs, focusing on different improvements. For example, [12,13,23], contributed towards the runtime enhancement, [27] explored a new novel proposal move called "radical restructure", which changes the structure of the tree (T), without changing the number of leaves or the partition of observations into leaves. Moreover, [4] proposed different criteria for accepting or rejecting the T on MCMC, such as posterior probability, marginal likelihood, residual sum of squares, and misclassification rates. The general approach to improving the proposal function in Monte Carlo methods has been introduced previously. However, in the context of Bayesian DTs, there is still enough space for exploration.

Our contribution is then:

- We describe for the first time a novel algorithm inspired by the Evolutionary Algorithms to improve the proposal function of the Bayesian DTs that uses an inherently parallel algorithm, a Sequential Monte Carlo (SMC) sampler, to generate samples.

2 Bayesian Decision Trees

A DT operates by descending a tree T. The process of outputting a classification probability for a given datum starts at a root node (see Fig. 1). At each non-leaf node, a decision as to which child node to progress to is made based on the datum and the parameters of the node. This process continues until a leaf node is reached. At the leaf node, a node-specific and datum-independent classification output is generated.

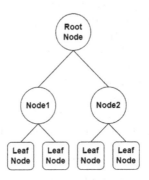

Fig. 1. Decision Tree

Our model describes the conditional distribution of a value for Y given the corresponding values for x, where x is a vector of predictors and Y the corresponding values that we predict. We define the tree to be T such that the function of the non-leaf nodes is to (implicitly) identify a region, A, of values for x for which $p(Y|x \in A)$ can be approximated as independent of the specific value of $x \in A$, i.e. as $p(Y|x) \approx p(Y|\phi_j, x \in A)$. This model is called a probabilistic classification tree, according to the quantitative response Y.

For a given tree, T, we define its depth to be $d(T)$, the set of leaf nodes to be $L(T)$ and the set of non-leaf nodes to be $\bar{L}(T)$. The T is then parameterised by: the set of features for all non-leaf nodes, $k_{\bar{L}(T)}$; the vector of corresponding thresholds, $c_{\bar{L}(T)}$; the parameters, $\phi_{L(T)}$, of the conditional probabilities associated with the leaf nodes. This is such that the parameters of the T are $\theta(T) = [k_{\bar{L}(T)}, c_{\bar{L}(T)}, \phi_{L(T)}]$ and $\theta(T)_j$ are the parameters associated with the jth node of the T, where:

$$\theta(T)_j = \begin{cases} [k_j, c_j] & j \in \bar{L}(T) \\ \phi_j & j \in L(T). \end{cases} \tag{1}$$

Given a dataset comprising N data, $Y_{1:N}$ and corresponding features, $x_{1:N}$, and since DTs are specified by T and $\theta(T)$, a Bayesian analysis of the problem proceeds by specifying a prior probability distribution, $p(\theta(T), T)$ and associated likelihood, $p(Y_{1:N}|T, \theta(T), x_{1:N})$. Because $\theta(T)$ defines the parametric model for T, it will usually be convenient to adopt the following structure for the joint probability distribution of N data, $Y_{1:N}$, and the N corresponding vectors of predictors, $x_{1:N}$:

$$p(Y_{1:N}, T, \theta(T)|x_{1:N}) = p(Y_{1:N}|T, \theta(T), x_{1:N})p(\theta(T), T) \tag{2}$$
$$= p(Y_{1:N}|T, \theta(T), x_{1:N})p(\theta(T)|T)p(T) \tag{3}$$

which we note is proportional to the posterior, $p(T, \theta(T)|Y_{1:N}, x_{1:N})$, and where we assume

$$p(Y_{1:N}|T, \theta(T), x_{1:N}) = \prod_{i=1}^{N} p(Y_i|x_i, T, \theta(T)) \tag{4}$$

$$p(\theta(T)|T) = \prod_{j \in T} p(\theta(T)_j|T) \tag{5}$$

$$= \prod_{j \in T} p(k_j|T)p(c_j|k_j, T) \tag{6}$$

$$p(T) = \frac{a}{(1 + d(T))^\beta} \tag{7}$$

Equation 4 describes the product of the probabilities of every data point, Y_i, being classified correctly given the datum's features, x_i, the T structure, and the features/thresholds, $\theta(T)$, associated with each node of the T. At the jth node, Eq. 6 describes the product of possibilities of picking the k_jth feature and corresponding threshold, c_j, given the T structure. Equation 7 is used as the prior for the T. This prior is recommended by [5] and three parameters specify this prior: the depth of the T, $d(T)$; the parameter, a, which acts as a normalisation constant; the parameter, $\beta > 0$, which specifies how many leaf nodes are probable, with larger values of β reducing the expected number of leaf nodes. β is crucial as this is the penalizing feature of our probabilistic T which prevents an algorithm that uses this prior from over-fitting and allows convergence to occur [21]. Changing β allows us to change the prior probability associated with "bushy" trees, those whose leaf nodes do not vary too much in depth.

An exhaustive evaluation of Eq. 2 over all trees will not be feasible, except in trivially small problems, because of the sheer number of possible trees.

Despite these limitations, Bayesian algorithms can still be used to explore the posterior. Such algorithms simulate a chain sequence of trees, such as:

$$T_0, T_1, T_2, ..., T_n \qquad (8)$$

which converge in distribution to the posterior, which is itself proportional to the joint distribution, $p(Y_{1:N}|T, \theta(T), x_{1:N})p(\theta(T)|T)p(T)$, specified in Eq. 2. We choose to have a simulated sequence that gravitates toward regions of the higher posterior probability. Such a simulation can be used to search for high-posterior probability trees stochastically. We now describe the details of algorithms that achieve this and how they can be implemented.

2.1 Stochastic Processes on Trees

To design algorithms that can search the space of trees stochastically, we first need to define a stochastic process for moving between trees. More specifically, we consider the following four kinds of move from one T to another:

- Grow (G): we sample one of the leaf nodes, $j \in L(T)$, and replace it with a new node with parameters, k_j and a c_j, which we sample uniformly from their parameter ranges.
- Prune (P): we sample the jth node (uniformly) and make it a leaf.
- Change (C): we sample the jth node (uniformly) from the non-leaf nodes, $\bar{L}(T)$, and sample k_j and a c_j uniformly from their parameter ranges.
- Swap (S): we sample the j_1th and j_2th nodes uniformly, where $j_1 \neq j_2$, and swap k_{j_1} with k_{j_2} and c_{j_1} with c_{j_2}.

We note that there will be situations (e.g. pruning from a T with a single node) when some moves cannot occur. We assume each 'valid' move is equally likely, which makes it possible to compute the probability of transition from one T, to another, T', which we denote $q(T', \theta(T')|T, \theta(T))$.

3 Our Approach on Evolutionary Algorithms

Evolutionary Algorithms (EA) mimic living organisms' behavior, using natural mechanisms to solve problems [2]. In our approach, the optimisation problem is represented as a multi-dimensional space on which the population lives (in our case, the population is the total number of trees). Each location on the solution space corresponds to a feasible solution to the optimisation problem. The optimal solution is found by identifying the optimal location in the solution space.

Pheromones play a crucial role in evolutionary algorithms as they are used for communication among the population [7, 16]. When a member of the population moves, it receives pheromones from the other member of the population and uses them to determine its next move. Once all members of the population reach new locations, they release pheromones; the concentration and type of pheromones released depend on the objective function or fitness value at that location. The solution space is the medium for transmitting pheromones, allowing individuals to receive and be affected by the pheromones released by other individuals, creating a global information-sharing model.

The population will gradually gain a rough understanding of global information through their movements, which can significantly benefit the optimisation process. In our approach, the EA can use the solution space as a memory to record the best and worst solutions produced in each iteration. Once the positioning stage is finished, all pheromones on the solution space are cleared. To guide the optimisation process, the most representative extreme solutions are selected from the recorded solutions, and the corresponding locations are updated with permanent pheromones. Unlike permanent pheromones, temporary pheromones only affect the movements of trees in the next iteration.

4 Methods

4.1 Conventional MCMC

One approach is to use a conventional application of Markov Chain Monte Carlo to DTs, as found in [12].

More specifically, we begin with a tree, T_0 and then at the ith iteration, we propose a new T' by sampling $T' \sim q\left(T', \theta(T')|T_i, \theta(T_i)\right)$. We then accept the proposed T' by drawing $u \sim U([0,1])$ such that:

$$T_{i+1} = \begin{cases} T' & u \leq \alpha(T'|T) \\ T_i & u > \alpha(T'|T) \end{cases} \tag{9}$$

where we define the acceptance ratio, $\alpha(T',T)$ as:

$$\alpha(T'|T) = \frac{p(Y_{1:N}|T, \theta(T), x_{1:N})}{p(Y_{1:N}|T', \theta(T'), x_{1:N})} \frac{q\left(T, \theta(T)|T', \theta(T')\right)}{q\left(T', \theta(T')|T, \theta(T)\right)} \tag{10}$$

This process proceeds for n iterations. We take the opportunity to highlight that this process is inherently sequential in its nature.

4.2 Evolutionary Algorithm in Bayesian Decision Trees

Initializing Population. The initial population plays a crucial role in the solutions' quality. An initial population with a good mix of diversity and a substantial number of trees can help improve the optimisation performance of the algorithm. To create a diverse initial population, a random method is often employed. This approach helps to ensure that the algorithm can perform a global search efficiently, is easy to use, and has a diverse set of individuals in the initial population.

The population size of trees is invariable, denoted as n. The location of every T_i in the D - dimensional space can be described as : $T = (T_1^1, T_2^2, \ldots, T_n^d, \ldots, T_N^D)$. According to the value of our objective function $p(Y1 : N, T, \theta(T)|x1 : N)$, the fitness value of location d and T_i can be measured. By comparing the current location of each T_i in the initial population, the optimal location and the worst location in the initial population were obtained, and the value of the objective function of the optimal location in the initial population was recorded.

Positioning Stage. In the positioning stage, in our use case, trees release permanent and temporary pheromones. The solution space records the locations of the terrible solution and the excellent solution produced by each iteration. While all trees move to the new location, the pheromones are updated differently, which will be discussed in Sect. 4.2. The process in the positioning stage is shown in Fig. 2. In our case, the possible movements of the T_i are those described in Sect. 2.1. When the proposed move is the *Grow*, we search for possible solutions in a higher dimensional space; when the proposed move is the *Prune*, we explore for possible solutions in lower dimensional space; and when the proposed moves are *Change* and *Swap*, we search for solutions on the current dimensional space.

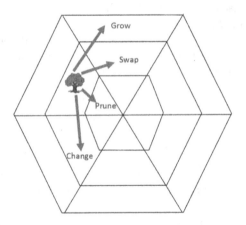

Fig. 2. Positioning Stage

Permanent Pheromones. *Permanent pheromones* have persistent effects. On each iteration, we evaluate each T_i to compute the value α_n (see Eq. 10). If α_n is greater than a uniform number between $[0, 1]$, we store the T_i and the stochastic move associated with *positive exploration* and *effective moves*, respectively. If α_n is less than a uniform number between $[0, 1]$ we store the T_i and the stochastic move associated, on *negative explartion* and *innefective moves* respectively (*negative explartion* and *innefective moves* will be used on *temporary pheromones*). We repeat the procedure above for all trees. We then calculate the *permanent pheromones* given the *effective moves*. *Permanent pheromones* is a single list with 4 real numbers, representing the possibilities of each one of the 4 stochastic moves to be selected on the next iteration. We update the *permanent pheromones* by adding to each possibility on the *permanent pheromones*, the number of times each move is in the *effective moves* list, and we then divide each element of the list by the sum of the *effective moves* (we normalise to add up to 1). We have also designed a mechanism to avoid having a biased *permanent pheromones* list. For example, in the first stages of the algorithm, it is common the *Grow* move to be a more useful proposal compared to the *Prune*. In such cases, if a specific move has more than 80% possibilities to be chosen, we re-weight the list by setting the dominant move having 40% possibilities to be selected and the rest having 20% possibilities to be chosen on the next iteration.

Temporary Pheromones. *Temporary pheromones* can only affect the movements of trees in the next iteration. We discussed above how we end up with a list called *effective moves* and a list called *ineffective moves*. When the iteration ends, *temporary pheromones* clear, and each T_i will release new *temporary pheromones*, which will be recorded. Algorithm 1 shows how we produce and store pheromones.

4.3   Sequential Monte Carlo with EA

We are considering a Sequential Monte Carlo (SMC) sampler [6] to handle the problem of sampling the trees from the posterior. After we have collected all the useful information from the movement of the trees, we now need to sample using the pheromones produced. As the Algorithm 2 shows, there are three possible sampling techniques where one has a subcategory. We choose the sampling technique by drawing a uniform number between $[0, 1]$. At this point, we need to specify that on each iteration, we draw a new uniform number, so each T_i has the possibility to sample with a different strategy.

The first sampling technique uses the *temporary pheromones*, and it has a possibility of 45% to be chosen. As mentioned earlier, *temporary pheromones* are produced during the previous iteration. This sampling technique has a sub-category, where the samples classified in *positive exploration* use a different sampling technique from those listed on *negative exploration*. If the T_i is listed in *positive exploration*, we pick a stochastic move m from the list with the *ineffective moves* uniformly. We then remove all the identical m from the *ineffective moves* and sample uniformly with equal probabilities from the remaining *ineffective moves*. If the T_i is in *negative exploration*, we pick uniformly a stochastic move m from the *effective moves* list to sample the particular T_i.

The second sampling technique uses the *permanent pheromones*, and it has a possibility of 45% to be chosen. As mentioned earlier, *permanent pheromones* are updated dynamically on every iteration, considering all the previous iterations. We sample the T_i using the possibilities in the list we describe in subsection *permanent pheromones* in Sect. 4.2.

The last sampling technique is straightforward. We use a list with stochastic moves, where each move has a uniform probability of being chosen. This sampling technique is unaffected by the pheromones. The main reason for including this technique is to ensure our algorithm is not biased, as this is the most common way of sampling in Bayesian DTs.

Algorithm 1. Pheromones Production Stage

Initialise n trees(T)
Sample trees $[T_0, T_1, ..., T_n]$
Initialise initial possibilities $= [p(G), p(P), p(C), p(S)] = [0.25, 0.25, 0.25, 0.25]$
Initialise permanent pheromones $= [p(G), p(P), p(C), p(S)]$ ▷ permanent pheromones
Initialise *positive exploration* list
Initialise *effective moves* list   ▷ Temporary Pheromones
Initialise *negative exploration* list
Initialise *ineffective moves* list   ▷ Permanent Pheromones
iterations $= 10$
for $(i \leq iterations, i + +)$ **do**
    Evaluate trees $[T_0, T_1, ..., T_n]$
    Store their acceptance probability $[\alpha_0, \alpha_1, ..., \alpha_n]$
    for $(s \leq n, s + +)$ **do**
        Draw a uniform number $u_1 \sim u[0, 1]$
        if $\alpha_s > u_1$ **then**
            append T_s to list *positive exploration*
            append $T_s move$ to list *effective moves*
        else
            append T_s to list *negative expolation*
            append $T_s move$ to list *ineffective moves*
        end if
    end for
    update *permanent pheromones* list given the *effective moves* list
end for

Algorithm 2. SMC with EA

---

for $i \leq n, i + +$ **do**
    Draw a uniform number $u_2 \sim u[0,1]$
    if $u_2 \leq 0.45$ **then**
        if T_i *in positive exploration* **then**
            pick uniformly a move m from *ineffective moves*
            Remove every identical m from the *ineffective moves*
            Sample T_i uniformly given the updated *ineffective moves*
        end if
        if T_i *in negative exploration* **then**
            Pick uniformly a move m from *effective moves*
            Sample T_i by applying the selected move m
        end if
    else if $u_2 > 0.45$ and $u_2 \leq 0.9$ **then**
        Sample T_i using *permanent pheromones*
    else
        Sample T_i using *initial possibilities*
    end if
end for
Empty *positive exploration*
Empty *effective moves*   ▷ Temporary Pheromones
Empty *negative exploration*
Empty *ineffective moves*   ▷ Permanent Pheromones

---

5 Experimental Setup and Results

To demonstrate the accuracy and the run time efficiency improvements we achieved through our proposed methods, we experiment on three publicly available datasets[1] listed in Table 1. We acknowledge that the size of the datasets we conduct the experiments is small. The main reason is to have a fair comparison between the MCMC algorithm and SMC, as experiments show [13] that the former struggles to converge on an adequate time on big datasets, compared to the latter. We also aim to show that SMC, an inherently parallel algorithm combined with EA, can be a great fit within the context of big data. For each dataset, we have multiple testing hypotheses. Firstly, we compare the SMC-EA with MCMC on 1000 iterations and 10 chains for MCMC and 10 T for the SMC-EA, 100 iterations and 100 chains for MCMC and 100 T for the SMC-EA, and 10 iterations with 1000 chains for MCMC and 1000 T for the SMC-EA.

    This section presents the experimental results obtained using the proposed methods with a focus on accuracy improvement and the ability of the SMC-EA to evolve smoothly in a very short period of iterations. We obtained the following results using a local HPC platform comprising twin SP 6138 processors (20 cores, each running at 2GHz) with 384GB memory RAM. We use the same hyper-parameters α and β for every contesting algorithm for testing purposes and a fair comparison and evaluation.

---

[1] https://archive.ics.uci.edu/ml/index.php.

Table 1. Datasets description

Dataset	Attributes	Instances
Heart Disease	75	303
Lung Cancer	56	32
SCADI	206	70

We tested both MCMC and SMC-EA with a 5-Fold Cross-Validation. Results indicate what is discussed in Sect. 4.2. When we initialise more trees, we introduce a more diverse set, which helps improve the algorithm's optimization performance. More specifically, SMC-EA with 1000 T and 10 iterations running SCADI dataset has an accuracy improvement of ∼2% and ∼ 6% compared to having 100 T with 100 iterations, and 10 T with 1000 iterations respectively. On the Heart Disease dataset SMC-EA with 1000 T and 10 iterations has an accuracy improvement of ∼3% and ∼4% compared to having 100 T with 100 iterations and 10 T with 1000 iterations respectively. On the Lung Cancer dataset SMC-EA with 1000 T and 10 iterations, has an accuracy improvement of ∼8% and ∼14% compared to having 100 T with 100 iterations, and 10 T with 1000 iterations respectively.

On the other hand, MCMC performs poorly when we have fewer iterations and more chains compared to SMC-EA. This is expected as MCMC needs adequate time to converge. When MCMC ran for more iterations, the algorithm made better predictions, and the accuracy achieved cannot be accepted as an acceptable threshold. SMC-EA has a ∼12% better predictive accuracy on the SCADI dataset compared to MCMC, ∼7% on Heart Disease, and ∼17% on Lung Cancer(see Tables 2, 3 and 4).

On the SMC-EA algorithm, trees do not have a leading T_i, and their movements are guided by interactions between them rather than one individual T dominating the group of trees. This helps to prevent individualism and stagnation in the population's evolution. Furthermore, a diverse population of trees is more effectively handled by the approach we suggest, as we avoid falling into local optimisation. The stochastic nature of the SMC-EA algorithm helps in escaping local optimisation and achieving global optimisation. Combining positive and negative feedback from the different pheromones can incorporate the benefits of successful solutions while mitigating the negative effects of poor solutions. SMC-EA algorithm fully uses all the information on the solution space, avoiding unnecessary waste or duplication of information. The EA algorithm updates the positions of all trees by using a combination of their current and past positions within the population. This allows the algorithm to maintain a history of information, preventing rapid jumps and leading to a smooth algorithm evolution.

Due to the small size of the datasets we are using to conduct this study, we can only show the effectiveness of our method in exploring the solutions space faster. However, previous studies [13] have shown that the SMC DT algorithm can improve the runtime compared to MCMC DT by up to a factor of 343. We are

Table 2. SCADI dataset

Chains_Trees	Iterations	MCMC	SMC-EA
10	1000	85	90.2
100	100	63	94.2
1000	10	57	96.6

Table 3. Heart Disease dataset

Chains_Trees	Iterations	MCMC	SMC-EA
10	1000	75.7	78.9
100	100	73.5	79.2
1000	10	67.7	82.7

Table 4. Lung Cancer dataset

Chains_Trees	Iterations	MCMC	SMC-EA
10	1000	70.1	73.9
100	100	69.7	79.1
1000	10	69.1	87.2

optimistic that we can achieve the same results, as SMC and EA are inherently parallel algorithms. The main bottleneck of the SMC-EA algorithm is when we evaluate a big number of trees; for example, see the test case of 10 iterations and 1000 T. We can overcome this problem by the distributed implementation, as we can distribute the trees on many nodes and evaluate the trees concurrently.

6   Conclusion

Our study has shown that by combining two novel algorithms, the SMC and EA can tackle major problems on Bayesian Decision Tress and open the space for more research. According to our experimental results, our novel approach based on Sequential Monte Carlo and Evolutionary Algorithms explores the solution space with at least 100 times fewer iterations than a standard probabilistic algorithm, for example, Markov Chain Monte Carlo. As discussed in Sect. 5, we managed to tackle the problem of the naive proposals, and we suggest a method that takes advantage of the communication between the trees. We proposed a sophisticated method to propose new samples based on EA and minimised the burn-in period through SMC.

As we already mentioned, both SMC and EA are inherently parallel algorithms with existing parallel implementations [22, 25, 26]. We plan to extend this study by parallelising the SMC-EA, adding more testing scenarios with larger datasets, and showing improvement in run time.

References

1. Basse, G., Smith, A., Pillai, N.: Parallel Markov chain Monte Carlo via spectral clustering. In: Gretton, A., Robert, C.C. (eds.) Proceedings of the 19th International Conference on Artificial Intelligence and Statistics. Proceedings of Machine Learning Research, Cadiz, Spain, 09–11 May 2016, vol. 51, pp. 1318–1327. PMLR (2016)
2. Binitha, S., Sathya, S.S., et al.: A survey of bio inspired optimization algorithms. Int. J. Soft Comput. Eng. **2**(2), 137–151 (2012)
3. Blanes, S., Casas, F., Sanz-Serna, J.M.: Numerical integrators for the hybrid Monte Carlo method. SIAM J. Sci. Comput. **36**(4), A1556–A1580 (2014)
4. Chipman, H.A., George, E.I., McCulloch, R.E.: Bayesian cart model search. J. Am. Stat. Assoc. **93**(443), 935–948 (1998)
5. Chipman, H.A., George, E.I., McCulloch, R.E.: BART: Bayesian additive regression trees. Ann. Appl. Stat. (2010)
6. Del Moral, P., Doucet, A., Jasra, A.: Sequential Monte Carlo samplers. J. Roy. Stat. Soc.: Ser. B (Stat. Methodol.) **68**(3), 411–436 (2006)
7. Dorigo, M., Stützle, T.: Ant colony optimization: overview and recent advances. In: Gendreau, M., Potvin, J.Y. (eds.) Handbook of Metaheuristics. International Series in Operations Research & Management Science, vol. 272, pp. 311–351. Springer, Cham (2019). https://doi.org/10.1007/978-3-319-91086-4_10
8. Douc, R., Guillin, A., Marin, J.-M., Robert, C.P.: Convergence of adaptive mixtures of importance sampling schemes. Ann. Stat. **35**(1), 420–448 (2007)
9. Drousiotis, E., et al.: Probabilistic decision trees for predicting 12-month university students likely to experience suicidal ideation. In: Maglogiannis, I., Iliadis, L., MacIntyre, J., Dominguez, M. (eds.) AIAI 2023. IFIP Advances in Information and Communication Technology, vol. 675, pp. 475–487. Springer, Cham (2023). https://doi.org/10.1007/978-3-031-34111-3_40
10. Drousiotis, E., Pentaliotis, P., Shi, L., Cristea, A.I.: Capturing fairness and uncertainty in student dropout prediction – a comparison study. In: Roll, I., McNamara, D., Sosnovsky, S., Luckin, R., Dimitrova, V. (eds.) AIED 2021, Part II. LNCS (LNAI), vol. 12749, pp. 139–144. Springer, Cham (2021). https://doi.org/10.1007/978-3-030-78270-2_25
11. Drousiotis, E., Shi, L., Maskell, S.: Early predictor for student success based on behavioural and demographic indicators. In: Cristea, A.I., Troussas, C. (eds.) ITS 2021. LNCS, vol. 12677, pp. 161–172. Springer, Cham (2021). https://doi.org/10.1007/978-3-030-80421-3_19
12. Drousiotis, E., Spirakis, P.G.: Single MCMC chain parallelisation on decision trees. In: Simos, D.E., Rasskazova, V.A., Archetti, F., Kotsireas, I.S., Pardalos, P.M. (eds.) LION 2022. LNCS, vol. 13621, pp. 191–204. Springer, Cham (2023). https://doi.org/10.1007/978-3-031-24866-5_15
13. Drousiotis, E., Spirakis, P.G., Maskell, S.: Parallel approaches to accelerate Bayesian decision trees. arXiv preprint arXiv:2301.09090 (2023)
14. Duane, S., Kennedy, A.D., Pendleton, B.J., Roweth, D.: Hybrid Monte Carlo. Phys. Lett. B **195**(2), 216–222 (1987)
15. Le Brazidec, J.D., Bocquet, M., Saunier, O., Roustan, Y.: Quantification of uncertainties in the assessment of an atmospheric release source applied to the autumn 2017. Atmos. Chem. Phys. **21**, 13247–13267 (2021)
16. Kalivarapu, V., Foo, J.-L., Winer, E.: Improving solution characteristics of particle swarm optimization using digital pheromones. Struct. Multidiscip. Optim. **37**(4), 415–427 (2009)

17. Liu, J.S., Liang, F., Wong, W.H.: The multiple-try method and local optimization in metropolis sampling. J. Ame. Stat. Assoc. **95**(449), 121–134 (2000)
18. Marinari, E., Parisi, G.: Simulated tempering: a new Monte Carlo scheme. EPL (Europhys. Lett.) **19**(6), 451 (1992)
19. Mykland, P., Tierney, L., Bin, Yu.: Regeneration in Markov chain samplers. J. Am. Stat. Assoc. **90**(429), 233–241 (1995)
20. Robert, C.P., Elvira, V., Tawn, N., Wu, C.: Accelerating MCMC algorithms. Wiley Interdisc. Rev.: Comput. Stat. **10**(5), e1435 (2018)
21. Ročková, V., Saha, E.: On theory for BART. In: The 22nd International Conference on Artificial Intelligence and Statistics. PMLR (2019)
22. Shukla, U.P., Nanda, S.J.: Parallel social spider clustering algorithm for high dimensional datasets. Eng. Appl. Artif. Intell. **56**, 75–90 (2016)
23. Taddy, M.A., Gramacy, R.B., Polson, N.G.: Dynamic trees for learning and design. J. Am. Stat. Assoc. **106**(493), 109–123 (2011)
24. Taylor, D., Bright, J.-A., Buckleton, J.: Interpreting forensic DNA profiling evidence without specifying the number of contributors. Forensic Sci. Int. Genet. **13**, 269–280 (2014)
25. Varsi, A., Maskell, S., Spirakis, P.G.: An o (logn) fully-balanced resampling algorithm for particle filters on distributed memory architectures. Algorithms **14**(12), 342 (2021)
26. Varsi, A., Taylor, J., Kekempanos, L., Knapp, E.P., Maskell, S.: A fast parallel particle filter for shared memory systems. IEEE Signal Process. Lett. **27**, 1570–1574 (2020)
27. Yuhong, W., Tjelmeland, H., West, M.: Bayesian cart: prior specification and posterior simulation. J. Comput. Graph. Stat. **16**(1), 44–66 (2007)

Towards Tackling MaxSAT by Combining Nested Monte Carlo with Local Search

Hui Wang[1], Abdallah Saffidine[2], and Tristan Cazenave[1(✉)]

[1] LAMSADE, University Paris Dauphine - PSL, Paris, France
`tristan.cazenave@lamsade.dauphine.fr`
[2] The University of New South Wales, Sydney, Australia

Abstract. Recent work proposed the UCTMAXSAT algorithm to address Maximum Satisfiability Problems (MaxSAT) and shown improved performance over pure Stochastic Local Search algorithms (SLS). UCTMAXSAT is based on Monte Carlo Tree Search but it uses SLS instead of purely random playouts. In this work, we introduce two algorithmic variations over UCTMAXSAT. We carry an empirical analysis on MaxSAT benchmarks from recent competitions and establish that both ideas lead to performance improvements. First, a nesting of the tree search inspired by the Nested Monte Carlo Search algorithm is effective on most instance types in the benchmark. Second, we observe that using a static flip limit in SLS, the ideal budget depends heavily on the instance size and we propose to set it dynamically. We show that it is a robust way to achieve comparable performance on a variety of instances without requiring additional tuning.

1 Introduction

Maximum Satisfiability (MaxSAT) problem is an extension of Boolean Satisfiability (SAT) problem. For MaxSAT, the task is to find a truth value assignment for each literal which satisfies the maximum number of clauses [12]. Stochastic Local Search (SLS) algorithms like WalkSat [15] and Novelty [19] are well studied to solve MaxSAT problems. These methods can not find a provable optimal solution but are usually used to search for an approximate optimal solution especially for larger problem instances. However, SLS algorithms are easy to get stuck in a local optimal solution and it's hard for them to escape. Thus, it's important to find an effective way to get rid of the local optimal solution. As a well-known successful method to address this exploration-exploitation dilemma, Monte Carlo Tree Search (MCTS) with UCT formula [4] is an ideal algorithm to deal with MaxSAT problems.

MCTS has shown impressive performance on game playing (including perfect information games and imperfect information games) [8,11,25], probabilistic single-agent planning [23], as well as most of problems which can be formed as a sequential decision making process, also know as Markov Decision Process (MDP) [3]. Based on the UCT formula, MCTS can address the exploration

© The Author(s), under exclusive license to Springer Nature Switzerland AG 2023
M. Sellmann and K. Tierney (Eds.): LION 2023, LNCS 14286, pp. 332–346, 2023.
https://doi.org/10.1007/978-3-031-44505-7_23

and exploitation dilemma in a theoretically sound way because UCT provides a state-of-the-art way to build the search tree based on the previous search records (including the node visited count and the node estimate values of the visit). Typically, the estimate method of the leaf node in the search tree is a random rollout policy. However, in a lot of applications, many other rollout policies are created to improve the accuracy of the leaf node value estimation. For MaxSAT problem, UCTMAXSAT (simply written as UCTMAX in the following parts) employs SLS algorithms to estimate the node value [12].

However, UCTMAX only runs MCTS for the root node to build a search tree until the time out, which may not sufficiently use the advantage of UCT reported by the Nested Monte Carlo Tree Search (NMCTS) [2]. NMCTS runs MCTS from root to the end or the time out. For each step, after performing the MCTS, it chooses the best assignment value for the current step and then enters into the next step and performs the MCTS again. In addition, UCTMAX employs a fixed flip limit for SLS algorithms. But in a UCT-style SLS, the number of the unassigned variables (literals below the search tree frontier are unassigned) will decreases along with the search tree deepens. Therefore, we design a novel computation called *Dynamic SLS*, see Eq. 2, for Monte Carlo methods used in this paper. The experimental results show that for most of the MaxSAT instances[1], the Dynamic SLS way is more robust than the fixed way used for UCTMAX to achieve comparable performance on a variety of instances without extra tuning. Besides, the results show that the NMCTS is better than the UCTMAX on most instances with moderate improvement.

Moreover, Nested Monte Carlo Search (NMCS) method [5] and its variants [6,7] have been successfully applied to master many NP-hard combinatorial optimization problems, like Morpion Solitaire [9], and achieve impressive performance [5,27]. However, NMCS has not been investigated to deal with MaxSAT problems. Therefore, this paper further studies the effectiveness of NMCS (also using Dynamic SLS as the state estimate) for MaxSAT.

Overall, the main contribution of this paper can be summarized as follows:

1. We examine various Monte Carlo Search techniques for the domain of MaxSAT, especially rollout policies and high-level searches. Through an extensive empirical analysis, we establish that (a) Purely random or heuristic-based rollouts are weaker than a Stochastic Local Search policy. (b) An MCTS-based search is weaker than Nested MCTS, especially in larger instances. NMCTS with WalkSat is weaker than NMCS, but is stronger with Novelty.
2. We introduce Dynamic SLS, a new rollout policy that dynamically computes the flip budget available for a stochastic local search. We demonstrate that Monte Carlo algorithms building on Dynamic SLS achieve comparable performance on standard MaxSAT benchmarks with previously existing Monte Carlo approaches without extra tuning.

---

[1] The instances are from *ms_random* benchmark: http://www.maxsat.udl.cat/15/benchmarks/index.html.

The rest of the paper is structured as follows. Before introducing preliminaries of this work in Sect. 3, we present an overview of the most relevant literature in Sect. 2. Then we present Dynamic SLS based Monte Carlo methods in Sect. 4. Thereafter, we illustrate the orientation experiments on a group of MaxSAT instances to finalize the structure of our proposed methods in Sect. 5. Then the full length experiments are presented in Sect. 6. Finally, we conclude our paper and discuss future work.

2   Related Work

There are a lot of solvers created to master MaxSAT problems [1,13,14,18]. Generally, these solvers can be categorized into two different types, i.e. *complete* solvers and *incomplete* solvers. Complete solvers can provide provable the best solution for the problem. Incomplete solvers start from a random assignment and continue to search for a better solution according to some strategies. Typically4 Stochastic Local Search algorithms like WalkSat [15] and Novelty [19] are well studied on MaxSAT [16,21]. These *incomplete* solvers suffer from an exploration-exploitation dilemma. And MCTS has shown successful performance of dealing with this dilemma [4]. Therefore, Tompkins et al. implemented an experimentation environment for mastering SAT and MaxSAT, called UCBMAX [24]. Furthermore, Goffinet et al. proposed UCTMAX algorithm to enhance the performance of SLS [12]. However, UCTMAX only performs UCT search once from the root, which may not sufficiently use the power of MCTS comparing to run UCT search for each step until to the terminal node or time out, which is known as Nested Monte Carlo Tree Search [2]. In addition to MCTS and NMCTS, NMCS [5] and its variations [6,7,22] also perform well especially for single agent NP-hard combinatorial problems, like Morpion Solitaire [9], where they achieve the best record which has not yet been improved even employing deep learning techniques [10,27]. Therefore, in this paper, we firstly employ NMCTS and NMCS to master MaxSAT problems with SLS methods.

3   Preliminaries

3.1   MaxSAT

In MaxSAT, like SAT, the problem is specified by a propositional formula described in conjunctive normal form (CNF) [20]. But unlike SAT which the aim is to find a truth assignment to satisfy all clauses, MaxSAT is just to find a truth assignment to satisfy the maximum number of clauses. For a set of Boolean variables $V = \{v_1, v_2, v_3, ..., v_i\}$, a literal l_j is either a variable v_j or its negation $\neg v_j$, $1 \leq j \leq i$. A clause is a disjunction of literals (i.e., $c_i = l_1 \vee l_2 \vee, ..., \vee l_j$). A CNF formula F is a set of clauses as conjunctive normal form (i.e., $F = c_1 \wedge c_2 \wedge, ..., \wedge c_i$). MaxSAT instances written as CNF can be easily found in our tested benchmark.

3.2   Heuristics

In order to test the different rollout policies for Monte Carlo Methods, here we present 3 simple heuristics that commonly used for MaxSAT.

1. H1 is the heuristic which assigns the value *from the first variable to the last variable* and H1 sets 0 for a variable that its positive value occurs more times than its negative value in all clauses.
2. H2 is the heuristic which, for each step, assigns the *variable* first which occurs the most times and H2 sets 0 for a variable that its positive value occurs more times than its negative value in all clauses.
3. H3 is the heuristic which, for each step, assigns the *literal* first which occurs the most times and H3 sets 0 for a variable that its positive value occurs more times than its negative value in all clauses.

3.3   Stochastic Local Search

Based on [12], in this paper, we also investigate two well-studied Stochastic Local Search (SLS) algorithms to deal with MaxSAT problem, namely WalkSat and Novelty.

Algorithm 1. Walksat

1: **function** WALKSAT(s)
2:     assignment←INITASSIGNMENT()
3:     **while** fliptimes $< f$ **do**
4:         **if** RANDOM() $< \epsilon_1$ **then**
5:             $v \leftarrow$ random variable
6:         **else**
7:             $v \leftarrow$ best unassigned variable
8:         assignment←flip(v)
     return assignment

WalkSat. As it can be seen in Algorithm 1, the idea of WalkSat is to initialize a random assignment (basic version) or according to the current found best solution (enhanced version) for each variable. Then an unsatisfied clause is selected. Further step is to select a variable to flip which has the highest bonus after flipping in the selected unsatisfied clause. The bonus is the change of the number of satisfied clauses after flipping the variable.

Novelty. Novelty is similar to WalkSat. The first step is also to initialize a random assignment (basic version) or according to the current found best solution (enhanced version). But differently, for each variable in all unsatisfied clauses, its bonus is computed. Then in order to avoid flipping in a dead loop, a variable which has the highest bonus but not selected in the most recent flipping is selected to flip. Simply, after line 7 in Algorithm 1, we add **If** $v = v_f$ **and random()** $< 1 - \epsilon_2$ **then** $v \leftarrow v_s$. v_f is the most recent flipped variable and v_s is the second best unassigned variable.

3.4 Monte Carlo Tree Search

Algorithm 2. Monte Carlo Tree Search

1: **function** MCTS(s)
2:     Search(s)
3:     $\pi_s \leftarrow$ normalize($Q(s,\cdot)$)
4:     **return** π_s
5: **function** SEARCH(s)
6:     **if** s is a terminal state **then**
7:         $v \leftarrow v_{end}$
8:         **return** v
9:     **if** s is not in the Tree **then**
10:         Add s to the Tree, initialize $Q(s,\cdot)$ and $N(s,\cdot)$ to 0
11:         Run rollout policy and get the solution score $v_{rollout}$
12:         $v \leftarrow v_{rollout}$
13:         **return** v
14:     **else**
15:         Select an action a with highest UCT value
16:         $s' \leftarrow$ getNextState(s, a)
17:         $v \leftarrow$ Search(s')
18:         $Q(s,a) \leftarrow \frac{N(s,a)*Q(s,a)+v}{N(s,a)+1}$
19:         $N(s,a) \leftarrow N(s,a) + 1$
20:     **return** v

According to [26,28,29], a recursive MCTS pseudo code is given in Algorithm 2. For each search, the rollout value is returned (or the game termination score). For each visit of a non-leaf node, the action with the highest UCT value is selected to investigate next [4]. After each search, the average win rate value $Q(s,a)$ and visit count $N(s,a)$ for each node in the visited trajectory is updated correspondingly. The UCT formula is as follows:

$$U(s,a) = Q(s,a) + c\sqrt{\frac{ln(N(s,\cdot))}{N(s,a)+1}} \tag{1}$$

The Nested Monte Carlo Tree Search (Due to the high computation, we only investigate level 1 for NMCTS in this paper) calls MCTS for each step of the assignment process.

3.5 Nested Monte Carlo Search

According to [5], the Nested Monte Carlo Search algorithm employs nested calls with rollouts and the record of the best sequence of moves with different levels. The basic level only performs random moves. Since a nested search may obtain worse results than a previous lower level search, recording the currently found

Algorithm 3. Nested Monte Carlo Search

```
1: function NMC(s, level)
2:     chosenSeq←[], bestScore← −∞, bestSeq←[]
3:     while s is not terminal do
4:         for each m in legalMoves(s) do
5:             s' ← PerformMove(s, m)
6:             if level = 1 then
7:                 (score, seq) ← run rollout policy
8:             else
9:                 (score, seq) ← NMC(s', level-1)
10:            highScore ← highest score of the moves from s
11:            if highScore > bestScore then
12:                bestScore ← highScore
13:                chosenMove ← m associated with highScore
14:                bestSeq ← seq associated with highScore
15:            else
16:                chosenMove ← first move in bestSeq
17:                bestSeq ← remove first move from bestSeq
18:            s ← perform chosenMove to s
19:            chosenSeq ← append chosenMove to chosenSeq
20:     return (bestScore, chosenSeq);
```

best sequence and following it when the searches result in worse results than the best sequence is important. Therefore, we present the pseudo code for the basic Monte Carlo Search algorithm as Algorithm 3. In order to estimate the leaf nodes from themselves instead of their children, we further test a variant of NMCS, named ZNMCS (Zero Nested Monte Carlo Search), where in Algorithm 3, line 4 is changed to for $i = 0, i < t, i + +$ do, in our experiments, $t = 10$. In addition, line 5 has been removed. And line 9 is changed to (score, seq)← ZNMCS(s, level-1).

4   Dynamic SLS Based Monte Carlo Methods

This section proposes the Dynamic SLS method with MCTS and NMCS. Since the number of the unassigned variables decreases as the search tree deepens, we propose a Dynamic SLS to avoid redundant flips and enlarge search tree to improve the performance within a fixed time budget. The flip limit (written as f) is simply computed according to the following Equation:

$$f = w \times u \tag{2}$$

w is a weight number, u is the number of the unassigned variables which can be flipped. Considering MCTS, in the search tree, the variables, upon the leaf nodes, have already been assigned to a value, so they can not be flipped anymore. We also tested several exponent values powered by u and finally found exponent equals 1 is the best.

In this work, we insert Dynamic SLS to replace rollout policy for MCTS (line 11 in Algorithm 2) and NMCS (line 7 in Algorithm 3, same to ZNMCS). In addition, according to [12], it is reported that using square number of the score is the best for UCTMAX, so in this work, for MCTS, we also replace the value calculation in line 7 and line 12 in Algorithm 2 to $v = pow(v_{end}, 2)$ and $v = pow(v_{dsls}, 2)$.

5   Orientation Experiments

5.1   Trial with Different Rollout

There are several ways to estimate the state value for Monte Carlo methods. One typical way is to simply run random simulations to get approximate values. In addition, for MaxSAT, there are many well designed heuristics to assign the truth values, based on the assignment, a proper value can be obtained. Besides, there are also several well studied SLS algorithms which can be applied to estimate the state value. Therefore, in order to determine which way is the best for the state estimate function, we use different ways to work together with NMCTS and NMCS to process our test setting (50 different instances, 70 variables each). The NMCTS simulation is set as 100. Time cost for each run is 50 s. each setting runs 10 repetitions. The results are shown in Table 1. We see that the heuristics all outperform random rollout, H3 is better than H2 and H2 is better than H1. Importantly, SLS methods perform significantly the best. So we adopt WalkSat and Novelty as the rollout policies for the further experiments. In addition, WalkSat for NMCS is better than NMCTS, but NMCTS with Novelty is the best.

Table 1. Results for Max3Sat Instances (70 variables) Using Different Rollout Policies for MCTS, NMCS. Results are average number of unsatisfied clauses on tested group instances, same to the following results.

Method	NMCTS	NMCS		
Level	–	playout	level 1	level 2
Random	81.4	125.2	80.8	80.5
H1	56.1	70.0	54.4	53.7
H2	55.1	69.5	54.6	53.8
H3	53.2	64.4	52.2	52.2
WalkSat	47.9	52.0	**47.4**	**47.7**
Novelty	**47.7**	**51.9**	48.8	49.0

5.2   UCTMAX vs NMCTS

Since [12] only investigated the UCTMAX with one time MCTS from the root until the time out. However, it does not perform an action to enter next state and run UCT again like game playing. To this end, the NMCTS [2] method should be further investigated. We let the MCTS simulation as a fixed value (set as 100) so that each step will stop and get a search tree. Based on this search tree, a best action can be found and performed to enter to next state. Then it runs another UCT process until the time out or the termination. The results show that the NMCTS performs clearly better than the UCTMAX way. In order to enlarge the result difference for different settings, we use larger instances (50 instances, each has 140 variables. [12] also used 140 as the test instance size, but they only tested on one instance, we test on 50 different instances with this size to reduce the noise.) for this experiment and the following orientation experiments (Fig. 1).

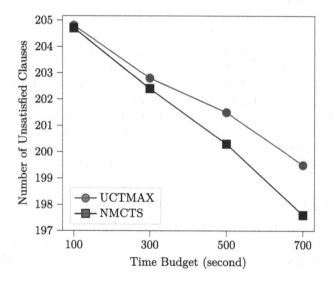

Fig. 1. Comparison of UCTMAX with NMCTS. NMCTS outperforms UCTMAX on 50 instances which has 140 variables each. For both UCTMAX and NMCTS, the f is set as 2000 which is reported as the best.

5.3   Current Global Best Solution

Based on [12] and [5], we know that it is the key to keep the global best solution (the best of the local solutions from all steps) found so far and initialize the SLS algorithms with this global best solution. We still do not know whether it is also important in our Nested Monte Carlo Methods with SLS. Therefore, we design different combinations to show the importance.

The results are shown in Table 2, we see that with a small time budget (100 s), for NMCTS, keeping the global best records has shown the advantage, and initializing based on the global best records is also better than not but with small improvements. For NMCS, with 100 s, although we still find that keeping the

Table 2. Impact of Random variable initialization and of keeping the global best solution on the performance of NMCTS and NMCS. Fixed number of flips (2000), 50 instances, 140 variables each.

Keep Global Initialization	No		Yes		
	Rand	Best	Rand	Local	Best
Time Budget	100 s				
NMCTS	221.2	220.8	204.8	205.1	**204.6**
NMCS	198.8	199.1	199.3	198.8	**198.7**
Time Budget	300 s				
NMCTS	219.9	219.8	202.9	202.9	**202.6**
NMCS	195.3	195.6	195.3	195.6	**193.1**

global best records and initializing with them is the best, but it's not very significant. However, we see a clear improvement with larger time budget (300 s). The reason that different initialization does not differ too much might be that the flip limit is set too big so even if it is initialized from random, it can also reach a global record level after flipping. From this experiment, we can conclude that keeping the global best records and initializing based on them for SLS (in this case, it is WalkSat) are both important to the nested search. NMCS works better than NMCTS with WalkSat on 140 variables instances.

5.4 Probabilistic SLS Initialization

In order to further investigate the contribution of initializing WalkSat based on the global best solution found so far, we adopt the simplest but commonly used way to balance the exploration and exploitation, ϵ-greedy, to initialize the assignment.

From Fig. 2, we see that $\epsilon = 0.1$ performs best, which further shows the best initialization way is to set literal assignment based on the best solution found so far but with a small randomness to initialize randomly. Thus, our following experiments are done with the ϵ as 0.1.

5.5 Fixed Flip Limits vs Dynamic Flip Limits

Goffinet et al. [12] used the fixed flip limits, which we found can be implemented in a dynamical way. Therefore, in this section, we test different w values (from 0.5 to 25, but finally we only present results of $w \in \{1, 2, 4\}$ as they are better) for dynamic flip limits calculation equation (see Eq. 2). And we found generally for both NMCTS and NMCS with different budget, $w = 2$ is the best (only the result of 300 s is weaker for NMCTS). In addition, we test fixed flip limit with 2000 (which is reported the best for UCTMAX tuned on a single instance) and 140 (same as the average flip limit for each step with $w = 2$). We found that with a fixed flip limit as 2000 is the worst and smaller limits increase the performance

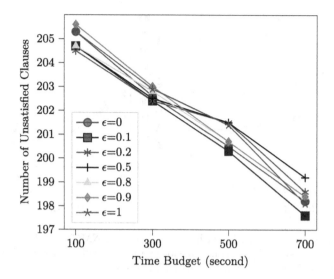

Fig. 2. Initializing Walksat Based on ϵ-greedy for NMCTS on 50 instances with 140 variables each, $\epsilon = 0$ means initializing WalkSat totally based on the global best solution. $\epsilon = 0.1$ means there is 10% probability to take a random initialization for the literal, and so on. The ϵ equals 0.1 is the best.

which shows that for Nested Monte Carlo methods, allocating time cost for relatively more steps contributes more.

Intuitively, even if a fine tuned fixed flip limit is found for a type of instances, it is not really applicable to set as the best for other instances. However, it is obviously that along with the increasing of sizes, the flip limit should also be larger. In order to test this assumption, we proposed the dynamic SLS and showed it works well for the category 140. Therefore, in order to show the adaptation of our Dynamic SLS method, after tuning the w for Dynamic SLS, we further test the best value we get for other larger instances which have 180 and 200 variables respectively, and compare the results with the fixed flip limit way (the best value is 140 for instances which have 140 variables). The results are presented in Fig. 4. We see that $2u$ achieves better performance for both 180 and 200 variables categories, showing that our Dynamic SLS is more adaptive to other instances. Therefore, no redundant extra tuning cost is needed.

6   Experiments on Benchmark

In this section, we will show the experimental results on tested benchmark instances with aforementioned SLS based different Monte Carlo methods. The benchmark consists of 383 instances categorized by different numbers of variables. And for each category, there are a bunch of instances with different numbers of clauses (Fig. 3).

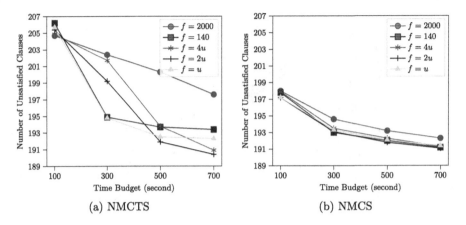

(a) NMCTS   (b) NMCS

Fig. 3. Comparison of Fixed SLS with Dynamic SLS for NMCTS and NMCS. In order to keep the w consistent for all runs, considering the overall results, we decide setting the weight w for Dynamic SLS flip limits as 2 is the best.

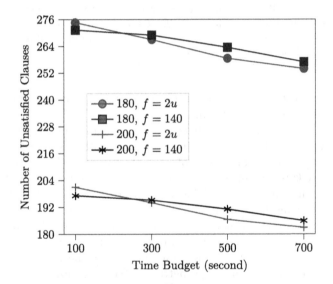

Fig. 4. Examples: Comparison of $2u$ and 140 flips for instances which have 180 and 200 variables respectively. NMCTS with Dynamic SLS is better than fixed flip limit on both 180 and 200 variables type, showing that our Dynamic SLS is more adaptive to other instances with different variable numbers.

Table 3. Results of MaxSat Instances Using WalkSat based UCTMAX, NMCTS, ZNMCS and NMCS respectively, with 300 s budget each run, 10 repetitions each.

Benchmark		Max_Walksat						Known Optimal Solution
Vars	Instances	UCTMAX	NMCTS	ZNMCS		NMCS		
			$m = 100$	round = 5, level 1	round = 1, level 2	round = 5, level 1	round = 1, level 2	
70	50	47.7	47.8	**47.1**	47.2	**47.1**	47.7	46.8
80	50	27.3	27.4	**27.1**	**27.1**	**27.1**	27.5	26.9
120	50	223.3	219.1	219.2	**218.7**	219.0	221.0	196.1
140	50	201.8	199.2	194.0	**193.1**	195.3	195.9	184.8
160	42	257.7	256.4	246.1	**243.1**	**243.1**	246.6	227.6
180	44	248.2	247.4	237.7	235.9	**235.4**	238.5	220.6
200	49	195.7	195.2	186.2	**184.5**	184.9	187.6	171.0
250	24	**7.7**	**7.7**	8.2	8.6	8.5	8.7	5.5
300	24	9.3	**9.1**	9.8	10.2	10.1	10.5	6.3

Table 4. Results of MaxSat Instances Using Novelty based UCTMAX, NMCTS, ZNMCS and NMCS respectively, with 300 s budget each run, 10 repetitions each.

Benchmark		Max_Novelty						Known Optimal Solution
Vars	Instances	UCTMAX	NMCTS	ZNMCS		NMCS		
			$m = 100$	round = 5, level 1	round = 1, level 2	round = 5, level 1	round = 1, level 2	
70	50	**47.1**	47.4	48.6	48.0	47.9	47.9	46.8
80	50	**27.4**	27.8	28.9	28.4	28.2	28.2	26.9
120	50	**212.7**	212.8	213.6	213.2	213.1	213.2	196.1
140	50	**185.7**	**185.7**	186.6	186.0	186.1	186.1	184.8
160	45	228.9	**228.8**	229.8	229.2	229.1	229.3	227.6
180	44	222.4	**222.2**	223.2	222.4	222.5	222.6	220.6
200	49	173.2	173.2	173.8	**173.1**	**173.1**	173.3	171.0
250	24	11.6	**11.2**	12.3	13.0	12.7	13.1	5.5
300	24	14.4	**14.0**	14.7	15.3	15.0	15.4	6.3

From Table 3, we can see that with WalkSAT, Nested Monte Carlo methods perform better than UCTMAX. For smaller instances like 70 and 80 variables categories, ZNMCS and NMCS level 1 perform the best, and ZNMCS level 2 achieves similar scores. Interestingly, for categories from 120 to 200, the best performance is achieved by ZNMCS level 2. And for largest instances, NMCTS is the best. These results confirm that the high level nesting of Monte Carlo methods may lead to worse performance.

From Table 4, we still see that for Novelty, NMCTS performs the best for larger instances. But differently, for the small instances, UCTMAX achieves best scores. Only for type 200, ZNMCS achieves the best and the scores do not vary too much. Importantly, it is clear that for most instances, Comparing with

WalkSat, Novelty achieves better scores which are much more close to the known optimal solutions, which also shows that a better SLS estimate method achieves better performance together with Nested Monte Carlo. This also leads to that the improvements of NMCTS for Novelty are smaller than that for WalkSat, but we still see a possibility of increasing improvements along with the increasing of the instances sizes, which we should further investigate in future work.

In addition, the type 250 and 300 variables instances are different from others since their clauses are much more easy to be satisfied. In these cases, we find that the NMCTS performs much stably the best.

Therefore, for both WalkSat and Novelty, we can conclude that the nesting search improves the performance of Monte Carlo methods, especially for nesting the MCTS while dealing with larger instances and employing the better SLS method.

7 Conclusion and Future Work

In this paper, we first investigated different rollout policies (random, heuristics, SLS) for different Nested Monte Carlo methods, including NMCTS and NMCS to deal with MaxSAT problem. We found that heuristics are better than random, but SLS is the best rollout policy to work with Monte Carlo methods in the domain of MaxSAT. In addition, we confirmed that also for Nested Monte Carlo methods, SLS methods should also record the global best records and initialize assignment based on the found current best record. In order to further balance the exploration and exploitation, we employed ϵ-greedy and found a proper ϵ value as 0.1 to randomly initialize the assignment for SLS, which improves the way that [12] initialized assignment fully based on the best record. The full benchmark experimental results show that for both WalkSat and Novelty based Monte Carlo methods, the nested tree search outperforms UCTMAX (Novelty in particularly performs better on larger instances), and NMCS with WalkSat also outperforms UCTMAX and even NMCTS. Therefore, we can conclude that nested search is important to deal with MaxSAT problems, especially for tree search on larger instances.

In the future, one way is to apply more powerful SLS algorithms together with Nested Monte Carlo methods like CCLS [17]. Besides, further investigation to find a light computation way for employing high level nested search is promising, especially for larger MaxSAT instances.

References

1. Ansótegui, C., Gabas, J.: WPM3: an (in) complete algorithm for weighted partial MaxSAT. Artif. Intell. **250**, 37–57 (2017)
2. Baier, H., Winands, M.H.: Nested Monte Carlo Tree Search for online planning in large MDPs. In: ECAI, vol. 242, pp. 109–114 (2012)
3. Brechtel, S., Gindele, T., Dillmann, R.: Probabilistic MDP-behavior planning for cars. In: 2011 14th International IEEE Conference on Intelligent Transportation Systems (ITSC), pp. 1537–1542. IEEE (2011)

4. Browne, C.B., et al.: A survey of Monte Carlo Tree Search methods. IEEE Trans. Comput. Intell. AI Games **4**(1), 1–43 (2012)
5. Cazenave, T.: Nested Monte-Carlo search. In: Twenty-First International Joint Conference on Artificial Intelligence (2009)
6. Cazenave, T.: Generalized nested rollout policy adaptation. In: Cazenave, T., Teytaud, O., Winands, M.H.M. (eds.) MCS 2020. CCIS, vol. 1379, pp. 71–83. Springer, Cham (2021). https://doi.org/10.1007/978-3-030-89453-5_6
7. Cazenave, T., Teytaud, F.: Application of the nested rollout policy adaptation algorithm to the traveling salesman problem with time windows. In: Hamadi, Y., Schoenauer, M. (eds.) LION 2012. LNCS, pp. 42–54. Springer, Heidelberg (2012). https://doi.org/10.1007/978-3-642-34413-8_4
8. Cowling, P.I., Ward, C.D., Powley, E.J.: Ensemble determinization in Monte Carlo Tree Search for the imperfect information card game magic: the gathering. IEEE Trans. Comput. Intell. AI Games **4**(4), 241–257 (2012)
9. Demaine, E.D., Demaine, M.L., Langerman, A., Langerman, S.: Morpion Solitaire. Theory Comput. Syst. **39**(3), 439–453 (2006)
10. Doux, B., Negrevergne, B., Cazenave, T.: Deep reinforcement learning for Morpion Solitaire. In: Browne, C., Kishimoto, A., Schaeffer, J. (eds.) ACG 2021. LNCS, vol. 13262, pp. 14–26. Springer, Cham (2022). https://doi.org/10.1007/978-3-031-11488-5_2
11. Gelly, S., Silver, D.: Combining online and offline knowledge in UCT. In: Proceedings of the 24th International Conference on Machine Learning, pp. 273–280 (2007)
12. Goffinet, J., Ramanujan, R.: Monte-Carlo tree search for the maximum satisfiability problem. In: Rueher, M. (ed.) CP 2016. LNCS, vol. 9892, pp. 251–267. Springer, Cham (2016). https://doi.org/10.1007/978-3-319-44953-1_17
13. Heras, F., Larrosa, J., Oliveras, A.: MiniMaxSAT: an efficient weighted MaxSAT solver. J. Artif. Intell. Res. **31**, 1–32 (2008)
14. Ignatiev, A., Morgado, A., Marques-Silva, J.: RC2: an efficient MaxSAT solver. J. Satisfiability Boolean Model. Comput. **11**(1), 53–64 (2019)
15. Kautz, H., Selman, B., McAllester, D.: Walksat in the 2004 SAT competition. In: Proceedings of the International Conference on Theory and Applications of Satisfiability Testing (2004)
16. Kroc, L., Sabharwal, A., Gomes, C.P., Selman, B.: Integrating systematic and local search paradigms: a new strategy for MaxSAT. In: Twenty-First International Joint Conference on Artificial Intelligence (2009)
17. Luo, C., Cai, S., Wu, W., Jie, Z., Su, K.: CCLS: an efficient local search algorithm for weighted maximum satisfiability. IEEE Trans. Comput. **64**(7), 1830–1843 (2014)
18. Martins, R., Manquinho, V., Lynce, I.: Open-WBO: a modular MaxSAT Solver'. In: Sinz, C., Egly, U. (eds.) SAT 2014. LNCS, vol. 8561, pp. 438–445. Springer, Cham (2014). https://doi.org/10.1007/978-3-319-09284-3_33
19. Menai, M.E., Batouche, M.: Efficient initial solution to extremal optimization algorithm for weighted MAXSAT problem. In: Chung, P.W.H., Hinde, C., Ali, M. (eds.) IEA/AIE 2003. LNCS (LNAI), vol. 2718, pp. 592–603. Springer, Heidelberg (2003). https://doi.org/10.1007/3-540-45034-3_60
20. Morgado, A., Heras, F., Liffiton, M., Planes, J., Marques-Silva, J.: Iterative and core-guided MaxSAT solving: a survey and assessment. Constraints **18**(4), 478–534 (2013)

21. Pelikan, M., Goldberg, D.E.: Hierarchical BOA solves Ising spin glasses and MAXSAT. In: Cantú-Paz, E., et al. (eds.) GECCO 2003. LNCS, vol. 2724, pp. 1271–1282. Springer, Heidelberg (2003). https://doi.org/10.1007/3-540-45110-2_3

22. Rosin, C.D.: Nested rollout policy adaptation for Monte Carlo Tree Search. In: Twenty-Second International Joint Conference on Artificial Intelligence (2011)

23. Seify, A., Buro, M.: Single-agent optimization through policy iteration using Monte Carlo Tree Search. arXiv preprint arXiv:2005.11335 (2020)

24. Tompkins, D.A.D., Hoos, H.H.: UBCSAT: an implementation and experimentation environment for SLS algorithms for SAT and MAX-SAT. In: Hoos, H.H., Mitchell, D.G. (eds.) SAT 2004. LNCS, vol. 3542, pp. 306–320. Springer, Heidelberg (2005). https://doi.org/10.1007/11527695_24

25. Wang, H., Emmerich, M., Plaat, A.: Assessing the potential of classical Q-learning in general game playing. In: Atzmueller, M., Duivesteijn, W. (eds.) BNAIC 2018. CCIS, vol. 1021, pp. 138–150. Springer, Cham (2019). https://doi.org/10.1007/978-3-030-31978-6_11

26. Wang, H., Emmerich, M., Preuss, M., Plaat, A.: Analysis of hyper-parameters for small games: iterations or epochs in self-play? arXiv preprint arXiv:2003.05988 (2020)

27. Wang, H., Preuss, M., Emmerich, M., Plaat, A.: Tackling Morpion Solitaire with AlphaZero-like ranked reward reinforcement learning. In: 2020 22nd International Symposium on Symbolic and Numeric Algorithms for Scientific Computing (SYNASC), pp. 149–152. IEEE (2020)

28. Wang, H., Preuss, M., Plaat, A.: Warm-Start AlphaZero self-play search enhancements. In: Bäck, T., et al. (eds.) PPSN 2020. LNCS, vol. 12270, pp. 528–542. Springer, Cham (2020). https://doi.org/10.1007/978-3-030-58115-2_37

29. Wang, H., Preuss, M., Plaat, A.: Adaptive warm-start MCTS in AlphaZero-like deep reinforcement learning. In: Pham, D.N., Theeramunkong, T., Governatori, G., Liu, F. (eds.) PRICAI 2021. LNCS (LNAI), vol. 13033, pp. 60–71. Springer, Cham (2021). https://doi.org/10.1007/978-3-030-89370-5_5

Relational Graph Attention-Based Deep Reinforcement Learning: An Application to Flexible Job Shop Scheduling with Sequence-Dependent Setup Times

Amirreza Farahani[1]([✉])[iD], Martijn Van Elzakker[2][iD], Laura Genga[1][iD],
Pavel Troubil[2][iD], and Remco Dijkman[1][iD]

[1] School of Industrial Engineering, Eindhoven University of Technology, Eindhoven, The Netherlands
{A.Farahani,L.Genga,R.M.Dijkman}@tue.nl
[2] Delmia R&D, Dassault Systèmes, 's-Hertogenbosch, The Netherlands
{Martijn.Vanelzakker,Pavel.Troubil}@3ds.com

Abstract. This paper tackles a manufacturing scheduling problem using an Edge Guided Relational Graph Attention-based Deep Reinforcement Learning approach. Unlike state-of-the-art approaches, the proposed method can deal with machine flexibility and sequence dependency of the setup times in the Job Shop Scheduling Problem. Furthermore, the proposed approach is size-agnostic. We evaluated our method against standard priority dispatching rules based on data that reflect a realistic scenario, designed on the basis of a practical case study at the Dassault Systèmes company. We used an industry-strength large neighborhood search based algorithm as benchmark. The results show that the proposed method outperforms the priority dispatching rules in terms of makespan, obtaining an average makespan difference with the best tested priority dispatching rules of 4.45% and 12.52%.

Keywords: Flexible Job Shop Scheduling · Optimization · Deep Reinforcement Learning

1 Introduction

In this paper, we propose an Edge Features Guided Relational Graph Attention-based Deep Reinforcement Learning (ERGAT-DRL) method to address a practical Flexible Job Shop Scheduling Problem with Sequence-dependent Setup Times (FJSP-SDST). FJSP-SDST is a more complex version of the well-known Job Shop Scheduling Problem (JSP), generalized with flexible machines and sequence-dependent setup times.

The Job shop Scheduling Problem (JSP) is one of the most-studied combinatorial optimization problems (COP), with many industrial applications [14,27]. In the JSP, a set of jobs have to be processed on a set of given machines. Each

M. Sellmann and K. Tierney (Eds.): LION 2023, LNCS 14286, pp. 347–362, 2023.
https://doi.org/10.1007/978-3-031-44505-7_24

job is composed of a set of operations, which are assigned to eligible machines with the aim of optimizing, e.g., the makespan or the tardiness [8]. When multiple alternative machines are available for an operation, the JSP is extended to the Flexible Job shop Scheduling Problem (FJSP) [25,27].

Another important aspect of many real-world scheduling problems are the setup times. In many real-world domains, such as e.g., pharmaceutical, or automobile manufacturing, setup operations (e.g, cleaning up or changing tools) are not only required between jobs but also heavily influenced by the immediately preceding operation on the same machine [21]. 'Sequence dependency' indicates that the magnitude of setup times is determined by the current operation as well as the direct previous operation processed on the same machine. The incorporation of setup times significantly impacts the applicability, dependability, and performance of scheduling methods. However, most studies in this area focus on operation time, with setup times assumed to be zero or constant [27]. The combination of these characteristics defines the FJSP-SDST addressed in this paper. This problem is considered a strongly NP-hard problem, and finding exact solutions to FJSP-SDST in practice is often impossible [9,15,19].

In addition to finding a high quality (i.e., approximately optimal) solution, there are other crucial factors in a practical scheduling problem: (1) *computational efficiency*, i.e, the computational time needed to find the solution, (2) *dynamicity and unexpected events*, i.e, the capability of of dealing with stochastic demands or events that make the practical scheduling environment non-deterministic, and (3) *size agnosticism*, i.e., the ability to deal with scheduling problems of varying numbers of jobs, operations, and machines without the need to update parameters for new instances. Exact solution methods, such as mathematical and constraint programming, are often impractical as they suffer from prohibitively long computational times, and they cannot deal with dynamic and unexpected events [7]. Meta-heuristics and approximate algorithms, such as evolutionary algorithms, are able to find reasonable solutions with less computational effort than exact solution methods. However, they are still expensive and do not guarantee that the found solution is globally optimal [3,13]. Moreover, solutions found via this group of methods often depend on a set of parameters, which may change for different problem instances or configurations [2]. Also, the design and set up of these methods often require expert knowledge. Priority dispatching rules (PDRs) are another solution approach for the FJSP-SDST problem. They assign sequentially dispatched individual operations to machines based on greedy rules. Compared with meta-heuristics and exact solution methods, they are sometimes considered more efficient and practical because of low computational time, capability to deal with dynamicity and unexpected events, and consistent approach for any problem size. Hence, they are often employed for real-life scheduling problems. However, PDRs sacrifice solution quality for efficiency. Also, they require substantial domain knowledge. Hence, usually, they have no guarantee of optimality and lack adaptability to specific situations [20].

To overcome these limitations, data-driven approaches have received a lot of attention in the last decade. Particularly, Deep Reinforcement Learning (DRL) approaches have been applied to different scheduling problems in several domains and industries [10,18,23]. Thanks to their capability to exploit data (i.e., simulated or historical solved problems) and deep neural networks' ability to capture the global patterns [17], DRL methods can provide high-quality schedules with promising computational efficiency, and they are considered in literature a good choice for non-deterministic environments [10,18,23]. The main challenge for DRL methods is that neural network dimensions have to be updated for new instances. Previous work suggested to overcome this issue by using Graph Neural Networks (GNN), which can process graphs of varying sizes. GNN-based methods are currently one of the best approaches for practical optimization problems [31]. Also, there are a few papers that used GNN transformers incorporated with DRL. They have proven that GNN-based approaches can satisfy size agnosticism requirements and make the DRL model more generic for scheduling problems [22,24,28,32].

Despite the capability and advantages of GNN transformers incorporated with DRL approaches, the presence of sequence-dependent setup times, in addition to machine flexibility, raises some challenges in different aspects of designing a DRL-based method. None of the previous Markov Decision Processes (MDPs) and proposed GNN transformers can simultaneously support machine flexibility and sequence-dependent setup time. Also, to the best of our knowledge, none of the existing related research tackled FJSP-SDST via DRL methods.

To fill this gap, this paper proposes a novel representation of FJSP-SDST as an MDP, based on a customized disjunctive graph where a weighted relational graph encodes the states. Furthermore, we propose an Edge Guided Relational Graph Attention-based Deep Neural Network (ERGAT) scheme that takes as input the state graph (i.e., all nodes, edges, and their features) and generates feature embeddings of the nodes and edges in the weighted relational graph that can support machine flexibility and sequence-dependent setup time simultaneously. Based on this scheme, we employ an Advantage Actor-Critic (A2C) to train our size-agnostic model to learn the policy of dispatching operations FJSP-SDST.

We evaluated our method against 1) an industry-strength large neighborhood search based algorithm used as a 'benchmark' and 2) standard priority dispatching rules. Data for our evaluation reflect a realistic scenario designed on the basis of a case study at the Dassault Systèmes company. The experiments show that while maintaining high computational efficiency, DRL methods can learn a high-quality policy for the FJSP-SDST problem under analysis.

Besides the methodological novelty, the proposed method also has good practical value. Its neural architecture is size agnostic and trained policy can be applied to solve instances of varying sizes, not only the sizes used in training. More importantly, the trained policy can rapidly solve large-scale FJSP-SDST instances and outperform the best tested PDRs in terms of makespan obtaining an average makespan difference from the benchmark within 4.45% and 12.52%. In particular, in instances with 400 operations (i.e., the largest size), our proposed method improved makespan by 36.37% compared to best tested PDRs, which is a promising improvement in practical scheduling. While our approach compares favorably with the PDR, there is still a significant quality gap with the benchmark solutions.

The rest of this paper is organized as follows. The next section introduces related works. The Preliminaries section provides formal definitions used throughout the paper. Our approach is then introduced in the Method section, and we describe its computational evaluation in the Experiments and Results section. Finally, the last section draws conclusions and delineates future work.

Table 1. GNN-based DRL methods for scheduling problems

Reference	COP	Graph	GNN
[24,32]	JSP	disjunctive graphs	GNN
[22]	HFSP	multi-graph	GCN - Attention-based
[28]	FJSP	disjunctive graphs	Heterogeneous GNN
Our Method	FJSP - SDST	disjunctive graphs	ERGAT

2   Related Work

In this section, we review previous work which also uses DRL and GNN transformers. We classify it based on: (1) the Scheduling Combinatorial Optimization problem (COP) it solves, (2) the type of graph it uses, and (3) the type of GNN it uses. Table 1 shows the comparison. Previous work dealing with scheduling problems usually relies on a disjunctive or multi-graph representation to encode the state graphs. This graph integrates the operations information via node features and relations between operations as edges [22,24,32]. However, none of these representations support the machine flexibility and sequence-dependent setup times aspects of scheduling. Their suggested GNN transformers do not support graphs with edge features and are limited to node features. To the best of our knowledge, there is only one exception, but that considers only machine flexibility aspect of problem. [28] proposed an integrated approach, which combines operation selection and machine assignment as one decision based on a multi-node types disjunctive graph (i.e., operations and machines as two different node types) and considers processing times as edge features for FJSP and applying a Heterogeneous GNN transformer to support multiple node types and single-dimensional edge features. In general, sequence-dependent setup times and machine flexibility make the problems much more complex in graph representation and decision-making. Hence, the main weak point of these approaches is that their proposed MDPs or GNNs cannot be applied to FJSP-SDST and are mainly compatible with more straightforward problems (e.g., JSP, FJSP). The attention-based GNN transformer used in [22,28] still cannot support sequence-dependent setup times in addition to machine flexibility. In contrast, our method is able to incorporate

sequence dependency of setup times as edge features in addition to machine flexibility via a customized disjunctive multi-relational graph with only one node type
which decreases complexity of multi node types graph. Furthermore, it employs
a novel Edge features-guided Relational Graph Attention-based transformer that
can support multi-dimensional edge features, allowing more information to be
incorporated into the disjunctive graph and different aspects of the scheduling
problem to be addressed.

3   Preliminaries

In this section, we first provide a formal definition of the Flexible Job Shop
Scheduling Problem with Sequence-dependent Setup Times (FJSP-SDST).
Then, we define our proposed graph structural properties of the FJSP-SDST
problem.

3.1   Flexible Job Shop Scheduling with Dynamic Setup Times

Let us consider a set J of n jobs and a set M of m machines. Each job $i \in J$
consists of a sequence of n_i consecutive operations, where the jth operation of
job i, denoted by o_{ij}, can be processed on any machine among a subset $m_{ij} \subseteq M$
of eligible (allowed) machines. Each machine can only process one operation at a
time. For each operation o_{ij}, let p_{ij} be its processing time on each machine, which
is the multiplication of quantity of o_{ij} and processing speed of this operation on
a machine respectively denoted as q_{ij} and v_{ij}. Note that we assume that the
allowed machines for each operation activity type (e.g., sorting, or matching)
are homogeneous and processing times (duration) are machine independent. We
define setup times dependent on the operation sequence and machine. A setup
time $d_{ij,i'j'}$ is incurred when operation j of job i (i.e., o_{ij}) and operation j' of
job i' (i.e., $o_{i'j'}$) are processed subsequently on a machine. Note that $d_{ij,i'j'}$ is
only defined if there exist operations o_{ij} and $o_{i'j'}$ such that exist in $m_{ij} \cap m_{i'j'}$.
Also, the setup time of sequence o_{ij} to $o_{i'j'}$ on a machine might be different from
the setup time of sequence $o_{i'j'}$ to o_{ij} on the same machine. The starting time
s_{ij} of an operation o_{ij} on a machine depends on: 1) completion time c_{ij-1} of
the previous operation of the same job, 2) completion time $c_{i'j'}$ of the previous
operation scheduled on the same machine, 3) setup time $d_{i'j',ij}$. In other words,
an operation cannot start earlier than the previous operation of the same job
terminates. Also, an operation cannot start before the previous operation on
the same machine terminates and the machine is setup again. The following
expressions represent the discussed constraints:

$$p_{ij} = q_{ij} * v_{ij} \tag{1}$$
$$s_{ij} = max(c_{ij-1}, c_{i'j'} + d_{i'j',ij}); \qquad c_{ij} = s_{ij} + p_{ij} \tag{2}$$

Our goal consists in determining a sequence of assignments of operations
to eligible machines which minimize the makespan $C_{max} = max_{ij}\{c_{ij}\}$ while
satisfying the constraints on the starting times.

3.2 Graph Structural Properties

To study the structural properties of the FJSP-SDST, we design a weighted relational graph based on the concept of the disjunctive graph to cover machine flexibility and sequence-dependent setup times aspects simultaneously. The main graph components and notation are illustrated in Fig. 1. This graph is inspired by a standard disjunctive model for JSP, proposed schemes for FJSP [11] and sequence-dependent setup times [4]. This graph is formally denoted as $G = (V, E^1, E^2, E^3, W_V, W_E)$. The set $V = \{o_{ij} \mid \forall i \in J, 1 \leq j \leq n_i\} \cup \{0, *\}$ contains all operations as well as dummy start and completion operations, which are respectively denoted as 0 and $*$. The set can be decomposed into subsets containing operations of a single job. Each subset represents all the operations of one particular job, i.e, job i indexed consecutively by the set of operations $\{o_{i1}, o_{i2}, ..., o_{in_i}\}$. We have three different edge types or relations. E^1 is a set of *conjunction arcs*, i.e., directed edges representing the precedence (job routing) constraints between every pair of consecutive operations of the same job (i.e., $o_{ij} \rightarrow o_{ij+1} \mid \forall i \in J, 1 \leq j < n_i$). They are illustrated via black arrows in Fig. 1. Set E^2 contains *disjunctive arcs*, i.e., weighted directed edges between pairs of operations with the same activity type that have to be executed on the same set of allowed machines. They are illustrated with dashed arrows in Fig. 1. These edges are weighted by sequence-dependent setup times between the origin and the source of the edge to represent the sequence-dependent setup times aspect of FJSP-SDST. E^3 contains the edges between operations that are planned directly after each other on the same machine and edges between the last operation scheduled on each machine in the current status of scheduling and all unplanned operations that could be scheduled next on that machine. E^3 is similar to E^2, with the difference that edges in this set are weighted by a binary edge indicator, which represents the existence of an edge between each pair of operations planned right after each other on the same set of machines. These edges are illustrated with solid blue arrows in Fig. 1. W_V is the weight on the nodes which is represented as a feature vector, that includes characteristics (e.g., size, color), activity type, quantity, process speed, earliest starting time, completion time, and status (i.e., a binary variable to identify whether an operation has been scheduled already or not) of the operation o_{ij}. W_E is the weight on the edges which is represented as an edge vector $W_E(o_{ij}, o_{i'j'}, E^r)$, where E^r represents the relation type. It means W_E maps a single-dimensional feature as a weight to a directed edge with relation type E^r. In particular, edges with relation type E^2 map to sequence-dependent setup times $d_{ij,i'j'}$ and E^1, E^3 map to binary values that represent the existence of the edge.

4 Method

In this paper, we propose an Edge Features Guided Relational Graph Attention-based Deep Reinforcement Learning (ERGAT-DRL) algorithm for solving the FJSP-SDST problem. This algorithm solves the FJSP-SDST problem as a sequential decision-making problem, which iteratively takes a dispatching action.

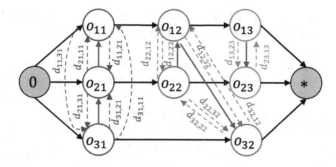

Fig. 1. FJSP-SDST problem - disjunctive graph

In this action, an operation is selected and then assigned to one of the eligible alternative machines, based on the first available machine heuristic. Figure 2 summarizes our proposed method. We discuss the various components in detail in the following subsections.

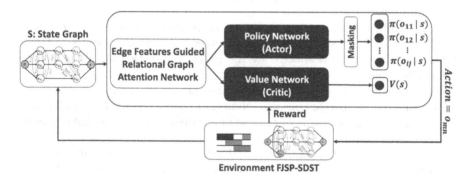

Fig. 2. Overview of proposed method for FJSP-SDST

4.1 Markov Decision Process Formulation

We formulate the FJSP-SDST as a MDP, defined by the following elements.

The **State** S is a weighted relation graph G (see Sect. 3) representing the current status of the scheduling environment and the partial solution of the problem. It consists of all operations with their attributes (i.e., characteristics, quantity, process speed, earliest starting time and completion time, and their status) and the relationships operations, eligibility of the machines, and sequence-dependent setup times. In particular, the state describes which operations have been scheduled already and which ones not yet.

The set of **Actions** A, consists of all operations (i.e., $\{o_{ij} \mid \forall i \in J, 1 \leq j \leq n_i\}$). By selecting an action, we dispatch an operation o_{ij} and assign it to the first available machine from the set m_{ij}. The dispatching is encoded in a node feature for this operation as a starting time that is the earliest time to start execution.

Note that not all actions are allowed in each state because of the problem constraints. First, operations have to be selected respecting their order within respective jobs, i.e., o_{ij+1} cannot be selected before o_{ij}. Second, an operation cannot be selected twice, i.e., it cannot be re-scheduled.

The **Transition** in this MDP is a multi-step process. First, based on the selected action, we assign the selected operation to the first available eligible machine. Then the state graph G is updated according to the following steps:

- Update the status of the selected operation based on selected action o_{ij} to 'planned operation' (i.e., already assigned to a machine) via the 'status' node feature.
- Update the starting time s_{ij} based on the availability of the assigned machine (i.e., completion time $c_{i'j}$ of the previous operation of the machine, plus sequence-dependent setup time $d_{i'j',ij}$) and the completion of the previous operation c_{ij-1} of the same job, and update the completion time c_{ij} based on quantity q_{ij} and process speed v_{ij}, by updating the corresponding node features (see Eq. 2).
- Update the earliest starting time and completion time node features of all other operations that have not been planned yet (i.e., not selected yet) and whose starting time is affected by the selected operation o_{ij}. This includes following operations of the same job (i.e., o_{ik} where $j \leq k < n_i$) and all operations allowed on the selected machine and their neighbors (i.e., operations of the same job of these operations). Note that these values may be different in the next iteration and only at the end of dispatching are the actual scheduled values determined.
- Update edges with type E^3. First, add a dispatching sequence edge from the selected operation o_{ij} to all unplanned operations that could be scheduled next on that machine. Second, remove dispatching sequence edges of the previous operation at the same machine (i.e. $o_{i'j'}$) that connected $o_{i'j'}$ to the other unplanned operations, except o_{ij}.

This transition process happens iteratively until all operations have been planned. Figure 3 presents a simplified example of an MDP transition. Starting

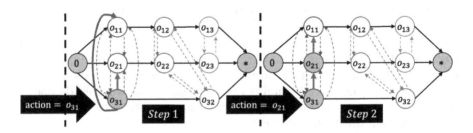

Fig. 3. A simplified example of an MDP transition

with the initial state graph, Step 1 selects o_{31} and assigns it to the first available machine in the set of eligible machines (i.e., dashed red edges). This updates the node features corresponding to the status, earliest starting time, and completion time of o_{31}, as well as the earliest starting and completion times of all nodes affected by o_{31} assignment. Additionally, a sequence dispatching edge (i.e., solid blue edges) is added from o_{31} to all unplanned operations that could be scheduled next on that machine (i.e., o_{21}, o_{11}). Step 2 similarly transitions by selecting o_{21} and additionally removing the sequence dispatching edge (i.e., solid blue edges) from previous selected operation o_{31} to the other unplanned operation (i.e. o_{11}).

In this paper, the training process is guided by cumulative immediate rewards to minimize makespan. These immediate **Rewards** are defined as makespan (i.e., $C_{max} = max_{ij}\{c_{ij}\}$) differences between two states that are processed after each other. Note that the approach can be used also in combination with different performance indicators to define the rewards.

4.2 Edge Features Guided Relational Graph Attention Network

To encode our weighted relational graph into a GNN, we follow the construction of the Relational Graph Attention Network (RGAT) [6], and Edge Features Guided Graph Attention Networks (EGAT) [16] to design an Edge Features Guided Relational Graph Attention Network (ERGAT) that supports multi-relational graphs with multidimensional real-valued features. RGAT and EGAT are extensions of the graph attention layer (GAT) proposed in [30]. RGAT extended GAT to the relational setting to encode graphs with different edge types (relations). EGAT incorporates edge features to indicate the relationship among nodes that can handle multidimensional real-valued edge features.

Network architecture overview, The input of the ERGAT layer is a weighted relational graph G with $|V|$ nodes and $|R|$ different relations. In this paper we consider $R = E^1, E^2, E^3$ (i.e., $|R| = 3$). We feed it to the network as two main components:

- Tensor of node features X, which is $X = [x_1 x_2 ... x_N] \in \mathbb{R}^{|V| \times F}$, where $x_i \in \mathbb{R}^F, \forall i \in V$ is a F dimensional feature vector with real-valued features (in this paper $F = 7$) of the ith node.
- Tensor of edge features E, Which is $E \in \mathbb{R}^{|V| \times |V| \times P \times E}$, where $e_{ij}^{(r)} \in E_P^{(r)}, \forall i, j \in V, \forall r \in |E|$ is a P dimensional feature vector of the edge connecting the ith to j^{th} nodes with relation type r (in this paper $P = 1$).

The output of the proposed layer is $X' = [x_1' x_2' ... x_N'], X' \in R^{|V| \times F'}$ which is a transformed node features matrix of X, where $x_i' \in \mathbb{R}^{F'}, \forall i \in V$ is a F' dimensional transformed feature vector of the ith node.

Network Operators. In ERGAT, we primarily use the operators which are employed in RGAT [6] with some advancements to make it compatible with edge guided approach and support real-valued edge features. The attention logits $a_{ij}^{(r)}$ of this suggested layer are additive attention logits for real-valued edge features which are calculated for each relation type and constructed query, key kernels and edge features to specify how the values, i.e. the intermediate representations $W_1^{(r)} x_i$ will combine to produce the updated node representations x_i'. A separate query kernel $Q^{(r)} \in \mathbb{R}^{F' \times D}$ and key kernel $K^{(r)} \in \mathbb{R}^{F' \times D}$ project the intermediate representations (i.e., $W_1^{(r)} x_i$) into query and key representations of dimensionality D, where $W_1^{(r)}, W_2^{(r)}$ are the learnable parameters of each relation r, (W is an $F' \times F$ matrix) of a shared linear transformation.

The following expressions determine additive attention logits for real-valued edge features:

$$q_i^{(r)} = W_1^{(r)} x_i . Q^{(r)}; \quad k_i^{(r)} = W_1^{(r)} x_i . K^{(r)} \tag{3}$$

$$a_{ij}^{(r)} = \text{LeakReLU}(q_i^{(r)} . k_j^{(r)} + W_2^{(r)} . e_{ij}^{(r)}) \tag{4}$$

The attention coefficients for each relation type r are then obtained using an across-relation attention mechanism, which calculates across node neighborhoods regardless of relation r [6].

$$\alpha_{ij}^{(r)} = \underset{j,r}{\text{softmax}}(a_{ij}^{(r)}) \frac{\exp(a_{ij}^{(r)})}{\sum_{r' \in E} \sum_{k \in \eta_i^{(r')}} \exp(a_{ik}^{(r')})} \tag{5}$$

By implementing a single probability distribution over the different representations $W_1^{(r)} x_j$ for nodes j in the neighborhood of node i, this mechanism encodes the prior that relation importance is a local property of the graph. Explicitly for any node i and all $r, r' \in E$, all $j \in \eta_i^{(r)}, k \in \eta_i^{(r')}$ yield competing attention $\alpha_{ij}^{(r)}$ and $\alpha_{ik}^{(r')}$ with sizes depending on their corresponding representations $W_1^{(r)} x_j$ and $W_1^{(r')} x_k$ [6].

Figure 4 shown an illustration of the ERGAT layer in four steps. Step 1: the intermediate representations for node i are combined with the intermediate representations for nodes in its neighborhood under each relation r type using edge features $e_{ij}^{(r)}$, to form each logit $a_{ij}^{(r)}$. Step 2: a softmax is taken across all logits independent of relation type to form the attention coefficients $\alpha_{ij}^{(r)}$. Step 3: these attention coefficients construct a weighted sum over the nodes in the neighborhood for each relation. Step 4: these are then aggregated and passed through a nonlinearity to produce the updated representation for node i.

4.3   Deep Reinforcement Learning

In this paper, we use an Advantage Actor Critic (A2C) algorithm to train our agent. The actor refers to the stochastic policy network $\pi(o_{ij}|s)$ to decide which operation (or node) to chose. The critic estimates the value function. This could be the action-value or state-value $V(s)$ to tell the actor how good its operation was and how it should adjust. In this paper we use an state-value

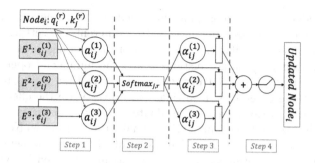

Fig. 4. Illustration of the ERGAT layer

approach. The stochastic policy $\pi(o_{ij}|s)$ is parameterized as a GNN with trainable parameter θ. Hence, it enables learning strong dispatching rules and size-agnostic generalization.

5 Experiments and Results

This section discusses the experiments we carried out to test our method. We first introduce the experimental settings and the tested competitors. Then, we discuss the obtained results.

5.1 Experimental Settings

Dataset. This experiment is designed on the basis of a practical case study at the Dassault Systèmes company. We used data with properties of an industrial manufacturing scheduling with the Job shop scheduling problem setting in the presence of machine flexibility and sequence-dependent setup times. These data include the following features: the number of jobs (ranging between 16 and 160), operations (with 2 to 3 operations per job and 40 to 400 operations in total), and machines (with 1 to 3 machines per operation and up to 6 machines in total), temporal properties, quantity, characteristics (e.g., size, length, type) of the operations, plus the sets of allowed machines per operation, and set of sequence-dependent setup times. We assume that the processing time of machines allowed for one specific type of activity are the same and depend on the operation's characteristics, quantity, and process speed of operation (see Eq. 1).

Training Parameters. We did hyperparameter tuning on the number of episodes (with options 500, 1000, 2000, 3000), and number of nodes per hidden layer (128, 256, 512). Also, we used some constant hyperparameters for training learning rate (1e-4), number of hidden layers (2), and discount factor 0.99. The learning method converged after 2000 episodes of one scheduling instance per episode. Note that the starting state of each episode is different from other episodes.

In each episode, all operations must be scheduled. For the validation of the model, we tested 125 unseen instances with 2 different size settings: 1) smaller instances, i.e., almost similar size configuration to training instances (i.e., less than 40 operations), 2) Larger instances, i.e., the number of operations is up to 10 times more than trained instances (i.e., more than 40 up to 400). The goal of using these different size settings is to investigate the effect of the size on the schedulers' performance and evaluate the size-agnosticism aspect of the model. We initialize an Edge Guided Relational Graph Attention Neural network as a graph transformer with 2 hidden layers of 512 hidden dimensions, ReLU activator, and Adam optimizer. For A2C algorithm, we used individual Fully-connected Feedforward Neural Networks with Backpropagation and a hidden layer as Policy Network (Actor) and Value Network (Critic). The remaining parameters are initialized according to PyTorch's default parameters. The agent and the simulation model are executed on a machine with an Intel(R) Core(TM) i7 Processor CPU @ 2.80 GHz and 16 GB of RAM, no graphics module is used for training the neural network.

Tested Competitors. We used an industry-strength algorithm based on a large neighborhood search as a 'benchmark', which has been provided by the Dassault Systèmes Delmia R&D team. Dassault Systèmes - DELMIA Quintiq provides companies with solutions to plan and optimize complex production value networks, optimize intricate logistics operations, and plan and schedule large, geographically diverse workforces. The Delmia Quintiq software was used as an industry-strength method for various planning and scheduling studies [1,12,29].

Furthermore, we reviewed existing literature [26] and had extensive discussions with experts from Dassault on how their scheduler currently works to select suitable PDRs to compare against our method. The selected rules were First Come, First Served (FCFS), (2) Shortest Processing Time (SPT), and (3) Shortest Setup Time (SIMSET). We also tested a random assignment.

5.2   Results

Methods Comparison. We tested the above-discussed methods in multiple experiments with different problem-size settings: smaller instances (i.e., up to 40 operations) and larger instances (i.e., up to 400 operations) that are respectively reported in Table 2. We evaluate their performance based on the average of the makespans and the average computational times. These evaluation measures are computed over 125 different test (unseen) instances.

The ERGAT-DRL method outperforms the tested PDRs in all the tested settings, obtaining an average makespan difference with the SPT (i.e., the best PDR) of 4.45% and 12.52% for smaller instances and larger instances, respectively. In particular, in instances with 400 operations (i.e., largest size), our proposed method improved the makespan by 36.37% compared to SPT, which is a promising improvement in practical scheduling. Even though our model never trains on instances with more than 40 operations and 50 problem instances, none of the tested instances have a similar number of operations to training instances.

While our approach compares favorably with the PDR, there is still a significant quality gap with the benchmark solutions. We believe this gap can be reduced by incorporating the DRL agent into a Monte Carlo tree search (MCTS) approach and improving the machine assignment heuristic (i.e., instead of using the first available machine heuristic, employing other heuristics or an additional DRL agent) used in our approach.

Table 2 also shows that, as expected, the PDRs rules are overall much faster than the competitors. However, as the number of operations increases from 40 to 400, the average computational time of PDRs increases by about 270 times. For the ERGAT-DRL method and benchmark these differences are around 68 and 30 times, respectively. It is worth noting that the performance of the ERGAT-DRL algorithm are comparable to the ones of the benchmark method, which is already a promising result. The company algorithm is a mature, optimized, and well-tested algorithm that actively provides services to customers, while our ERGAT-DRL algorithm still has a lot of potential to improve performance. For instance, our proposed DRL method was tested on regular CPU-based laptops (details of processors discussed above). Based on experimental results reported in the literature, employing GPUs can decrease the computational time 4 to 5 times [5]. Taking this into account, by optimizing the code, there is a potential to be 2–3 times more computationally efficient than the benchmark while performing substantially better in terms of results than existing PDRs.

Table 2. Average results over different size instances

Method	Smaller instances (up to 40)		Larger instances (up to 400)	
	Makespan (hrs)	time (s)	Makespan (hrs)	time (s)
Random	55.42	0.0818	283.79	22.54
SIMSET	55.37	0.083	303.76	**22.15**
FCFS	52.00	**0.0817**	250.45	22.29
SPT	51.74	0.0824	254.51	22.17
ERGAT-DRL	**49.95**	6.615	**234.43**	453.34
Benchmark	40.23	≈10	160.22	≈300

6   Conclusions and Future Work

This paper investigated the application of Graph Neural Network transformers incorporated in DRL in a practical Flexible Job Shop Scheduling Problem with Sequence-dependent Setup Times. The experimental results showed that the proposed approach consistently outperformed the tested PDRs in terms of makespan, obtaining an average makespan difference with the best tested PDR solution within 4.45% and 12.52%. These differences were more evident in larger instances; in the largest case, we achieved 36.37% of makespan improvement. The proposed architecture also effectively generalized to larger-sized problems and benchmarks unseen in training. Overall, these results show how the use of Edge Guided Relational Graph Attention Neural Network transformers incorporated in DRL can significantly decrease makespan in complex scheduling problem, thus suggesting that the use of these techniques can indeed bring significant practical advantages in the manufacturing domain with dynamic and non-deterministic environments that deal with large problem instances.

Nevertheless, our method presents some limitations that we plan to address in future work. The current version of the method was designed and tested to support single-objective optimization problems (e.g., makespan) rather than multi-objective optimization. Furthermore, we only considered machine flexibility and sequence-dependent setup times in this paper. In future work, we intend to extend our model to Flexible Assembly (i.e., incorporating relationships between different jobs) to increase the method's generality and applicability in real-world problems. In addition, we plan to incorporate dynamicity and uncertainty into the scheduling problem. Finally, we plan to investigate the performance improvement in terms of solution quality by incorporating the DRL agent into the Monte Carlo tree search (MCTS) approach and improving the machine assignment heuristic in our proposed method via other methods (e.g., heuristics or another DRL agent) rather than using the first available machine heuristic. Also we aim to improve the efficiency of the code to improve computational time.

References

1. Behnke, D., Geiger, M.J.: Test instances for the flexible job shop scheduling problem with work centers. Arbeitspapier/Research Paper/Helmut-Schmidt-Universität, Lehrstuhl für Betriebswirtschaftslehre, insbes. Logistik-Management (2012)
2. Bianchi, L., Dorigo, M., Gambardella, L.M., Gutjahr, W.J.: A survey on metaheuristics for stochastic combinatorial optimization. Nat. Comput. **8**, 239–287 (2009)
3. Blum, C., Roli, A.: Metaheuristics in combinatorial optimization: overview and conceptual comparison. ACM Comput. Surv. (CSUR) **35**(3), 268–308 (2003)
4. Brucker, P., Thiele, O.: A branch & bound method for the general-shop problem with sequence dependent setup-times. Operations-Research-Spektrum **18**(3), 145–161 (1996)
5. Buber, E., Banu, D.: Performance analysis and CPU vs GPU comparison for deep learning. In: 2018 6th International Conference on Control Engineering & Information Technology (CEIT), pp. 1–6. IEEE (2018)
6. Busbridge, D., Sherburn, D., Cavallo, P., Hammerla, N.Y.: Relational graph attention networks. arXiv preprint arXiv:1904.05811 (2019)
7. Chen, B., Matis, T.I.: A flexible dispatching rule for minimizing tardiness in job shop scheduling. Int. J. Prod. Econ. **141**(1), 360–365 (2013)
8. Cheng, R., Gen, M., Tsujimura, Y.: A tutorial survey of job-shop scheduling problems using genetic algorithms-I. representation. Comput. Ind. Eng. **30**(4), 983–997 (1996)

9. Cheng, T.E., Gupta, J.N., Wang, G.: A review of flowshop scheduling research with setup times. Prod. Oper. Manag. **9**(3), 262–282 (2000)

10. Cunha, B., Madureira, A.M., Fonseca, B., Coelho, D.: Deep reinforcement learning as a job shop scheduling solver: a literature review. In: Madureira, A.M., Abraham, A., Gandhi, N., Varela, M.L. (eds.) HIS 2018. AISC, vol. 923, pp. 350–359. Springer, Cham (2020). https://doi.org/10.1007/978-3-030-14347-3_34

11. Dauzère-Pérès, S., Paulli, J.: An integrated approach for modeling and solving the general multiprocessor job-shop scheduling problem using tabu search. Ann. Oper. Res. **70**, 281–306 (1997)

12. Dupláková, D., Teliškova, M., Török, J., Paulišin, D., Birčák, J.: Application of simulation software in the production process of milled parts. SAR J. **1**(2), 42–46 (2018)

13. Gao, K., Cao, Z., Zhang, L., Chen, Z., Han, Y., Pan, Q.: A review on swarm intelligence and evolutionary algorithms for solving flexible job shop scheduling problems. IEEE/CAA J. Autom. Sinica **6**(4), 904–916 (2019)

14. Gao, L., Zhang, G., Zhang, L., Li, X.: An efficient memetic algorithm for solving the job shop scheduling problem. Comput. Ind. Eng. **60**(4), 699–705 (2011)

15. Garey, M.R., Johnson, D.S., Sethi, R.: The complexity of flowshop and jobshop scheduling. Math. Oper. Res. **1**(2), 117–129 (1976)

16. Gong, L., Cheng, Q.: Adaptive edge features guided graph attention networks. arXiv preprint arXiv:1809.02709, vol. 2, pp. 811–820 (2018)

17. Hornik, K., Stinchcombe, M., White, H.: Multilayer feedforward networks are universal approximators. Neural Netw. **2**(5), 359–366 (1989)

18. Kayhan, B.M., Yildiz, G.: Reinforcement learning applications to machine scheduling problems: a comprehensive literature review. J. Intell. Manufact. 1–25 (2021). https://doi.org/10.1007/s10845-021-01847-3

19. Laguna, M.: A heuristic for production scheduling and inventory control in the presence of sequence-dependent setup times. IIE Trans. **31**(2), 125–134 (1999)

20. Mönch, L., Fowler, J.W., Mason, S.J.: Production Planning and Control for Semiconductor Wafer Fabrication Facilities: Modeling, Analysis, and Systems, vol. 52. Springer, New York (2012). https://doi.org/10.1007/978-1-4614-4472-5

21. Naderi, B., Zandieh, M., Balagh, A.K.G., Roshanaei, V.: An improved simulated annealing for hybrid flowshops with sequence-dependent setup and transportation times to minimize total completion time and total tardiness. Expert Syst. Appl. **36**(6), 9625–9633 (2009)

22. Ni, F., et al.: A multi-graph attributed reinforcement learning based optimization algorithm for large-scale hybrid flow shop scheduling problem. In: Proceedings of the 27th ACM SIGKDD Conference on Knowledge Discovery & Data Mining, pp. 3441–3451 (2021)

23. Panzer, M., Bender, B.: Deep reinforcement learning in production systems: a systematic literature review. Int. J. Prod. Res. **60**(13), 4316–4341 (2022)

24. Park, J., Chun, J., Kim, S.H., Kim, Y., Park, J.: Learning to schedule job-shop problems: representation and policy learning using graph neural network and reinforcement learning. Int. J. Prod. Res. **59**(11), 3360–3377 (2021)

25. Rossi, A.: Flexible job shop scheduling with sequence-dependent setup and transportation times by ant colony with reinforced pheromone relationships. Int. J. Prod. Econ. **153**, 253–267 (2014)

26. Sharma, P., Jain, A.: Performance analysis of dispatching rules in a stochastic dynamic job shop manufacturing system with sequence-dependent setup times: Simulation approach. CIRP J. Manuf. Sci. Technol. **10**, 110–119 (2015)

27. Shen, L., Dauzère-Pérès, S., Neufeld, J.S.: Solving the flexible job shop scheduling problem with sequence-dependent setup times. Eur. J. Oper. Res. **265**(2), 503–516 (2018)
28. Song, W., Chen, X., Li, Q., Cao, Z.: Flexible job-shop scheduling via graph neural network and deep reinforcement learning. IEEE Trans. Industr. Inf. **19**(2), 1600–1610 (2022)
29. van der Hoek, T. Optimization of crude oil operations scheduling and product blending and distribution scheduling within oil refineries (2014)
30. Velickovic, P., Cucurull, G., Casanova, A., Romero, A., Lio, P., Bengio, Y., et al.: Graph attention networks. Stat **1050**(20), 10–48550 (2017)
31. Wu, Z., Pan, S., Chen, F., Long, G., Zhang, C., Philip, S.Y.: A comprehensive survey on graph neural networks. IEEE Trans. Neural Netw. Learn. Syst. **32**(1), 4–24 (2020)
32. Zhang, C., Song, W., Cao, Z., Zhang, J., Tan, P.S., Chi, X.: Learning to dispatch for job shop scheduling via deep reinforcement learning. Adv. Neural. Inf. Process. Syst. **33**, 1621–1632 (2020)

Experimental Digital Twin for Job Shops with Transportation Agents

Aymen Gannouni[(✉)] [iD], Luis Felipe Casas Murillo[iD], Marco Kemmerling[iD], Anas Abdelrazeq[iD], and Robert H. Schmitt[iD]

Information Management in Mechanical Engineering WZL-MQ/IMA, RWTH Aachen University, Aachen, Germany
`aymen.gannouni@ima.rwth-aachen.de`
`https://cybernetics-lab.de/`

Abstract. Production scheduling in multi-stage manufacturing environments is subject to combinatorial optimization problems, such as the Job Shop Problem (JSP). The transportation of materials when assigned to mobile agents, such as Automated Guided Vehicles (AGVs), results in a Job Shop Problem with Transportation Agents (JSPTA). The transportation tasks require routing the AGVs within the physical space of the production environment. Efficient scheduling of production and material flow is thus crucial to enable flexible manufacturing systems.

Neural combinatorial optimization has evolved to solve combinatorial optimization problems using deep Reinforcement Learning (RL). The key aim is to learn robust heuristics that tackle the trade-off of optimality versus time complexity and scale better to dynamic changes in the problem. The present simulation environments used to train RL agents for solving the JSPTA lack accessibility (e.g. use of proprietary software), configurability (e.g. changing shop floor layout), and extendability (e.g. implementing other RL methods).

This research aims to address this gap by designing an Experimental Digital Twin (EDT) for the JSPTA. It represents an RL environment that considers the physical space for the execution of production jobs with transportation agents. We created our EDT using a simulation tool selected based on requirement analysis and tested it with a customized state-of-the-art neural combinatorial approach against two common Priority Dispatching Rules (PDRs).

With a focus on the makespan, our findings reveal that the neural combinatorial approach outperformed the other PDRs, even when tested on unseen shop floor layouts. Furthermore, our results call for further investigation of multi-agent collaboration and layout optimization. Our EDT is a first step towards creating self-adaptive manufacturing systems and testing potential optimization scenarios before transferring them to real-world applications.

Keywords: Job Shop Scheduling with Transportation · Neural Combinatorial Optimization · Experimental Digital Twins · Reinforcement Learning · Flexible Manufacturing Systems

© The Author(s), under exclusive license to Springer Nature Switzerland AG 2023
M. Sellmann and K. Tierney (Eds.): LION 2023, LNCS 14286, pp. 363–377, 2023.
https://doi.org/10.1007/978-3-031-44505-7_25

1   Introduction

Many multi-stage production processes are complex by nature and require a high degree of planning and monitoring. One example is assembly lines, where raw and semi-finished products are processed, transported, and assembled over multiple stations. Today's assembly products are highly customizable and can be individually shaped to customer needs. To build such products, so-called "lot-size one manufacturing" is crucial to produce a wide range of unique products at scale. The goal is to make manufacturing systems so flexible that even newly defined products can be ad-hoc integrated into production lines.

As good as the vision of "lot-size one manufacturing" is, the practical implementation poses various challenges, especially in production planning and control. For instance, assembly environments involve tackling Combinatorial Optimization Problems (COPs), such as the Job Shop Problem (JSP). This problem is known to be NP-hard, which is exponentially expensive given its time complexity when solved with brute-force approaches. Exact methods and meta-heuristics constitute a widely-used option for solving COPs. However, these are often impractical in reality, especially under dynamic conditions. The dynamics in production are present at many levels: from the mobility of production actors (both human and machine) to unexpected events (e.g. machine breakdown) or global disruptions (e.g. COVID-19). Therefore, the need for resilient approaches to solving COPs is continuously increasing.

Neural Combinatorial Optimization (NCO) has evolved in recent years as a potential approach to satisfy the need for robust heuristics. NCO uses deep Reinforcement Learning (RL) for solving COPs, including the JSP. Some RL environments have been developed for NCO approaches to the JSP with Transportation Agents (JSPTA). Still, many of these environments lack accessibility (e.g. use of proprietary software), configurability (e.g. change of shop floor layout), and extendability (e.g. out-of-the-box implementation of other RL methods). Meanwhile, Digital Twins (DT) are defined in the context of manufacturing as virtual representations of production systems that can run on different simulations for various purposes with synchronization between the virtual and real system [14]. By omitting the synchronization between the virtual and real system with the sole purpose of experimenting with optimization scenarios in the virtual representation of the system, digital twins are experimentable and therefore called Experimental Digital Twins (EDTs) [17].

To address the current lack of accessible, configurable, and extendable environments for studying NCO approaches to the JSPTA, we designed an EDT that enables the training and testing of RL-based approaches considering the physical space of the JSP and its material flow with transportation agents. We created our EDT using a simulation tool selected based on requirement analysis and tested it with a state-of-the-art neural combinatorial approach against two priority dispatching rules using an elaborated design of experiment.

This paper will cover in Sect. 2 the related works of current environments available for NCO approaches to solve the JSPTA. Section 3 describes the requirement analysis, development tool selection, and the design of the simulation environment. Moreover, it illustrates the experimental setup for training the RL agents and testing the overall EDT. Section 4 shows the results of our EDT evaluation, where the implications and limitations are discussed. Finally, this paper concludes with our contribution and future work in Sect. 5.

2   Related Work

This section presents the related work to our EDT for the JSPTA in the following. First, Feldkamp et al. [4] implemented a Double Deep Q-learning Network (DDQN) within a discrete event simulation model of a modular production system. The simulator used was created in Siemens Plant Simulation [19] and consisted of a grid of workstations spread throughout the shop floor and a variable number of AGVs. The jobs contained priority charts with partially ordered tasks. The DDQN model was evaluated against two PDRs based on lead times. The results showed that the DDQN model outperformed the PDRs [4].

Hu et al. [8] proposed also a DDQN model for AGV real-time scheduling within a flexible shop floor. The simulation was built within the Tecnomatix [20] platform where no AGV collisions were simulated, which simplifies the AGV navigation and impacts the JSPTA solutions in layouts with limited traversable areas. Furthermore, the AGVs possessed different speeds, and the jobs were directly split into sub-tasks providing no explicit job descriptions for the agents. The tasks were generated randomly and the task arrival interval was based on a normal distribution. The DDQN model was compared with two alternative RL methods, Q-learning and SARSA, and five different PDRs. The testing was performed within 4 different layout configurations and focused on delay ratios and makespan. The proposed DDQN model outperformed all other methods, including Q-learning and SARSA [8].

Malus et al. [12] implemented a multi-agent RL approach based on the TD3 policy gradient algorithm [5] for order dispatching with a fleet of AGVs. The RL model was trained using a fast, physically inaccurate representation of the real-world environment and then validated with a more accurate environment developed using Gazebo. The RL agents were compared against two PDRs. With a focus on completion times, the RL agents outperformed the PDRs [12].

Other relevant works that use DTs as tools for improving JSPTA solutions include Torterelli et al. [21] where a deep RL-based algorithm was implemented for an assembly line with resource constraints. Huang et al. [9] demonstrated how the DT concept can be implemented to improve flexible systems' design. Zhang et al. [24] proposed a pipeline for solving the simultaneous scheduling and layout problems innate to flexible manufacturing systems through simulation. To validate their pipeline they implemented it for a production system simulation with 12 workstations, 6 industrial robots, and 1 AGV.

The current related simulations and digital twins show a lack of accessibility (use of proprietary software, such as Siemens Plant Simulation), configurability (changing number of e.g. jobs, stations, agents), and extendability (implementation of new RL methods) for the applications of JSPTA. The experimentation with RL-based approaches for solving the JSPTA is currently bound to the adaptability of previous works. We aim to fill this gap with a highly-configurable EDT that enables the investigation of neural combinatorial approaches with different scenarios for the JSPTA. In the subsequent section, we describe the design of our EDT and the requirement analysis done for selecting a suitable simulation tool.

3   EDT Design

3.1   Tool Selection

To select the proper tool, we analyzed functional and non-functional requirements for building our EDT. On the one hand, functional requirements primarily serve the core functionalities, such as the components of the JSPTA: jobs, workstations, and transportation agents. The EDT has to simulate the relevant physics within the job shop, such as the routing of the transportation agents. Moreover, the environment should allow the training and testing of RL agents. Trained agents should be testable also in different shop floor configurations. On the other hand, non-functional requirements represent features that increase the usability of the EDT, especially its *accessibility, configurability*, and *extendability*. Accessibility focuses on making our EDT available to a larger audience and limiting possible entry barriers for the usage (e.g. need for a commercial license). While configurability enables the experimentation with different JSPTA components (e.g. number of agents or workstations, shop floor layout), and extendability targets the addition of new features to the EDT.

After studying reviews of simulation environments (particularly [3], [10], [11]), we identified **MuJoCo** [13], **Gazebo** [6] and **Unity** [22] as suitable candidates for building the EDT. Table 1 presents the major requirements for selecting the simulation tool used to build the EDT. Additionally, Table 1 shows our qualitative assessment of how each tool fulfills the defined requirements over three degrees: not fulfilling, partially fulfilling, and fulfilling.

Considering the core functional requirements in Table 1, Gazebo and MuJoCo show stronger capabilities in physics simulation when compared to Unity (*see FR.1.*). Among other factors, this is due to the usage of Gazebo and MuJoCo for robotic applications and hence their superiority in modeling contact forces and their behaviors. All of the assessed tools allow the training and testing of RL agents (*see FR.2.*). When it comes to the visualization capabilities, Unity has a superior range of features, especially regarding the graphical theme customization of simulation environments (*see FR.3.*).

Given the non-functional requirements in Table 1, we identified various strengths and shortcomings between the tools. Gazebo and MuJoCo are very accessible thanks to their open-source character. Moreover, MuJoCo fully supports the wrapping to OpenAI Gyms [1], which eases the manipulation of the environments and the application of already implemented RL methods such as Stable-Baselines3 [16]. Meanwhile, Unity is stronger than Gazebo and MuJoCo in terms of configurability and extendability, especially when combined with Unity ML agents [10]. Thanks to its powerful visualization capabilities, Unity

Table 1. Requirement Analysis and Simulation Tools Comparison

Category	Requirement	Gazebo	MuJoCo	Unity
	Functional Requirements (FRs)			
Core	FR.1. Physics simulation	●	●	◐
	FR.2. RL agent training & testing	●	●	●
	FR.3. Visualization capabilities	◐	◐	●
	Non-functional Requirements (NFRs)			
Accessibility	NFR.1. OpenAI Gym wrapper	◐	●	◐
	NFR.2. Open-source platform	●	●	◐
Configurability	NFR.3. Configure JSPTA	◐	◐	●
	NFR.4. Customize shop floor layout	◐	◐	●
Extendability	NFR.5. User-controlled Agents	◐	◐	●
	NFR.6. Mixed player types	○	○	●

Legend: ○ not fulfilling; ◐ partially fulfilling; ● fulfilling

allows the customization of shop floors, which enables the impact investigation of changing shop floor layouts in the JSPTA. Additionally, by providing high flexibility in the graphical design of the JSPTA components, Unity can simulate real-world scenarios (e.g. specific domain of production, plant size, and layout). Furthermore, Unity makes it possible to study mixed agents where humans can control agents with their input and thus interact with other RL agents.

To select an overall favorable tool for building the EDT for the JSPTA, we compromised over the trade-off between accessibility (favoring Gazebo and especially MuJoCo) on the one hand, and both configurability and extendability (favoring Unity) on the other hand. The accessibility aspect is crucial for the adoption of our EDT. Unity is a tool that can be used at no cost and is an enabler for experimenting with many relevant questions around the adaptability of RL-based approaches to the JSPTA. The combination of Unity and its ML agents' library opens the door to investigating multiple RL methods, hence its large potential for extendability. Therefore, we chose Unity to build the simulation environment of the JSPTA and Unity ML agents for implementing a neural combinatorial optimization approach.

3.2 JSPTA Environment Components

Figure 1 illustrates the created environment to model the JSPTA and all its components: products (jobs), workstations (machines), transportation agents (AGVs), geofences (restricted areas for AGVs), and a delivery station. The products and workstations are represented respectively by spheres and rectangular volumes with an input feed and an output feed each. Both products and workstations are color-coded to their current operation and workstation type respectively. Products can be grabbed by AGVs, which are represented by black cylinders. Once a product is loaded, it appears above the AGV and moves with it towards its next processing workstation. Once an AGV brings the semi-finished product to its target workstation's input, the semi-finished product is loaded onto the workstation, where the corresponding operation is performed. When the specified operation processing time ends, the processed product is placed at the workstation's output feed. This process continues until the product is finished and can be delivered. All finished products are delivered at a delivery station represented in light blue at the bottom right corner. The shop floor is represented as a customizable rectangular platform with surrounding boundaries. Moreover, geofences are implemented as rectangular boundaries represented in red. The AGVs are not allowed to enter the geofences, which aims at modeling safety areas in factories, where human workers can be endangered by possible collisions.

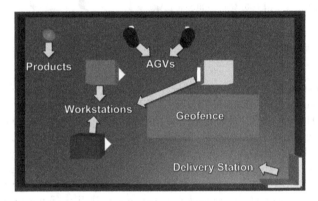

Fig. 1. Components of the JSPTA

3.3   Neural Combinatorial Optimization Approach

To test the capabilities of our simulation environment, we adapted a state-of-the-art RL approach and compared it with other PDRs. The work of Zhang et al. [23] showed how their RL approach outperformed different PDRs in solving the JSP. The approach uses a disjunctive graph representation of the JSP and a Graph Neural Network (GNN)-based scheme in order to solve multiple sizes of the JSP. This state-of-the-art NCO approach was selected thanks to its accessibility and outperformance of a wide range of PDRs. However, it could not be directly used in the context of the JSPTA. Therefore, some customizations of the original approach were necessary. First, the original approach deals with the JSP, meaning that the physical space of the job shop affecting the transportation times was not taken into account. Second, the actions described within the approach were for a single job scheduler. However, in case of the JPSTA, the scheduling involves the AGVs transporting and thus executing the jobs in their scheduled order. Consequently, the Markov Decision Process (MDP) formulation of the original approach is modified as in the following:

- **State S**: The states of the environment were described by two values per job operation o_{ij}, which stands for operation i of job j. The first value $I(o_{ij}, s_t)$ is a binary indicator that is equal to one if the operation is completed at s_t, otherwise, it is equal to zero. The second value $C_{LB}(o_{ij}, s_t)$ is the lower bound of the estimated completion time of operation o_{ij} at state s_t. Therefore, the lower bound of operation o_{ij} is computed as $C_{LB}(o_{ij}, s_t) = C_{LB}(o_{ij-1}, s_t) + p_{ij} + m_{ij}$, which is equivalent to the addition of the lower bound for the precedence constraint $C_{LB}(o_{ij-1}, s_t)$, its processing time p_{ij}, and the estimated movement time of the product to the required workstation m_{ij}. If the job has been completed, then the lower bound is equal to the completion time. Finally, two states were added per job describing the relative position of the AGV to the underlying product. In case of multiple AGVs, the environment state includes the last actions of the other AGVs within the system.
- **Action A^k**: The agents within our environment are the AGVs. Therefore, the MDP formulation is of multiple agents where a_t^k is the action of agent k at time step t. The actions represent the allocation of operations for each AGV. Given that every job has at most one operation that can be worked on at a time step t, the action space of each AGV consists of the number of jobs to be executed and an additional action to return to start position.
- **State Transition $P_{ss'}^{a^k}$**: When an agent performs an action a_t^k, the AGV is allocated to the corresponding job and moves using Unity's navigation system NavMesh. First, the AGV moves towards the corresponding product within the shop floor. Then, it picks up the product and places it onto the corresponding workstation, which triggers the next step of the AGV agent. If a job is allocated when the corresponding product is being processed by a workstation, the AGV will go to the workstation output location and wait for the processing to complete to take the product to the next workstation. If the operation is already allocated to a different AGV, the agent will transition directly to the next step.
- **Reward R**: the original reward was designed to minimize the total makespan of the JSP. The reward function is the quality difference between the partial solutions corresponding to subsequent states s_t and s_{t+1}. $R(a_t, s_t) = H(s_t) - H(s_{t+1})$, where $H(s_t)$ is the quality measure of the JSP solution at state s_t. The quality measure was defined as the lower bound of the makespan C_{max} at time step s_t, computed as $H(s_t) = max_{i,j}\{C_{LB}(o_{ij}, s_t)\}$.

We used Proximal Policy Optimization [18] as in Zhang et al. [23], which is available as an RL method within the Unity ML-Agents Toolkit. However, modifications to the underlying neural architecture of Zhang et al. [23] were essential to speed up the training process and consequently the experimentation process. For the problem representation, Zhang et al. [23] used a GNN that is size-invariant but computationally expensive. Therefore, we skipped the GNN for the problem representation by adjusting the input and output layers of the action selection and value prediction Multi Layer Perceptron (MLP) networks resulting in individual configurations for each problem size.

3.4 Experimental Setup

For experimenting with different JSPTA configurations, we developed a Design of Experiment (DoE) that spans over the following representation $n \times m \times k$: with n the number of jobs, m the number of workstations, and k the number of agents (AGVs). For speedy experimentation, we focused on a small set of values for the number of jobs, machines, and agents. Our DoE space consists of the combinations over $3, 6, 9$ for the number of both jobs and machines and $1, 2, 3$ for the number of agents (AGVs). This resulted in the combinations in Table 2. Our experimentation of NCO for the JSPTA was based on a set of particular configurations (colored in blue in Table 2) to target interesting aspects as in the following:

- Performance with agent scalability (6×3×1, 6×3×2, 6×3×3)
- Performance with job scalability (3×3×2, 6×3×2, 9×3×2)
- Performance with JSP scalability (3×3×2, 6×6×2, 9×9×2)
- Performance with different layouts (6×3×2* with 3 layouts: $L1$, $L2$, and $L3$)

Table 2. Design of Experiment for the JSPTA

n (jobs) ×	m (workstations)			× k (agents)
	3	6	9	
3	3×3×1	3×6×1	3×9×1	
6	6×3×1	6×6×1	6×9×1	1
9	9×3×1	9×6×1	9×9×1	
3	3×3×2	3×6×2	3×9×2	
6	6×3×2*	6×6×2	6×9×2	2
9	9×3×2	9×6×2	9×9×2	
3	3×3×3	3×6×3	3×9×3	
6	6×3×3	6×6×3	6×9×3	3
9	9×3×3	9×6×3	9×9×3	

Legend: used in EDT evaluation. * with different layouts

Each of the JSPTA configurations was tested within 5 different instances, which were similarly created as in Taillard benchmarks [2]. The same set of hyperparameters was used to train every RL agent. The MLP networks of the action selector had two hidden layers with 512 neurons, whereas the MLP networks of the value predictor had two hidden layers with 256 neurons. Given a configuration $n \times m \times k$, $2n + 2m(n+1) + k - 1$ and $n + 1$ neurons were needed for the input and output layers respectively for the MLPs. For scalable training, our shop floor was duplicated 25 times within the training environment and 4 copies of the environment were run in parallel for a total of 100 shop floor copies. For every JSPTA configuration, the RL agents were trained using the 5 different JSP instances. To measure the performance of the different approaches (RL-based vs. PDRs), we defined the optimality gap as a metric calculated using the following Eq. 1.

$$
\text{Optimality Gap} = \frac{\text{JSPTA Makespan} - \text{JSP-Optimum}}{\text{JSP-Optimum}} \tag{1}
$$

To our best knowledge, no solver could be used to calculate the optimum of a JSPTA instance, therefore we used the optimum of the JSP part of our instances, as if the agents would instantly transport the products. The optima of the derived JSP instances were calculated using Google OR Tools [7]. For training the RL agents, we used a computational unit that consisted of an Nvidia Titan X (Pascal) with 12 GB of memory GPU and an Intel Core i7-6850K CPU with 12 cores. The code of our EDT is publicly available under https://github.com/aymengan/EDT_JSPTA.

4   EDT Evaluation

4.1   Testing Results

The trained agents were tested and evaluated against two PDRs: namely Shortest Processing Time (SPT) and Longest Processing Time (LPT). The optimality gap was computed for each method and each instance based on Eq. 1. The JSP optimum was computed with Google OR-Tools outside the simulation environment and therefore did not include the AGV traveling times. In contrast, the RL agents and PDRs were implemented within the simulation environment, where the makespans include transportation and processing times. Every JSPTA configuration was tested with 5 different JSP instances, which were solved 10 times, resulting in 50 data points per evaluation method (RL vs. SPT vs. LPT).

Performance with Agent Scalability. Figure 2 shows the results of the performance with agent scalability involving the following JSPTA configurations: $6 \times 3 \times 1$, $6 \times 3 \times 2$, $6 \times 3 \times 3$, and their corresponding layouts respectively in Fig. 2c, Fig. 2d, Fig2e. Considering the resulted makespans in Fig. 2a the RL agents outperformed both the SPT and LPT in every configuration. While the trend of the makespan is decreasing with more agents for the PDRs, Fig. 2a shows that using two AGVs was better than three AGVs for the RL agents. Furthermore, Fig. 2b shows that the distribution of optimality gaps was more stable for SPT than LPT and RL agents, where it increased, especially from 2 AGVs ($6 \times 3 \times 2$) to 3 AGVs ($6 \times 3 \times 3$). This increase might be due to the fact that the excess of agents resulted in more competitiveness and less occupancy per agent.

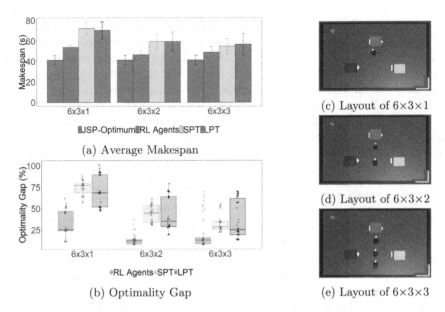

(a) Average Makespan

(b) Optimality Gap

(c) Layout of 6×3×1

(d) Layout of 6×3×2

(e) Layout of 6×3×3

Fig. 2. Testing results with increasing number of agents

Performance with Job Scalability. Figure 3 shows the results of the performance with job scalability involving the following JSPTA configurations: 3×3×2, 6×3×2, 9×3×2, and their corresponding layout in 3c. Considering the resulted makespans in Fig. 3a the RL agents outperformed both the SPT and LPT in every configuration. While the makespans are increasing with more jobs for all methods, Fig. 3b shows that the distribution of optimality gaps was stably decreasing for the RL agents in comparison with SPT and LPT. This stability in decrease for the RL agents might be explained by the increase of their learning efficiency with more jobs.

Performance with JSP Scalability. Figure 4 shows the results of the performance with JSP scalability (both job and machine) involving the following JSPTA configurations: 3×3×2, 6×6×2, 9×9×2, and their corresponding layouts respectively in Fig. 4c, Fig. 4d, Fig. 4e. Considering the resulted makespans in Fig. 4a the RL agents outperformed both the SPT and LPT in every configuration. While the makespans are increasing for all methods at a higher rate with larger JSP sizes, Fig. 4b shows that the increase of optimality gaps was at a lower pace for the RL agents in comparison with the PDRs. The higher increase in makespans with the scalability of the JSP is due to the combinatorial nature of the problem and its higher complexity, particularly for larger sizes.

Performance with Different Layouts. Figure 5 shows the results of the performance on 6×3×2 with three different layouts with increasing geofence areas as

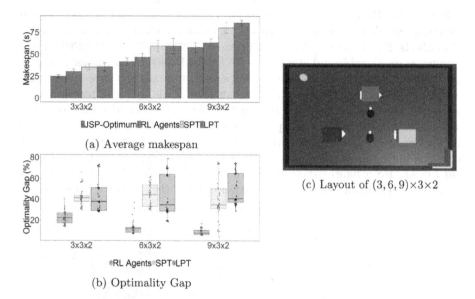

(a) Average makespan

(b) Optimality Gap

(c) Layout of $(3, 6, 9) \times 3 \times 2$

Fig. 3. Testing results with increasing number of jobs

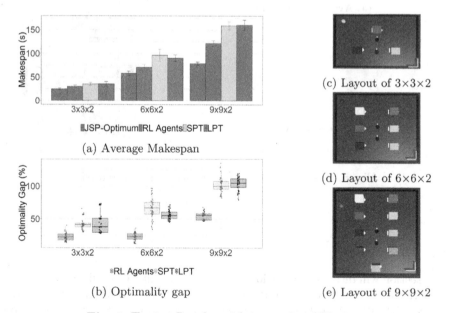

(a) Average Makespan

(b) Optimality gap

(c) Layout of $3 \times 3 \times 2$

(d) Layout of $6 \times 6 \times 2$

(e) Layout of $9 \times 9 \times 2$

Fig. 4. Testing Results with increasing JSP size

represented in Fig. 5c, Fig. 5d, and Fig. 5e. Considering the resulted makespans in Fig. 5a the RL agents outperformed both the SPT and LPT in every layout configuration. While the makespans are increasing with the geofence areas for all methods, Fig. 5b shows that the increase of optimality gaps and their distribution was more stable and at a lower rate for the RL agents in comparison with the PDRs. The higher increase in makespans with larger geofence areas might be due to the fact that the mobility of agents is increasingly restricted with more geofences. Thus, more efforts are needed to avoid geofences and collisions, which adds overhead time to the makespan. In order to study the generalization of our RL approach in other unseen settings, we tested trained RL agents in unseen layouts resulting in so-called non-native agents represented in orange on Fig. 5a and Fig. 5b. Both figures show that although non-native RL agents could not match the performance of their native counterparts, they still outperformed the PDRs.

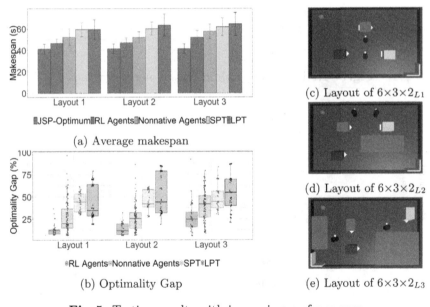

(a) Average makespan

(b) Optimality Gap

(c) Layout of $6 \times 3 \times 2_{L1}$

(d) Layout of $6 \times 3 \times 2_{L2}$

(e) Layout of $6 \times 3 \times 2_{L3}$

Fig. 5. Testing results with increasing geofence area

4.2   Discussion

This section will discuss the previously presented results. Overall, the RL agents outperformed the PDRs in all 12 configurations that investigated the performance with JSPTA scalability focusing on agents, jobs, JSP sizes, and shop floor layouts.

First, the results in Fig. 2 showed that having more agents increases efficiency in general, but not necessarily for the RL agents. This requires more investigation of whether RL approaches have more overhead dealing with the coordination of multiple agents than PDRs, which urges the study of collaborative multi-agent RL approaches. Second, the results in Fig. 3 and Fig. 4 showed how the makespans increased for all methods at higher rates for the JSP scalability in comparison to the job scalability. This calls for the analysis of machine scalability (e.g. testing the JSPTA configurations $9\times3\times2$, $9\times6\times2$, $9\times9\times2$) to study the impact of job vs. machine scalability on the overall JSP scalability. In terms of solution quality, the RL agents outperformed PDRs in Fig. 3b thanks to decreasing distributions of optimality gaps, which suggests that the RL agents learned a more efficient heuristic with more jobs. Moreover, the increase of makespan with JSP scalability in 4a is at a lower pace for the RL agents in comparison to the PDRs all while achieving better solution quality with denser distributions of optimality gaps as represented in Fig. 4b. Third, the results in Fig. 5 showed that the layout of the shop floor affects the makespan. The bigger the area of geofences in the layout the higher the makespan for all methods with the RL agents outperforming the PDRs, especially in solution quality (see distributions of optimality gaps in Fig. 5b). One of the interesting insights was not only the ability of RL agents of generalizing to unseen layouts but also their outperformance of PDRs. Limiting the mobility space for AGVs seems to add more overhead due to the avoidance of collisions and geofences reinforcing the urge to study collaborative multi-agent RL approaches.

In summary, our EDT fulfilled all core requirements that could be fulfilled by Unity (see Table 1), especially the ability to train and test RL agents and visualize their behaviors within the environment. Moreover, the configurability of our EDT was showcased through the testing of various JSPTA configurations. The extendability requirements focused on mixed agent settings that were out of the scope of this work, but these are not bound to implementation thanks to the ability of Unity to have user-controlled agents and enable their interaction with RL agents using Unity ML Agents.

In our work, we made choices that compromised over different trade-offs (e.g. accessibility vs. usability during tool selection). To speed up the experimentation and training process, we skipped for the problem representation the usage of GNNs, which were implemented in the original approach of Zhang et al. [23]. It is important to note that the GNNs are not bound to implementation. Furthermore, to our best knowledge, no solver could be used for calculating the optimal solutions for our JSPTA configurations. Thus, we used the JSP optima solved by Google OR Tools [7] when calculating the optimality gaps. The lack of such solvers represents an important subject of research that needs to be tackled. Finally, we relied on Unity's internal NavMesh for the routing of the agents, which calls for the integration and investigation of agent routing strategies. Against the backdrop of more high-fidelity production virtualization, other underlying simulation tools can be tested (e.g. Nvidia Omniverse [15]) to further narrow the gap between simulation and real-world scenarios.

5   Conclusion

Current simulation environments and digital twins for the job shop with transportation agents lack configurability and limit the investigation of neural combinatorial approaches to planning optimization in production and its physical space. To fill this gap, we designed an extendable Experimental Digital Twin allowing for the study of RL-based approaches for solving the JSPTA. We developed our EDT using Unity as a tool that we selected based on a requirement analysis. We customized the approach of Zhang et al. [23] to the JSPTA using the Unity ML-agents Toolkit, which allowed the implementation of RL agents based on proximal policy optimization as an RL method. Using a design of experiment, we evaluated our EDT by analyzing the performance of RL agents in comparison with two priority dispatching rules: shortest processing time and longest processing time. Our evaluation measured the makespan focusing on scalability with agents, jobs, JSP size, and shop floor layouts. The results showed that our RL-based approach outperformed the other PDRs. Moreover, we showed the ability of neural combinatorial approaches to generalize to unseen layouts, which implies its adaptability to dynamic changes in the problem. Furthermore, our EDT allowed the visualization of trained agents' behavior during testing. The future work consists of testing other RL methods, investigating multi-agent collaboration, and incorporating agent routing strategies.

Acknowledgement. Funded by the Deutsche Forschungsgemeinschaft (DFG, German Research Foundation) under Germany's Excellence Strategy - EXC-2023 Internet of Production - 390621612

References

1. Brockman, G., et al.: Openai gym. https://arxiv.org/pdf/1606.01540v1
2. Taillard, E.: Benchmarks for basic scheduling problems **64**, 278–285 (1993)
3. Erez, T., Tassa, Y., Todorov, E.: Simulation tools for model-based robotics: comparison of bullet, havok, mujoco, ode and physx, pp. 4397–4404 (2015). https://doi.org/10.1109/ICRA.2015.7139807
4. Feldkamp, N., Bergmann, S., Strassburger, S.: Simulation-based deep reinforcement learning for modular production systems. In: 2020 Winter Simulation Conference (WSC). IEEE (2020). https://doi.org/10.1109/WSC48552.2020.9384089
5. Fujimoto, S., van Hoof, H., Meger, D.: Addressing function approximation error in actor-critic methods. https://arxiv.org/pdf/1802.09477v3
6. Gazebo: simulation tool (2023). https://gazebosim.org/home
7. Google: or tools (2022). https://developers.google.com/optimization
8. Hu, H., Jia, X., He, Q., Fu, S., Liu, K.: Deep reinforcement learning based AGVs real-time scheduling with mixed rule for flexible shop floor in industry 4.0. Comput. Ind. Eng. **149**(2), 106749 (2020). https://doi.org/10.1016/j.cie.2020.106749
9. Huang, S., Mao, Y., Peng, Z., Hu, X.: Mixed reality-based digital twin implementation approach for flexible manufacturing system design (2022). https://doi.org/10.21203/rs.3.rs-1265753/v1
10. Juliani, A., et al.: Unity: a general platform for intelligent agents (2020). https://arxiv.org/pdf/1809.02627v2.pdf

11. Körber, M., Lange, J., Rediske, S., Steinmann, S., Glück, R.: Comparing popular simulation environments in the scope of robotics and reinforcement learning (2021). https://arxiv.org/pdf/2103.04616v1.pdf
12. Malus, A., Kozjek, D., Vrabič, R.: Real-time order dispatching for a fleet of autonomous mobile robots using multi-agent reinforcement learning. CIRP Ann. 69(1), 397–400 (2020). https://doi.org/10.1016/j.cirp.2020.04.001
13. MuJoCo: simulation tool (2022). https://mujoco.org/
14. Negri, E., Fumagalli, L., Macchi, M.: A review of the roles of digital twin in cps-based production systems. Procedia Manuf. 11, 939–948 (2017). https://doi.org/10.1016/j.promfg.2017.07.198
15. Nvidia: Omniverse (2023). https://www.nvidia.com/en-us/omniverse/
16. Raffin, A., Hill, A., Gleave, A., Kanervisto, A., Ernestus, M., Dormann, N.: Stable-baselines3: reliable reinforcement learning implementations. J. Mach. Learn. Res. 22(268), 1–8 (2021). https://jmlr.org/papers/v22/20-1364.html
17. Schluse, M., Priggemeyer, M., Atorf, L., Rossmann, J.: Experimentable digital twins-streamlining simulation-based systems engineering for industry 4.0. IEEE Trans. Ind. Inform. 14(4), 1722–1731 (2018). https://doi.org/10.1109/TII.2018.2804917
18. Schulman, J., Wolski, F., Dhariwal, P., Radford, A., Klimov, O.: Proximal policy optimization algorithms (2017). https://doi.org/10.48550/ARXIV.1707.06347, https://arxiv.org/abs/1707.06347
19. Simulation, S.P.: Simulation tool (2023). https://plant-simulation.de/
20. Tecnomatix: Simulation tool (2023). https://www.plm.automation.siemens.com/global/en/products/tecnomatix/
21. Tortorelli, A., Imran, M., Delli Priscoli, F., Liberati, F.: A parallel deep reinforcement learning framework for controlling industrial assembly lines. Electronics 11(4), 539 (2022). https://doi.org/10.3390/electronics11040539
22. Unity: simulation tool (2023). https://unity.com
23. Zhang, C., Song, W., Cao, Z., Zhang, J., Tan, P.S., Chi, X.: Learning to dispatch for job shop scheduling via deep reinforcement learning. In: Larochelle, H., Ranzato, M., Hadsell, R., Balcan, M.F., Lin, H. (eds.) Advances in Neural Information Processing Systems, vol. 33, pp. 1621–1632. Curran Associates, Inc. (2020). https://proceedings.neurips.cc/paper/2020/file/11958dfee29b6709f48a9ba0387a2431-Paper.pdf
24. Zhang, Z., Wang, X., Wang, X., Cui, F., Cheng, H.: A simulation-based approach for plant layout design and production planning. J. Ambient. Intell. Humaniz. Comput. 10(3), 1217–1230 (2018). https://doi.org/10.1007/s12652-018-0687-5

Learning to Prune Electric Vehicle Routing Problems

James Fitzpatrick[1]([✉])[iD], Deepak Ajwani[2][iD], and Paula Carroll[1][iD]

[1] School of Business, University College Dublin, Dublin, Ireland
james.fitzpatrick1@ucdconnect.ie, paula.carroll@ucd.ie
[2] School of Computer Science, University College Dublin, Dublin, Ireland
deepak.ajwani@ucd.ie

Abstract. Electric vehicle variants of vehicle routing problems are significantly more difficult and time-consuming to solve than traditional variants. Many solution techniques fall short of the performance that has been achieved for traditional problem variants. Machine learning approaches have been proposed as a general end-to-end heuristic solution technique for routing problems. These techniques have so far proven flexible but don't compete with traditional approaches on well-studied problem variants. However, developing traditional techniques to solve electric vehicle routing problems is time-consuming. In this work we extend the learning-to-prune framework to the case where exact solution techniques cannot be used to gather labelled training data. We propose a highly-adaptable deep learning heuristic to create high-quality solutions in reasonable computational time. We demonstrate the approach to solve electric vehicle routing with nonlinear charging functions. We incorporate the machine learning heuristics as elements of an exact branch-and-bound matheuristic, and evaluate performance on a benchmark dataset. The results of computational experiments demonstrate the usefulness of our approach from the point of view of variable sparsification.

Keywords: Vehicle Routing Problem · Electric Vehicle · Machine Learning

1 Introduction

Road transport constitutes the highest proportion of overall transport green house gas (GHG) emissions, with carbon dioxide as the main green house gas. Passenger cars and light commercial vehicles account for 12% and 25% of carbon dioxide emissions in the European Union (EU). To meet obligations set forth by the Paris Agreement, the EU aims to promote the use of clean vehicles and develop public transport and electric vehicle (EV) charging infrastructure [17].

There is a rapidly-expanding family of EV-related vehicle routing problems some of which may yet have no effective solution techniques or problem-specific knowledge to be leveraged. Such problems are in general more difficult to solve

M. Sellmann and K. Tierney (Eds.): LION 2023, LNCS 14286, pp. 378–392, 2023.
https://doi.org/10.1007/978-3-031-44505-7_26

than their traditional counterparts. This makes it difficult to formulate and to solve emerging EV routing problems variants at the scale and speed necessary for practical applications.

Running times of solution techniques for Combinatorial Optimisation Problems (COPs) are generally a function of the number of variables. Variable fixing techniques aim to identify a subset of the decision variables of a problem formulation that can be fixed a-priori, reducing the effective size of the instance to be solved. Unfortunately most existing approaches for variable fixing either require lengthy development times or rely on the ability to collect optimal solutions. In recent years the learning-to-prune framework has emerged as a general variable-fixing tool for COPs. The essence of this framework is a machine learning (ML) classifier that is trained to predict decision variable values. The advantage of this is that the trained classifier can quickly perform inference to fix a large number of the decision variables of a problem instance at test time. Classifiers used in this way are called sparsifiers and may allow many solution methods to be used to solve the resultant problem with a reduced number of variables in a reasonable computational time, which may not be possible otherwise.

In this work we propose an alternative learning-to-prune framework that relaxes the requirement for an exact solution labelling technique. Instead we require only that any (relatively) quick heuristic exists to solve a given problem variant, and that it can be leveraged to produce many different feasible solutions for any individual problem instance. We use this to develop a more general learning-to-prune framework and use it to obtain solutions for some electric vehicle routing problems (E-VRPs). Based on observations of the effectiveness, we draw conclusions about how this approach should be extended for other difficult optimisation problems.

In particular, the contributions of our work are as follows:

- We extend the learning-to-prune framework to problems for which obtaining optimal solutions for labels is impractical, but where we can collect partial information about edges that appear in good solutions (pseudo-labels).
- We extend machine learning heuristics to solve the electric vehicle routing problem with non-linear charging function.
- We present analysis of a set of computational experiments to demonstrate the effectiveness of this sparsifier approach.

The paper format is as follows: first, in Sect. 2 we give an introduction to the problem we attempt to solve, the related literature, issues associated with the current approaches and a high-level overview of how we attempt to address these and the contributions we make. Then, in Sect. 3 we explain in detail the approach that we take to solve these problems, explaining the results in Sect. 4. Finally, we discuss these results in Sect. 5, draw conclusions and briefly mention proposed future works.

2 Related Literature

In the following sections we present an overview of the relevant literature.

2.1   The Electric Vehicle Routing Problem

In the traditional capacitated VRP (CVRP), a graph $G = (V, E)$ underlies a particular problem instance. The goal is for a fleet of vehicles to depart from a depot node, visiting and servicing customer nodes, before returning to a depot node. The capacity of a vehicle limits the number of customers that may be serviced in a route. The goal is usually to minimise the cumulative distance travelled by all vehicles. E-VRPs are a broad class of vehicle routing problems for which some or all of the vehicles under consideration are EVs. Vehicles may not run out of battery charge (become stranded) along a route and may detour to recharge at a charging station (CS). The time to charge is non-negligible and depends on many factors such as the CS technology type, and the EV battery state of charge (SoC) upon arrival. The set of nodes V is heterogeneous, containing depot node(s) D, customer nodes I, and recharging stations F.

In many variants the duration of the routes is minimised, with each route constrained to a maximum duration. Such constraints typically make routing problems very difficult to solve. To better model the battery constraints, increasingly intricate variants of the problem consider more accurate approximations to the charging time. Initial approaches considered charge times to be static, or linear [5,6,20], while more recent works model the charge time as a piecewise linear function of the SoC upon arrival, yielding the Electric Vehicle Routing Problem with Nonlinear Charging (E-VRP-NL). [9,11,16] This renders the problem yet more challenging.

2.2   End-to-End Machine Learning Heuristics

The success of ML in recent years has prompted research on whether ML can be used to derive a principled approach to solving COPs for which we have limited success with other techniques, but for which we have a wealth of data. Several of these works directly tackle routing problems [3,12,22], acting as a construction heuristic by taking a problem instance as input and returning a solution as output. Such approaches have become known as end-to-end learning approaches in the literature. All of these approaches fall victim to poor generalisation for different problem sizes than those encountered at training-time. However, they are highly-adaptable to new problem variants. This adaptability is achieved via a masking scheme, preventing infeasible solutions, which is usually easy to modify. Some attempts have been made to extend the generalisability and size of problems for which end-to-end methods are effective, but little effort has been made to adapt them to more intricate, highly constrained problem types. The work of [15] is the only work that considers electric vehicles, to the knowledge of the authors, but here charging times are considered as linear functions of the state of charge. Such simplifications are problematic and can yield infeasible solutions in practice [16].

2.3 Learning to Prune

The learning-to-prune framework has been applied successfully to a range of COPs enabling the identification of best known solutions for some problem instances [4,7,14,21]. So far, the learning-to-prune approach has relied on the availability of optimal solutions to provide labels for training problem instances. In particular, the observed values of the decision variables in the optimal solutions are used as targets for the ML classifier. The values predicted by the classifier are the values that the decision variables are fixed to, if the confidence in the prediction is sufficiently high. In the traditional setting, we assume that we can solve a set of small problem instances in reasonable computational time, and that information from these optimal solutions will allow us to make inferences about solutions of unseen, possibly large, problem instances. Unfortunately, in some cases, lengthy exact solve times prevent us from solving a sufficient number of smaller problem instances to optimality within a reasonable time frame. In the case of the E-VRP-NL, even solving problems with ten customers to optimality is prohibitive. This makes it difficult to gather optimal solutions for a large enough number of instances for ML purposes.

We trade optimality for expediency, producing a large number of sub-optimal, relatively-high quality solutions in a reasonable computation time. We sample a set of unique solutions, combining the information from each solution to estimate the importance of different edges. This allows us to create a set of labels for the variable-pruning classifier in a manner similar to weak supervision methods. We show that with this approach we can extend the learning-to-prune framework to the E-VRP-NL.

3 Methodology

In Sect. 3.1 we present a branch-and-bound matheuristic solver composed of a learned construction heuristic, a hand-crafted construction heuristic and a learned sparsifying heuristic. We discuss the details of the construction heuristics in Sect. 3.2 and the details of the sparsifying heuristic in Sect. 3.4. To train the sparsifying heuristic we collect a set of target "pseudolabels", which is explained in Sect. 3.3.

3.1 Pruning Matheuristic Methodology

We formulate the E-VRP-NL problem as a MILP using the improved arc-based formulation of [9], an extension of the Miller-Tucker-Zemlin formulation of the capacitated VRP. It has a stronger linear relaxation than the originally proposed formulation of [16] while remaining relatively straightforward to implement and understand. In this formulation a binary variable is associated with each edge along with time and battery energy tracking variables. Fixing the value of one edge variable permits fixing all three such variables associated with an edge.

We present the pseudocode of our approach in Algorithm 1. In the first two lines we use the deep learning heuristic described in Sect. 3.2 and the initial construction heuristic *montoyaConstructionHeuristic* of [16] to compute heuristic solutions. The solution with the smaller objective function value $z(\hat{S})$ is stored as the best current solution in lines four through seven. In lines eight to ten we sparsify the problem by using the sparsifying heuristic to fix a large fraction of the decision variables associated with the edges. The classifier $h(\cdot; theta)$ takes an edge embedding q_e for the edge e and produces a probability that the associated variables should be fixed. In lines eleven to twelve we ensure feasibility by releasing any of the decision variables associated with the edges of the two identified solutions. Next, an exact solver is invoked to solve the resulting reduced problem until a global time limit (3600 s) is reached. The total time allowed for optimisation includes the solving time of both heuristics, the pruning time and the solving time of the branch-and-bound solver.

Algorithm 1. Pruning Matheuristic

1: Input: Problem P, Threshold η, Time limit T.
2: $S_1 = montoyaConstructionHeuristic(P)$
3: $S_2 = deepLearningHeuristic(P)$
4: **if** $z(S_1) \leq z(S_2)$ **then**
5:     $\hat{S} \leftarrow S_1$
6: **else**
7:     $\hat{S} \leftarrow S_2$
8: **for** each edge $e \in E$ **do**:
9:     **if** $h(q_e; \theta) < \eta$ **then**
10:        Fix decision variables associated with e.
11: **for** each edge e in S_1 and S_2 **do**
12:        Release decision variables associated with e.
13: Warm-start branch-and-bound solver with solution $\hat{S}$.
14: Solve until time limit T is reached, output solution S_3.
15: **if** $z(S_3) \leq z(\hat{S})$ **then**
16:     $\hat{S} \leftarrow S_3$
17: Output: Best found solution $\hat{S}$.

3.2 Deep Learning Heuristic

A solution to a routing problem can be viewed as a sequence of nodes. Taking this view, many construction heuristics follow a two-step process: first select one (feasible) node then insert the selected node somewhere into the sequence (partial solution) of S until S forms a complete solution. This process relies on an effective heuristic for selecting the next node for insertion. In this work we follow [12] and subsequent approaches by adopting a transformer-based neural network. This approach uses the neural network as a scoring mechanism to determine which node should be selected and appends it to a solution. Nodes that are estimated to be more likely to produce a good solution if selected next will have higher scores. Once a node is selected, it is appended to the end of the partial solution. When trained, neural network models of this kind can be used to rapidly sampling of many high-quality solutions. For the E-VRP-NL, however, a secondary decision must be made at CS nodes: how much time should be spent charging? This requires architectural changes that are outlined in the next section. Our neural network heuristic becomes a three-step process: first select a feasible node, then if it is a CS node select a charging duration, then append that node to the partial solution.

Network Architecture. The neural network consists of an encoder network and a decoder network. The encoder embeds some low-dimensional feature representation of the nodes into a static high-dimensional latent space representation containing information about each node and how they relate to the other nodes in the graph. The training process encourages this transformation to produce an embedding for each node that will help the decoder network to decide which node should be selected next. The initial features for the set of depot nodes, the set of customer nodes and the set of station nodes are as follows:

$$x_i = \begin{cases} [r_i^1, r_i^2] & \text{if } i \in D \text{ or } i \in I, \\ [r_i^1, r_i^2, w_i] & \text{if } i \in F. \end{cases} \tag{1}$$

where r_i^1 and r_i^2 are the standardised coordinates of node i on the Cartesian plane and w_i is a weight associated with the charging station technology:

$$w_i = \begin{cases} 0.0 & \text{if } i \text{ is a slow charging station,} \\ 0.5 & \text{if } i \text{ is a normal charging station,} \\ 1.0 & \text{if } i \text{ is a fast charging station.} \end{cases} \tag{2}$$

The first step is to transform each x_i to a space with of dimensionality m. We use linear transformations to ensure the encoding of each node is of the same dimension. For each of the disjoint subsets D, I, F of V there is a unique transformation (weights are node shared). The embedded representations are then transformed by three attention layers in sequence, exactly as in [12]. We then compute a graph embedding q as the mean of each of the node embeddings. This produces an m-dimensional embedding q_i for each node and one, q, that characterises the entire graph. The context embedding consists of the embedding of the previously visited node, the normalised SoC and the normalised time remaining in the given route.

We must make both categorical decisions (the node we select), and continuous decisions (how much time we spend at it). We spend exactly g_i units of time at each customer, the service time for customer i. There is no utility to waiting at any node, since time windows are absent. Spending too much or too little time at a CS can negatively affect the solution, possibly rendering it infeasible. We therefore modify the decoder to produce two outputs for each node: the probability estimate that node i is the best node to insert next and the amount of time that should be spent there.

We parameterise a probability density over the range of possible values that the continuous variable may assume. The possible charge times lie in the range $[l_i, u_i] \in \mathbb{R}_+ \cup \{0\}$. However, the partial solution restricts the choice of feasible charge times to the range $[\alpha_i^t, \beta_i^t]$, where $\alpha_i^t \geq l_i$ and $\beta_i^t \leq u_i$. We must ensure that the probability that any values outside the effective range is zero, so we model the probability density as a normal distribution and clip the outputs. The clipping acts as the masking does for the node selection. We modify the neural network to have an additional linear transformation that maps each dynamic node embedding and the context embedding to a two-dimensional vector, representing the mean μ_i and variance σ_i of the distribution $\hat{p}(\cdot; \mu_i(\theta), \sigma_i(\theta))$ for node i. This results in a neural network with two outputs: the probability p^t that we choose a node i next and the time t_i that should be spent at each node.

Training the Deep Learning Heuristic. Inference is carried out with repeated decoding steps until the selected nodes and charge times form a feasible solution. For a given parameter set θ of the neural network, greedy, deterministic inference can be performed by selecting at each time step t the node and charge time with the highest probability and probability density, respectively, $\max_i\{p_i^t\}$. To perform stochastic inference, we sample nodes according to their approximated probabilities and times from their parameterised distributions.

We compute the cost $\hat{z}(S)$ of a complete solution S as a quantity measured in hours. It comprises of three terms, the traditional driving times plus the service times, $routeDuration(S)$, plus two penalties. The penalties include the number of times a station was visited $countStationVisits(S)$ and the number of times the depot was visited. $countDepotVisits(S)$ in a solution. These discourage the neural network from visiting charging stations unnecessarily or from visiting the depot many times and visiting too few charging stations.

$$\hat{z}(S) = routeDuration(S) + countStationVisits(S) + countDepotVisits(S) \quad (3)$$

Similar to [12], we train the neural network using the REINFORCE algorithm. Since the objective function value includes the time spent charging, it is a function of the parameters of the neural network, so the loss must be modified as:

$$\nabla \mathcal{L}(S|\theta) = \mathbb{E}_{p_\theta}[\nabla_\theta(\hat{z}(S) - b(S)) \log p_\theta(S)], \quad (4)$$

where $b(S)$ is the objective of the solution produced by the baseline model (see [12] for full details).

Once improvements in solution quality have not been observed for Γ epochs, training is stopped. Training first takes place on generated problem instances with 10 customers and 2 charging stations. Problem instances are generated by throwing points onto the unit square. Charging station nodes are generated to ensure the feasibility of the problem instances. After the exponential baseline is traded out, a variety of randomly generated problem instances are produced at each batch, with $|I| = \{9, 10, 11, 12\}$ customers and $|F| = \{0, 1, 2, 3\}$ charging stations. Each batch contains problems with the same number of customers and charging stations. The Adam optimiser is used for performing gradient computations, with a learning rate of 1×10^{-4}. Each batch contained 400 problem instances and each epoch consisted of 50 batches. Training terminates after 150 epochs, even if improvements were still occurring. The baseline model is replaced after a paired t-Test check at the end of each epoch, just as with [12]. If a baseline replacement occurred, then the test set was replaced with a newly-generated test set.

3.3   Constructing Pseudo-labels

The traditional learning to prune approach involves computing optimal solutions and using the decision variable values in those solutions as targets in a classification problem. This is not possible for many difficult routing problems. We assume there exists a solution technique $\mathcal{M}$ that can be used to sample, for each problem instance $P_i \in P$, a set of m unique solutions $S_i = \{S_i^1, ..., S_i^m\}$. Each problem contains s_i continuous decision variables and p_i integer decision variables, which can be arranged (in no particular order) into a tuple $x_i = (x_i^1, ..., x_i^{s_i}, x_i^{s_i+1}, ..., x_i^{s_i+p_i})$. Each solution $S_i^j \in S_i$ is represented by some value $x_i^{l,j}, l \in \{1, ..., s_i + p_i\}$ to each of the decision variables in x_i and is associated with some objective function value z_i^j.

In minimisation problems, lower values of z_i^j are representative of better solutions. When we sample the m solutions for each problem, When we construct the pseudo-labels for a given problem instance P_i, the higher-quality solutions should contribute more. A weighting ρ_i^j is computed for each solution using a softmax function:

$$\rho_i^j = 1 - \frac{e^{\hat{z}_i^j}}{\sum\limits_{j=1}^{m} e^{\hat{z}_i^j}}, \tag{5}$$

where $\hat{z}_i^j = z_i^j / \min\limits_{k \in \{1,...,m\}} z_i^k$. Using these weightings, we compute the pseudo-labels as:

$$\gamma_i^l = \sum_{j=1}^{m} \rho_i^j x_i^{l,j}. \tag{6}$$

This provides the label for each variable for each problem instance. This can then be thought of as weak supervision. Next, a feature vector $q_i^l \in \mathbb{R}^r$ must be computed for each variable so that we may construct the labelled dataset $\mathcal{D} = \left\{ \bigcup\limits_{l=1}^{s_i+p_i} \{(q_i^l, \gamma_i^l)\} \right\}_{i=1}^{n}$. The values of γ_i^l are not, in general, integral. We may pose the variable fixing problem as a regression task and train a classifier $h(\cdot; \theta)$ to approximate the mapping between the feature space and the space of the pseudo-labels. In practice we find it more effective to interpret the problem as a classification problem, to fix some variables and to leave others free for optimisation.

3.4   Pruning as Classification

For each problem size in the benchmark dataset ($I = \{10, 20, 40, 80, 160, 320\}$) three hundred new problem instances were generated as training data for the sparsification task. For each of these problem instances, one hundred unique heuristic solutions were identified and used to construct pseudo-labels. The aim of the sparsifying classifier is then to separate the edges that we think are unlikely to be in a good solution from those that we think are likely to belong to a good solution. These threshold values are presented in Table 1. Three classification models were chosen for evaluation, inspired by previous works invoking the learning-to-prune framework: a linear support vector machine, a random forest classifier and a logistic regression model [7,21]. These models were chosen for their relative explainability and quick inference times. This is important because the larger problems may require hundreds of thousands of variables. Each classifier was trained with two hundred of the generated training instances of each size, with fifty instances retained for testing and validation. The positive class samples are far fewer than the negative samples in these datasets. To balance the positive and negative classes and avoid class imbalance, samples from the negative class in the selected training instances were randomly under-sampled.

Table 1. Thresholds determined for each classifier for the pruning/classification task.

model	$|I| = 10$	$|I| = 20$	$|I| = 40$	$|I| = 80$	$|I| = 160$	$|I| = 320$
$lSVM$	0.05	0.05	0.01	0.01	0.01	0.01
RFC	0.05	0.05	0.05	0.01	0.01	0.01
LR	0.05	0.05	0.01	0.01	0.01	0.01

3.5   Computational Setup

Experiments were performed using Python. For training the deep learning heuristic, PyTorch was used [18]. Training for the pruning classifier was carried out using SciKit-Learn [19]. All graph operations were performed using the NetworkX package [10]. The Xpress optimisation suite [2] with default settings was used for MILP solving, using the Python wrapper to interact with its functionalities. Concorde [1] was used to solve TSP problem instances and the frvcpy [13] package was used to solve Fixed Route Vehicle Charging Problems. Training and solving were performed on a machine with an AMD EPYC 7281 16-Core Processor with 32 threads. The machine runs Ubuntu 20.04 with a total memory of 96GB and a total L3 cache of 32MB. Training of the classifier was carried out with the Scikit-Learn package [19].

4 Results

First we discuss the training regime for the deep learning heuristic, then how it performs as a stand-alone heuristic. We then compare the solutions we obtain to those we get from the standard initial construction heuristic of [16]. We then discuss these in the context of the sparsification task. Finally with finish with a discussion of the results of the final solve using the reduced problem instances. All results presented are obtained by evaluating the techniques presented against the benchmark dataset introduced in [16], where Best Known Solution objective values are obtained from [8].

4.1 Training the Deep Learning Heuristic

The running time of the deep learning heuristic depends on the length of the solution that it produces. More visits to stations and depots result in more passes of the decoder, which requires a new masking procedure each time. In the early stages one of the major obstacles that slows down training is repeated spurious visits to charging stations and one-customer tours. Using the depot-visit and station-visit penalties in the cost function, observations of solutions throughout the training process indicates that visits of these kinds are discouraged first. Once this has occurred, it generally seems that a second training regime begins, where the task is focused on making cleverer choices about the routes to make the solution duration shorter. For this reason, bootstrapping during training makes it easier to train the network: starting with problem instances with a small number of customers and a small number of charging stations allows the model to learn to be parsimonious with station and depot visits earlier on. Training can then be tuned by introducing batches of larger problems into the process and phasing out the batches of smaller instances (Table 2).

It is desirable that the heuristic produces not only produces solutions with as small an objective value as possible, but that it also can be used to sample a variety of good solutions. Many of the solutions are significantly worse than the greedy solution, while some are better. T In general, we observe that solutions with fewer visits to station nodes and depot nodes produce better objective function values. This is not particularly surprising, but it does identify some of the traps that are encountered at training time. To discourage spurious visits to charging stations and early returns to the depot, penalty terms were added to the cost function. Often, early in training, we observe that the gradient descent method pushes the learning in one of these directions, either forcing many visits to the stations and few to the depot, or vice versa. Care must be taken with the penalty weights to prevent this. This is illustrated in Fig. 1.

Table 2. Mean objective function ratio of of best sampled solution and greedy solution obtained from deep learning heuristic to the best known solution, as well as the mean time required to produce a greedy solution for each fixed $|I|$ of the benchmark set.

$|I|$	$\langle z_{100}/z_{BKS}\rangle$	$\langle z_{\text{greedy}}/z_{BKS}\rangle$	$t(s)$
10	1.0016	1.0104	0.05
20	1.0057	1.0167	0.11
40	1.0105	1.0301	0.29
80	1.0562	1.0856	0.64
160	1.0590	1.1017	0.98
320	1.0934	1.2201	2.01

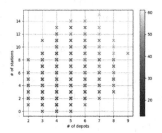

Fig. 1. Scatter plot of mean behaviour for sampled solutions with the corresponding number of depots and charging stations included. The intensity of the marker on the plot indicates the solution quality: blue is better. (Color figure online)

The deep learning heuristic models do not generalise well to larger problem sizes. This means that for any given problem size, there must exist a trained heuristic that has been tailored to problems of around that size. This is because solution quality degrades sharply as the size of the problem diverges from that encountered during training, greatly affecting the solution quality. This has severe negative impacts both on the quality of the warm-start solutions and the quality of the pseudo-labels that can be produced for the sparsification task. In general, it also leads to longer running times, since depots and stations will be visited too often.

4.2   The Pruning Classification Model

The three models that were investigated for pruning where the linear support vector machine (lSVM), the random forest classifier (RFC) and the logistic regression model (LR). Each of these models can be evaluated relatively quickly, since inference is quick, though it is clear that the random forest classifier, even when forced to remain small, is the slowest. As observed with [7], the lSVM model affords the best performance for similar computational cost to that of the LR model, so it is chosen as the classifier for pruning in the following sections. We first evaluate the performance of the classifiers in terms of the pruning rate p and the false-negative rate ϕ achieved. We present results, for brevity, as means over sets of benchmark instances for which the number of customers remains constant. That is $\langle p \rangle_{10}$ is the mean pruning rate observed over all instances with ten customers. Similarly $\langle \phi \rangle_{40}$ is the mean false negative rate observed over the set of all problem instances with forty customers.

Table 3. Pruning rate achieved by each model for each subset of the benchmark problem instances, where $|I|$ is fixed.

model	$\langle p \rangle_{10}$	$\langle p \rangle_{20}$	$\langle p \rangle_{40}$	$\langle p \rangle_{80}$	$\langle p \rangle_{160}$	$\langle p \rangle_{320}$
$lSVM$	0.76	0.81	0.85	0.90	0.94	0.97
RFC	0.76	0.79	0.82	0.86	0.92	0.94
LR	0.69	0.73	0.79	0.85	0.91	0.94

It is particularly necessary in the case of the E-VRP-NL that a high pruning rate is achieved, especially for particularly large problem instances, which may have hundreds of thousands of variables and constraints. The number of edges pruned must not only grow with the size of the problem, but so too must the fraction of edges pruned. For small problems, we prune a relatively small fraction of the edges, as can be seen in Table 3. We can see that the mean pruning rate for all of the benchmark instances with $|I| = 10$ is 0.76 with the lSVM. On the other hand, for the case where $|I| = 320$, we see that this grows to 0.97.

[7] observe that the false negative rate ϕ acts as a proxy measure for the success of the pruner. The authors note that there is a strong correlation between

Table 4. False negative rate achieved by each model for each subset of the benchmark problem instances, where $|I|$ is fixed.

model	$\langle\phi\rangle_{10}$	$\langle\phi\rangle_{20}$	$\langle\phi\rangle_{40}$	$\langle\phi\rangle_{80}$	$\langle\phi\rangle_{160}$	$\langle\phi\rangle_{320}$
$lSVM$	0.06	0.03	0.02	0.02	0.02	0.02
RFC	0.08	0.05	0.04	0.04	0.03	0.03
LR	0.09	0.08	0.07	0.06	0.05	0.05

the false negative rate of the TSP pruner and the degradation in the solution quality following problem reduction. We consider false negatives to be any edges that were selected for pruning, but carried a pseudo-label value greater than 0.01. Table 4 further indicates that the $lSVM$ achieves the lowest false negative rate across all of the benchmark instances, indicating that it prunes the most variables in general at the least detriment to the quality of the solution that can be obtained from the reduced problem.

4.3 Pruning Then Optimising

Initial objective function values, especially for the smaller problem instances are relatively close to the best known values for the benchmark problem instances, as seen in Table 5. For the larger problem instances, a significant gap exists between these values. Smaller reduced problems can generally be solved to optimality within one minute, whereas solve times for larger problems grows quite quickly with the number of customers. For problems with up to forty customers, we can solve the reduced problems to optimality. Large problems are not always solveable within the time limit of one hour, though in most cases the problems with eighty customers are solveable. The gaps for the problem instances with 160 and 320 customers are relatively large; most problems could not be solved within the given time limit. For the largest class of problems, not a single reduced problem could be solved to optimality, despite the high degree of pruning.

Table 5. Mean performance statistics across fixed problem sizes for reduced problems solved with a branch-and-bound solver.

$\|I\|$	$\langle z_{\text{initial}}/z_{BKS}\rangle$	$\langle z_{\text{final}}/z_{BKS}\rangle$	$t(s)$	$Gap(\%)$
10	1.0104	1.0002	46	0.00
20	1.0163	1.0024	322	0.00
40	1.0289	1.0142	561	0.00
80	1.0832	1.0305	756	1.03
160	1.0999	1.0568	1203	2.38
320	1.1705	1.0930	3600	6.12

5   Conclusions and Discussion

Although deep learning techniques cannot yet guarantee optimality nor any measure of quality of their solutions, they remain a promising research direction.

First, as we have demonstrated, the ML techniques are highly flexible and adaptable to new problem variants, especially to cases where no other approaches have yet been developed, a property especially important in practical settings where constraints and requirements may change regularly.

Furthermore, the deep learning approaches can take advantage of dedicated hardware for neural networks, increasing the speed at which inference can be made. Swapping computationally expensive search procedures for lengthy training phases, these heuristics take advantage of learned information from training problem instances to perform relatively well later on for unseen test instances.

The RL approach allows us to apply the learning-to-prune framework to problems where collecting optimal labels is not practical, even offline. Our computational experiments show that learning to prune achieves high levels of sparsification for the highly-constrained and challenging E-VRP-NL problem. This demonstrates that our ML framework can generate sufficient training data to train pruning classifiers.

Similar to the observations of [15] we note that charging decisions made by the deep learning heuristic can be sub-optimal. This is a result of two related issues. In some cases the neural network does not select charging station nodes optimally. This leads to over-charging or under-charging, or visiting a slow charging station even if a faster one is nearby.

It is clear that the use of pruning approaches to reduce the number of variables in a linear programming formulation of routing-type problems of this complexity is insufficient. Even with an extremely high pruning rate, solve times can be unacceptably long. This suggests that more work is needed to aid branch-and-bound solvers if these techniques are to be extended for problems that have formulations with weak relaxations. Although there has been research in that direction, learning-to-cut approaches have shown limited success to date.

Acknowledgement. This publication has emanated from research supported in part by a grant from Science Foundation Ireland under Grant number 18/CRT/6183. For the purpose of Open Access, the author has applied a CC BY public copyright licence to any Author Accepted Manuscript version arising from this submission.

References

1. Applegate, D.L., Bixby, R.E., Chvátal, V., Cook, W.J.: The Traveling Salesman Problem: A Computational Study. Princeton University Press, Princeton (2006)
2. Ashford, R.: Mixed integer programming: a historical perspective with Xpress-MP. Ann. Oper. Res. **149**(1), 5 (2007)
3. Bello, I., Pham, H., Le, Q.V., Norouzi, M., Bengio, S.: Neural combinatorial optimization with reinforcement learning. arXiv preprint arXiv:1611.09940 (2016)

4. Tayebi, S.R.D., Ajwani, D.: Learning to prune instances of k-median and related problems. In: Algorithm Engineering and Experiments (ALENEX). ACM-SIAM (2022)
5. Desaulniers, G., Errico, F., Irnich, S., Schneider, M.: Exact algorithms for electric vehicle-routing problems with time windows. Oper. Res. **64**(6), 1388–1405 (2016)
6. Felipe, Á., Ortuño, M.T., Righini, G., Tirado, G.: A heuristic approach for the green vehicle routing problem with multiple technologies and partial recharges. Transp. Res. Part E: Logist. Transp. Rev. **71**, 111–128 (2014)
7. Fitzpatrick, J., Ajwani, D., Carroll, P.: Learning to sparsify travelling salesman problem instances. In: Stuckey, P.J. (ed.) CPAIOR 2021. LNCS, vol. 12735, pp. 410–426. Springer, Cham (2021). https://doi.org/10.1007/978-3-030-78230-6_26
8. Froger, A., Jabali, O., Mendoza, J.E., Laporte, G.: The electric vehicle routing problem with capacitated charging stations. Transp. Sci. **56**(2), 460–482 (2022)
9. Froger, A., Mendoza, J.E., Jabali, O., Laporte, G.: Improved formulations and algorithmic components for the electric vehicle routing problem with nonlinear charging functions. Comput. Oper. Res. **104**, 256–294 (2019)
10. Hagberg, A., Swart, P., Chult, D.S.: Exploring network structure, dynamics, and function using NetworkX. Technical report, Los Alamos National Lab. (LANL), Los Alamos, NM (United States) (2008)
11. Kancharla, S.R., Ramadurai, G.: Electric vehicle routing problem with non-linear charging and load-dependent discharging. Expert Syst. Appl. **160**, 113714 (2020)
12. Kool, W., Van Hoof, H., Welling, M.: Attention, learn to solve routing problems! arXiv preprint arXiv:1803.08475 (2018)
13. Kullman, N.D., Froger, A., Mendoza, J.E., Goodson, J.C.: frvcpy: an open-source solver for the fixed route vehicle charging problem. INFORMS J. Comput. **33**(4), 1277–1283 (2021)
14. Lauri, J., Dutta, S.: Fine-grained search space classification for hard enumeration variants of subset problems. In: The Thirty-Third AAAI Conference on Artificial Intelligence, AAAI, pp. 2314–2321. AAAI Press (2019)
15. Lin, B., Ghaddar, B., Nathwani, J.: Deep reinforcement learning for the electric vehicle routing problem with time windows. IEEE Trans. Intell.Transp. Syst. **23**, 11528–11538 (2021)
16. Montoya, A., Guéret, C., Mendoza, J.E., Villegas, J.G.: The electric vehicle routing problem with nonlinear charging function. Transp. Res. Part B: Methodol. **103**, 87–110 (2017)
17. Parliament, E.: Directive (EU) 2019/1161 of the European parliament and of the council of 20 June 2019 amending directive 2009/33/EC on the promotion of clean and energy-efficient road transport vehicles. Off. J. Eur. Union **62**, 116–130 (2019)
18. Paszke, A., et al. PyTorch: an imperative style, high-performance deep learning library. In: Advances in Neural Information Processing Systems, vol. 32 (2019)
19. Pedregosa, F., et al.: Scikit-learn: machine learning in python. J. Mach. Learn. Res. **12**, 2825–2830 (2011)
20. Schneider, M., Stenger, A., Goeke, D.: The electric vehicle-routing problem with time windows and recharging stations. Transp. Sci. **48**(4), 500–520 (2014)
21. Sun, Y., Ernst, A., Li, X., Weiner, J.: Generalization of machine learning for problem reduction: a case study on travelling salesman problems. OR Spect. **43**(3), 607–633 (2021)
22. Vinyals, O., Fortunato, M., Jaitly, N.: Pointer networks. In: Advances in Neural Information Processing Systems, vol. 28 (2015)

Matheuristic Fixed Set Search Applied to Electric Bus Fleet Scheduling

Raka Jovanovic[1]([✉])[iD], Sertac Bayhan[1], and Stefan Voß[2][iD]

[1] Qatar Environment and Energy Research Institute (QEERI),
Hamad bin Khalifa University, PO Box 5825, Doha, Qatar
rjovanovic@hbku.edu.qa
[2] Institute of Information Systems, University of Hamburg,
Von-Melle-Park 5, 20146 Hamburg, Germany

Abstract. In recent years, there has been an increasing growth in the number of electric vehicles on the road. An important part of this process is the electrification of public transport with the use of electric buses. There are several differences between scheduling an electric or diesel bus fleet to cover a public transport timetable. The main reason for this is that electric buses have a shorter range and need to be charged during operating hours. The related optimization problems are often modeled using mixed-integer programming (MIP). An issue is that standard MIP solvers usually cannot solve problem instances corresponding to real-world applications of the model within a reasonable time limit. In this paper, this is addressed by extending the fixed set search to a matheuristic setting. The conducted computational experiments show that the new approach can be applied to much larger problems than the basic MIP. In addition, the proposed approach significantly outperforms other heuristic and metaheuristic methods on the problem of interest for problem instances up to a specific size.

Keywords: Matheuristic · electric buses · fleet scheduling

1 Introduction

Recent years have seen a massive rise in the worldwide adoption of electric vehicles (EVs). Governments have offered substantial incentives for EV adoption to reach their net-zero emission targets. Researchers are exploring ways to optimize the use of EVs by constructing better charging infrastructure and leveraging the charging flexibility of EVs to address challenges posed by the increased reliance on renewable energy sources [10,21]. The use of electric buses (EBs) in public transport has surpassed that of personal EVs. In China, for instance, there are over 400,000 EBs in use, representing more than 14% of all public transport buses. A study of the growth of EBs in Beijing, China can be found in [25]. The use of EBs offers numerous benefits such as reduced air and noise pollution in urban areas. They are also crucial for reaching net-zero goals and

M. Sellmann and K. Tierney (Eds.): LION 2023, LNCS 14286, pp. 393–407, 2023.
https://doi.org/10.1007/978-3-031-44505-7_27

offer a cost-effective public transportation alternative compared to diesel buses (DBs) [4,23,27]. This is partly due to the high daily mileage of EBs, leading to a significant decrease in operational costs as electricity is used instead of diesel, which quickly covers the initial investment. A high adoption rate of EVs can also be seen in car-hailing and taxi services for the same reason. There is also an increasing research interest in efficiently planning charging infrastructure for large fleets and optimizing smart charging schedules [3,6,28].

The adoption of EBs brings new operational difficulties, such as limited driving range and prolonged charging times, compared to DBs. The scheduling of DBs in urban public transportation is well studied through the vehicle scheduling problem (VSP) [2,5]. In the VSP, the objective is to optimize the allocation of trips to vehicles based on a cost function, often related to the number of buses used. However, incorporating the constraints of EBs' range and charging schedule makes this problem more challenging and is typically addressed through the electric vehicle scheduling problem (E-VSP) [22] and its variations. These variations incorporate elements similar to the traditional VSP for DBs, such as single [9,18,26,30] and multiple [19,29] depot scenarios. Studies have analyzed the scheduling of pure EB fleets [9,18,26] and mixed fleets of EBs and DBs [29,30]. A noteworthy investigation, found in [26], assesses the robustness of scheduling under varying traffic conditions. Most of these models are solved through mixed-integer programming (MIP) [9,18,26,30], while limited research has explored the use of heuristic methods, such as a combination of a greedy algorithm and simulated annealing [30], the greedy randomized adaptive search procedure (GRASP) [11] or genetic algorithms [29].

In the general case, mathematical models for problems of this type often need a high level of detail to well represent the related real-world problems. Because of this, to a large extent, the implementation of metaheuristics becomes highly complex. On the other hand, problem instances corresponding to real-world systems are usually of large size, and generally cannot be solved using MIP within a reasonable amount of time. In this paper, an approach that combines MIP and a heuristic approach is proposed. Methods of this type are frequently called matheuristics, for which a recent review can be found in [1]. The proposed method avoids the complexity of a versatile metaheuristic implementation and manages to solve larger problems than the corresponding MIP. The proposed method combines MIP and the fixed set search (FSS) metaheuristic. The FSS is a novel population-based metaheuristic that has previously been successfully applied to solve the traveling salesman problem [14], the power dominating set problem [17], machine scheduling [15], the minimum weighted vertex cover problem [13,16], covering location with interconnected facilities problem [20] and the clique partitioning problem [12]. The inspiration for the FSS is the fact that generally high-quality solutions, for a specific problem instance, have many of the same elements in them. The idea is to generate new solutions that contain such elements. In essence, these elements, the fixed set, are included in the solutions that will be generated and the computational effort is dedicated to complete the partial solution or, in other words, "filling in the gaps". The proposed matheuristic FSS (MFSS) is applied to a variation of the single-depot E-VSP where charging time depends on the battery state proposed in [11]. The conducted computational experiments compare the performance of the MFSS to the use of the MIP and the GRASP metaheuristic.

The paper is organized as follows. Section 2 provides the outline of the problem being solved and how it is modeled. This is followed by a section dedicated to the graph formulation of bus scheduling problems. The next section presents the MFSS for the problem of interest. Section 6 is dedicated to the presentation of the conducted computational experiments. The paper is finalized with some concluding remarks.

2   Model Outline

In this section, an outline of the model proposed in [11] is presented. The purpose of the model is to determine the minimum number of EBs required to serve all bus routes in a timetable. This is achieved by assigning a collection of trips (from the timetable) to a group of EBs. Each trip is defined by its starting point, endpoint, duration, and start time. Note that the trips do not consider intermediate stops. The model operates under the following assumptions: scheduling of EBs to trips takes place within a predetermined time frame, for a set of stations, and a single depot. Additionally, it is assumed that the distances between each station and the depot are known, and that EBs can only be recharged at the depot. The assumption is that all buses start and finish their day at the depot. An EB has a defined driving range and must have its battery fully recharged before it can be scheduled. This means that an EB can only be used if it has enough battery power. It is also assumed that all EBs travel at a constant speed, which allows for the calculation of battery usage, proportional to trip duration, to be simplified by converting distance to duration.

For an EB to undertake a trip from a particular origin, it must arrive prior to the trip's starting time. The objective of the proposed model is to evaluate the number of EBs with varying ranges required to fulfill a public transportation timetable. The following points describe the model's specifications:

- There are N locations, represented by the set $\mathcal{L} = \{1, ..., N\}$.
- There is one single depot $D \in \mathcal{L}$.
- The distance between any two locations $i, j \in \mathcal{L}$ is known and equal to d_{ij}.
- The scheduling takes place over P time periods, represented by the set $\mathcal{P} = \{1, ..., P\}$.
- A timetable trip is a 4-tuple (o, d, s, l) where $o, d \in \mathcal{L}$ and $s, l \in \mathcal{P}$ represent the origin-destination pairs and start time with duration, respectively. The set $\mathcal{T}$ represents the timetable comprising of such trips. $e = s + l$ denotes the completion time of each trip.
- An EB has a fixed battery range R, and all EBs are assumed to have the same range.
- The objective is to determine the minimum number of EBs needed to perform all trips in the timetable.

- An EB can only be recharged at the depot.
- An EB is fully charged at the start of the first time period.
- There is a charging rate γ representing how much the battery is charged in one time period.
- The battery of an EB is always charged to full capacity.

3   Graph Formulation

The following text provides a graph representation for scheduling buses. This concept is used in several papers, e.g. [11, 26]. It focuses solely on assigning buses to trips in the schedule. Subsequent sections will delve into the extension of this model to incorporate battery capacity and charging, providing a more comprehensive understanding.

Let us describe the problem using a directed graph $G = (V, E)$. Every trip $t \in T$ is represented by a node $t \in V$. Additionally, the set of nodes V includes two special nodes, D_s and D_e, which correspond to the start (departure from the depot) and end (arrival at the depot) of all buses. The notation V_t will be used to denote the set of all nodes that correspond to trips in the timetable. The distance between two nodes, or trips, i and j, is calculated as the distance between the destination i_d of trip i and the origin j_o of trip j as follows.

$$d_{ij} = \hat{d}_{i_d j_o} \tag{1}$$

An edge (i, j) belongs to the edge set E if it is feasible to carry out trip j immediately after trip i. The set of edges E can be formally defined as follows.

$$E_n = \{(i, j) \mid i, j \in \mathcal{L} \wedge (s_j - e_i - d_{ij} \geq 0)\} \tag{2}$$
$$E_s = \{(D_s, i) \mid i \in \mathcal{L}\} \tag{3}$$
$$E_e = \{(i, D_e) \mid i \in \mathcal{L}\} \tag{4}$$
$$E = E_n \cup E_s \cup E_e \tag{5}$$

The set E_n consists of all the trip pairs (i, j) where the start time of trip j (s_j) is equal to or greater than the sum of the completion time e_i of trip i and the travel time between the destination of trip i (d_{ij}) and the origin of trip j. The start trip node D_s is connected to all other nodes via the edge set E_s. All trip nodes are also connected to the end node D_e using the edge set E_e. E represents the complete set of edges in the graph. It is important to note that, as defined, the directed graph G does not contain any cycles. Each edge $(i, j) \in E$ has an associated length property.

The schedule for all the buses is represented as a directed subgraph $S = (V, \tilde{E})$ of G. An edge (i, j) signifies that a bus performs trip j immediately after trip i. Let us examine the implications of this solution for the problem at hand. The vertex set of graph S is the same as that of G, meaning that all trips are performed by some EB. Each trip node $i \in V_t$ will have exactly one incoming and one outgoing edge with respect to some node $x \in V$, because a trip can only be directly performed before or after one other trip. The number of edges in the form (D_s, i) is equal to the number of buses required to complete the timetable. As a result, a single bus's schedule will be a directed path of the form $(D_s, \ldots, D_e)$. An illustration of how a trip timetable with known distances between locations and a bus schedule is converted to the graph formulation can be seen in Fig. 1.

Timetable and distances between destinations and origins of trips

Trip	Start Time	End Time	Orig./Dest.	a	b	c	d	Depot
a	320	340	a	0	20	10	30	10
b	390	420	b	25	0	40	10	15
c	420	460	c	15	40	0	30	20
d	440	470	d	25	10	35	0	10
			Depot	15	15	20	15	0

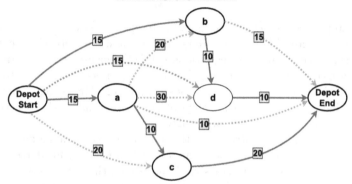

Schedule in graph presentation

Fig. 1. Illustration of the conversion of a timetable with known distances between origins and destinations between trips. The directed graph has a node for each trip and edges only connect trips that can be sequentially done. The set of gray dotted edges corresponds to precedence relations that do not appear in the bus schedule. The set of red edges (bottom) is a schedule of EBs. It contains the two buses with schedules $(Depot, a, c, Depot)$ and $(Depot, b, d, Depot)$. (Color figure online)

4   Mathematical Model

In this section, the MIP for the problem of interest proposed in [11] based on the presented formulations is given. The model uses the following input parameters:

- The distance between trip i and trip j is designated by d_{ij} and is calculated as outlined above.
- Each trip i has a starting time s_i and a duration of d_i. For ease of notation, the completion time of trip i is defined as $e_i = s_i + d_i$.

- The rate of battery charging per time period is denoted by γ.
- The battery usage per time period is represented by α.
- The battery capacity of all EBs is the same and is equal to G_B.

The scheduling of EBs and the battery-related restrictions are established in the model through the following decision variables:

- For each trip pair (i,j) in the edge set E, there is a binary variable x_{ij} which indicates whether an EB performs trip j immediately after trip i.
- For each trip i in the node set V, the variable b_i denotes the battery level of the EB when it reaches the starting point of trip i before embarking on it.
- For each trip node i in V, a binary variable c_i is defined, indicating whether the EB that carries out trip i will recharge its battery after completing it.

The goal of the model is to minimize the number of electric buses required to complete all trips in the timetable. This objective can be represented mathematically as the number of edges originating from the start node D_s as follows.

$$\text{Minimize} \sum_{i \in V_t} x_{D_s i}. \tag{6}$$

The final step in constructing the model is to specify its constraints, which are outlined in (9)–(20). These constraints can be separated into three categories. The first set of constraints, in equations (9)–(11), ensures that all trips in the timetable are completed by some EB. These constraints do not take into account the range of the EBs. The next set of constraints, (12)–(18), deals with the state of the EB battery and thus the range of the EBs. Finally, the constraints in (19)–(20) regulate the charging of the EBs.

The battery-related constraints in the model have conditional parts. To clarify the constraints, two types of auxiliary variables are defined based on the values of x_{ji} and c_j. Although not strictly necessary, these variables improve the readability of the constraints. They are specified in the following equations:

$$v_{ji} = 1 - x_{ji} + c_j \tag{7}$$
$$w_{ji} = 2 - x_{ji} - c_j \tag{8}$$

This variable, v_{ji}, given in Eq. (7), identifies the situation where an EB performs trip j before trip i and the EB is not charged after trip j. In this scenario, the value of v_{ji} would be equal to 0. The variable w_{ji}, given in Eq. (8), recognizes the state when an EB performs trip j before trip i and is charged after trip j. In this case, the value of w_{ji} is equal to 0. With these auxiliary variables, the constraints of the proposed model can be defined as follows:

$$\sum_{(i,j) \in E} x_{ij} = 1 \qquad i \in V_t, \tag{9}$$

$$\sum_{(i,j) \in E} x_{ij} = 1 \qquad j \in V_t, \tag{10}$$

$$\sum_{i \in V_t} x_{D_s i} = \sum_{i \in V_t} x_{i D_e} \tag{11}$$

$$b_i - \alpha d_i \geq \alpha d_{i D_e} \quad i \in V_t \tag{12}$$

$$b_i \leq M v_{ji} + b_j - \alpha(d_j + d_{ji}) \quad i \in V_t, j \in V, \tag{13}$$

$$b_i \geq -M v_{ji} + b_j - \alpha(d_j + d_{ji}) \quad i \in V_t, j \in V, \tag{14}$$

$$b_i \leq M w_{ji} + G_B - \alpha d_{D_s i} \quad i \in V_t, j \in V, \tag{15}$$

$$b_i \geq -M w_{ji} + G_B - \alpha d_{D_s i} \quad i \in V_t, j \in V, \tag{16}$$

$$b_{D_s} = G_B, \tag{17}$$

$$b_i \leq G_B \quad i \in V_t, \tag{18}$$

$$s_i \geq -M w_{ji} + e_j + d_{j D_e} + d_{D_s i}$$
$$+ \frac{1}{\gamma}(G_B - (b_j - d_j - d_{j D_e})) \quad i, j \in V_t, \tag{19}$$

$$0 \leq c_j \leq 1 \quad j \in V_t. \tag{20}$$

Equations (9) and (10) ensure that a single EB performs one trip after and before a given timetable trip i, respectively. It is important to note that both starting the day at the depot and ending it at the depot are considered trips as well (pull-out and pull-in). Equation (11) ensures that the same number of EBs depart from the starting depot as those that arrive at the ending depot.

The following set of constraints focuses on the state of charge of the battery. Constraints (12) set the minimum value of battery load before a trip i in the timetable can be performed. The battery level must be greater than or equal to the amount of charge required to complete the trip and reach the depot afterwards. Constraints (13) and (14) are conditional constraints that are implemented with the use of a large value M. They only take effect if $v_{ji} = 0$, meaning that an EB performs trip i after trip j and does not charge inbetween. In this scenario, the state of the battery will be equal to the state of the battery after trip j (b_j) minus the charge used to complete trip j and travel to the starting point of trip i. Similarly, the constraints in (15) and (16) are conditional constraints for when an EB charges between trips j and i, as specified by the variable w_{ji}. In this case, the battery state before performing trip i is equal to the full battery charge (G_B) minus the charge required to move from the depot to the starting point of trip i. It should be noted that a high enough value of M for (13)–(16) is $2G_B$ (twice the capacity of the EB battery). The constraint in (17) ensures that an EB leaves the starting depot with a fully charged battery. Finally, the constraints in (18) are used to enforce the battery capacity limits before the start of any trip in the timetable.

The conditional constraints outlined in (19) determine when it is possible to charge the battery between trips j and i. These constraints are tied to time intervals, meaning that charging can only occur if there is enough time between the completion of trip j at period e_j and the start of trip i (s_i) to travel from the endpoint of trip j to the depot, then from the depot to the starting point of trip i, and to fully charge the battery. It is important to note that a suitable value for the large parameter M in this constraint is $2P$.

5 Matheuristic Fixed Set Search

In this section the proposed MFSS is presented. Its main idea is to use the MIP presented in the previous section as a component of an algorithm that can solve problem instances of a certain size on which the MIP cannot be effectively applied. Occasionally this size refers to a corridor such as in [24]. The main assumption is that the MIP is highly efficient to solve problems up to a specific size. The FSS method utilizes the fact that many high quality solutions to a given combinatorial optimization problem share common elements. By integrating some of these elements into newly created solutions, the FSS focuses its computational efforts on finding optimal or nearly optimal solutions within the corresponding subset of the solution space. The set of such common elements is referred to as the "fixed set" in the further text. The FSS achieves this by incorporating a learning mechanism into the GRASP. This section introduces the MFSS, which extends this idea to a matheuristic setting. The MFSS comprises multiple components, such as representing the solution as a subset of a ground set of elements, defining techniques for generating fixed sets, generating initial feasible solutions, implementing the learning mechanism, and defining a method for completing a solution from a fixed set.

5.1 Fixed Set

In this subsection, we describe the approach for generating fixed sets for the problem of interest. In order to apply the FSS algorithm, it is necessary to represent a solution as a subset of the ground set of elements. As described in Sect. 3, a solution of an EB scheduling problem is equal to a subset of the set of edges E of the graph G which is the ground set of elements.

The subsequent step involves establishing a procedure for generating multiple fixed sets F with a controllable size (cardinality) $|F|$, which can be used to produce feasible solutions of equal or better quality than the previously generated ones. The following definitions are introduced: $\mathcal{S}_n = \{S_1, .., S_n\}$ denotes the set of n best solutions produced in the preceding steps of the algorithm. A base solution $B \in \mathcal{S}_n$ is a solution chosen randomly from the top n solutions. If a fixed set satisfies $F \subset B$, it can be utilized to generate a feasible solution of at least the same quality as B, with F having the freedom to include any number of elements of B. The main objective is to create F such that it includes frequently occurring elements in a group of high-quality solutions. We define $\mathcal{S}_{kn}$ as the set of k randomly chosen solutions out of the n best solutions, $\mathcal{S}_n$. The function $C((i,j), S)$, where S is a solution and (i,j) is an element (edge), is defined as 1 if $(i,j) \in S$ and 0 otherwise. Using $C((i,j), S)$, the number of times an element (i,j) occurs in $\mathcal{S}_{kn}$ can be counted with the function:

$$O((i,j), \mathcal{S}_{kn}) = \sum_{S \in \mathcal{S}_{kn}} C((i,j), S) \tag{21}$$

We can define $F \subset B$ as the set of elements (i,j) that have the highest value of $O((i,j), \mathcal{S}_{kn})$. In the generation of the fixed set the edges corresponding to the movements of EBs from/to the depot, having the form $(D_s, i)/(i, D_e)$, are never used since they highly constrain the quality of the generated solution. Additionally, we can define the function $F = Fix(B, S_{kn}, Size)$ as the fixed set produced for a base solution B and a set of solutions S_{kn} with a size of $Size$ elements.

The original FSS uses a randomized greedy algorithm with a pre-selected set of elements to diversify the generation of solutions. However, in the MFSS, diversification is partly achieved in the method for generating the fixed set. Specifically, we utilize the way ties are resolved. Suppose the last element (i, j) to be added to the fixed set F has the value of $O((i, j), S_{kn}) = f$. In the general case, there may be multiple elements with the same value of this function. Let us assume that there are $l > Size$ elements $(i, j) \in B$ that have a value greater than or equal to f for the function $O((i, j), S_{kn})$. We use the notation $\hat{F}$ to represent the set of such elements. In this approach, when there are multiple elements with the same function value, the function $Fix(B, S_{kn}, Size)$ returns $\hat{F}$ with $l - Size$ randomly selected elements removed. It is important to note that a removed element (i, j) may not necessarily have the lowest value of $O((i, j), S_{kn})$.

5.2   Integer Program Use

The proposed matheuristic approach aims to take advantage of the fact that for many problems, using an IP solver can be highly efficient up to a specific instance size. However, in the general case, standard IP solvers spend most of their time searching for the optimal solution and proving its optimality, while being able to find high-quality solutions for that instance size at a relatively low computational cost.

The proposed approach aims to reduce the computational cost of solving an integer programming (IP) problem by utilizing fixed sets. Fixing the values of some decision variables can be an effective way to achieve this, and fixed sets can naturally facilitate the process. To achieve this, the following set of constraints can be added to the IP model presented in (9)–(20):

$$x_{ij} = 1 \qquad (i, j) \in F \tag{22}$$

Equation (22) guarantees that any element (i, j) of the fixed set F is a part of the newly generated solution.

The use of the objective function given in (6) has a disadvantage that a wide range of solutions have an equivalent objective value. In real-world EBs scheduling problems, generally, the most relevant factor is the number of used EBs but their total travel length is of high importance. In addition, we have observed that schedules with a lower total travel length have a higher potential of being improved to ones using a lower number of EBs. Because of this, the MFSS uses an objective that extends the one given in (6) by considering the total travel length of all EBs as follows.

$$\text{Minimize} \quad \Lambda \sum_{i \in V_t} x_{D_s i} + \sum_{(i,j) \in E} d_{ij} x_{i,j}. \tag{23}$$

In (23), the constant $\Lambda > \sum_{(i,j)\in E} d_{ij}$ is used to give a higher importance to the number of used EBs in the solution. The value of the objective function is equal to the sum of the number of used EBs scaled by Λ and the total distance the EB need to travel between two consecutive trips.

Let us define $IPS(F,t)$ as a function that solves the IP using the objective function given in (23) and the constraints defined by (9)–(20) and (22) with a maximal computational time of t. In the practical utilization of IP solvers, it is often advantageous to supply an initial, high-quality incumbent solution S for a "warm start", as this can eliminate portions of the search space in the branch-and-cut algorithm, potentially resulting in smaller branch-and-cut trees. The proposed method exploits this approach to enhance the performance of the IP solver. We define the function $IPS(F,t,S)$ as an extension of the IPS function that includes an initial incumbent solution S.

5.3 Learning Mechanism

Let us start with an overview of the MFSS approach. We begin with a population $\mathcal{P} = \{S_1, \ldots, S_{|P|}\}$ that consists of randomly generated solutions. The population is then iteratively improved using the following procedure: First, a random base solution B is selected from the set of n best solutions, denoted by $\mathcal{S}_n$. Additionally, a set $\mathcal{S}_{kn}$ is created by selecting k solutions from the set of n best solutions $\mathcal{S}_n$. Note that the solution quality is measured based on the objective given in (23). A fixed set F is constructed using B and $\mathcal{S}_{kn}$, and a new solution S' is generated using the IP algorithm with additional constraints related to the fixed set F. The resulting solution S' is then added to the population of solutions $\mathcal{P}$. It is worth noting that $\mathcal{P}$ is a set of unique solutions, so duplicates are not included. This process is repeated for the new population of solutions, and further details about the implementation of this method are provided in the subsequent text.

A simple method is used for generating initial feasible solutions. To be exact, schedules for individual EBs are generated without consideration of battery recharging using the following procedure. An EB starts from the depot and trips are iteratively added to its schedule. At each iteration, a random trip that has not yet been assigned to any EB and it is reachable from the current trip i and after completing it there is sufficient charge to reach the depot is added to the schedule of the EB. If there is no such trip, the EB returns to the depot and scheduling for a new EB is started. This procedure is repeated until all the trips in the timetable are assigned to some EB. During the iterative procedure, which implements the learning mechanism, several factors must be taken into account. Initially, let us focus on the computational cost since the computational expense of the IPS may be high. One issue to consider is that for a low-quality fixed set, even if the IPS is given an extended execution time, it is unlikely to acquire high-quality solutions. In contrast, for a high-quality fixed set, it is reasonable to allow the IPS an extended execution time since it explores a portion of the solution space that contains high-quality solutions. As more solutions are generated, the quality of solutions in the population increases, and the quality of the fixed sets improves accordingly. Therefore, it is reasonable to allow only short computational times for the IPS during the early iterations of the algorithm and longer times during the later ones. To implement this strategy, the proposed method starts with an initial allowed computational time t_{init} for the IPS and increases it as the algorithm begins to stagnate. Specifically, when no new solution is generated among the best n solutions in the last $StagMax$ iterations, the algorithm is deemed stagnant.

Algorithm 1. Pseudocode for MFSS

1: **Parameters:** Initial population size N_{pop}, n best solutions to be used, subproblem size $Free_{max}$, initial computational time for IP t_{init}
2:
3: Generate initial population $\mathcal{P}$
4: $Stag \leftarrow 0$, $t \leftarrow t_{init}$
5: **while** Not Time Limit Reached **do**
6:     Select random $k \in [k_{min}, k_{max}]$
7:     Select random $B \in \mathcal{S}_n$
8:     Generate random $\mathcal{S}_{kn}$
9:     $Size = InnerEdges(B) - Max_{Free}$
10:    $F \leftarrow Fix(B, \mathcal{S}_{kn}, Size)$
11:    $S \leftarrow IPS(F, t, B)$
12:    Check if S is a new best solution
13:    $\mathcal{P} = \mathcal{P} \cup \{S\}$
14:    **if** $\mathcal{S}_n$ has changed **then**
15:        $Stag = 0$
16:    **else**
17:        $Stag = Stag + 1$
18:    **end if**
19:    **if** $Stag \geq MaxStag$ **then**
20:        $Stag = 0$
21:        $t \leftarrow 2t$
22:    **end if**
23: **end while**

This procedure is better understood by observing the pseudocode given in Algorithm 1. The first step is generating the initial population of N_{pop} solutions using the previously described method for generating feasible solutions. In the main loop, firstly a random value of the size of the set $\mathcal{S}_{kn}$ is selected from the interval $[k_{min}, k_{max}]$. The value of k changes in subsequent iterations to increase the diversity of generated fixed sets. Next, the random base solution B and set $\mathcal{S}_{kn}$ are selected from the set of the best n solutions. The size of the fixed set $Size$ is set to the total number of inner edges, i.e., the ones not going to or from the depot, of B minus the maximal allowed number of free edges $Free_{max}$. Next, a fixed set F is generated using the function $Fix(B, \mathcal{S}_{kn}, Size)$. A new solution S is acquired using the function $IPS(F, t, B)$, for the fixed set F, with a time limit t and an initially incumbent solution B. The solution S is added to the population of solutions $\mathcal{P}$ and it is checked if S is the new best solution. In case that the set $\mathcal{S}_n$ has not changed, the stagnation counter is increased by 1, otherwise it is set to 0. After this, it is checked if stagnation has occurred, and if so, the allowed computational time for IPS is doubled. This procedure is repeated until a time limit is reached.

6 Results

In this section, the results of the conducted computational experiments are presented. Their objective is to evaluate the effectiveness of the MFSS. This is done in comparison to the MIP model and the GRASP algorithm proposed in [11]. The MIP and MFSS approaches have been implemented in ILOG CPLEX for C#.NET through Concert Technology. The same code has been used for the GRASP algorithm as in [11], which has been implemented in C#. The computational experiments have been performed on a personal computer running Windows 10 having Intel(R)Xeon(R) Gold 6244 CPU @3.60 GHz with 128 GB memory.

The methods are evaluated on randomly generated test instances using the same method as in [11] for which a short outline is given as follows. A set of N random locations is selected from a rectangle and one of them is selected as the depot. The distance measure is Euclidean. The scheduling takes place within a 24-hour time frame, with each period being equal to one minute, resulting in a total of 1440 periods. A bus route is described by its starting point (designated as l_o), its destination (designated as l_d), the start time (t_s), the end time (t_e), the trip duration (d), and the frequency of departures (f). The two points, start and destination, are randomly selected from the set of locations. The values of the other parameters related to routes are randomly selected from predefined intervals. A problem instance has been generated for a minimal number of trips N. To be specific, bus routes are generated based on randomly selected parameters, one by one until the total number of trips in them is at least N. In the conducted computational experiments the minimal number of trips ranged from 30 to 200. For each problem size 10 different instances are used.

The GRASP algorithm uses the same set of parameters as in [11]. In the case of the MFSS, the following parameter values are used $N_{pop} = 20$, $n = 20$, $t_{init} = 0.5s$, and $Free_{max} = 20$. These values have been selected empirically for the used computer. In the case of the MIP method for all instances a time limit of 1200 s is used. The number of seconds allowed for the execution of the GRASP and the MFSS is four times the minimal number of trips used for generating the instance. For the GRASP in the case of the largest instance this results in more than 50,000 iterations, so it is not likely that it could produce better results.

In the conducted computational experiments problem instances with two EB settings are evaluated. In the first one the battery capacity G_B of an EB is 250 and the charging rate γ is 1. In the second one the battery capacity is 300 and the charging rate is 1.5. Each of the methods performed a single run on each of the instances. In Table 1 average solution quality and average time needed to find the best solution can be seen for each of the problem sizes.

Table 1. Comparison of the proposed MFSS, GRASP and MIP for synthetic problem instances of different sizes and EB properties.

Minimal number of trips	Average solution			Average time [s]		
	MFSS	GRASP	MIP	MFSS	GRASP	MIP
$G_B = 250$, $\gamma = 1$ and $\alpha = 1$						
30	6.40	6.50	6.60	21.76	1.26	926.62
40	7.80	8.10	8.50	34.19	1.97	1205.58
50	9.30	9.80	10.80	51.96	11.58	1204.01
75	13.00	13.60	16.80	151.87	42.85	1209.93
100	17.30	18.50	25.90	214.05	12.12	1219.23
150	25.90	26.90	31.90	442.57	80.54	1218.41
200	35.30	35.40	43.60	661.98	171.51	1207.67
$G_B = 300$, $\gamma = 1.5$ and $\alpha = 1$						
30	5.40	5.60	5.90	17.59	1.54	1022.36
40	6.30	6.90	7.30	33.37	2.68	1207.76
50	7.80	8.10	9.30	60.10	15.16	1207.81
75	11.30	11.60	14.60	87.43	0.69	1213.98
100	15.00	15.30	20.40	166.90	6.56	1217.17
150	22.00	22.60	25.90	379.05	46.37	1226.91
200	30.90	29.30	35.70	730.62	122.97	1217.01

The first thing that can be observed from these results is that both GRASP and MFSS significantly outperform MIP when solution quality is considered. In addition, the MIP needed significantly more time to find such solutions than the other two methods. Overall the MFSS had a significantly better performance than the GRASP for instances of less than 200 trips. It had a better average solution than GRASP in all 12 such test groups. The MFSS has a higher advantage in solution quality in case of EBs with lower range and slower charging rate, where a higher number of buses is needed to perform all the trips. From our observations, the performance of the MFSS drops for instances of 200 trips and higher, due to the fact that solving the IPS becomes inefficient. To be exact, the IPS manages to find only low quality solutions within the available time period for the subproblem. In all the instances, the GRASP has a significantly lower execution time. In most cases the GRASP finds the best solution within the first few iterations. On the other hand, the MFSS had a significantly lower convergence speed but managed to frequently find better solutions. It is important to point out that from the results presented in [11], the GRASP algorithm can be applied to much larger instances.

7 Conclusion

In this paper, the problem of scheduling a fleet of EBs to satisfy a public transport timetable has been addressed. A matheuristic approach based on the FSS has been proposed and applied to an existing model for the problem of interest. The conducted computational experiments show that the new approach significantly outperforms the direct application of the MIP with standard solvers. In addition, it is highly competitive to existing metaheuristic approaches, specifically to the GRASP for problem instances up to a specific size. The main advantage of the method is that it can, potentially, be applied to other mathematical models for EB fleet scheduling problems like the ones presented in [9,18,26,30]. Another direction of future research is the adaptation of the method to solve very large-scale problem instances and adapting it to a multi-objective setting. This would also allow to look into more general integrated vehicle and crew scheduling problems such as from [7,8] and adding additional uncertainty. Finally, due to the varying usability of EBs under practical circumstances, an uncertain number of available buses needs to be considered.

References

1. Boschetti, M.A., Maniezzo, V.: Matheuristics: using mathematics for heuristic design. 4OR **20**(2), 173–208 (2022)
2. Desaulniers, G., Hickman, M.D.: Public transit. In: Handbooks in Operations Research and Management Science, vol. 14, pp. 69–127 (2007)
3. Dreier, D., Rudin, B., Howells, M.: Comparison of management strategies for the charging schedule and all-electric operation of a plug-in hybrid-electric bi-articulated bus fleet. Public Transp. **12**, 363–404 (2020)
4. Feng, W., Figliozzi, M.: Vehicle technologies and bus fleet replacement optimization: problem properties and sensitivity analysis utilizing real-world data. Public Transp. **6**(1–2), 137–157 (2014)
5. Foster, B.A., Ryan, D.M.: An integer programming approach to the vehicle scheduling problem. J. Oper. Res. Soc. **27**(2), 367–384 (1976)
6. Gallo, J.B., Bloch-Rubin, T., Tomić, J.: Peak demand charges and electric transit buses. US Department of Transportation, Technical report (2014)
7. Ge, L., Kliewer, N., Nourmohammadzadeh, A., Voß, S., Xie, L.: Revisiting the richness of integrated vehicle and crew scheduling. Public Transp. 1–27 (2022)
8. Ge, L., Nourmohammadzadeh, A., Voß, S., Xie, L.: Robust optimization for integrated vehicle and crew scheduling based on uncertainty in the main inputs. In: The Fifth Data Science Meets Optimisation Workshop at IJCAI-22 (2022). https://sites.google.com/view/ijcai2022dso/?pli=1
9. Janovec, M., Koháni, M.: Exact approach to the electric bus fleet scheduling. Transp. Res. Proc. **40**, 1380–1387 (2019)
10. Jovanovic, R., Bayhan, S., Bayram, I.S.: A multiobjective analysis of the potential of scheduling electrical vehicle charging for flattening the duck curve. J. Comput. Sci. **48**, 101262 (2021)
11. Jovanovic, R., Bayram, I.S., Bayhan, S., Voß, S.: A GRASP approach for solving large-scale electric bus scheduling problems. Energies **14**(20), 6610 (2021)

12. Jovanovic, R., Sanfilippo, A.P., Voß, S.: Fixed set search applied to the clique partitioning problem. Eur. J. Oper. Res. **309**(1), 65–81 (2023)
13. Jovanovic, R., Sanfilippo, A.P., Voß, S.: Fixed set search applied to the multi-objective minimum weighted vertex cover problem. J. Heuristics **28**, 481–508 (2022)
14. Jovanovic, R., Tuba, M., Voß, S.: Fixed set search applied to the traveling salesman problem. In: Blesa Aguilera, M.J., Blum, C., Gambini Santos, H., Pinacho-Davidson, P., Godoy del Campo, J. (eds.) HM 2019. LNCS, vol. 11299, pp. 63–77. Springer, Cham (2019). https://doi.org/10.1007/978-3-030-05983-5_5
15. Jovanovic, R., Voß, S.: Fixed set search application for minimizing the makespan on unrelated parallel machines with sequence-dependent setup times. Appl. Soft Comput. **110**, 107521 (2021)
16. Jovanovic, R., Voß, S.: Fixed set search applied to the minimum weighted vertex cover problem. In: Kotsireas, I., Pardalos, P., Parsopoulos, K.E., Souravlias, D., Tsokas, A. (eds.) SEA 2019. LNCS, vol. 11544, pp. 490–504. Springer, Cham (2019). https://doi.org/10.1007/978-3-030-34029-2_31
17. Jovanovic, R., Voss, S.: The fixed set search applied to the power dominating set problem. Expert. Syst. **37**(6), e12559 (2020)
18. van Kooten Niekerk, M.E., van den Akker, J., Hoogeveen, J.: Scheduling electric vehicles. Public Transp. **9**(1–2), 155–176 (2017)
19. Li, L., Lo, H.K., Xiao, F.: Mixed bus fleet scheduling under range and refueling constraints. Transp. Res. Part C Emerg. Technol. **104**, 443–462 (2019)
20. Lozano-Osorio, I., Sánchez-Oro, J., Martínez-Gavara, A., López-Sánchez, A.D., Duarte, A.: An efficient fixed set search for the covering location with interconnected facilities problem. In: Di Gaspero, L., Festa, P., Nakib, A., Pavone, M. (eds.) Metaheuristics, pp. 485–490. Springer, Cham (2023). https://doi.org/10.1007/978-3-031-26504-4_37
21. Meyer, D., Wang, J.: Integrating ultra-fast charging stations within the power grids of smart cities: a review. IET Smart Grid **1**(1), 3–10 (2018)
22. Perumal, S.S., Lusby, R.M., Larsen, J.: Electric bus planning & scheduling: a review of related problems and methodologies. Eur. J. Oper. Res. **301**(2), 395–413 (2022)
23. Quarles, N., Kockelman, K.M., Mohamed, M.: Costs and benefits of electrifying and automating bus transit fleets. Sustainability **12**(10), 3977 (2020)
24. Sniedovich, M., Voß, S.: The corridor method: a dynamic programming inspired metaheuristic. Control. Cybern. **35**(3), 551–578 (2006)
25. Song, Z., Liu, Y., Gao, H., Li, S.: The underlying reasons behind the development of public electric buses in China: the Beijing case. Sustainability **12**(2), 688 (2020)
26. Tang, X., Lin, X., He, F.: Robust scheduling strategies of electric buses under stochastic traffic conditions. Transp. Res. Part C Emerg. Technol. **105**, 163–182 (2019)
27. Topal, O., Nakir, İ: Total cost of ownership based economic analysis of diesel, CNG and electric bus concepts for the public transport in Istanbul city. Energies **11**(9), 2369 (2018)
28. Vuelvas, J., Ruiz, F., Gruosso, G.: Energy price forecasting for optimal managing of electric vehicle fleet. IET Electr. Syst. Transp. **10**, 401–408 (2020)
29. Yao, E., Liu, T., Lu, T., Yang, Y.: Optimization of electric vehicle scheduling with multiple vehicle types in public transport. Sustain. Cities Soc. **52**, 101862 (2020)
30. Zhou, G., Xie, D., Zhao, X., Lu, C.: Collaborative optimization of vehicle and charging scheduling for a bus fleet mixed with electric and traditional buses. IEEE Access **8**, 8056–8072 (2020)

Class GP: Gaussian Process Modeling for Heterogeneous Functions

Mohit Malu[1,3]([✉]), Giulia Pedrielli[2], Gautam Dasarathy[1], and Andreas Spanias[1,3]

[1] School of ECEE, Arizona State University, Tempe, AZ 85281, USA
{mmalu,giulia.pedrielli,gautamd,spanias}@asu.edu
[2] SCAI, Arizona State University, Tempe, AZ 85281, USA
[3] SenSIP Center, Tempe, USA

Abstract. Gaussian Processes (GP) are a powerful framework for modeling expensive black-box functions and have thus been adopted for various challenging modeling and optimization problems. In GP-based modeling, we typically default to a stationary covariance kernel to model the underlying function over the input domain, but many real-world applications, such as controls and cyber-physical system safety, often require modeling and optimization of functions that are locally stationary and globally non-stationary across the domain; using standard GPs with a stationary kernel often yields poor modeling performance in such scenarios. In this paper, we propose a novel modeling technique called Class-GP (Class Gaussian Process) to model a class of heterogeneous functions, i.e., non-stationary functions which can be divided into locally stationary functions over the partitions of input space with one active stationary function in each partition. We provide theoretical insights into the modeling power of Class-GP and demonstrate its benefits over standard modeling techniques via extensive empirical evaluations.

Keywords: Gaussian process · Black-box modeling · Heterogeneous function · Non-stationary function modeling · Optimization

1 Introduction

Many modern day science and engineering applications, such as machine learning, hyperparameter optimization of neural networks, robotics, cyber-physical systems, etc., call for modeling techniques to model black-box functions. Gaussian Process (GP) modeling is a popular Bayesian non-parametric framework heavily employed to model expensive black-box functions for analysis such as prediction or optimization [20]. Traditionally, GP models assume stationarity of the underlying, unknown, function. As a result a unique covariance kernel (with constant hyperparameters) can be used over the entire domain. However, many real-world systems such as cyber-physical, natural, recommendation can only be characterized by locally stationary but globally non stationary functions. Breaking the assumption underlying the stationary kernel can deteriorate the quality of predictions generated by the GP.

M. Sellmann and K. Tierney (Eds.): LION 2023, LNCS 14286, pp. 408–423, 2023.
https://doi.org/10.1007/978-3-031-44505-7_28

Many studies in the literature tackle this problem. We can classify these studies in to three categories:

1. *Locally stationary and partition based approaches*: The work by Gramacy et al., [6] is one of the first ones to tackle the modeling of heterogeneous functions by partitioning the input space with tree-based algorithms and using independent stationary GPs to model the underlying function. Kim et al., [10] and Pope et al., [19] propose Voronoi tessellations based partitioning of the input space. Candelieri et al., [2] extends the work [6] by overcoming the modeling limitation of axis aligned partitions by using a support vector machine (SVM) based classifiers at each node of the tree. Fuentes et al., [4] uses an alternative kernel which is modeled as a convolution of fixed kernel with independent stationary kernel whose parameters vary over the sub regions of the space.

2. *GP's with non-stationary kernels*: The studies [5,17,18] use non-stationary kernels to model the heterogeneous function, [8] uses non stationary kernels with a input dependent kernel parameters and further models the parameter functions with a smooth stationary Gaussian process. However, the use of non stationary kernel makes these methods complex and computationally intractable.

3. *Warping or spatial deformation based methods*: Methods in [15,21] map the original input space to a new space where the process is assumed to be stationary, [3] uses composite GP to warp the input space.

For many engineering systems, the structure of the non-stationariety is known or can be evaluated. As an example, vehicle automatic transmission will exhibit switching behaviors, with a discrete and finite number of switches (gears changes). When a specific behavior (gear) is exercised, the system exhibits smooth state dynamics and the metrics associated to the system that we are interested in monitoring/predicting maintain such smoothness. The work [16], has considered the case of identifying unsafe system-level behaviors for Cyber-Physical Systems without exploiting any information about, for example, switching dynamics.

In this paper, we make a first step in the direction of improving analysis of systems that are characterized by a discrete and finite number of "classes" of behaviors. Notice that a single class may be represented by disconnected sets of the input space. In particular, given an input, we assume the class and the closest class that the system can switch to, can be both evaluated. Under this scenario, we extend the existing partition based methods to a family of heterogeneous functions adding the class information. We model the homogeneous behavior classes by imposing that the GPs learnt within the subregions of the input space that belong to the same class have same kernel hyperparameters. These functions are often encountered in many real world applications, where we can access the class information by learning a classifier using the features. To the best of our knowledge, we present a first tree-based algorithm with information sharing across non-contiguous partitions that belong to same homogeneous behaviour class to better learn the GP models. Our contributions include:

- A novel Class GP framework to model heterogeneous function with class information.

- Theoretical analysis - we compute uniform error bounds for our framework and compare it with the error bounds achieved by standard GP.
- Empirical analysis - we provide extensive empirical evaluations of the model and compare it with other modeling techniques.

The rest of the paper is organized as follows: Sect. 2 gives a formal introduction to the problem and notations used in the paper followed by a brief overview of Gaussian process modeling and classification tree algorithm in Sect. 3 and introducing the Class GP framework in Sect. 4. Section 5 provides theoretical insights for Class-GP algorithm followed by details of experimental setup and corresponding results in Sect. 6. Section 7 gives conclusion over the performance of Class-GP as compared to other methods and insights on future work. Finally, paper ends with an appendix in Sect. A.

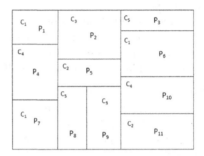

Fig. 1. Class-partition space with axis aligned partitions (P_j) and classes (C_i)

2   Problem Setup and Notation

Let $\mathcal{X} \subseteq \mathbb{R}^d$ be a compact space with p axis aligned partitions $\{\mathcal{X}_j\}_{j=1}^p$ and each partition $j \in [p]$ is assigned a class label $i \in [m]$ i.e., $\mathbb{C}(j) = i$, we call this space as *class-partition* space. This paper, models a family of non-stationary functions f defined over class-partition space, $f : \mathcal{X} \to \mathbb{R}$, such that f boils down to a stationary functions g_j's over each partition $j \in [p]$ where each g_j's are sampled from a Gaussian process with a continuous stationary kernel κ_j i.e., $g_j \sim \mathcal{GP}(\mu_j(.), \kappa_j(.,.))$. For notational convenience, and *w.l.o.g*, we assume the prior mean $\mu_j = 0$. Further, partitions j_1, j_2 that belong to same class i have same covariance kernel i.e., $\kappa_{j_1} = \kappa_{j_2} = \kappa_i$. Let $\mathcal{C}_i$ denote all the partitions with class label i, i.e., $\mathcal{C}_i = \cup_{\{j:\mathbb{C}(j)=i\}}\mathcal{X}_j$, this can be visualized with the help of an example as shown in Fig. 1. The function f is formally given as follows:

$$f(\mathbf{x}) = \sum_{j=1}^{p} \mathbb{1}\{\mathbf{x} \in \mathcal{X}_j\}g_j(\mathbf{x}) \tag{1}$$

Note: For consistency we denote partitions with a subscript j and classes with subscript i, owing to this notation any variable with subscript i or j would refer to class or partition variable respectively.

2.1   Observation Model

Evaluating the function at any point $\mathbf{x}$ in the input space reveals the following information: function evaluation y, the class label z of the partition to which the point belongs and the tuple distance $w =$ (distance from closest boundary, feature index). We denote that training data set $\mathcal{D} = \{\mathbf{x}_n, y_n, z_n, w_n\}_{n=1}^{N}$ where N is number of training data points. Also, $\mathbf{X} = [\mathbf{x}_1, \dots, \mathbf{x}_n]^T$, $\mathbf{y}, \mathbf{z}$ are the vector of corresponding evaluations, classes respectively and $\mathbf{W}$ is a list of tuples of distance and feature index along which the distance is measured.

3   Background

This section gives a brief overview of Gaussian Process modeling and the classification tree algorithm used in the Class-GP framework.

3.1   Gaussian Process Modeling

Gaussian process (GP) modeling is a popular statistical framework to model non-linear black box functions f due to its analytical tractability of posteriors. With in this framework the function, $f : \mathcal{X} \subseteq \mathbb{R}^d \to \mathbb{R}$, being modeled is assumed the to be a distributed as per a Gaussian process prior, formally written as follows:

$$f \sim \mathcal{GP}(\mu(.), \kappa(., .)),$$

GP is completely given by its mean $\mu(.)$ and covariance kernel $\kappa(., .)$, where for convenience, and without loss of generality, the mean function $\mu(.)$ is set to 0. The choice of the covariance kernel depends on the degree of smoothness warranted by the function being modeled and is defaulted to stationary kernels such as squared exponential (SE) kernel or Matérn kernel. Functions modeled within this framework are typically assumed to be *stationary* i.e., function can be modeled using a same stationary covariance function over the entire input space.

Learning a GP model involves computing the posterior conditioned on the observed data and learning kernel hyperparameters. Let $\mathcal{D}_n$: $\{(\mathbf{x}_1, y_1) \dots (\mathbf{x}_n, y_n)\}$ be the sampled training data of the objective function f. The posterior of the function f conditioned on the training data $\mathcal{D}_n$ is given by a Gaussian distribution i.e., $f(\mathbf{x}) | \mathcal{D}_n \sim \mathcal{N}(\mu_n(\mathbf{x}), \sigma_n^2(\mathbf{x}))$, where the mean $\mu_n(\mathbf{x})$ and covariance $\sigma_n^2(\mathbf{x})$ are given as follows:

$$\mu_n(\mathbf{x}) = k^T K^{-1} \mathbf{y} \quad \text{and} \quad \sigma_n^2(\mathbf{x}) = \kappa(\mathbf{x}, \mathbf{x}) - k^T K^{-1} k \tag{2}$$

Here, $\mathbf{y}$ is the vector of noise free observations, k is a vector with $k_p = \kappa(\mathbf{x}, \mathbf{x}_p)$. The matrix K is such that $K_{p,q} = \kappa(\mathbf{x}_p, \mathbf{x}_q)$ $p, q \in \{1, \dots, n\}$.

The hyperparameters of the model are learnt by maximising the log marginal likelihood which is given as follows:

$$\log p(\mathbf{y}|\mathbf{X}, \theta) = -\frac{1}{2}\mathbf{y}^T K^{-1} \mathbf{y} - \frac{1}{2} \log |K| - \frac{n}{2} \log 2\pi \tag{3}$$

and $\theta* = \arg\max_\theta \log p(\mathbf{y}|\mathbf{X}, \theta)$, this optimization problem is solved using off the shelf non convex optimization packages such as Dividing Rectangles (DIRECT) [9], LBGFS [12], CMA-ES [7]. For a detailed treatment of Gaussian process modeling we refer readers to [20] and [22].

3.2   Classification Tree Algorithm

A classification tree/decision tree classifier is a binary tree algorithm which yield axis aligned partitions of the input space by recursively partitioning the input space on one dimension (feature) at a time. The tree is learnt from the training data and the predictions for any input $\mathbf{x}$ is given by traversing the learnt tree from root node to a leaf node. Notice that each leaf node corresponds to a partition of the input space. We use CART algorithm to grow/learn the tree. During training at each the goal is to select the best feature and splitting threshold that minimizes the weighted impurity metric of the children nodes. Most of the tree based algorithms typically use Gini index as the impurity metric to grow the tree, which is given as follows:

$$Gini\ index = 1 - \sum_{i=1}^{n}(p_i)^2 \tag{4}$$

where p_i is the probability of a given class at a given node. The recursion is continued until one of the stopping criterion's is met or no further improvement in the impurity index is achievable. Typical choice of stopping criterion's include maximum depth of the tree, minimum number of samples in a leaf, maximum number of leaf/partitions. For more detailed overview on classification tree please refer the work by [1] and [13].

4   Class-GP Framework

In this section, we introduce a framework to model the family of heterogeneous functions as defined in Sect. 2. Within this framework we solve two sub-problems: 1. Learning partitions of the input space using closest boundary information $\mathbf{W}$ along with class information $\mathbf{z}$ and, 2. Training a Gaussian Process within each partition such that GP's of the partitions that share same class label learn the same set of hyperparameters. Current framework considers both noise less and noisy function evaluations. Further, this framework can also be extended to other partitioning methods that can use closest boundary information.

4.1   Learning Partitions

To learn the partitions of the input space we use decision tree algorithm tailored for the current framework. The algorithm learns the tree is 2 steps: While in both the steps we use recursive binary splitting to grow the tree, in the first step we use closest boundary information to find the best feature index and splitting threshold that maximize reduction of Gini index (or any other impurity metric) until all the closest boundary information $\mathbf{W}$ is exhausted, in the second step the best feature index and splitting threshold that maximize reduction of Impurity metric (Gini index) are selected from available training data at each node as in the CART algorithm Sect. 3.2. The nodes are recursively split until a stopping criterion is satisfied. In our proposed framework we default to Gini index as impurity metric and, minimum number of samples at the leaf node and max depth are used as stopping criterion's.

4.2 Gaussian Process in Each Partition

The partitions of the input space learnt from the decision tree algorithm Sect. 4.1 is used to divide the training data set $\mathcal{D}$ into partition based subsets $\mathcal{D}_j$ with n_j data points for all $j \in [p]$. For each partition $\mathcal{X}_j$ underlying stationary function g_j is modeled using a zero mean Gaussian Process prior with continuous stationary kernel $\kappa_j(.,.)$ and subset of training data $\mathcal{D}_j$ is used to learn/train the model. The function modeling and training in each partition is similar to that of a standard Gaussian process regression problem with one exception of learning the hyperparameters of the partition GPs. The posterior of partition GP conditioned on the partition training data is given by $y(\mathbf{x})|\mathbf{x}, j, \mathcal{D}_j \sim \mathcal{N}(\mu_{j,n_j}(\mathbf{x}), \sigma^2_{j,n_j}(\mathbf{x}))$ where mean $\mu_{j,n_j} = \mathbb{E}(y|\mathbf{x}, j, \mathcal{D}_j)$ and variance σ^2_{j,n_j} are given as follows:

$$\mu_{j,n_j}(\mathbf{x}) = \mathbf{k}_j^T K_j^{-1} \mathbf{y}_j \quad \text{and} \quad \sigma^2_{j,n_j}(\mathbf{x}) = \kappa_j(\mathbf{x}, \mathbf{x}) - \mathbf{k}_j^T K_j^{-1} \mathbf{k}_j$$

where $\mathbf{y}_j$ is the vector of observations in given partition j, $\mathbf{k}_j$ is the vector with $k_j^{(s)} = \kappa_j(\mathbf{x}, \mathbf{x}_s)$. The matrix K_j is such that $K_j^{(s,t)} = \kappa_j(\mathbf{x}_s, \mathbf{x}_t)$ where $s, t \in \{1, \ldots, n_j\}$. Note the superscripts here represent the components of the vector and matrix.

Learning hyperparameters in a standard GP typically involves finding the set of hyperparameters that maximize the log marginal likelihood of the observations 3, where as in the current framework we are required to find the set of hyperparameters that maximizes the log marginal likelihood across all the partition within a class. We propose a novel method to learn of the hyperparameters. A new class likelihood is formed by summing the log marginal likelihoods of all partition GPs for a given class and class kernel hyperparameters are learnt by maximizing new likelihood. The formulation of the class-likelihood function assumes that the data from different partitions are independent of each other and this reduces the computational complexity of learning the hyperparameters significantly while still taking data from all the partitions in the class.

The extensive empirical results show that this new class likelihood reduces the modelling error and provides with the better estimates of the hyperparameters, intuitively this makes sense as we have more data points to estimate hyperparameters even though all the data points do not belong to the same partition. The new class likelihood function for a given class i is given as follows:

$$\mathcal{L}_i(\mathbf{y}_i|\mathbf{X}_i, \theta_i) = \sum_{\{j:\mathbb{C}(j)=i\}} \log p(\mathbf{y}_j|\mathbf{X}_j, \theta_i) = -\frac{1}{2}\mathbf{y}_i^T K_i^{-1} \mathbf{y}_i - \frac{1}{2}\log|K_i| - \frac{n_i}{2}\log 2\pi$$

where $\theta_i^* = \arg\max_{\theta_i} \mathcal{L}_i(\mathbf{y}_i | \mathbf{X}_i, \theta_i)$, K_i is the block diagonal matrix with blocks of K_j's for all $\{j : \mathbb{C}(j) = i\}$, $n_i = \sum_{\{j:\mathbb{C}(j)=i\}} n_j$, and $\mathbf{y}_i$'s is the vector formed by $\mathbf{y}_j$ for all $\{j : \mathbb{C}(j) = i\}$.

5   Class-GP Analysis

In this section, we provide formal statement for probabilistic uniform error bounds for Class GP framework, the results are the extension of the results from [11]. To state our theorem we first introduce the required assumptions on the unknown function $f = \sum_{j=1}^{p} \mathbb{1}\{x \in \mathcal{X}_j\} g_j$ over the input space with p partitions.

A0: g_j 's in each partition are continuous with Lipschitz constant L_{g_j}.

A1: g_j 's in each partition are sampled from a zero mean Gaussian process with known continuous kernel function κ_j on the compact set $\mathcal{X}_j$.

A2: Kernels κ_j 's are Lipschitz continuous with Lipschitz constant L_{κ_j}

Theorem 1. Consider an unknown function $f : \mathcal{X} \to \mathbb{R}$ which induces p partitions on the input space, and is given as $f = \sum_{j=1}^{p} \mathbb{1}\{x \in \mathcal{X}_j\} g_j$ obeying **A0:,A1:** and **A2:**. Given $n_j \in \mathbb{N}$ noisy observations $\mathbf{y}_j$ with i.i.d zero mean Gaussian noise in a given partition $j \in [p]$ the posterior mean (μ_{n_j}) and standard deviation (σ_{n_j}) of the Gaussian Process conditioned on $\mathcal{D}_j = \{\mathbf{X}_j, \mathbf{y}_j\}$ of the partition are continuous with Lipschitz constant $L_{\mu_{n_j}}$ and modulus of continuity $\omega_{\sigma_{n_j}}$ on $\mathcal{X}_j$ such that

$$L_{\mu_{n_j}} \leq L_{\kappa_j} \sqrt{n_j} \|\widehat{\mathbf{K}}_j^{-1} \mathbf{y}_j\| \tag{5}$$

$$\omega_{\sigma_{n_j}}(r) \leq \sqrt{2r L_{\kappa_j} \left(1 + n_j \|\widehat{\mathbf{K}}_j^{-1}\| \max_{\mathbf{x}, \mathbf{x}' \in \mathcal{X}_j} \kappa_j(\mathbf{x}, \mathbf{x}')\right)} \tag{6}$$

where $\widehat{\mathbf{K}}_j = (\mathbf{K}_j + \eta^2 \mathbf{I}_{n_j})$.
Moreover, pick $\delta_j \in (0, 1), r \in \mathbb{R}_+$ and set

$$\beta_j(r) = 2\log\left(\frac{M(r, \mathcal{X}_j)}{\delta_j}\right) \tag{7}$$

$$\gamma_j(r) = (L_{\mu_{n_j}} + L_{g_j})r + \sqrt{\beta(r)}\omega_{\sigma_{n_j}} \tag{8}$$

then the following bound on each partition holds with probability $1 - \delta_j$

$$\left|g_j(\mathbf{x}) - \mu_{n_j}(\mathbf{x})\right| \leq \sqrt{\beta_j(r)}\sigma_{n_j}(\mathbf{x}) + \gamma_j(r), \forall\, \mathbf{x} \in \mathcal{X}_j \tag{9}$$

and the following bound on the entire input space holds with probability $1 - \delta$ where $\delta = \sum_{j=1}^{p} \mathbb{1}\{x \in \mathcal{X}_j\}\delta_j$ i.e.,

$$|f(\mathbf{x}) - \mu_n(\mathbf{x})| \leq \sqrt{\beta(r)}\sigma_n(\mathbf{x}) + \gamma(r), \forall\, \mathbf{x} \in \mathcal{X} \tag{10}$$

Corollary 1. *Given problem setup defined in the Theorem 1 the following bound on L_1 norm holds with probability $1 - \delta$*

$$\|f - \mu_n\|_1 \leq \zeta r^d \sum_{j=1}^{p} M(r, \mathcal{X}_j) \left(\sqrt{\beta_j(r)} \sigma_{n_j}(\mathbf{x}) + \gamma_j(r) \right) \tag{11}$$

where ζ is a non negative constant, $\delta = \sum_{j=1}^{1} \delta_j$ and $\delta_j = 1 - M(r, \mathcal{X}_j) e^{-\beta_j(r)/2}$.

6   Numerical Results

In this section, we compare the performance of Class-GP framework with other baselines Partition-GP and Standard-GP over the extensive empirical evaluations on both noisy and noiseless simulated dataset. Performance of the models is evaluated using the mean square error (MSE) metric. Brief overview of the baseline models is given below:

Standard GP: In this framework, we use a single GP with continuous stationary kernel across the entire space to model the underlying function.

Partition GP: This framework is similar to that of Treed Gaussian process framework [6] with additional class information. We learn the partitions of the input space using the class information followed by modeling the function in each partition individually, i.e., the hyperparameters in each partition are learnt independently of other partition data of the same class.

6.1   Synthetic Data and Experimental Setup

Synthetic data for all the experiments is uniformly sampled from input space $\mathcal{X} = [-10, 10]^d$ where, d is the dimension of the input space. Partitions p are generated by equally dividing the input space and each partition $j \in [p]$ is assigned a class label $i \in [m]$ to forms a checkered pattern. We use Gini index as the node splitting criterion to learn the tree and, squared exponential (SE) covariance kernel for all GPs to model the underlying function in each learnt partition. Each model is evaluated and compared across different functions as given below:

1. Harmonic function:

$$f(\mathbf{x}) = \sum_{j=1}^{p} \mathbb{1}\{\mathbf{x} \in \mathcal{X}_j\}(\cos \mathbf{x}^T \omega_{\mathbb{C}(j)} + b_j)$$

2. Modified Levy function (only for d = 2):

$$f(\mathbf{x}) = \sum_{j=1}^{p} \mathbb{1}\{\mathbf{x} \in \mathcal{X}_j\} \left(\sin^2(\pi v_1) + \sum_{k=1}^{d-1} (v_k - 1)^2 \left(1 + 10 \sin^2(\pi v_k + 1) \right) \right.$$

$$\left. + (v_d - 1)^2 (1 + \sin^2(2\pi v_d)) + b_j \right), \quad \text{where} \quad v_k = \left(1 + \frac{x_k - 1}{4} \right) \omega_{\mathbb{C}(j),k}$$

3. Modified Qing function (only for d = 2):

$$f(\mathbf{x}) = \sum_{j=1}^{p} \mathbb{1}\{\mathbf{x} \in \mathcal{X}_j\} \left(\sum_{k=1}^{d} \left((x_k \omega_{\mathbb{C}(j),k})^2 - i \right)^2 + b_j \right)$$

4. Modified Rosenbrock function (only for d = 2):

$$f(\mathbf{x}) = \sum_{j=1}^{p} \mathbb{1}\{\mathbf{x} \in \mathcal{X}_j\} \left(\sum_{k=1}^{d-1} [100 \left(v_{k+1} - v_k^2 \right)^2 + (1 - v_k)^2] + b_j \right)$$

here $v_k = x_k \omega_{\mathbb{C}(j),k}$.

where the frequency vector $\omega_i = i * \omega$ depends on the class i, $\omega_{\mathbb{C}(j),k}$ is the kth component. Further, vector ω is sampled from a normal distribution, constant b_j (intercepts) depends on the partition j and each b_j is sampled from a normal distribution.

Following parameters are initialized for each simulation: dimension d, training budget N, number of partitions p, number of classes m, frequency vector ω, Constant (intercept) vector $\mathbf{b}$, maximum depth of the tree, minimum samples per node, initial kernel hyperparameters θ. For a fixed set of initialized parameters, 50 independent simulation runs for each baseline are performed to compute a confidence interval over the model performance, training data is re-sampled for each run.

To analyze the effects and compare the model performance with baselines each parameter i.e., number of partitions (p), number of classes (m), training

Table 1. Parameter initialization across simulations

Parameters	Values
Number of Classes (c)	2, 4, 6, 8
Number of Partitions (p)	4, 16, 36, 64
Training Budget (N)	50, 500, 1000
Dimension (d)	2, 3

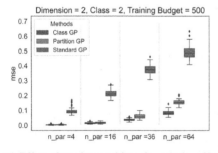 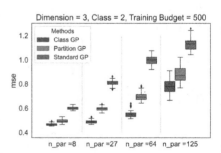

(a) Effect of varying partitions for noiseless 2D harmonic function evaluation.   (b) Effect of varying partitions for noiseless 3D harmonic function evaluation.

Fig. 2. Effect of the number of partitions.

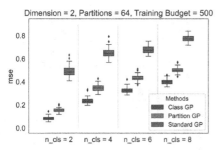 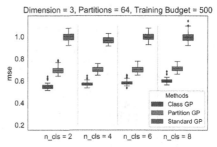

(a) Effect of varying classes for noiseless 2D harmonic function evaluation.

(b) Effect of varying classes for noiseless 3D harmonic function evaluation.

Fig. 3. Effect of number of classes

budget (N) and, dimension (d), is varied while keeping the others constant. Various initialization of the parameters for the simulations are shown in the Table 1. The performance measure metric (MSE) for each model across all the simulations is evaluated on uniformly sampled test data set of fixed size i.e., 5000 data points.

Effects of each parameter on the model performance is analyzed below:

1. Effect of number of partitions (p): For a fixed set of initial parameters as the number of partitions (p) is increased the performance of all the models deteriorates as seen in Fig. 2(a) and 2(b). Notice that, the performance of Class-GP is superior or at least as good as Partition-GP owing to the fact that, while keeping the training budget constant and increasing the number of partitions leads to decrease in number of points per partition available to learn the hyperparameters of each GP independently in each partition for Partition-GP, whereas, since the number of classes (m) remain constant, number of points available to learn hyperparameters of GPs remain same for Class-GP because of the new likelihood function. The only information we lose on is the correlation information which leads to the small deterioration in the model performance. Also, when the training budget (N) is small or close to number of partitions (p), Standard-GP outperforms both class-GP and Partition-GP. This due to the insufficient data to learn all the partitions of the input space leading to sharp rise in MSE of Class-GP and Partition-GP.

2. Effect of number of classes (m): Increasing the number of classes (m) while keeping other parameters fixed does not effect the performance of the models when the training budget (N) is significantly high because of the large training data available to learn the underlying model, whereas when the training budget is moderate the reduction in the performance of all the models is observed, as seen in the Fig. 3(a) and 3(b), with the increasing classes, while keeping to the trend of performance between modeling methods. This is observed due to following reasons: For Class-GP with the increasing classes number of data points to learn the hyperparameters decreases resulting in reduction in the performance, where as for Partition-BO even though the number of

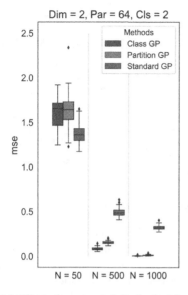
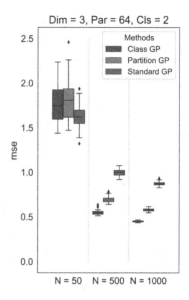

(a) Effect of varying training budget for noiseless 2D harmonic function.

(b) Effect of varying training budget for noiseless 3D harmonic function.

Fig. 4. Effect of training budget

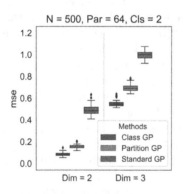
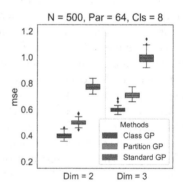

(a) Effect of varying dimension for noiseless harmonic function with 2 classes.

(b) Effect of varying dimension for noiseless harmonic function with 2 classes.

Fig. 5. Effect of dimension

data points per partition remain same we observe reduction in performance due to the fact that modeling of high frequency functions (which increase as the number of classes increase) require larger data points and, whereas for the Standard-GP the reduction in performance is because more functions are being modeled with a single GP.

3. Effect of Training Budget (N): Increasing the training budget N has an obvious effect of improvement in the performance of models as seen in Fig. 4(a) and 4(b) owing to the fact that GP's learns the model better with more training data points, but the drawback of increasing the data points is the computational complexity of the model increases. Also, Notice that the gain in performance of Standard-GP is not as significant as Class-GP or Partition-GP because single GP is used to model the heterogeneous function.

4. Effect of Dimension (d): An increase in the problem's dimensionality increases the number of data points required to model the underlying function. This is also observed in the performance of the models as shown in Fig. 5(a), and 5(b) i.e., with the increase in the number of dimensions, model performance decreases, while the other parameters are fixed.

We also evaluate the model's performance over noisy data sets for various parameter initialization, but due to the constraint of space, we only display a subset of the results in the tabular format. Table 2 shows each model's average MSE (not normalized) over 50 independent runs. It can be observed that when the number of partitions is low, the performance of Class GP is as good as Partition GP, whereas when the partitions increase, Class GP outperforms other methods. The full code used to perform simulations can be found at following Github repository.

Table 2. Model performance comparison in presence of noise

Parameters				Average MSE over 50 runs		
Functions	Training Budget	Classes	Partitions	Class GP	Partition GP	Standard GP
Harmonic	500	6	4	1.35	1.37	1.51
			64	1.68	1.9	1.85
		8	4	1.35	1.37	1.53
			64	1.7	1.95	1.95
Levy	500	6	4	2.97 e2	2.97 e2	5.72 e2
			64	1.15 e3	1.44 e3	1.56 e3
		8	4	2.97 e2	2.97 e2	5.83 e2
			64	3.36 e3	3.38 e3	4.02 e3
Qing	500	6	4	1.58 e3	1.58 e3	8.45 e7
			64	1.79 e9	2.74 e9	9.45 e10
		8	4	1.58 e3	1.58 e3	8.45 e7
			64	3.25 e10	5.24 e10	8.87 e11
Rosenbrock	500	6	4	1.51 e5	1.52 e5	1.85 e10
			64	9.28 e11	9.42 e11	2.24 e13
		8	4	1.51 e5	1.52 e5	1.85 e10
			64	6.13 e12	1.45 e13	1.97 e14

7   Conclusion and Future Work

This paper presents a novel tree based Class GP framework which extends the existing partition based methods to a family of heterogeneous functions with access to class information. We introduced a new likelihood function to exploit the homogeneous behaviour of the partitions that belong to same class, leading to enhanced the performance of GP models allowing to learn the hyperparameters across the entire class instead of individual partitions. Furthermore, we establish a tailored tree algorithm suitable for current framework that uses the closest boundary information to learn the initial tree. We also provide some theoretical results in terms of the probabilistic uniform error bounds and bounds on L_1 norm. Finally, we conclude with extensive empirical analysis and clearly show the superior performance of Class GP as compared other baselines. Extension of the Class GP modeling framework to optimization, scaling to higher dimensions [14] and extensive theoretical analysis of the algorithm with practical error bounds are some promising venue to be explored in the future work.

Acknowledgements. This work is supported in part by National Science Foundation (NSF) under the awards 2200161, 2048223, 2003111, 2046588, 2134256, 1815361, 2031799, 2205080, 1901243, 1540040, 2003111, 2048223, by DARPA ARCOS program under contract FA8750-20-C-0507, Lockheed Martin funded contract FA8750-22-9-0001, and the SenSIP Center.

A   Appendix

Proof sketch for the Theorem 1 follows along the lines of the proof of Theorem 3.1 in [11]. We get probabilistic uniform error bounds for GPs in each partitions $j \in [p]$ from [11] and we use per partition based bounds to bound the over all function and to derive bound on L_1 norm. The proof for the theorem and corollary given as follows:

Proof. 1 Following bounds on each partition holds with probability $1 - \delta_j$

$$\left| g_j(\mathbf{x}) - \mu_{n_j}(\mathbf{x}) \right| \le \sqrt{\beta_j(r)} \sigma_{n_j}(\mathbf{x}) + \gamma_j(r), \forall\, \mathbf{x} \in \mathcal{X}_j \tag{12}$$

where $\beta_j(r)$ and $\gamma_j(r)$ are given as follows

$$\beta_j(r) = 2 \log \left(\frac{M(r, \mathcal{X}_j)}{\delta_j} \right) \tag{13}$$

$$\gamma_j(r) = (L_{\mu_{n_j}} + L_{g_j}) r + \sqrt{\beta(r)} \omega_{\sigma_{n_j}} \tag{14}$$

Now to bound the entire function lets look at the difference $|f(\mathbf{x}) - \mu_n(\mathbf{x})|$.

$$|f(\mathbf{x}) - \mu_n(\mathbf{x})| = \left| \sum_{j=1}^{p} \mathbb{1}\{x \in \mathcal{X}_j\}(g_j(\mathbf{x}) - \mu_{n_j}(\mathbf{x})) \right| \tag{15}$$

$$= \sum_{j=1}^{p} \mathbb{1}\{x \in \mathcal{X}_j\} \left| g_j(\mathbf{x}) - \mu_{n_j}(\mathbf{x})) \right| \tag{16}$$

$$\leq \sum_{j=1}^{p} \mathbb{1}\{x \in \mathcal{X}_j\} \left(\sqrt{\beta_j(r)}\sigma_{n_j}(\mathbf{x}) + \gamma_j(r) \right), \forall \, \mathbf{x} \in \mathcal{X}_j \tag{17}$$

The last inequality (17) follows from (12) and holds with probability $1-\delta$, where $\delta = \sum_{j=1}^{p} \mathbb{1}\{x \in \mathcal{X}_j\}\delta_j$.

Now, redefining $\sum_{j=1}^{p} \mathbb{1}\{x \in \mathcal{X}_j\} \left(\sqrt{\beta_j(r)}\sigma_{n_j}(\mathbf{x}) \right) = \sqrt{\beta(r)}\sigma_n(\mathbf{x})$ and $\sum_{j=1}^{p} \mathbb{1}\{x \in \mathcal{X}_j\}\gamma_j(r) = \gamma(r)$, we have the result. □

The proof for the Corollary 1 uses the high confidence bound 10 and is given as follows:

Proof. We know that L_1 norm is given by

$$\|f(\mathbf{x}) - \mu_n(\mathbf{x})\|_1 = \mathrm{E}[|f(\mathbf{x}) - \mu_n(\mathbf{x})|] \tag{18}$$

$$= \int |f(\mathbf{x}) - \mu_n(\mathbf{x})|\, d\mu \tag{19}$$

$$= \int \left| \sum_{j=1}^{p} \mathbb{1}\{x \in \mathcal{X}_j\}(g_j(\mathbf{x}) - \mu_{n_j}(\mathbf{x})) \right| d\mu \tag{20}$$

$$= \sum_{j=1}^{p} \int \mathbb{1}\{x \in \mathcal{X}_j\} \left| (g_j(\mathbf{x}) - \mu_{n_j}(\mathbf{x})) \right| d\mu \tag{21}$$

$$= \sum_{j=1}^{p} \int_{\mathcal{X}_j} \left| (g_j(\mathbf{x}) - \mu_{n_j}(\mathbf{x})) \right| d\mu \tag{22}$$

$$\leq \zeta r^d \sum_{j=1}^{p} M(r, \mathcal{X}_j) \left(\sqrt{\beta_j(r)}\sigma_{n_j}(\mathbf{x}) + \gamma_j(r) \right) \quad \text{holds w.p } 1 - \delta \tag{23}$$

where $\delta = \sum_{j=1}^{p} \delta_j$ and $\delta_j = 1 - M(r, \mathcal{X}_j)\exp(-\beta_j(r)/2)$. □

References

1. Breiman, L., Friedman, J.H., Olshen, R.A., Stone, C.J.: Classification and Regression Trees. Routledge, Milton Park (2017)

2. Candelieri, A., Pedrielli, G.: Treed-gaussian processes with support vector machines as nodes for nonstationary Bayesian optimization. In: 2021 Winter Simulation Conference (WSC), pp. 1–12. IEEE (2021)
3. Davis, C.B., Hans, C.M., Santner, T.J.: Prediction of non-stationary response functions using a Bayesian composite gaussian process. Comput. Stat. Data Anal. **154**, 107083 (2021)
4. Fuentes, M., Smith, R.L.: A new class of nonstationary spatial models. Technical report, North Carolina State University, Department of Statistics (2001)
5. Gibbs, M.N.: Bayesian Gaussian processes for regression and classification. Ph.D. thesis, Citeseer (1998)
6. Gramacy, R.B., Lee, H.K.H.: Bayesian treed gaussian process models with an application to computer modeling. J. Am. Stat. Assoc. **103**(483), 1119–1130 (2008)
7. Hansen, N., Ostermeier, A.: Completely derandomized self-adaptation in evolution strategies. Evol. Comput. **9**(2), 159–195 (2001)
8. Heinonen, M., Mannerström, H., Rousu, J., Kaski, S., Lähdesmäki, H.: Nonstationary gaussian process regression with hamiltonian monte carlo. In: Gretton, A., Robert, C.C. (eds.) Proceedings of the 19th International Conference on Artificial Intelligence and Statistics. Proceedings of Machine Learning Research, Cadiz, Spain, vol. 51, pp. 732–740. PMLR (2016)
9. Jones, D.R., Perttunen, C.D., Stuckman, B.E.: Lipschitzian optimization without the lipschitz constant. J. Optim. Theory Appl. **79**(1), 157–181 (1993)
10. Kim, H.M., Mallick, B.K., Holmes, C.C.: Analyzing nonstationary spatial data using piecewise gaussian processes. J. Am. Stat. Assoc. **100**(470), 653–668 (2005)
11. Lederer, A., Umlauft, J., Hirche, S.: Uniform error bounds for gaussian process regression with application to safe control. In: Advances in Neural Information Processing Systems, vol. 32 (2019)
12. Liu, D.C., Nocedal, J.: On the limited memory BFGS method for large scale optimization. Math. Program. **45**(1), 503–528 (1989)
13. Loh, W.Y.: Classification and regression trees. Wiley Interdiscip. Rev. Data Mining Knowl. Discov. **1**(1), 14–23 (2011)
14. Malu, M., Dasarathy, G., Spanias, A.: Bayesian optimization in high-dimensional spaces: a brief survey. In: 2021 12th International Conference on Information, Intelligence, Systems & Applications (IISA), pp. 1–8. IEEE (2021)
15. Marmin, S., Ginsbourger, D., Baccou, J., Liandrat, J.: Warped gaussian processes and derivative-based sequential designs for functions with heterogeneous variations. SIAM/ASA J. Uncertain. Quantif. **6**(3), 991–1018 (2018)
16. Mathesen, L., Yaghoubi, S., Pedrielli, G., Fainekos, G.: Falsification of cyberphysical systems with robustness uncertainty quantification through stochastic optimization with adaptive restart. In: 2019 IEEE 15th International Conference on Automation Science and Engineering (CASE), pp. 991–997. IEEE (2019)
17. Paciorek, C.J., Schervish, M.J.: Spatial modelling using a new class of nonstationary covariance functions. Environmetrics Official J. Int. Environ. Soc. **17**(5), 483–506 (2006)
18. Paciorek, C.J.: Nonstationary Gaussian processes for regression and spatial modelling. Ph.D. thesis, Carnegie Mellon University (2003)
19. Pope, C.A., et al.: Gaussian process modeling of heterogeneity and discontinuities using voronoi tessellations. Technometrics **63**(1), 53–63 (2021)
20. Rasmussen, C.E.: Gaussian processes in machine learning. In: Bousquet, O., von Luxburg, U., Rätsch, G. (eds.) ML -2003. LNCS (LNAI), vol. 3176, pp. 63–71. Springer, Heidelberg (2004). https://doi.org/10.1007/978-3-540-28650-9_4

21. Schmidt, A.M., O'Hagan, A.: Bayesian inference for non-stationary spatial covariance structure via spatial deformations. J. Roy. Stat. Soc. Ser. B (Stat. Methodol.) **65**(3), 743–758 (2003)
22. Schulz, E., Speekenbrink, M., Krause, A.: A tutorial on gaussian process regression: modelling, exploring, and exploiting functions. J. Math. Psychol. **85**, 1–16 (2018)

Surrogate Membership for Inferred Metrics in Fairness Evaluation

Melinda Thielbar[1], Serdar Kadıoğlu[1,2]($\boxtimes$)(iD), Chenhui Zhang[1], Rick Pack[1], and Lukas Dannull[1]

[1] AI Center, Fidelity Investments, Boston, USA
{melinda.thielbar,serdar.kadioglu,chenhui.zhang,rick.pack,
lukas.dannull}@fmr.com
[2] Department of Computer Science, Brown, Providence, USA

Abstract. As artificial intelligence becomes more embedded into daily activities, it is imperative to ensure models perform well for all subgroups. This is particularly important when models include underprivileged populations. Binary fairness metrics, which compare model performance for protected groups to the rest of the model population, are an important way to guard against unwanted bias. However, a significant drawback of these binary fairness metrics is that they require protected group membership attributes. In many practical scenarios, protected status for individuals is sparse, unavailable, or even illegal to collect. This paper extends binary fairness metrics from deterministic membership attributes to their surrogate counterpart under the probabilistic setting. We show that it is possible to conduct binary fairness evaluation when exact protected attributes are not immediately available but their surrogate as likelihoods is accessible. Our inferred metrics calculated from surrogates are proved to be valid under standard statistical assumptions. Moreover, we do not require the surrogate variable to be strongly related to protected class membership; inferred metrics remain valid even when membership in the protected and unprotected groups is equally likely for many levels of the surrogate variable. Finally, we demonstrate the effectiveness of our approach using publicly available data from the Home Mortgage Disclosure Act and simulated benchmarks that mimic real-world conditions under different levels of model disparity.

Keywords: Fairness Metrics · Surrogate Modelling · Responsible AI

1 Introduction

Algorithmic decision-making has become ubiquitous, turning fairness evaluation into a crucial cornerstone of Responsible AI. Evaluating whether models are fair (equally performant) for different groups is the least that responsible science shall offer. For that purpose, various metrics that measure fairness in different ways have been proposed [5]. However, all of the standard fairness metrics require that group membership is known. Unfortunately, in many practical

M. Sellmann and K. Tierney (Eds.): LION 2023, LNCS 14286, pp. 424–442, 2023.
https://doi.org/10.1007/978-3-031-44505-7_29

scenarios this is not feasible. Fairness concerns are often about discrimination based on information many people consider private, e.g., race, religion, gender, and that are legally protected in domains including housing [23], credit systems [8], and human resources [24]. In practice, this information is often hard to obtain, limited, and possibly illegal to collect. This renders the most common fairness metrics invalid for many practical applications [2]. We argue that the lack of membership data should not exempt models from fairness evaluation. While immediate information may not be available, there is often some associated data to be gathered. Other researchers have proposed using this type of side information to predict class membership [6,12] and showed how to estimate boundaries for fairness based on this incomplete information [12,22]. All of the existing methods require a predictive model for protected class membership.

In this paper, we show how to utilize surrogate membership information for fairness evaluation *without attempting to predict class membership for individuals* thus, extending the scope of fairness metrics to scenarios where fairness testing was impossible before. More broadly, our approach offers a generalization of the existing work from the deterministic information to its probabilistic counterpart.

This paper also extends our previous work on surrogate ground truth generation [20] for the analogous scenario where ground truth labels are unavailable, yet another dependency of existing fairness metrics. In combination, our joint work addresses the two severe limitations of the existing binary fairness literature. We first alleviate the requirement for ground truth labels, as shown in [20], and in this paper, we address the need for the protected membership attribute.

2 Problem Definition

Let us start with a description of our setting that consists of the Probabilistic Membership Problem and then illustrate it with an example:

Definition 1 (Probabilistic Membership Problem (PMP)). *Consider a population, X, with individuals, $x \in X$, that is divided into two cohorts by a class membership attribute $A \in \{\top, \bot\}$ such that $\top$ and $\bot$ represent protected and unprotected membership, respectively. Let $X = X^\top \cup X^\bot$.*

For practical reasons, e.g., privacy concerns, the protected information of each individual remains unknown, i.e., $x \overset{?}{\in} \{X^\top, X^\bot\}$, but there exists a surrogate grouping so that membership in the surrogate group reveals the probability of being protected, i.e. $P_z(x \in X^\top) \; \forall x \in z$. Note that $P_z(x \in X^\top) = 1 - P_z(x \in X^\bot)$ and every individual belongs to exactly one surrogate group, $\exists! \, z \in Z \land x \in z, \forall x$.

Consider a binary classification model trained on historical data to make predictions about individuals, $\mathrm{ML}(x)$. We would like to evaluate this model against unwanted bias between $X^\top$ and $X^\bot$. Let m be a metric about this model, e.g., statistical parity, true positive rate, false positive rate, etc.

The goal of the Probabilistic Membership Problem (PMP) is to estimate the disparity in the model metric m between the protected and unprotected cohorts, *i.e.*, $m(X^\top) - m(X^\perp)$. *Contrary to its deterministic counterpart, in PMP, the protected attribute of individuals remains unknown. Instead, a probability* $P_z(x \in X^\top)$ *at the group level,* $\forall x \in z$ *and* $\forall z \in Z$, *is known.*

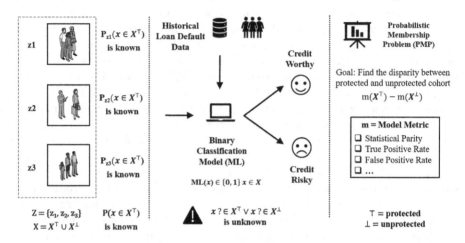

Fig. 1. Illustrative PMP example on Credit Loan Default prediction across a population X with protected, $X^\top$, and unprotected, $X^\perp$, cohorts. Further, there are 3 surrogate groups $Z = \{z_1, z_2, z_3\}$, e.g., zip codes. The probability of being in the protected cohort is known within each surrogate group. However, the protected attribute of each individual remains unknown. The goal of PMP is to find the model disparity between the protected and unprotected cohorts, $m(X^\top) - m(X^\perp)$, for a given model metric, m.

Figure 1 illustrates PMP using the Credit Loan Default prediction as an example. Let us demonstrate how $PMP\langle X, Z, P_z(x \in X^\top), m\rangle$ captures practical fairness scenarios in various domains.

This example considers the classical setting for predicting successful credit applications. For that purpose, a binary classification model is trained on the historical loan behavior of customers to predict who is credit-worthy in the future. There are two cohorts in the population X; protected, $X^\top$, and unprotected, $X^\perp$. The protected membership can be based on any attribute A such as gender, race, age, or marital status that are legally protected against discrimination. The model metric, m, measures the disparity of the machine learning model between these two cohorts.

Imagine $A \in \{\top, \perp\}$ denotes race as in white and non-white. As defined in PMP, we do not have access to such personal information of individuals, e.g., due to privacy constraints. The absence of confidential protected attributes is often the case in reality, and unfortunately, *all existing binary fairness evaluation metrics that require protected membership information becomes invalid in these cases* [2]. This is the gap we address in our paper.

The primary motivation behind PMP, and our paper, is that the absence of protected attributes should not jeopardize the evaluation of machine learning models against fairness metrics, here m, to surface potential unwanted bias.

As a remedy, we assume access to a surrogate variable, Z, e.g., the zip code of the population that provides the likelihood of protected membership, $P_z(x \in X^\top)$, at the *group level* for individuals in the same zip code area, $x \in z$. Here we have three zip codes where the probability of white and non-white cohorts is known, e.g., gathered from the publicly available Census data. The goal of PMP is to leverage this surrogate zip code information to find the model disparity $m(X^\top) - m(X^\perp)$ between white and non-white cohorts to conduct fairness evaluation.

To address PMP, in the next section (Sect. 3), we show that, if Z is available and the calculation for m can be expressed as an *arithmetic mean*, then we can infer the model metric disparity, i.e., $m(X^\top) - m(X^\perp)$, under standard statistical conditions. We call these estimates *inferred metrics* obtained from surrogate membership. Then, in (Sect. 4), we utilize inferred metrics for fairness evaluation. Without our proposed approach to infer these metrics, binary fairness evaluation would not be possible when protected membership is absent. This is the primary contribution of our paper as detailed next.

3   Solving PMP with Surrogate Membership

In the following, we describe our approach for calculating the inferred metrics for the PMP leveraging surrogate membership.

Let m be a model measure that can be expressed as an arithmetic mean, and let $m(X^\top) - m(X^\perp)$ be the model fairness disparity metric we would like to estimate. Then, by the linearity property of expectation [21], the model measure for each level of Z, denoted by m_z, can be approximated by a linear combination of the model measures for groups $X^\top$ and $X^\perp$ weighted by the population proportions of each group within z:

$$m_z = P_z(x \in X^\top)m(X^\top) + P_z(x \in X^\perp)m(X^\perp) \qquad (1)$$

In our toy example in Fig. 1, we assumed to know $P_z(x \in X^\top)$, $P_z(x \in X^\perp)$, and m_z without error, i.e., we measured the entire population. This would allow us to solve group-level metrics arithmetically using a system of equations.

In practice, we do not have access to the entire population; hence our model metrics cannot be exact. As a result, there will be some error within each m_z. Accordingly, we express m_z with an error term as:

$$m_z = P_z(x \in X^\top)m(X^\top) + P_z(x \in X^\perp)m(X^\perp) + e_z \qquad (2)$$

where each e_z remains unknown.

The addition of e_z means we can no longer solve Eq. 2 as a system of linear equations as in Eq. 1. Therefore, we need an optimization solution that will allow us to estimate $m(X^\top)$ and $m(X^\perp)$ with the minimum error. To achieve that, let us re-write the Eq. 2 into a form that lends itself to this kind of estimation.

Remember that we have two groups: protected, $\top$, and unprotected, $\perp$, and each individual is classified into exactly one group. Then $P_z(x \in X^\top) = 1 - P_z(x \in X^\perp)$, and we can re-write Eq. 2 as:

$$m_z = P_z(x \in X^\top)m(X^\top) + (1 - P_z(x \in X^\top))m(X^\perp) + e_z \qquad (3)$$

$$m_z = m(X^\perp) + (m(X^\top) - m(X^\perp))P_z(x \in X^\top) + e_z \qquad (4)$$

The critical insight behind our approach is to replace the unknown $m(X^\top)$ and $m(X^\perp)$ with parameters from Linear Regression:

$$m_z = \beta_0 + \beta_1 P_z(x \in X^\top) + e_z \qquad (5)$$

where $\beta_0 = m(X^\perp)$, and $\beta_1 = m(X^\top) - m(X^\perp)$.

With this transformation, notice how β_1 neatly captures the disparity of the model metric between the two cohorts.

For linear relationships as described in Eq. 5, the method of Ordinary Least Squares (OLS) is the standard estimation technique for β_0 and β_1 [4]. Under the following assumptions, the Gauss-Markov theorem states that ordinary least squares estimators for β_0 and β_1 are unbiased and have minimum variance:

1. The error terms e_z must have an expected value of zero given the value $P_z(x \in X^\top)$, i.e. $E[e_z|P_z(x \in X^\top)] = 0$. In our case, this condition is met by assuming m can be expressed as an arithmetic mean. This allows us to write m_z as a linear function of the population values of $m(X^\top)$, $m(X^\perp)$, and $P_z(x \in X^\top)$.
2. The error terms e_z must be iid. In our case, this assumption is met if we assume that $P(m_z, Z)$ and $P(A, Z)$ are independent draws from their respective marginal distribution, where $A \in \{\top, \perp\}$ is the unknown class membership. This assumption is a relaxed version of the independence assumption described in [18].
3. Ordinary least squares requires that the variance of e_z be constant for all z. In our case, this assumption is violated because the error of a mean varies with the number of observations unless we observe exactly the same number of individuals in each category z. Therefore, we relax the equal variances assumption by using an alternative estimator called Weighted Ordinary Least Squares (WOLS) [17]. The weight for each z is the number of observations for that level of Z. We denote this value as n_z. This method of WOLS and its properties were first proposed by [1], and we refer to [17] for a detailed description of WOLS for regression on means and the justification for using n_z as the weight.

To summarize, in this section, we made the connection between our metric m in PMP and the β parameters in WOLS. This connection allows us to leverage the WOLS estimator to infer the metrics we are interested in; precisely, $m(X^\top)$ and $m(X^\perp)$. Overall, this allows us capture the disparity in the model metric,

$m(X^\top) - m(X^\perp)$, between the protected and unprotected group for fairness evaluation. In the next section, we show how well-known fairness metrics can be neatly calculated given the inferred disparity metric.

4   From PMP to Fairness Evaluation

Many fairness metrics have been developed in the literature, see, e.g., [5] for an overview. In this paper, we consider the following standard metrics:

$$Statistical\ Parity = P(ML(x) = 1|x \in X^\top) - P(ML(x) = 1|x \in X^\perp)$$
$$Equal\ Opportunity = TPR_\top - TPR_\perp$$
$$Predictive\ Equality = FPR_\top - FPR_\perp$$
$$Average\ Odds = (Predictive\ Equality + Equal\ Opportunity)\ /\ 2$$

where TPR is the true positive rate, FPR is the False Positive Rate, and $ML(x)$ is the predicted class.

These are well-known fairness metrics that are commonly available in fairness packages such as IBM AI FAIRNESS 360 [3] and FIDELITY JURITY [20], among others. Considering statistics based on the TPR and FPR allows us to examine whether the inferred metrics are equally performant for fairness metrics calculated on different parts of the confusion matrix. Considering Average Odds shows that inferred metrics that are sums or differences of other inferred metrics and/or inferred metrics multiplied by constants are unbiased.

4.1   Fairness Metrics as Functions of Arithmetic Means

Our approach for solving PMP from (Sect. 3) requires inferred metrics be expressed as arithmetic means. Here we show that this holds for standard fairness metrics.

We first study Statistical Parity and make the observation that probabilities are estimated by summing the number of individuals who are classified into the positive case for each group, e.g.;

$$P(ML(x) = 1 \mid x \in X^\top) = \frac{1}{|X^\top|} \sum_{x \in X^\top} I(ML(x) = 1) \qquad (6)$$

This is the *arithmetic mean* of the indicator function $ML(x) = 1$ The probability of being predicted positive is, therefore, a suitable metric m that can be expressed as an arithmetic mean. Consequently, we can use surrogate membership for PMP to infer $m(X^\top)$ and $m(X^\perp)$. The observation in equation (6) that probabilities can be expressed as arithmetic means allows us to calculate the other fairness statistics in our list.

Next, we consider Equal Opportunity, which is the difference between the true positive rates for the protected and unprotected groups. The true positive rate is calculated as follows:

$$TPR = \frac{True\ Positives}{True\ Positives + False\ Negatives} = \frac{\sum I\big(ML(x) = 1 \wedge Y = 1\big)}{\sum I(Y = 1)}$$

where Y is the binary label for the model $Y \in \{0,1\}$ and I is the indicator function.

If we divide the numerator and denominator of this equation by N, the total number of individuals, the calculation is unchanged, but TPR becomes an expression based on probabilities:

$$TPR = \frac{\frac{1}{N}\sum I\big(ML(x) = 1 \wedge Y = 1\big)}{\frac{1}{N}\sum I((Y = 1)} = \frac{P(ML(x) = 1 \wedge Y = 1)}{P(Y = 1)}$$

As with statistical parity, each of these probabilities can be expressed as an arithmetic mean of an indicator function. That means we can calculate them as inferred metrics. This idea allows us to infer a host of fairness metrics that are based on the *confusion matrix* for a binary classifier.

4.2 Bootstrap Estimation

Even when fairness metrics can be expressed as arithmetic mean, one caveat still remains. As also pointed out in [16], while the expected values of the estimated probabilities are equal to their true population values, values that are functions of these probabilities are not guaranteed to have their true expectation. We solve this issue by using a resampling technique known as the *bootstrap*, where we draw multiple samples from our data with repetition and calculate a value for our statistic of interest for each sample. We then report the mean estimate from these samples as our bootstrapped estimate. The advantage of this method is two-fold. First, it allows us to produce a more robust estimate for our inferred statistics. Second, the variation within bootstrapped samples can be used to estimate the error incurred from inferred metrics instead of actual class values. We refer to [11] for a detailed treatment of the bootstrap methodology and the proof for its guarantee to yield true expected value.

5 Related Work

This section presents an overview of the previous research on solving PMP and how our method contributes to the literature. First, the lack of class membership data and the problems it poses for fairness calculations have been well-documented [2]. The idea of using data related to the unknown class to infer class fairness metrics is also present in the literature and researchers have approached this problem from various angles. In [12], the authors use Census data to build a predictive model for race and use the prediction as if it were the true value for protected class membership. Using a prediction from a model leads to some error in fairness metric calculation, and in [22], the authors develop a method to estimate bounds on fairness metrics that are estimated with imputed race, building on [15], which presents a general method for estimating uncertainty for machine learning models. In general, these models rely on strong predictors for class membership. When class membership is not well-predicted by the model, there is more uncertainty in the fairness metrics [22].

Our method differs from the existing work in that we do not attempt to assign a predicted class membership to *individuals*. Therefore our approach does not rely on having "good" predictors or suffer from its lack thereof. Instead, we require that a discrete surrogate variable exists, which may have many levels where protected and unprotected class membership are equally likely. The novelty of our approach is to *bypass* predicting protected attributes for individuals via inferred metrics calculated at the surrogate group level. As demonstrated in our experiments, our solution yields good results in practice, producing fairness metrics close to oracle values.

The method developed in [6] is the closest to our approach, where the authors calculate fairness statistics as a weighted average of the probabilities of protected class membership. The probabilities are assumed to come from a predictive model. The resulting fairness statistic can be expressed in a form close to the estimator we develop here. Our estimator exhibits better properties than [6] as it is guaranteed to be *unbiased* under the regularity conditions described earlier. We present a detailed proof of our unbiased guarantee in Appendix A.

6   Experiments

To demonstrate the effectiveness of our approach when solving PMP in practice, we consider two specific questions:

Q1: How do our inferred metrics from surrogate membership compare to an oracle that produces exact fairness evaluation using deterministic membership? When the inferred disparity metric is used, how does fairness evaluation change, and are binary fairness metrics still within the same range as in the exact results?

Q2: How robust are our inferred metrics from surrogate membership under different scenarios with varying disparity conditions?

To answer each question, we start with an overview of the dataset and the modelling setup. We then present numerical results with discussions.

6.1   [Q1] Performance Against Oracle

Home Mortgage Disclosure Act Dataset: To evaluate our method against an oracle with known membership attributes, we need a dataset that reveals protected information about individuals. The Home Mortgage Disclosure Act (HMDA) dataset [14] fits our experimental purposes neatly. First, this publicly available dataset provides information on the self-reported race of 1.3 million mortgage applicants. Second, it also contains zip code data, which we use as our surrogate variable. We use the US Census's American Community Survey (ACS) to acquire the probability of protected status based on race. Zip code estimates for race percentages were assigned based on the state and census tract with the highest ratio of residential addresses according to the 2018 Q4 HUD-USPS Crosswalk file [9]. We drop 300K individuals who do not have valid zip codes and drop 9,460 zip codes that have less than 30 individuals.

Table 1. Comparison of fairness metrics on HMDA dataset obtained by the Oracle, using deterministic protected attribute, and our Inferred Metrics, using probabilistic protected attribute based on the surrogate variable.

Fairness Metric	Oracle Value	Inferred Value	Difference |Oracle - Inferred|	Oracle in range	Inferred in range
Statistical Parity	−0.166	−0.190	0.024	✓	✓
Equal Opportunity	−0.085	−0.110	0.025	✓	✓
Predictive Equality	−0.102	−0.094	0.008	✓	✓
Average Odds	−0.093	−0.102	0.007	✓	✓

HMDA Model: To test our method on the HMDA data, we need a binary classifier to evaluate fairness metrics. To that end, we use the R package glmnet to fit a predictive model for whether a loan originated (1=Yes and 0=No) on approximately 1 million home purchase mortgage applications, excluding refinance and reverse mortgages. Let us stress that our goal is to calculate fairness metrics and not to design the best predictor on this data. That said, generalized linear models are desirable in loan applications thanks to their interpretability. Our focus is on fairness statistics to capture the model disparity between the protected (non-white) and the non-protected (white) group. The Oracle uses the exact labels from the loan applicant's self-reported race. Using zip code as the surrogate variable, our inferred metrics are calculated, as discussed in (Sect. 3). Both approaches utilize the same generalized linear model as their model predictions.

Numerical Results on HMDA Dataset: Table 1 compares the fairness statistics on the HMDA data by the Oracle, using self-reported race, and the Inferred metrics, calculated with zip code as the surrogate. It presents the actual and the inferred value, their difference, and whether the resulting fairness evaluation remains in ideal ranges according to the 80/20 rule [19].

As shown in Table 1, for Predictive Equality and Average Odds, the difference between the oracle fairness statistics calculated with actual race and the inferred values is negligible, the equivalent of a rounding error. However, for Statistical Parity and Equal Opportunity, the inferred metrics are somewhat different from the oracle values. Nevertheless, when checking for ideal ranges for fairness evaluation, inferred metrics lead us to the same conclusion as the oracle. We conjecture that the cases where the inferred and actual fairness metrics are slightly different are most likely due to *omitted variable bias* that are associated with model fairness, such as location and race. This bias affects other methods also for calculating fairness when the true protected status is unknown [6, 22]. Later on, in Practical Considerations section (Sect. 7), we discuss these issues together with their potential fixes.

Table 2. Simulation rates for $\langle FPR, FNR, P(ML(X) = 1) \rangle$ in the synthetic benchmarks between protected, $\top$, and unprotected, $\bot$, groups.

Scenario	False Positive Rate		False Negative Rate		Positive Rate	
	Protected	Unprotected	Protected	Unprotected	Protected	Unprotected
Fair	0.20	0.20	0.10	0.10	0.20	0.20
Slightly Unfair	0.10	0.20	0.35	0.10	0.10	0.20
Moderately Unfair	0.10	0.30	0.45	0.10	0.10	0.30
Very Unfair	0.10	0.30	0.65	0.10	0.10	0.40
Extremely Unfair	0.05	0.20	0.65	0.10	0.10	0.50

6.2 [Q2] Robustness Under Different Fairness Scenarios

While the results from HMDA are promising, it is possible that the HMDA data and the glmnet model produce fairness statistics that are inherently close to the inferred metrics. Therefore, to run a controlled experiment, we need synthetic data to study how inferred metrics perform under a variety of scenarios.

Synthetic Data: For synthetic benchmarks, we consider a surrogate variable Z with 3,800 levels and 20 to 50 individuals per level. The probability of protected class membership is set between 0.01 and 0.999, with a distribution skewed toward small probabilities of being in the protected group. The resulting hypothetical population hosts 126,000 individuals. These values were chosen so the synthetic samples would have less favorable characteristics for inferred metrics than the HMDA data. Later on, in Practical Considerations section (Sect. 7), we discuss how the characteristics of Z affect the estimations and share best practices.

The Simulation Process: In HDMA experiments, we used the features of individuals to train a glmnet model for binary prediction. Here, we do not have access to features for model building; instead, based on the population characteristics, we *simulate the results* of a binary classifier with controlled unfairness. Then, given the simulation results, which are precisely the confusion matrices, we calculate the Oracle and the Inferred fairness metrics.

This simulation process enables us to study how inferred metrics perform under different scenarios ranging from a *fair model* where the classifier produces the same results for both cohorts to an *extremely unfair model* where the classifier highly favors the unprotected cohort. A classifier can be unfair by:

Algorithm 1. The Simulation Procedure

Input: *False Positive & Negative Rates* $\mathcal{FPR}_\top, \mathcal{FNR}_\top, \mathcal{FPR}_\bot, \mathcal{FNR}_\bot$
Input: *Positive Rates* $\mathcal{P}(Y = 1 \mid x \in X^\top), \mathcal{P}(Y = 1 \mid x \in X^\bot)$
Output: *The disparity of the model metric m*

Step 1. Calculate the confusion matrix, $\mathcal{CM}^\mathcal{A}$, probabilities for $\mathcal{A} \in \{\top, \bot\}$:
$\mathcal{P}(\mathcal{FN} \mid \mathcal{A}) \leftarrow (1 - \mathcal{P}(Y = 1 \mid \mathcal{A})) \times \mathcal{FNR}_\mathcal{A}$
$\mathcal{P}(\mathcal{FP} \mid \mathcal{A}) \leftarrow \mathcal{P}(Y = 1 \mid \mathcal{A}) \times \mathcal{FPR}_\mathcal{A}$
$\mathcal{P}(\mathcal{TN} \mid \mathcal{A}) \leftarrow (1 - \mathcal{P}(Y = 1 \mid \mathcal{A})) \times (1 - \mathcal{FPR}_\mathcal{A})$
$\mathcal{P}(\mathcal{TP} \mid \mathcal{A}) \leftarrow \mathcal{P}(Y = 1 \mid \mathcal{A}) \times (1 - \mathcal{FNR}_\mathcal{A})$
$\mathcal{P}(\mathcal{CM}^\mathcal{A}) \leftarrow \mathcal{P}(\mathcal{FP}|\mathcal{A}), \mathcal{P}(FN|\mathcal{A}), \mathcal{P}(\mathcal{TN}|\mathcal{A}), \mathcal{PTP}|\mathcal{A})$

Step 2. Assign each individual into a quadrant in the confusion matrix
for $\forall x \in \mathcal{X} \mid Z = z \wedge x \in z$ **do**
  $x_{is\_protected} \leftarrow$ assign with probability $\mathcal{P}_z(x \in X^\top)$
  $\mathcal{CM} \leftarrow \mathcal{CM}^\top$ **if** $x_{is\_protected}$ **else** $\mathcal{CM}^\bot$
  assign x to a quadrant in $\mathcal{CM}$ with probability **choice**$(\mathcal{P}(\mathcal{CM}))$
end for

Step 3. Calculate the model metric based on simulated confusion matrices
Oracle: $m(\mathcal{X}^\mathcal{A}) \leftarrow$ apply m to $\mathcal{CM}^\mathcal{A}$ for $\mathcal{A} \in \{\top, \bot\}$
Inferred: WOLS estimator for $m_z = \beta_0 + \beta_1 \mathcal{P}_z(x \in \mathcal{X}^\top) + e_z \forall z \in \mathcal{Z}$ to find β_1

Step 4. Return the model disparity
return $m(\mathcal{X}^\top) - m(\mathcal{X}^\bot)$

1. **FPR:** incorrectly classifies unprotected group members into the positive case more often, i.e., the difference in false positive rate.
2. **FNR:** incorrectly classifies protected individuals into the negative case more often, i.e., the difference in the false negative rate.
3. **P(Y=1):** The other degree of freedom stems from the bias in the target variable, where positive outcomes for the protected group are observed more rarely than the unprotected group, i.e., the rate at which target is positive.

Table 2 presents the values of $\langle FPR, FNR, P(Y = 1) \rangle$ that jointly determine the probability of an individual being classified into one of the four quadrants in the confusion matrix within each fairness scenario. Notice that the protected and unprotected groups are subject to different rates depending on the unfairness level we want to simulate. The settings depict unfairness ranges that practitioners are likely to encounter, ranging from 0.1 to 0.55. The statistics are set to favor the unprotected group since, by symmetry, the reverse case is the same calculation but of the opposite sign. We also simulate a fair model as a baseline, where the confusion matrices are the same for the protected and unprotected groups.

Algorithm 1 presents the details of our simulations for Oracle and Inferred values. Conceptually, in a fair model, individuals from the protected and unprotected groups are classified into four quadrants of the confusion matrix at the same rate. Contrarily, in an unfair model, individuals are classified into the four quadrants at different rates, resulting in two different confusion matrices.

Numerical Results on Synthetic Dataset: Table 3 presents the simulation results for five scenarios using the settings from Table 2 and the procedure in Algorithm 1 across all fairness metrics. As before, we compare the Oracle, which calculates the actual disparity from the confusion matrix, with our inferred metrics that leverage the surrogate information. The results are averaged over 30 runs for robustness, and we also report the standard deviation (σ). The main takeaways from these numerical results are as follows.

Table 3. Simulation Results from the Oracle and Inferred metrics, averaged over 30 runs, on various unfairness settings.

Fairness Metrics	Fair		Slightly Unfair		Moderately Unfair		Very Unfair		Extremely Unfair	
	Oracle	Inferred (σ)	Oracle	Inferred (σ)	Setting	Inferred (σ)	Oracle	Inferred (σ)	Oracle	Inferred (σ)
Statistical Parity	0.0	−0.002 (.005)	−0.19	−0.184 (.005)	−0.34	−0.335 (.004)	−0.415	−0.414 (.005)	−0.47	−0.470 (.005)
Equal Opportunity	0.0	−0.001 (.009)	−0.25	−0.250 (.013)	−0.35	−0.348 (.023)	−0.55	−0.544 (.027)	−0.54	−0.544 (.029)
Predictive Equality	0.0	−0.001 (.006)	−0.10	−0.098 (.005)	−0.20	−0.200 (.004)	−0.20	−0.200 (.005)	−0.15	−0.150 (.004)
Average Odds	0.0	0.000 (.006)	−0.18	−0.174 (.006)	−0.28	−0.274 (.011)	−0.38	−0.376 (.013)	−0.35	−0.347 (.015)

First, in accordance with our simulation design, we observe that the metric disparity gets worse (larger values) from fair to unfair scenarios. Second, in scenarios in Table 2, we deliberately set low positive rates for the protected group. Rates with small denominators are inherently less stable than rates with larger denominators [10]. When comparing standard deviations, this is why the Equal Opportunity, where the denominator is the number of positive cases, has a higher standard deviation than the other statistics. Finally, and most importantly, our inferred metrics closely follow the oracle values. This holds across *all metrics in all scenarios*, demonstrating the effectiveness of our approach.

7   Practical Considerations

Finally, we discuss practical issues such as confounding factors and the characteristics of the surrogate variable, Z, with their potential fixes.

7.1   Omitted Variable Bias

Our approach for inferred metrics might suffer in scenarios where variables that are correlated with both the surrogate and the model metric can cause bias in the estimated disparity. The same issue also occurs in [6,18], which utilize surrogate class variables (referred to as *proxies*). This aspect has been well-studied in other applications of WOLS, and is known as *omitted variable bias, confounding effects*, and *spurious correlation* [25].

Suppose there is a variable C that is correlated with both the surrogate variable Z and our model metric m. Then our inferred metric is misspecified, and the true model is:

$$m_z = \beta_0 + \beta_1 P_z(x \in X^{\top}) + \beta_2 C_z + e_z \tag{7}$$

If this is the true model, but we use the inferred estimator shown in equation (5), the expected value of β_1 is no longer $m(X^\top) - m(X^\perp)$, but is instead:

$$m(X^\top) - m(X^\perp) + \rho_{Cp}\frac{\sigma_{Cm}}{\sigma_p}, \text{ where}$$

- ρ_{Cp} is the correlation between the population percentages for the z levels and the omitted variable.
- σ_{Cm} is the covariance between m_z and the omitted variable.
- σ_p is the variance of the population percentages.

The amount and direction of the bias depends on the magnitude and direction of the correlations with the omitted variable. For bias to occur in our inferred metric, the omitted variable must be correlated with the population percentages $P_z(x \in X^\top)$ and $P_z(x \in X^\perp)$ within groups of Z. Potential omitted variables must also be correlated with m_z to cause bias in our disparity estimates. For example, the home mortgage data has many variables related to household wealth. Previous studies indicate that those variables correlate with location and race [7]. Therefore, we expect those variables to cause omitted variable bias in our estimates for the disparity that use Prediction Rate (the probability of the model classifying a person as someone who should get a loan) but less bias for metrics that measure model accuracy for different groups, e.g., Predictive Equality. We observe this in the HMDA experiments in Table 1. The most significant difference between the actual value calculated using race and the inferred metrics are Statistical Parity and Equal Opportunity.

Addressing Confounding Effects: The best solution for omitted variable bias is to add the omitted variable to the model. In our experiments, variables that affect m_z are also the features that are used in the model. As such, we can calculate their mean for each level of Z just as easily as model metrics. We can then add them to the inferred metric model to estimate an unbiased difference solely because of class membership. However, adding additional variables to the model can introduce other problems, most likely *multicollinearity*. Multicollinearity occurs when model estimates become unstable because of linear dependencies between the predictors [13]. Let us also call out that for some applications, adding omitted variables to the models might not be possible because effects are unavailable or unmeasurable.

7.2   Characteristics of Z and $P_z(x \in X^\top)$

An essential characteristic of Z is how many levels it exhibits and whether $P_z(x \in X^\top)$ and $P_z(x \in X^\perp)$ are nearly equal. According to the properties of WOLS [1], the uncertainty in the inferred metrics $m(X^\top) - m(X^\perp)$ is inversely proportional to the number of levels in Z and the variation in $P_z(x \in X^\top)$. Ideally, our surrogate variable shall have many levels while $P_z(x \in X^\top)$ covers the range $(0, 1)$. The HMDA dataset and the synthetic benchmarks exhibit these properties in our experiments. For cases where there is less coverage, i.e., where most or all of the levels of Z have the same value for $P_z(x \in X^\top)$, estimates for $m(X^\top) - m(X^\perp)$ will be more uncertain.

Addressing Z Characterization: For surrogate variables that have few levels or relatively low variation in $P_z(x \in X^\top)$, the bootstrap method that we use to achieve robust and unbiased estimates will also allow us to estimate the uncertainty in $m(X^\top) - m(X^\perp)$. Because the bootstrap method requires repeated samples and repeated estimates of $m(X^\top) - m(X^\perp)$, the standard deviation in the inferred metrics across samples estimates the uncertainty in the estimate [11]. Estimating uncertainty in calculated fairness metrics and using those estimates to make fairness decisions are also discussed in [22].

8   Conclusions

In this paper we introduced the *Probabilistic Membership Problem (PMP)*, for practitioners to estimate a model's fairness with only access to the probability of protected class membership, and we showed how to use *inferred metrics* to solve it. We described the assumptions required for inferred metrics to be unbiased. We demonstrated the effectiveness of inferred metrics on publicly available data and on benchmark data that was simulated to have known amounts of model unfairness. We discussed practical implications for calculating these metrics and methods to handle situations when the assumptions for unbiased estimates are not met. We also discuss the need for additional research to assist in practical applications of inferred metrics, including the amount of omitted variable bias one can expect in practice and how to estimate uncertainty.

Overall, our inferred metrics are novel and practical contributions to the existing methodology for estimating model fairness. It alleviates the requirement for exact deterministic protected attributes for every individual to its relaxation to a surrogate variable that only reveals group-level protected attribute probabilities. This allows us to test models where fairness evaluation would have been impossible before. We hope our work motivates other researchers and practitioners not to overlook fairness evaluation that might influence the trustworthiness of algorithmic decisions.

A   Appendix - Comparison to Weighted Fairness Statistic

Our inferred metrics are similar in approach to an estimator described in [6]. In this section, we re-write our inferred metrics and the weighted estimator so they can be compared directly and present a mathematical argument for why the weighted estimator is biased toward 0 under the regularity conditions described.

First, our estimator $m(X^\top) - m(X^\perp)$ is derived from the WOLS estimator of the value β_1 from Eq. 5.

$$m_{wols}(X^\top) - m_{wols}(X^\perp) = \frac{\sum_z n_z(m_z - \bar{m})(P_z(x \in X^\top) - \bar{P}(x \in X^\top))}{\sum_z n_z(P_z(x \in X^\top) - \bar{P}(x \in X^\top))^2} \quad (8)$$

where $\bar{m}$ is the overall mean for the model metric and $\bar{P}(x \in X^\top)$ is the overall mean for the probability of being in the protected group.

The weighted estimator described in [6] is:

$$m_w(X^\top) - m_w(X^\perp) = \frac{\sum_x P_z(x \in X^\top)m(x)}{\sum_x P_z(x \in X^\top)} - \frac{\sum_x P_z(x \in X^\perp)m(x)}{\sum_x P_z(x \in X^\perp)} \quad (9)$$

where

- $m(x)$ is the value of the model metric for each individual (e.g. if the m is statistical parity, $m(x) = I(ML(x) = 1)$)
- $\sum_x$ indicates a sum over all N individuals for which we are calculating fairness metrics
- $P_z(x \in X^\top)$ is the probability that each individual is in the protected group given their surrogate class membership $z \in Z$.

In the proof below, we re-write these equations to show that they are the same except for one term in the denominator. Specifically, we re-write our inferred metric as:

$$m_{wols}(X^\top) - m_{wols}(X^\perp) = \frac{N\sum_x m(x)P_z(x \in X^\top) - \sum_x m(x)\sum_x P_z(x \in X^\top)}{N\sum_x P_z(x \in X^\top)^2 - \left(\sum_x P_z(x \in X^\top)\right)^2} \quad (10)$$

We re-write the weighted estimator from [6] as:

$$m_w(X^\top) - m_w(X^\perp) = \frac{N\sum_x m(x)P_z(x \in X^\top) - \sum_x m(x)\sum_x P_z(x \in X^\top)}{N\sum_x P_z(x \in X^\top) - \left(\sum_x P_z(x \in X^\top)\right)^2} \quad (11)$$

where $N = \sum_z n_z$ (N is the total number of individuals for which we are calculating fairness metrics).

Equation 10 and Eq. 11 are the same except for the first term in the denominator. We argue here that this difference implies that the weighted estimator is biased toward 0 under the conditions described in (Sect. 3). This means that the weighted estimator will show smaller differences between groups than are actually present in the data.

First, note that $P_z(x \in X^\top)$ is a probability, and therefore bounded between (0,1)

$$P_z(x \in X^\top) < 1 \implies N > \sum_x P_z(x \in X^\top) \implies N\sum_x P_z(x \in X^\top) > \left(\sum_x P_z(x \in X^\top)\right)^2$$

This means that the sign of the weighted estimator (whether it is negative or positive) is determined by the numerator of the equation.

Now, because $P_z(x \in X^\top)$ is a probability,

$$P_z(x \in X^\top)^2 < P_z(x \in X^\top) \forall x \implies \sum_x P_z(x \in X^\top) < \sum_x P_z(x \in X^\top)^2$$

This shows that the first term in the denominator is smaller for our inferred estimator, and therefore:

$$\left| \frac{N\sum_x m(x)P_z(x \in X^\top) - \sum_x m(x)\sum_x P_z(x \in X^\top)}{N\sum_x P_z(x \in X^\top)^2 - \left(\sum_x P_z(x \in X^\top)\right)^2} \right|$$
$$> \left| \frac{N\sum_x m(x)P_z(x \in X^\top) - \sum_x m(x)\sum_x P_z(x \in X^\top)}{N\sum_x P_z(x \in X^\top) - \left(\sum_x P_z(x \in X^\top)\right)^2} \right|$$

which means:

$$|m_{wols}(X^\top) - m_{wols}(X^\perp)| > |m_w(X^\top) - m_w(X^\perp)|$$

In (Sect. 3) we refer to a set of conditions where WOLS is unbiased that follows from [1] and the Gauss-Markov theorem. The weighted estimator is always smaller in absolute value and must therefore be biased toward 0 under the same conditions.

A.1   Re-Writing the Weighted Estimator

In order to compare the weighted estimator with our inferred estimator, we rewrite the weighted estimator for the case where there are two groups, and one surrogate variable Z that acts as a predictor. Now, $P_z(x \in X^\perp) = 1 - P_z(x \in X^\top)$, so that:

$$m_w(X^\top) - m_w(X^\perp) = \frac{\sum_x m(x)P_z(x \in X^\top)}{\sum_x P_z(x \in X^\top)} - \frac{\sum_x m(x)(1 - P_z(x \in X^\top))}{\sum(1 - P_z(x \in X^\top))}$$

Multiply each of these fractions to get a common denominator.

$$m_w(X^\top) - m_w(X^\perp) = \frac{\sum m(x)P_z(x \in X^\top)\sum(1 - P_z(x \in X^\top)) - \sum m(x)(1 - P_z(x \in X^\top))\sum P_z(x \in X^\top)}{\sum P_z(x \in X^\top)\sum(1 - P_z(x \in X^\top))}$$

Then, starting with the numerator, we expand the parentheses and distribute the sums, which gives the following:

$$N\sum m(x)P_z(x \in X^\top) - \sum m(x)P_z(x \in X^\top)\sum P_z(x \in X^\top)$$
$$- \sum m(x)\sum P_z(x \in X^\top) + \sum m(x)P_z(x \in X^\top)\sum P_z(x \in X^\top) \tag{12}$$

The second and fourth terms cancel, so that:

$$m_w(X^\top) - m_w(X^\perp) = \frac{N\sum_x m(x)P_z(x \in X^\top) - \sum_x m(x)\sum_x P_z(x \in X^\top)}{\sum_x(1 - P_z(x \in X^\top))\sum_x P_z(x \in X^\top)}$$

Following the same process for the denominator gives us the following form for the weighted estimator:

$$m_w(X^\top) - m_w(X^\perp) = \frac{N\sum_x m(x)P_z(x \in X^\top) - \sum_x m(x)\sum_x P_z(x \in X^\top)}{N\sum_x P_z(x \in X^\top) - \left(\sum_x P_z(x \in X^\top)\right)^2}$$

A.2   Re-Writing the Inferred Estimator

We can follow the same process to re-write the estimator for the difference between $m_{wols}(X^\top) - m_{wols}(X^\perp)$, and express our inferred fairness metric in terms of the individual values $m(x)$.

As before, start with the numerator, expand the terms in parentheses and distribute the sums, which gives us the following expression.

$$\sum_z n_z(m_z - \bar{m})(P_z(x \in X^\top) - \bar{P}(x \in X^\top)) = \sum_z n_z m_z \bar{P}(x \in X^\top)$$
$$- \bar{m} \sum_z n_z P_z(x \in X^\top)$$
$$- \bar{P}(x \in X^\top) \sum_z n_z m_z + \bar{m}\bar{P}(x \in X^\top) \sum_z n_z \tag{13}$$

Observe the following:

- We require m to be an arithmetic mean, therefore, $m_z = \frac{1}{n_z} \sum_z m(x)$, and $n_z m_z = \sum_z m_x$
- $\bar{m} = \frac{1}{N} \sum_x m(x)$
- $\bar{P}(x \in X^\top) = \frac{1}{N} \sum_x P(x \in X^\top) = \frac{1}{N} \sum_z n_z P_z(x \in X^\top)$

Taking each term in the numerator separately, we re-write them as:

- $\sum_z n_z m_z \bar{P}(x \in X^\top) = \sum_z n_z(m_z P_z(x \in X^\top)) = \sum_x m(x) P_z(x \in X^\top)$
- $\bar{m} \sum_z n_z P_z(x \in X^\top) = \frac{1}{N} \sum_x m(x) \sum_x P(x \in X^\top) = N\bar{m}\bar{P}(x \in X^\top)$
- $\bar{P} \sum_z n_z m_z = \frac{1}{N} \sum_x P(x \in X^\top) \sum_x m(x) = N\bar{m}\bar{P}(x \in X^\top)$
- $\bar{m}\bar{P}(x \in X^\top) \sum_z n_z = N\bar{m}\bar{P}(x \in X^\top)$

This lets us collect three of the four terms in the numerator and leaves us with:

$$m(X^\top) - m(X^\perp) = \frac{\sum_x m(x) P_z(x \in X^\top) - N\bar{m}\bar{P}(x \in X^\top)}{\sum_z n_z(P_z(x \in X^\top) - \bar{P}(x \in X^\top))^2} \tag{14}$$

For the denominator, we again expand the parentheses and collect the sums to give the following:

$$\sum_z n_z(P_z(x \in X^\top) - \bar{P}(x \in X^\top))^2 = \sum_z n_z P_z(x \in X^\top)^2$$
$$- 2\sum_z n_z P_z(x \in X^\top)\bar{P}(x \in X^\top) + \sum_z n_z \bar{P}(x \in X^\top)^2 \tag{15}$$

Again, taking each term separately, we simplify as follows:

- $\sum_z n_z P_z(x \in X^\top)^2 = \sum_z \sum_{x \in z} P_z(x \in X^\top)^2 = \sum_x P_z(x \in X^\top)^2$
- $-2\sum_z n_z P_z(x \in X^\top)\bar{P}(x \in X^\top) = -2N\bar{P}(x \in X^\top)^2$
- $\sum_z n_z \bar{P}(x \in X^\top)^2 = \bar{P}(x \in X^\top)^2 \sum_z n_z = N\bar{P}(x \in X^\top)^2$

This gives us the expression:

$$m_{wols}(x \in X^{\top}) - m_{wols}(X^{\perp}) = \frac{\sum_x m(x)p(x \in X^{\top}) - N\bar{m}\bar{P}}{\sum_x P_z(x \in X^{\top})^2 - N\bar{P}(x \in X^{\top})^2}$$

Multiplying the above fraction by $\frac{N}{N}$, gives us the form of the equation as written in (10).

$$m_{wols}(x \in X^{\top}) - m_{wols}(X^{\perp}) = \frac{N\sum_x m(x)p(x \in X^{\top}) - \sum_x m(x)\sum_x P_z(x \in X^{\top})}{N\sum_x P_z(x \in X^{\top})^2 - \left(\sum_x P_z(x \in X^{\top})\right)^2}$$

References

1. Aitken, A.C.: On least squares and linear combination of observations. Proc. R. Soc. Edinb. **55**, 42–48 (1936)
2. Andrus, M., Spitzer, E., Brown, J., Xiang, A.: What we can't measure, we can't understand: challenges to demographic data procurement in the pursuit of fairness. In: Proceedings of the 2021 ACM Conference on Fairness, Accountability, and Transparency, pp. 249–260 (2021)
3. Bellamy, R.K.E., et al.: AI Fairness 360: an extensible toolkit for detecting, understanding, and mitigating unwanted algorithmic bias (2018). https://arxiv.org/abs/1810.01943
4. Box, G.E.: Use and abuse of regression. Technometrics **8**(4), 625–629 (1966)
5. Caton, S., Haas, C.: Fairness in machine learning: a survey. arXiv preprint arXiv:2010.04053 (2020)
6. Chen, J., Kallus, N., Mao, X., Svacha, G., Udell, M.: Fairness under unawareness: assessing disparity when protected class is unobserved. In: Proceedings of the Conference on Fairness, Accountability, and Transparency, pp. 339–348 (2019)
7. Chenevert, R., Gottschalck, A., Klee, M., Zhang, X.: Where the wealth is: the geographic distribution of wealth in the united states. US Census Bureau (2017)
8. Department, U.F.R.: Federal fair lending regulations and statutes (2020). https://www.federalreserve.gov/boarddocs/supmanual/cch/fair_lend_over.pdf. Accessed 04 Sept 2020
9. Din, A., Wilson, R.: Crosswalking zip codes to census geographies. Cityscape **22**(1), 293–314 (2020)
10. Duris, F., et al.: Mean and variance of ratios of proportions from categories of a multinomial distribution. J. Stat. Distrib. Appl. **5**(1), 1–20 (2018)
11. Efron, B., Tibshirani, R.: Bootstrap methods for standard errors, confidence intervals, and other measures of statistical accuracy. Stat. Sci. **1**, 54–75 (1986)
12. Elliott, M.N., Morrison, P.A., Fremont, A., McCaffrey, D.F., Pantoja, P., Lurie, N.: Using the census bureau's surname list to improve estimates of race/ethnicity and associated disparities. Health Serv. Outcomes Res. Method. **9**(2), 69–83 (2009)
13. Farrar, D.E., Glauber, R.R.: Multicollinearity in regression analysis: the problem revisited. In: The Review of Economic and Statistics, pp. 92–107 (1967)
14. Federal Financial Institutions Examination Council: Home mortgage disclosure act snapshot national loan level dataset. Technical report, U.S. Government (2018). https://ffiec.cfpb.gov/data-publication/snapshot-national-loan-level-dataset/2018

15. Gal, Y., Ghahramani, Z.: Dropout as a Bayesian approximation: representing model uncertainty in deep learning. In: International Conference on Machine Learning, pp. 1050–1059. PMLR (2016)
16. Hays, W.: The algebra of expectations. In: Statistics, p. 630. CBS College Publishing, Holt Rhinehart Winston New York (1981)
17. Heitjan, D.F.: Inference from grouped continuous data: a review. Stat. Sci. **4**(2), 164–179 (1989)
18. Kallus, N., Mao, X., Zhou, A.: Assessing algorithmic fairness with unobserved protected class using data combination. Manage. Sci. **68**(3), 1959–1981 (2022)
19. of Labor, D.: Uniform guidelines on employee selection procedures (1978). https://uniformguidelines.com/questionandanswers.html. Accessed 05 Sept 2020
20. Michalský, F., Kadioglu, S.: Surrogate ground truth generation to enhance binary fairness evaluation in uplift modeling. In: 20th IEEE International Conference on ML and Applications, ICMLA 2021, USA, 2021, pp. 1654–1659. IEEE (2021)
21. Papoulis, A.: Expected value; dispersion; moments (1984)
22. Racicot, T., Khoury, R., Pere, C.: Estimation of uncertainty bounds on disparate treatment when using proxies for the protected attribute. In: Canadian Conference on AI (2021)
23. U.S. Department of Housing and Urban Development: Fair housing rights and obligations (2020). https://www.hud.gov/program_offices/fair_housing_equal_opp/fair_housing_rights_and_obligations. Accessed 04 Sept 2020
24. U.S. Equal Employment Opportunity Commission: Prohibited employment policies/practices (2020). https://www.eeoc.gov/prohibited-employment-policiespractices. Accessed 04 Sept 2020
25. VanderWeele, T.J., Shpitser, I.: On the definition of a confounder. Ann. Stat. **41**(1), 196 (2013)

The BeMi Stardust: A Structured Ensemble of Binarized Neural Networks

Ambrogio Maria Bernardelli[1]([✉])[iD], Stefano Gualandi[1][iD], Hoong Chuin Lau[2][iD], and Simone Milanesi[1][iD]

[1] Department of Mathematics, University of Pavia, Pavia, Italy
{ambrogiomaria.bernardelli01,simone.milanesi01}@universitadipavia.it,
stefano.gualandi@unipv.it
[2] School of Information Systems, Singapore Management University, Singapore, Singapore
hclau@smu.edu.sg

Abstract. Binarized Neural Networks (BNNs) are receiving increasing attention due to their lightweight architecture and ability to run on low-power devices, given the fact that they can be implemented using Boolean operations. The state-of-the-art for training classification BNNs restricted to few-shot learning is based on a Mixed Integer Programming (MIP) approach. This paper proposes the BeMi ensemble, a structured architecture of classification-designed BNNs based on training a single BNN for each possible pair of classes and applying a majority voting scheme to predict the final output. The training of a single BNN discriminating between two classes is achieved by a MIP model that optimizes a lexicographic multi-objective function according to robustness and simplicity principles. This approach results in training networks whose output is not affected by small perturbations on the input and whose number of active weights is as small as possible, while good accuracy is preserved. We computationally validate our model using the MNIST and Fashion-MNIST datasets using up to 40 training images per class. Our structured ensemble outperforms both BNNs trained by stochastic gradient descent and state-of-the-art MIP-based approaches. While the previous approaches achieve an average accuracy of 51.1% on the MNIST dataset, the BeMi ensemble achieves an average accuracy of 61.7% when trained with 10 images per class and 76.4% when trained with 40 images per class.

Keywords: Binarized neural networks · Mixed-integer linear programming · Structured ensemble of neural networks · Few-shot learning

1 Introduction

State-of-the-art Neural Networks (NNs) contain a huge number of neurons organized in several layers, and they require an immense amount of data for training

M. Sellmann and K. Tierney (Eds.): LION 2023, LNCS 14286, pp. 443–458, 2023.
https://doi.org/10.1007/978-3-031-44505-7_30

[11]. The training process is computationally demanding and is typically performed by stochastic gradient descent algorithms running on large GPU-based clusters. Whenever the trained (deep) neural network contains many neurons, also the network deployment is computationally demanding. However, in real-life industrial applications, GPU-based clusters are often unavailable or too expensive, and training data is scarce and contains only a few data points per class.

Binarized Neural Networks (BNNs) were introduced in [6] as a response to the challenge of running NNs on low-power devices. BNNs contain only binary weights and binary activation functions, and hence they can be implemented using only bit-wise operations, which are very power-efficient: such neural networks can be implemented using Boolean operations, eliminating the need for CPU usage. However, the training of BNNs raises interesting challenges for gradient-based approaches due to their combinatorial structure. In [22], the authors show that the training of a BNN performed by a hybrid constraint programming (CP) and mixed integer programming (MIP) approach outperforms the stochastic gradient approach proposed in [6] by a large margin, if restricted to a few-shot-learning context [23]. Indeed, the main challenge in training a NN by an exact MIP-based approach is the limited amount of training data that can be used since, otherwise, the size of the optimization model explodes. However, in [21], the hybrid CP and MIP method was further extended to integer-valued neural networks: exploiting the flexibility of MIP solvers, the authors were able to (i) minimize the number of neurons during training and (ii) increase the number of data points used during training by introducing a MIP batch training method.

We remark that training a NN with a MIP-based approach is more challenging than solving a verification problem, as in [1,4], even if the structure of the nonlinear constraints modeling the activation functions is similar. In NNs verification [10], the weights are given as input, while in MIP-based training, the weights are the decision variables that must be computed. Furthermore, recent works aim at producing compact and lightweight NNs that maintain acceptable accuracy, e.g., in terms of parameter pruning [19,26], loss function improvement [20], gradient approximation [17], and network topology structure [13].

Contributions. In this paper, we propose the BeMi[1] ensemble, a structured ensemble of classification-designed BNNs, where each single BNN is trained by solving a lexicographic multi-objective MIP model. Given a classification task over k classes, the main idea is to train $\frac{k(k-1)}{2}$ BNNs, where every single network learns to discriminate only between a given pair of classes. When a new data point (e.g., a new image) must be classified, it is first fed into the $\frac{k(k-1)}{2}$ trained BNNs, and later, using a Condorcet-inspired majority voting scheme [25], the most frequent class is predicted as output. Also, this method is similar to and generalizes the SVM-OVO approach [2], and it has not yet been applied within the context of neural networks, to the best of our knowledge. For training every single BNN, our approach extends the methods introduced in [22] and [21]. Our computational results using the MNIST and the Fashion-MNIST dataset show

---

[1] Acronym from the last names of the two young authors who had this intuition.

that the BeMi ensemble permits to use for training up to 40 data points per class, and permits reaching an accuracy of 78.8% for MNIST and 72.9% for Fashion-MNIST. In addition, thanks to the multi-objective function that minimizes the number of neurons, up to 75% of weights are set to zero for MNIST, and up to 50% for Fashion-MNIST.

Outline. The outline of this paper is as follows. Section 2 introduces the notation and defines the problem of training a single BNN with the existing MIP-based methods. Section 3 presents the BeMi ensemble, the majority voting scheme, and the improved MIP model to train a single BNN. Section 4 presents the computational results of the MNIST and Fashion-MNIST. Finally, Sect. 5 concludes the paper with a perspective on future works.

2 Binarized Neural Networks

In this section, we formally define a single BNN using the same notation as in [22], while, in the next section, we show how to define a structured ensemble of BNNs. The architecture of a BNN is defined by a set of layers $\mathcal{N} = \{N_0, N_1, \ldots, N_L\}$, where $N_l = \{1, \ldots, n_l\}$, and n_l is the number of neurons in the l-th layer. Let the training set be $\mathcal{X} := \{(\boldsymbol{x}^1, y^1), \ldots, (\boldsymbol{x}^t, y^t)\}$, such that $\boldsymbol{x}^i \in \mathbb{R}^{n_0}$ and $y^i \in \{-1, +1\}^{n_L}$ for every $i \in T = \{1, 2, \ldots, t\}$. The first layer N_0 corresponds to the size of the input data points $\boldsymbol{x}^k$.

The link between neuron i in layer N_{l-1} and neuron j in layer N_l is modeled by weight $w_{ilj} \in \{-1, 0, +1\}$. Note that whenever a weight is set to zero, the corresponding link is removed from the network. Hence, during training, we are also optimizing the architecture of the BNN. The activation function is the binary function

$$\rho(x) := 2 \cdot \mathbb{1}(x \geq 0) - 1, \tag{1}$$

that is, a sign function reshaped such that it takes ± 1 values. The indicator function $\mathbb{1}(p)$ outputs $+1$ if proposition p is verified, and 0 otherwise.

To model the activation function (1) of the j-th neuron of layer N_l for data point $\boldsymbol{x}^k$, we introduce a binary variable $u_{lj}^k \in \{0, 1\}$ for the indicator function $\mathbb{1}(p)$. To rescale the value of u_{lj}^k in $\{-1, +1\}$ and model the activation function value, we introduce the auxiliary variable $z_{lj}^k = (2u_{lj}^k - 1)$. For the first input layer, we set $z_{0j}^k = x_j^k$; for the last layer, we account in the loss function whether z_{Lh}^k is different from y_h^k. The definition of the activation function becomes

$$z_{lj}^k = \rho \left(\sum_{i \in N_{l-1}} z_{(l-1)i}^k w_{ilj} \right) = 2 \cdot \mathbb{1} \left(\sum_{i \in N_{l-1}} z_{(l-1)i}^k w_{ilj} \geq 0 \right) - 1 = 2u_{lj}^k - 1.$$

Notice that the activation function at layer N_l gives a nonlinear combination of the output of the neurons in the previous layer N_{l-1} and the weights w_{ilj} between the two layers. Section 3.3 shows how to formulate this activation function in terms of mixed integer linear constraints. The choice of a family of parameters $W := \{w_{ilj}\}_{l \in \{1,\ldots,L\}, i \in N_{l-1}, j \in N_l}$ determines the classification function $f_W : \mathbb{R}^{n_0} \to \{\pm 1\}^{n_L}$. The training of a neural network is the process of computing the family W such that f_W classifies correctly both the given training data, that is, $f_W(x^i) = y^i$ for $i = 1, \ldots, t$, and new unlabelled testing data.

The training of a BNN should target two objectives: (i) the resulting function f_W should generalize from the input data and be *robust* to noise in the input data; (ii) the resulting network should be *simple*, that is, with the smallest number of non-zero weights that permit to achieve the best accuracy. Deep neural networks are believed to be inherently robust because mini-batch stochastic gradient-based methods implicitly guide toward robust solutions [8,9,16]. However, as shown in [22], this is false for BNNs in a few-shot learning regime. On the contrary, MIP-based training with an appropriate objective function can generalize very well [21,22], but it does not apply to large training datasets, because the size of the MIP training model is proportional to the size of the training dataset. To generalize from a few data samples, the training algorithm should maximize the margins of the neurons. Intuitively, neurons with larger margins require larger changes to their inputs and weights before changing their activation values. This choice is also motivated by recent works showing that margins are good predictors for the generalization of deep convolutional NNs [7]. Regarding the simplicity objective, a significant parameter is the number of connections [15]. The training algorithm should look for a BNN fitting the training data while minimizing the number of non-zero weights. This approach can be interpreted as a simultaneous compression during training. Although this objective is challenged in [3], it remains the basis of most forms of regularization used in modern deep learning [18].

MIP-Based BNN Training. In [22], two different MIP models are introduced: the Max-Margin, which aims to train robust BNNs, and the Min-Weight, which aims to train simple BNNs. These two models are combined with a CP model into two hybrid methods HW and HA in order to obtain a feasible solution within a fixed time limit because otherwise, the MIP models fail shortly when the number of training data increases. We remark that two objectives, robustness and simplicity, are never optimized simultaneously. In [21], three MIP models are proposed that generalize the BNN approach to consider integer values for weights and biases. The first model, Max-Correct, is based on the idea of maximizing the number of corrected predicted images; the second model, Min-Hinge, is inspired by the squared hinge loss; the last model, Sat-Margin, combines aspects of both the first two models. These three models always produce a feasible solution but use the margins only on the neurons of the last level, obtaining, hence, less robust BNNs.

Gradient-Based BNN Training. In [6], a gradient descent-based method is proposed, consisting of a local search that changes the weights to minimize a square hinge loss function. Note that a BNN trained with this approach only learns -1 and $+1$ weights. An extension of this method to admit zero-value weights, called (GD_t), is proposed in [22], to facilitate the comparison with their approach.

3   The BeMi Ensemble

In this section, we present our structured ensemble of neural networks.

3.1   The BeMi Structure

If we define $\mathcal{P}(S)_m$ as the set of all the subsets of the set S that have cardinality m, and we name $\mathcal{I}$ the set of the classes of the classification problem, our structured ensemble is constructed in the following way.

- We set a parameter $1 < p \leq n = |\mathcal{I}|$.
- We train a BNN denoted by $\mathcal{N}_{\mathcal{J}}$ for every $\mathcal{J} \in \mathcal{P}(\mathcal{I})_p$.
- When testing a data point, we feed it to our list of trained BNNs, namely $(\mathcal{N}_{\mathcal{J}})_{\mathcal{J} \in \mathcal{P}(\mathcal{I})_p}$, obtaining a list of predicted labels $(\mathfrak{e}_{\mathcal{J}})_{\mathcal{J} \in \mathcal{P}(\mathcal{I})_p}$.
- We then apply a majority voting system.

Note that we set $p > 1$, otherwise our structured ensemble would have been meaningless. Whenever $p = n$, our ensemble is made of one single BNN. When $p = 2$, we are using a one-versus-one scheme.

The idea behind this structured ensemble is that, given an input $\boldsymbol{x}^k$ labelled $l\ (= y^k)$, the input is fed into $\binom{n}{p}$ networks where $\binom{n-1}{p-1}$ of them are trained to recognize an input with label l. If all of the networks correctly classify the input $\boldsymbol{x}^k$, then at most $\binom{n-1}{p-1} - \binom{n-2}{p-2}$ other networks can classify the input with a different label. With this approach, if we plan to use $r \in \mathbb{N}$ inputs for each label, we are feeding our BNNs a total of $p \times r$ inputs instead of feeding $n \times r$ inputs to a single large BNN. When $p = 2 \ll n$, it is much easier to train our structured ensemble of BNNs rather than training one large BNN.

3.2   Majority Voting System

After the training, we feed one input $\boldsymbol{x}^k$ to our list of BNNs, and we need to elaborate on the set of outputs.

Definition 1 (Dominant label). *For every $b \in \mathcal{I}$, we define*

$$C_b = \{\mathcal{J} \in \mathcal{P}(\mathcal{I})_p \mid \mathfrak{e}_{\mathcal{J}} = b\},$$

*and we say that a label b is a **dominant label** if $|C_b| \geq |C_l|$ for every $l \in \mathcal{I}$. We then define the set of dominant labels*

$$\mathcal{D} := \{b \in \mathcal{I} \mid b \text{ is a dominant label}\}.$$

Using this definition, we can have three possible outcomes.

(a) There exists a label $b \in \mathcal{I}$ such that $\mathcal{D} = \{b\} \implies$ our input is labelled as b.
(b) There exist $b_1, \ldots, b_p \in \mathcal{I}$, $b_i \neq b_j$ for all $i \neq j$ such that $\mathcal{D} = \{b_1, \ldots, b_p\}$, so $\mathcal{D} \in \mathcal{P}(\mathcal{I})_p \implies$ our input is labelled as $\mathfrak{e}_{\{b_1, \ldots, b_p\}} = \mathfrak{e}_{\mathcal{D}}$.
(c) $|\mathcal{D}| \neq 1 \wedge |\mathcal{D}| \neq p \implies$ our input is labelled as $z \notin \mathcal{I}$.

While case (a) is straightforward, we can label our input even when we do not have a clear winner, that is, when we have trained a BNN on the set of labels that are the most frequent (i.e., case (b)). Note that the proposed structured ensemble alongside its voting scheme can also be exploited for regular NNs.

Definition 2 (Label statuses). *In our labeling system, when testing an input seven different cases, herein called* **label statuses**, *can show up. Note that every input test will fall into one and only one label status.*

(s-0) *There exists exactly one dominant label and it is correct.*

(s-1) *There exist exactly p dominant labels $b_1, \ldots, b_p$ and $\mathfrak{e}_{\{b_1,\ldots,b_p\}}$ is correct.*

(s-2) *There exist exactly p dominant labels $b_1, \ldots, b_p$ and the correct one belongs to the set $\{b_1, \ldots, b_p\} - \mathfrak{e}_{\{b_1,\ldots,b_p\}}$.*

(s-3) *There exist exactly $\hat{p}$ dominant labels, $\hat{p} \neq 1 \wedge \hat{p} \neq p$, and one of them is correct.*

(s-4) *There exist exactly $\hat{p}$ dominant labels, $\hat{p} \neq 1 \wedge \hat{p} \neq p$, none of which are correct.*

(s-5) *There exist exactly p dominant labels $b_1, \ldots, b_p$, none of which are correct.*

(s-6) *There exists exactly one dominant label, but it is not the correct one.*

Example 1. Let us take $\mathcal{I} = \{bird, cat, dog, frog\}$ and $p = 2$. Note that, in this case, we have to train $\binom{4}{2} = 6$ networks:

$$\mathcal{N}_{\{bird, cat\}}, \quad \mathcal{N}_{\{bird, dog\}}, \quad \mathcal{N}_{\{bird, frog\}}, \quad \mathcal{N}_{\{cat, dog\}}, \quad \mathcal{N}_{\{cat, frog\}}, \quad \mathcal{N}_{\{dog, frog\}},$$

the first one distinguishes between *bird* and *cat*, the second one between *bird* and *dog*, and so on. A first input could have the following predicted labels:

$$\mathfrak{e}_{\{bird, cat\}} = bird, \qquad \mathfrak{e}_{\{bird, dog\}} = bird, \qquad \mathfrak{e}_{\{bird, frog\}} = frog,$$
$$\mathfrak{e}_{\{cat, dog\}} = cat, \qquad \mathfrak{e}_{\{cat, frog\}} = cat, \qquad \mathfrak{e}_{\{dog, frog\}} = dog.$$

We would then have

$$C_{bird} = \{\{bird, cat\}, \{bird, dog\}\}, \qquad C_{cat} = \{\{cat, dog\}, \{cat, frog\}\},$$
$$C_{dog} = \{\{dog, frog\}\}, \qquad C_{frog} = \{\{bird, frog\}\}.$$

In this case $\mathcal{D} = \{bird, cat\}$ because $|C_{bird}| = |C_{cat}| = 2 > 1 = |C_{dog}| = |C_{frog}|$ and we do not have a clear winner, but since $|\mathcal{D}| = 2 = p$, we have trained a network that distinguishes between the two most voted labels, and so we use its output as our final predicted label, labelling our input as $\mathfrak{e}_{\{bird, cat\}} = bird$. If *bird* is the right label we are in label status (s-1), if the correct label is *cat*, since $cat \in \{cat\} = \{bird, cat\} - \{bird\} = \{bird, cat\} - \mathfrak{e}_{\{bird, cat\}}$, we are in label status (s-2). Else we are in label status (s-5).

Example 2. Let us take $\mathcal{I} = \{0, 1, \ldots, 9\}$ and $p = 2$. Note that, in this case, we have to train $\binom{10}{2} = 45$ networks and that $|C_b| \leq 9$ for all $b \in \mathcal{I}$. Hence, an input could be labelled as follows:

$$C_0 = (\{0, i\})_{i=1,2,3,5,7,8}, \quad C_1 = (\{1, i\})_{i=5,6}, \quad C_2 = (\{2, i\})_{i=1,5,8},$$
$$C_3 = (\{3, i\})_{i=1,2,4,5}, \quad C_4 = (\{4, i\})_{i=0,1,2,5,6,7,9}, \quad C_5 = (\{5, i\})_{i=6,7},$$
$$C_6 = (\{6, i\})_{i=0,2,3,7}, \quad C_7 = (\{7, i\})_{i=1,2,3},$$
$$C_8 = (\{8, i\})_{i=1,3,4,5,6,7}, \quad C_9 = (\{9, i\})_{i=0,1,2,3,5,6,7,8}.$$

Visually, we can represent this labelling with the following scheme:

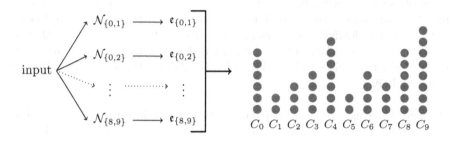

where we have omitted the name of each element of the set C_i for simplicity: for example, the dots above C_1 represent the sets $\{1, 5\}, \{1, 6\}$. Since $\mathcal{D} = \{9\}$, our input is labelled as 9. If 9 is the right label, we are in label status (s-0), if it is the wrong one, we are in label status (s-6). If instead $\hat{C}_j = C_j$, $j = 0, \ldots, 7$, and

$$\hat{C}_8 = (\{8, i\})_{i=1,3,4,5,6,7,9}, \quad \hat{C}_9 = (\{9, i\})_{i=0,1,2,3,5,6,7},$$

then $|\hat{\mathcal{D}}| = |\{4, 8, 9\}| = 3$, so that our input were labelled as -1. If the correct label is $4, 8$ or 9, we are in label status (s-3), else we are in label status (s-4). Lastly, if $\bar{C}_j = C_j$, $j \in \{0, 1, 2, 4, 5, 6, 7, 8\}$, and

$$\bar{C}_3 = (\{3, i\})_{i=1,2,4,5,9}, \quad \bar{C}_9 = (\{9, i\})_{i=0,1,2,5,6,7,8}$$

then $|\bar{\mathcal{D}}| = |\{4, 9\}| = 2 = p$ and since $\{4, 9\} \in \bar{C}_4$ our input is labelled as 4. If 4 is the correct label, we are in label status (s-1), if 9 is the correct label, we are in label status (s-2), else we are in label status (s-5). Note that in this example we used the notation $(\{j, i\})_{i=i_1,\ldots,i_n} = \{\{j, i_1\}, \ldots, \{j, i_n\}\}$, $j, i_1, \ldots, i_n \in \{0, \ldots, 9\}$ for brevity.

3.3   A Multi-objective MIP Model for Training BNNs

In this subsection, we present how each of single small BNN is trained with a multi-objective MIP model. For ease of notation, we denote with $\mathcal{L} := \{1, \ldots, L\}$ the set of layers and with $\mathcal{L}_2 := \{2, \ldots, L\}$, $\mathcal{L}^{L-1} := \{1, \ldots, L-1\}$ two of its subsets. We also denote with $\mathfrak{b} := \max_{k \in T, j \in N_0}\{|x_j^k|\}$ a bound on the values of the training data.

Training a BNN with a Multi-objective MIP Model. A few MIP models are proposed in the literature to train BNNs efficiently. In this work, to train a single BNN, we use a lexicographic multi-objective function that results in the sequential solution of three different MIP models: the Sat-Margin (S-M) described in [21], the Max-Margin (M-M), and the Min-Weight (M-W), both described in [22]. The first model S-M maximizes the number of confidently correctly predicted data. The other two models, M-M and M-W, aim to train a BNN following two principles: robustness and simplicity. Our model is based on a lexicographic multi-objective function: first, we train a BNN with the model S-M, which is fast to solve and always gives a feasible solution. Second, we use this solution as a warm start for the M-M model, training the BNN only with the images that S-M correctly classified. Third, we fix the margins found with M-M, and minimize the number of active weights with M-W, finding the lightest BNN with the robustness found by M-M.

Problem Variables. The critical part of our model is the formulation of the nonlinear activation function (1). We use an integer variable $w_{ilj} \in \{-1, 0, +1\}$ to represent the weight of the connection between neuron $i \in N_{l-1}$ and neuron $j \in N_l$. Variable u_{lj}^k models the result of the indicator function $\mathbb{1}(p)$ that appears in the activation function $\rho(\cdot)$ for the training instance x^k. The neuron activation is actually defined as $2u_{lj}^k - 1$. We introduce auxiliary variables c_{ilj}^k to represent the products $c_{ilj}^k = (2u_{lj}^k - 1)w_{ilj}$. Note that, while in the first layer, these variables share the same domain of the inputs, from the second layer on, they take values in $\{-1, 0, 1\}$. Finally, the auxiliary variables $\hat{y}^k$ represent a predicted label for the input x^k, and variable q_j^k are used to take into account the data points correctly classified. The procedure is designed such that the parameter configuration obtained in the first step is used as a warm start for the (M-M). Similarly, the solution of the second step is used as a warm start for the solver to solve (M-W). In this case, the margins lose their nature as decision variables and become deterministic constants derived from the solution of the previous step.

Sat-Margin (S-M) Model. We first train our BNN using the following S-M model.

$$\max \quad \sum_{k \in T} \sum_{j \in N_L} q_j^k \tag{2a}$$

$$\text{s.t.} \quad q_j^k = 1 \implies \hat{y}_j^k \cdot y_j^k \geq \frac{1}{2} \qquad\qquad \forall j \in N_L, k \in T, \tag{2b}$$

$$q_j^k = 0 \implies \hat{y}_j^k \cdot y_j^k \leq \frac{1}{2} - \epsilon \qquad\qquad \forall j \in N_L, k \in T, \tag{2c}$$

$$\hat{y}_j^k = \frac{2}{n_{L-1} + 1} \sum_{i \in N_{L-1}} c_{iLj}^k \qquad\qquad \forall j \in N_L, k \in T, \tag{2d}$$

$$u_{lj}^k = 1 \implies \sum_{i \in N_{l-1}} c_{ilj}^k \geq 0 \qquad\qquad \forall l \in \mathcal{L}^{L-1}, j \in N_l, k \in T, \tag{2e}$$

$$u_{lj}^k = 0 \implies \sum_{i \in N_{l-1}} c_{ilj}^k \le -\epsilon \qquad \forall l \in \mathcal{L}^{L-1}, j \in N_l, k \in T, \qquad (2f)$$

$$c_{i1j}^k = x_i^k \cdot w_{i1j} \qquad \forall i \in N_0, j \in N_1, k \in T, \qquad (2g)$$

$$c_{ilj}^k = (2u_{(l-1)j}^k - 1)w_{ilj} \qquad \forall l \in \mathcal{L}_2, i \in N_{l-1}, j \in N_l, k \in T, \qquad (2h)$$

$$q_j^k \in \{0, 1\} \qquad \forall j \in N_L, k \in T, \qquad (2i)$$

$$w_{ilj} \in \{-1, 0, 1\} \qquad \forall l \in \mathcal{L}, i \in N_{l-1}, j \in N_l, \qquad (2j)$$

$$u_{lj}^k \in \{0, 1\} \qquad \forall l \in \mathcal{L}^{L-1}, j \in N_l, k \in T, \qquad (2k)$$

$$c_{i1j}^k \in [-\mathfrak{b}, \mathfrak{b}] \qquad \forall i \in N_0, j \in N_1, k \in T, \qquad (2l)$$

$$c_{ilj}^k \in \{-1, 0, 1\} \qquad \forall l \in \mathcal{L}_2, i \in N_{l-1}, j \in N_l, k \in T. \qquad (2m)$$

The objective function (2a) maximizes the number of data points that are correctly classified. Note that ϵ is a small quantity standardly used to model strict inequalities. The implication constraints (2b) and (2c) and constraints (2d) are used to link the output $\hat{y}_j^k$ with the corresponding variable q_j^k appearing in the objective function. The implication constraints (2e) and (2f) model the result of the indicator function for the k-th input data. The constraints (2g) and the bilinear constraints (2h) propagate the results of the activation functions within the neural network. We linearize all these constraints with standard big-M techniques.

The solution of model (2a)–(2m) gives us the solution vectors $c_{\text{S-M}}$, $u_{\text{S-M}}$, $w_{\text{S-M}}$, $\hat{y}_{\text{S-M}}$, $q_{\text{S-M}}$. We then define the set

$$\hat{T} = \{k \in T \mid q_{j \text{ S-M}}^k = 1, \ \forall j \in N_L\}, \qquad (3)$$

of confidently correctly predicted images. We use these images as input for the next Max-Margin M-M, and we use the vector of variables $c_{\text{S-M}}$, $u_{\text{S-M}}$, $w_{\text{S-M}}$ to warm start the solution of M-M.

Max-Margin (M-M) Model. The second level of our lexicographic multi-objective model maximizes the overall margins of every single neuron activation, with the ultimate goal of training a robust BNN. Starting from the model S-M, we introduce the margin variables m_{lj}, and we introduce the following Max-Margin model.

$$\max \ \sum_{l \in \mathcal{L}} \sum_{j \in N_l} m_{lj} \qquad (4a)$$

$$\text{s.t.} \quad (2g)\text{–}(2m) \qquad \forall k \in \hat{T},$$

$$\sum_{i \in N_{L-1}} y_j^k c_{iLj}^k \ge m_{Lj} \qquad \forall j \in N_L, k \in \hat{T}, \qquad (4b)$$

$$u_{lj}^k = 1 \implies \sum_{i \in N_{l-1}} c_{ilj}^k \geq m_{lj} \qquad \forall l \in \mathcal{L}^{L-1}, j \in N_l, k \in \hat{T}, \qquad (4c)$$

$$u_{lj}^k = 0 \implies \sum_{i \in N_{l-1}} c_{ilj}^k \leq -m_{lj} \qquad \forall l \in \mathcal{L}^{L-1}, j \in N_l, k \in \hat{T}, \qquad (4d)$$

$$m_{lj} \geq \epsilon \qquad \qquad \qquad \forall l \in \mathcal{L}, j \in N_l. \qquad (4e)$$

Again, we can linearize constraints (4c) and (4d) with standard big-M constraints. This model gives us the solution vectors $\mathbf{c}_{\text{M-M}}, \mathbf{u}_{\text{M-M}}, \mathbf{w}_{\text{M-M}}, \mathbf{m}_{\text{M-M}}$. We then evaluate $\mathbf{v}_{\text{M-M}}$ as

$$v_{ilj_{\text{M-M}}} = |w_{ilj_{\text{M-M}}}| \qquad \forall l \in \mathcal{L}, i \in N_{l-1}, j \in N_l. \qquad (5)$$

Min-Weight (M-W) Model. The third level of our multi-objective function minimizes the overall number of non-zero weights, that is, the connection of the trained BNN. We introduce the new auxiliary binary variable v_{ilj} to model the absolute value of the weight w_{ilj}. Starting from the solution of model M-M, we fix $\hat{\mathbf{m}} = \mathbf{m}_{\text{M-M}}$, and we pass the solution $\mathbf{c}_{\text{M-M}}, \mathbf{u}_{\text{M-M}}, \mathbf{w}_{\text{M-M}}, \mathbf{v}_{\text{M-M}}$ as a warm start to the following M-W model:

$$\min \sum_{l \in \mathcal{L}} \sum_{i \in N_{l-1}} \sum_{j \in N_l} v_{ilj} \qquad \qquad (6a)$$

$$\text{s.t.} \quad (2g)\text{--}(2m) \qquad \qquad \forall k \in \hat{T},$$

$$\sum_{i \in N_{L-1}} y_j^k c_{iLj}^k \geq \hat{m}_{Lj} \qquad \qquad \forall j \in N_L, k \in \hat{T}, \qquad (6b)$$

$$u_{lj}^k = 1 \implies \sum_{i \in N_{l-1}} c_{ilj}^k \geq \hat{m}_{lj} \qquad \forall l \in \mathcal{L}^{L-1}, j \in N_l, k \in \hat{T}, \qquad (6c)$$

$$u_{lj}^k = 0 \implies \sum_{i \in N_{l-1}} c_{ilj}^k \leq -\hat{m}_{lj} \qquad \forall l \in \mathcal{L}^{L-1}, j \in N_l, k \in \hat{T}, \qquad (6d)$$

$$-v_{ilj} \leq w_{ilj} \leq v_{ilj} \qquad \qquad \forall l \in \mathcal{L}, i \in N_{l-1}, j \in N_l, \qquad (6e)$$

$$v_{ilj} \in \{0, 1\} \qquad \qquad \forall l \in \mathcal{L}, i \in N_{l-1}, j \in N_l. \qquad (6f)$$

Note that whenever v_{ilj} is equal to zero, the corresponding weight w_{ilj} is set to zero due to constraint (6e), and, hence, the corresponding link can be removed from the network.

Lexicographic multi-objective. By solving the three models S-M, M-M, and M-W, sequentially, we first maximize the number of input data that is correctly classified, then we maximize the margin of every activation function, and finally, we minimize the number of non-zero weights. The solution of the decision variables w_{ilj} of the last model M-W defines our classification function $f_W : \mathbb{R}^{n_0} \to \{\pm 1\}^{n_L}$.

4   Computational Results

We run three types of experiments to address the following questions.

- **Experiment 1:** How does our approach compare with the previous state-of-the-art MIP models for training BNNs in the context of few-shot learning?
- **Experiment 2:** How does the BeMi ensemble scale with the number of training images, considering two different types of BNNs?
- **Experiment 3:** How does the proposed approach perform on a different dataset, comparing the running time, the average gap to the optimal training MIP model, and the percentage of links removed?

Datasets. The experiments are performed on the MNIST dataset [12] for a fair comparison with the literature, and Fashion-MNIST [24] dataset. We test our results on 500 images for each class. For each experiment, we report the average over three different samples of images. The images are sampled uniformly at random in order to avoid overlapping between different experiments.

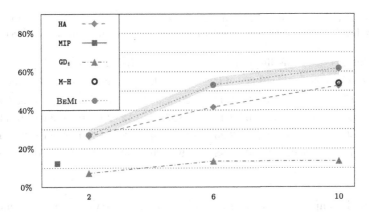

Fig. 1. Comparison of published approaches vs BeMi, in terms of accuracy over the MNIST dataset using few-shot learning with 2, 6, and 10 images per digit.

Implementation Details. We use Gurobi version 9.5.1 [5] to solve our MIP and ILP models. The parameters of Gurobi are left to the default values. All the MIP experiments were run on an HPC cluster running CentOS but using a single node per experiment. Each node has an Intel CPU with 8 physical cores working at 2.1 GHz and 16 GB of RAM. In all of our experiments, we fix the value $\epsilon = 0.1$. The source code will be available on GitHub in case of acceptance of this paper.

4.1 Experiment 1

The first set of experiments aims to compare the BeMi ensemble with the following state-of-the-art methods: the hybrid CP and MIP model based on Max-Margin optimization (HA) [22]; the gradient-based method GD_t introduced in [6] and adapted in [22] to deal with link removal; and the Min-hinge (M-H) model proposed in [21]. For the comparison, we fix the setting of [22], which takes from the MNIST up to 10 images for each class, for a total of 100 training data points, and which uses a time limit of $7\,200$ s to solve their MIP training models. In our experiments, we train the BeMi ensemble with 2, 6, and 10 samples for each digit. Since our ensemble has 45 BNNs, we leave for the training of each single BNN a maximum of $160\,$s (since $160 \times 45 = 7\,200$). In particular, we give a 75 s time limit to the solution of S-M, 75 s to M-M, and 10 s to M-W. In all of our experiments, whenever the optimum is reached within the time limit, the remaining time is added to the time limit of the subsequent model. We remark that our networks could be trained in parallel, which would highly reduce the wall-clock runtime. For the sake the completeness, we note that we are using $45 \times (784 \times 4 + 4 \times 4 + 4 \times 1) = 142\,020$ parameters (all the weights of all the 45 BNNs) instead of the $784 \times 16 + 16 \times 16 + 16 \times 10 = 12\,960$ parameters used in [22] for a single large BNN. Note that, in this case, the dimension of the parameter space is $3^{12\,960} (\cong 10^{6\,183})$, while, in our case, it is $45 \times 3^{3\,156} (\cong 10^{1507})$.

Table 1. Percentages of MNIST images classified as correct, wrong, or unclassified $(n.l.)$, and of label statuses from (s-0) to (s-6), for the architecture $\mathcal{N}_a = [784, 4, 4, 1]$.

Images per class	Classification			Label status						
	correct	wrong	$n.l.$	s-0	s-1	s-2	s-3	s-4	s-5	s-6
10	61.80	36.22	1.98	58.30	3.50	1.84	1.30	0.68	4.74	29.64
20	69.96	27.60	2.44	66.68	3.28	2.12	2.18	0.26	3.04	22.44
30	73.18	24.56	2.26	70.14	3.04	1.88	1.88	0.38	2.68	20.00
40	78.82	19.30	1.88	75.56	3.26	1.90	1.72	0.16	1.70	15.70

Figure 1 compares the results of our BeMi ensemble with four other methods: the hybrid CP-MIP approach HA [22]; the pure MIP model in [22], which can handle a single image per class; the gradient-based method GD_t, which is the version of [6] modified by [22]; the minimum hinge model M-H presented in [21], which report results only for 10 digits per class. We report the best results reported in the original papers for these four methods. The BeMi ensemble obtains an average accuracy of 61%, outperforms all other approaches when 6 or 10 digits per class are used, and it is comparable with the hybrid CP-MIP method when only 2 digits per class are used.

4.2   Experiment 2

This second set of experiments studies how our approach scales with the number of data points (i.e., images) per class, and how it is affected by the architecture of the small BNNs within the BeMi ensemble. For the number of data points per class we use $10, 20, 30, 40$ training images per digit. We use the layers $\mathcal{N}_a = [784, 4, 4, 1]$ and $\mathcal{N}_b = [784, 10, 3, 1]$ for the two architectures, Herein, we refer to Experiments $2a$ and $2b$ as the two subsets of experiments related to the architectures $\mathcal{N}_a$ and $\mathcal{N}_b$. In both cases, we train each of our 45 BNNs with a time limit of 290 s for model S-M, 290 s for M-M, and 20 s for M-W, for a total of 600 s (i.e., 10 min for each BNN).

Figure 2a shows the results for Experiments 2a and 2b: the dotted and dashed lines refer to the two average accuracies of the two architectures, while the colored areas include all the accuracy values obtained as the training instances vary. While the two architectures behave similarly, the best average accuracy exceeds 75% and it is obtained with the first architecture $\mathcal{N}_a$.

Table 1 reports the results for the BEMI ensemble where we distinguish among images classified as correct, wrong, or unclassified. These three conditions refer to different label statuses specified in Definition 2: the correct labels are the sum of the statuses (s-0) and (s-1); the wrong labels of statuses (s-2), (s-5), and (s-6); the unclassified labels ($n.l.$) of (s-3) and (s-4).

4.3 Experiment 3

In the third experiment, we replicate Experiments 2a and 2b with the two architectures $\mathcal{N}_a$ and $\mathcal{N}_b$, using the Fashion-MNIST dataset.

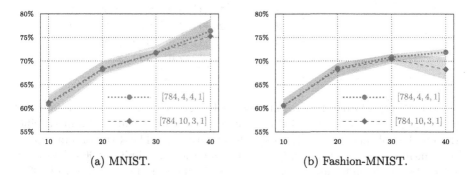

(a) MNIST.   (b) Fashion-MNIST.

Fig. 2. Average accuracy for the BEMI ensemble tested on two architectures, namely $\mathcal{N}_a = [784, 4, 4, 1]$ and $\mathcal{N}_b = [784, 10, 3, 1]$, using $10, 20, 30, 40$ images per class.

Figure 2b shows the results of Experiments 3a and 3b. As in Fig. 1, the dotted and dashed lines represent the average percentages of correctly classified images, while the colored areas include all accuracy values obtained as the instances vary. The first architecture is comparable with the second, with up to 30 training images per digit, while it is significantly better with 40 images. For the Fashion-MNIST, the best average accuracy exceeds 70%.

Table 2. Aggregate results for Experiments 2 and 3: the 4-th column reports the runtime to solve the first model S-M; *Gap (%)* refers to the mean and maximum percentage gap at the second MIP model M-M; *Links (%)* is the percentage of non-zero weights after the solution of models M-M and M-W.

Dataset	Layers	Images per class	Model S-M time (s)	Gap (%) mean	max	Links (%) (M-M)	(M-W)
MNIST	784,4,4,1	10	2.99	17.37	28.25	49.25	27.14
		20	5.90	19.74	24.06	52.95	30.84
		30	10.65	20.07	26.42	56.90	30.88
		40	15.92	18.50	23.89	58.70	29.42
	784,10,3,1	10	6.88	6.28	9.67	49.46	23.96
		20	17.02	7.05	8.42	53.25	26.65
		30	25.84	7.38	15.88	57.21	25.02
		40	44.20	9.90	74.16	59.08	24.22
F-MNIST	784,4,4,1	10	7.66	17.21	25.92	86.38	56.54
		20	14.60	22.35	28.00	93.18	57.54
		30	26.10	19.78	29.53	92.56	58.78
		40	39.90	22.71	75.03	93.13	64.61
	784,10,3,1	10	13.83	6.14	8.98	86.65	53.72
		20	26.80	7.84	9.59	93.57	51.03
		30	38.48	7.18	16.09	92.90	52.50
		40	64.52	12.10	55.19	93.57	55.67

Table 2 reports detailed results for all Experiments 2 and 3. The first two columns give the dataset and the architecture, and the third column specifies the number of images per digit used during training. The 4-th column reports the runtime for solving model S-M. Note that the time limit is 290 s; hence, we solve exactly the first model, consistently achieving a training accuracy of 100%. The remaining four columns give: *Gap (%)* refers to the mean and maximum percentage gap at the second MIP model (M-M) of our lexicographic multi-objective model, as reported by the Gurobi MIPgap attribute; *Links (%)* is the percentage of non-zero weights after the solution of the second model M-M, and after the solution of the last model M-W. The results show that the runtime and the gap increase with the size of the input set. However, for the percentage of removed links, there is a significant difference between the two datasets: for MNIST, our third model M-W removes around 70% of the links, while for the Fashion-MNIST, it removes around 50% of the links. Note that in both cases, these significant reductions show how our model is also optimizing the BNN architecture.

5   Conclusions

In this work, we have introduced the BeMi ensemble, a structured architecture of BNNs for classification tasks. Each network specializes in distinguishing between pairs of classes and combines different approaches already existing in the literature to preserve feasibility while being robust and lightweight. These features and the nature of the parameters are critical to enabling neural networks to run on low-power devices. In particular, such networks can be implemented using Boolean operations and do not need CPUs to run. The output of the BeMi ensemble is chosen by a majority voting system that generalizes the one-versus-one scheme. Notice that the BeMi ensemble is a general architecture that could be employed using other types of neural networks, for example, Integer-valued NNs [21]. A current limitation of our approach is the strong dependence on the randomly sampled images used for training. In future work, we plan to improve the training data selection by using a k-medoids approach, dividing all images of the same class into disjoint non-empty subsets and consider their centroids as training data. This approach should mitigate the dependency on the sampled training data points. We also plan to better investigate the scalability of our method with respect to the number of classes of the classification problem, varying the parameter p and training fewer BNNs, namely, one for every $\mathcal{J} \in \mathcal{Q} \subset \mathcal{P}(\mathcal{I})_p$, with $|\mathcal{Q}| << |\mathcal{P}(\mathcal{I})_p|$. In conclusion, we used the MNIST dataset to provide a fair comparison with existing literature. In future, we intend to investigate datasets more appropriate for the task of few-shot learning [14].

References

1. Anderson, R., Huchette, J., Ma, W., Tjandraatmadja, C., Vielma, J.P.: Strong mixed-integer programming formulations for trained neural networks. Math. Program. **183**(1), 3–39 (2020)
2. Bishop, C.M., Nasrabadi, N.M.: Pattern Recognition and Machine Learning. vol. 4, no. 4 (2006)
3. Domingos, P.: A few useful things to know about machine learning. Commun. ACM **55**(10), 78–87 (2012)
4. Fischetti, M., Jo, J.: Deep neural networks and mixed integer linear optimization. Constraints **23**(3), 296–309 (2018)
5. Gurobi Optimization, LLC: Gurobi optimizer reference manual (2022). https://www.gurobi.com
6. Hubara, I., Courbariaux, M., Soudry, D., El-Yaniv, R., Bengio, Y.: Binarized neural networks. In: Advances in Neural Information Processing Systems (NeurIPS), vol. 29, pp. 4107–4115 (2016)
7. Jiang, Y., Krishnan, D., Mobahi, H., Bengio, S.: Predicting the generalization gap in deep networks with margin distributions. In: International Conference on Learning Representations (ICLR) (2019)
8. Kawaguchi, K., Kaelbling, L.P., Bengio, Y.: Generalization in deep learning. arXiv:1710.05468 (2017)
9. Keskar, N.S., Mudigere, D., Nocedal, J., Smelyanskiy, M., Tang, P.T.P.: On large-batch training for deep learning: generalization gap and sharp minima. In: International Conference on Learning Representations (ICLR), vol. 5 (2017)
10. Khalil, E.B., Gupta, A., Dilkina, B.: Combinatorial attacks on binarized neural networks. In: International Conference on Learning Representations (ICLR) (2019)
11. LeCun, Y., Bengio, Y., Hinton, G.: Deep learning. Nature **521**(7553), 436–444 (2015)

12. LeCun, Y., Cortes, C., Burges, C.J.: The MNIST database of handwritten digits (1998). https://yann.lecun.com/exdb/mnist
13. Lin, X., Zhao, C., Pan, W.: Towards accurate binary convolutional neural network. In: Advances in Neural Information Processing Systems (NeurIPS), vol. 30 (2017)
14. Lorenzo, B., Bjorn, B., Luca, I., Joachim, D.: Image classification with small datasets: overview and benchmark. IEEE Access **10**, 49233–49250 (2022)
15. Moody, J.: The effective number of parameters: an analysis of generalization and regularization in nonlinear learning systems. In: Advances in Neural Information Processing Systems (NeurIPS), vol. 4, pp. 847–854 (1991)
16. Neyshabur, B., Bhojanapalli, S., McAllester, D., Srebro, N.: Exploring generalization in deep learning. In: Advances in Neural Information Processing Systems (NeurIPS), vol. 30, pp. 5947–5956 (2017)
17. Sakr, C., Choi, J., Wang, Z., Gopalakrishnan, K., Shanbhag, N.: True gradient-based training of deep binary activated neural networks via continuous binarization. In: 2018 IEEE International Conference on Acoustics, Speech and Signal Processing (ICASSP), pp. 2346–2350. IEEE (2018)
18. Schmidhuber, J.: Deep learning in neural networks: an overview. Neural Netw. **61**, 85–117 (2015)
19. Serra, T., Kumar, A., Ramalingam, S.: Lossless compression of deep neural networks. In: Hebrard, E., Musliu, N. (eds.) CPAIOR 2020. LNCS, vol. 12296, pp. 417–430. Springer, Cham (2020). https://doi.org/10.1007/978-3-030-58942-4_27
20. Tang, W., Hua, G., Wang, L.: How to train a compact binary neural network with high accuracy? In: Thirty-First AAAI Conference on Artificial Intelligence (2017)
21. Thorbjarnarson, T., Yorke-Smith, N.: Optimal training of integer-valued neural networks with mixed integer programming. PLoS ONE **18**(2), 1–17 (2023)
22. Toro Icarte, R., Illanes, L., Castro, M.P., Cire, A.A., McIlraith, S.A., Beck, J.C.: Training binarized neural networks using MIP and CP. In: Schiex, T., de Givry, S. (eds.) CP 2019. LNCS, vol. 11802, pp. 401–417. Springer, Cham (2019). https://doi.org/10.1007/978-3-030-30048-7_24
23. Vanschoren, J.: Meta-learning. In: Hutter, F., Kotthoff, L., Vanschoren, J. (eds.) Automated Machine Learning. The Springer Series on Challenges in Machine Learning, pp. 35–61. Springer, Cham (2019). https://doi.org/10.1007/978-3-030-05318-5_2
24. Xiao, H., Rasul, K., Vollgraf, R.: Fashion-MNIST: a novel image dataset for benchmarking machine learning algorithms. arXiv:1708.07747 (2017)
25. Young, H.P.: Condorcet's theory of voting. Am. Polit. Sci. Rev. **82**(4), 1231–1244 (1988)
26. Yu, X., Serra, T., Ramalingam, S., Zhe, S.: The combinatorial brain surgeon: pruning weights that cancel one another in neural networks. In: International Conference on Machine Learning (ICML), pp. 25668–25683 (2022)

Discovering Explicit Scale-Up Criteria in Crisis Response with Decision Mining

Britt Lukassen, Laura Genga⬚, and Yingqian Zhang$^{(\boxtimes)}$⬚

Eindhoven University of Technology, Eindhoven, The Netherlands
{L.Genga,yqzhang}@tue.nl

Abstract. In modern society, incidents such as road incidents or fires, occur daily, requiring institutions to develop appropriate management protocols to react quickly. When an incident requires coordination among different emergency services or it is estimated to have a significant impact on the population, *crisis management* processes are used. The overall aim of crisis management is to provide the right resources to manage the incident and return to a normal situation as soon as possible. However, the decision to scale up an incident to a crisis level is often left to the experience of operational commanders without explicit criteria or guidelines. In this research, we propose a framework combining data-driven decision-mining approaches with implicit knowledge formalization techniques to discover explicit criteria to support decision-makers in crisis response. We tested our approach in a case study at VRU, the safety region for the region Utrecht, in The Netherlands. The obtained results show that the approach has been able to extract criteria acknowledged by the decision-makers, which is the first step to developing appropriate guidelines to steer the decisional process of the incident scaling up.

Keywords: Crisis management · Process mining · Machine learning

1 Introduction

Daily, incidents such as road accidents or kitchen fires are reported at the emergency dispatch centres. Based on the nature and gravity of the incident on hand, emergency services (i.e., *disciplines*) are dispatched to manage the incident. Dispatch centres alert and direct, for instance, the fire brigade, the police, ambulance services, and the municipality, as well as a command and control structure for incident management. Larger scale command and control structures are dispatched when a routine incident evolves into a *crisis*, i.e., when an incident potentially leads to a significant impact on the population or when the incident management needs larger scale co-ordination in a process called *Scale-up*. Government institutions are in charge of developing and implementing appropriate *crisis management* processes to guarantee a quick response. The overall aim of crisis management is to provide the right resources to control the crisis and return to a normal situation as soon as possible. Crisis response processes

M. Sellmann and K. Tierney (Eds.): LION 2023, LNCS 14286, pp. 459–474, 2023.
https://doi.org/10.1007/978-3-031-44505-7_31

are characterized by being dynamic, highly knowledgeable, and unstructured [8]. Furthermore, there are critical factors such as time and information availability that must be considered when making decisions in crisis situations [9]. Usually, the decision to scale up an incident to a crisis level is often left to the experience of operational commanders without explicit criteria or guidelines. The subjective and unstructured nature of these decisions poses significant challenges in the management and the improvement of crisis management processes, which is a topic of utmost interest from institutions involved in incident management.

This research aims to answer the following question: *how to define explicit criteria to support decision-makers in determining the most appropriate multidisciplinary scale-up in crisis management?* To address this challenge, we investigate the feasibility of *Decision mining* approaches. The goal of this discipline consists in analyzing data recording past decisions taken by human actors both to create business rules and to check compliance to business rules and regulations [11]. Previous work has already proved the capabilities of these techniques to identify data patterns of interest in reaching a certain decision by applying machine-learning techniques. In order to incorporate implicit knowledge in decision criteria, not always available in historic data, we also leverage methods for knowledge acquisition from qualitative research, and we propose a framework able to keep into account both sources of information in a structured way.

To validate our approach, we carried out a case study at the Veiligheidsregio Utrecht (VRU), the safety region for the region Utrecht, in The Netherlands. VRU are responsible for developing and implementing policy plans to deal with incidents in the region, for collecting relevant information needed to control a crisis, and for incident response by the fire service. The VRU is interested in evaluating the decision process leading to a scale-up to different *GRIP* levels, where GRIP is the abbreviation for Coordinated Regional Incident Management Procedure. GRIP links the recommended level of scale-up to characteristics of the incident. However, decisions for scale-up in each specific case are made by humans and can therefore also be based on subjective criteria and human judgment. Results obtained by applying the proposed framework at VRU show that the proposed method was able to obtain decision models with good accuracy and highlighting criteria recognized and validated by the decision-makers at VRU. Furthermore, different decision points in the process have been identified, each of them with its specific set of criteria, to provide tailored support to the decision-makers.

The rest of the manuscript is organized as follows. Section 2 illustrates the proposed approach; Sect. 3 describes the case study; Sect. 4 discusses the data gathering; Sect. 5 illustrates the analysis of the VRU scale-up process; Sect. 6 discusses the application of machine learning techniques to analyze decision points; Sect. 7 provides an overview of related work; finally, Sect. 8 draws some conclusions and delineates future work.

2 Methodology

The proposed approach involves three main phases. The first one is the *business and data understanding*, which requires an exploration of all the steps involved in multidisciplinary incident managemnt scale-up and the gathering and preprocessing of the incident data. Note that determining criteria relevant to decision-makers requires not only knowledge of past decisions but also an insight into why these decisions were made; namely, we are interested in which criteria are important to different decision-makers, for which we need to collect appropriate implicit knowledge. In our approach, we have chosen to use a questionnaire to gather this knowledge, to consider the opinion of multiple decision-makers and to grasp a general view of their considerations from different perspectives. The output of the first phase corresponds to the results of the questionnaire and to a so-called *event log* storing all the information related to past instances of the incident management process. This log is used as input in the second phase of the methodology, i.e., the *decision mining* phase. Here we refer to the methodology proposed by Rozinat et al. [15], which in turn, involves two steps. The first one is the *process discovery* step, whose goal is to discover the actual multidisciplinary scale-up process. Understanding the process is a starting point for understanding the decision that has to be made. In our experiments, we compared the results obtained by different state-of-the-art algorithms to select the process model that best describes the process under analysis. For the selected process model, the decision points are identified, which are then further analyzed with *Machine-Learning* (ML) algorithms to identify data patterns for different decisions. The input features are designed based on the criteria highlighted by decision-makers, with the help of the available data gathered in previous steps. Finally, in the *evaluation* phase, explainability techniques are used to determine the most important features for the tested predictive models, which are then compared to the implicit knowledge insights of decision-makers. The results of this analysis represent the first step to drawing tailored, efficient guidelines to guide the decisional process of the human actors in the process.

3 Crisis Management Process at VRU

The crisis management process always starts with an incident notification at the emergency dispatch centre. The call is directed to a corresponding dispatcher, which assigns vehicles accordingly. While some incidents can be handled by the disciplines at the location, others require further coordination, for which several scale-up options are available, hereafter referred to as *GRIP levels*.

For each GRIP level, there is a designated operational team. The first two levels involve mostly stakeholders working at the operational level, while levels 3 and 4 involve actors from the organizational level, such as the municipality. Currently, each GRIP level includes a standard operational team that is alarmed. However, each incident has specific characteristics that determine who could be meaningful to include in the procedure.

As a result, decision-makers have several decision options to evaluate properly. They have to judge every situation with specific characteristics correctly and assign the most appropriate people, that have value to resolve the incident. For scale-up to the operational level (GRIP 1 and GRIP 2) the decision-makers are mostly operational commanders, which are often experts in their discipline but may not adequately considering the needs of other disciplines when making a decision. In addition, operators tend to lose themselves in the heat of the incident, losing the ability to zoom out. For scale-up to the organizational level (GRIP 3 and GRIP 4) the decision-makers are not directly involved at the incident location, but they are accountable for the operational execution and decisions. For scale-up to GRIP 3, this is the mayor of the municipality corresponding to the incident, and for scale-up to GRIP 4 this is the chairman of the VRU. These scale-ups only occurr when an incident has long-term effects. Therefore, decisions for organizational scale-up are made differently than decisions to operational scale-up, considering the impact on the specific municipality or region. Challenging is that they are not directly involved in the incident but are informed when needed.

4   Data Gathering

4.1   Historic Data

We extracted historic data sets which stored diverse information about the incidents and the deployed vehicles. In order to apply process discovery, an *event log* has to be generated by integrating the different data sets and creating activities based on the timestamps in the data. In an event log, each row represents a new activity with a timestamp. In our case, *Case ID* is the *Incident number*. One incident (aka, one case) can have multiple activities. From the available data sets, activities have been labeled by exploiting domain knowledge and combining it with the timestamps of an incident number. Besides, as many attributes as possible are added to represent the incident context in the event log.

According to the research scope, the aim is to analyze the decisions in multidisciplinary incidents. Therefore a selection is made of which incidents include the assistance of the fire brigade, police, and medical team. An excerpt of the created event log is shown in Table 1. The following attributes are considered in the event log: Classification criteria, Priority, Municipality, City, Object incident, Object report, Type location, Intervention type. The integration results in an event log with 18,349 activities and 6,866 unique multidisciplinary incidents.

4.2   Implicit Knowledge

Questionnaire Setup. The aim of this questionnaire is to gain insight into the criteria that are important to decision-makers. The group of decision-makers contacted for the questionnaire involves all people in position to scale-up to GRIP 1 and/or GRIP 2. In total approximately 150 people have been contacted to complete the questionnaire. We chose to use three real past crisis situations, asking the decision makers to make a decision on scaling up. One situation

Table 1. Example event log

Incident Number	Timestamp	Activity	textbfAttributes
201513899	2015–01–15 17:10:00	Start Incident	...
201513899	2015–01–15 17:10:00	Partner organization	...
201513899	2015–01–15 17:23:00	OvD Fire Brigade	...
201513899	2015–01–15 17:37:00	GRIP 1	...
201513951	2015–01–15 17:40:00	Start Incident	...
201513899	2015–01–15 17:55:00	IM	...
201513951	2015–01–15 18:54:00	OvD Fire Brigade	

requires extensive coordination but no GRIP. Another situation clearly needs scale-up to GRIP. The last situation is doubtful since it is not a regular GRIP but still requires some of the aspects of GRIP. All of these crisis situations have been retrieved from "Lessen uit crises en mini crises" [4,5,24] and discussed with the experts from the crisis management department.

In the questionnaire, each situation consists of two parts representing different moments in time. The first part consists of incident notification information followed by a set of questions. The first question asked to the decision-maker if a multidisciplinary scale-up is appropriate with the current information. Furthermore, an explanatory statement for this decision and possible game changers, are asked. Since more information is gathered during the incident that might change the decision, we included a second part as well. The second part starts by providing complementary information about the incident, gathered by the disciplines arrived at the incident location. The first question of the first part is repeated, to see if the judgement of the decision-makers changes with the additional incident information. In addition, the follow-up question asks the decision-maker to select points of attention from the situation description. In this question, the decision-maker has the possibility to summarize which criteria they think they consider. In the next question, criteria are provided based on themes in National Control Room System (LCMS), including Incident, Risks and Safety, Meteo, Victims/population, Environmental analysis, Communication, Services involved, and Missing information. These themes are covered and divided into several criteria each. The decision-makers have the possibility to check as many of the criteria they consider for this specific incident situation.

Questionnaire Results. The questionnaire was sent by email to approximately 150 people of which 71 responded. Figure 1 shows the count of the mentioned criteria extracted by the answers. The figure suggests that all criteria are considered and that their use heavily depends on the situation. The criteria found important by decision-makers in all three situations are: 'Incident location', 'Incident size', 'Incident type', '(number of) injured and injury classification', 'Sensitivity incident on social media' and 'Own disciplines involved'. These all have more than half of the votes in all three situations. Some other criteria that have an higher than average score are: 'Expected duration incident', 'Possible effects on people/material', 'Safe/unsafe area', 'Involved partners' and 'Duration of incident unknown'. These criteria are discussed with the VRU. They recognize these insights from practice. It is worth noting that these criteria are quite general and can be applied also to different crises.

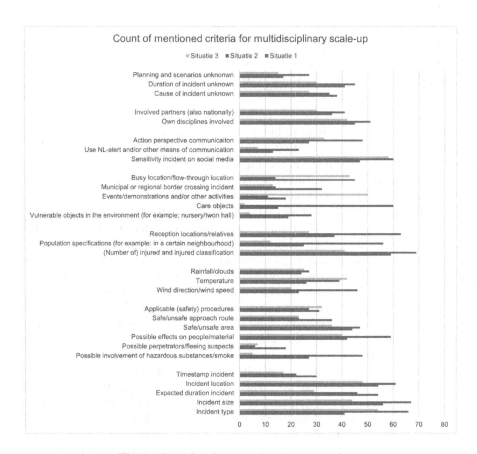

Fig. 1. Considered criteria by decision-makers

Analyzing the answers related to the criteria proposed by the decision-makers, we found that most of the answers were in line with the listed criteria, except for some answers related to specific incidents. We decided not to discard these criteria because of the narrow range of situations they apply to.

5   Process Discovery

In this section, we discuss the results obtained by process discovery techniques, from which we select the model that best fits the event log created at the previous step. This model will then be used to identify the decision points.

Before extracting the process model, we applied some pre-processing techniques to remove irrelevant data and outliers commonly used in process mining literature [7]. First we applied *selection* filters to remove traces showing outlier or noisy behaviours. For instance, we removed all traces where the first event was not 'Start incident'. Analyzing the event log with the domain experts, we found out that this was mostly due to recording errors. We also applied *aggregation* filters to rename specialized events and group them to higher-level groups. In particular, the events "OVD B","OVD P", and "OVD G" are renamed as one general event "OVD" (note that "OVD" stands for "Duty officer"). The reason is that all OVDs are there for the same purpose, to coordinate their units and ensure collaboration with other disciplines.

Once the event log has been cleaned, we need to select the process discovery algorithm to use. We investigated the results obtained by two miners frequently used in literature, i.e., the heuristic miner [23] and the inductive miner [10].

Both the miners have been fitted on our event log using the package PM4Py[1]. The discovered models are evaluated for their performance based on four well-known metrics within process mining; fitness, precision, simplicity, and generalization. *Fitness* quantifies how much of the observed behavior in the event log is captured by the process model [22]. *Precision* quantifies how much behavior exists in the process model that was not observed in the event log [22]. *Simplicity* quantifies the complexity of the model [22]. *Generalization* quantifies how well the model explains unobserved system behaviour [22].

Several approaches have been proposed to quantify these dimensions. In this research, we used fitness, and precision metrics based on *alignments* [1], while we used *token-based reply* [16] for generalization. These are metrics commonly used in literature and available in the python implementation PM4Py.

Table 2 shows the results obtained by applying the different miners on the event log. Overall, the inductive miner seems the best in balancing all four metrics. Therefore, the process model found with the inductive miner is chosen as the best-performing model to continue with machine learning. Note that the process model has also been validated with experts from VRU. Figure 2 shows the model obtained by the inductive miner, where we highlighted the decision points and named them with numbers. The flow visualized in this process model starts always with 'Start incident'. Next, the first decision point is identified with 1. At this point, a decision is made between assigning a partner organization (event: 'Instantie assigned') or not (invisible activity). Whether or not a partner is assigned, after these activities the flow merges again in a new decision point, identified with 2. Two choices are possible, with both an invisible activity as the first event. The lower edge goes directly to the event 'End incident'; in other

---

[1] https://pm4py.fit.fraunhofer.de/.

words, no scale-up is necessary. The other choice directly results in a new decision point, namely decision point 3. At this point, there are four outgoing arcs, all defining different choices in multidisciplinary scale-up. These four possible events are 'IM', 'CAC', 'GRIP 1' and 'OVD'. Respectively the second, third and fourth choices lead to the next decision point, number 4. While the event IM leads directly to decision point 5. From decision point 4, the choice for scale-up to GRIP 2 can be made. After which, this flow also enters decision point 5. Decision point 5 is the final decision point in this process model, which includes a loop back to decision point 3. The other option from decision point 5 is to enter the event 'End incident', after which the process is completed.

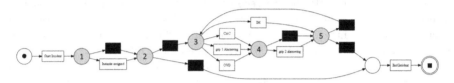

Fig. 2. Process model with inductive miner

The five identified decision points have been discussed with the VRU, according to which decision point 2 and decision point 5 are the most interesting. For the second decision point, the initial choice has two options, entering multidisciplinary scale up or 'routine' incident handling. Notice that decision point 3 follows immediately after the invisible activity of decision point 2. For VRU, this point is of interest because they are interested in what the difference is between incidents that require scale-up and those incidents that do not need scale-up. In decision point 5, there are two outgoing arcs with invisible activities, representing no further scale-up required and a loop back to decision point 3. The presence of a loop makes this decision point of special interest for experts at VRU. In particular, they would like to determine if there are differences in decisions made for process instances which executed the loop one time versus those executing the loop multiple times.

6   Decision Criteria Extraction

In Sect. 5, decision points 2 and decision point 5 are identified as interesting. Both decision points have two outgoing arcs followed by invisible activities. Therefore, we decided to model the decision mining problem for these decision points as binary classification problems. We labelled the invisible activities using domain knowledge. The lower arrow of Fig. 2 leaving decision point 2 does not include any activities until the 'End incident'. Therefore, this invisible activity can be labeled as a 'routine incident'. The upper arrow in Fig. 2 towards decision point 3, is the connection with all events that are defined as multidisciplinary scale-up. Therefore, this activity could be named 'Multi scale up required'. These labels are used for the classification of decision point 2.

Table 2. Model performance

	Fitness	Precision	Simplicity	Generalization
Heuristic Miner	0.91	0.99	0.51	0.80
Inductive Miner	1	0.83	0.65	0.94

For decision point 5, the upper arrow is a loop back to decision point 3, which identifies a request for additional multidisciplinary scale-up. Therefore, this invisible activity could be named 'Additional multi-scale up'. The lower arc leaving decision point 5 leaves towards 'End incident'. Therefore, the invisible activity can be named 'No additional multi scale up'. These labels are used for the classification of decision point 5.

To leverage information related to the activities executed before the decision point, we need to generate a set of *prefixes* for each decision point, as commonly done in predictive process monitoring literature [20]. A prefix is a subtrace involving all activities prior to a given position in the process execution. In our case, this corresponds to each decision point. In particular, we created three *buckets*, each involving prefixes corresponding to activities executed before the corresponding decision point. Note that we created two buckets for decision point 5, one involving prefixes with a single loop execution and the other one involving prefixes with multiple loop executions. For each bucket, a separate model is trained. Before the modelling start, the data need to be prepared. First, the target has to be defined. The first bucket is based on decision point 2; therefore, the labels 'Multi scale up required' and 'Routine incident' are possible targets. Bucket 2 and bucket 3 are based on decision point 5, with the labels 'Additional multi scale up' and 'No additional multi scale up'. From now on these buckets are referred to as decision moments, respectively decision moment 1, decision moment 2 and decision moment 3. Notice that for decision moment 1 the label 'Multi scale up required' occurs approximately for 33 % of the data in this data set. This distribution is similar for decision moment 3. While for decision moment 2 the imbalance is even larger. Here the label 'Additional multi scale up' occurs by approximately 15 % of the data in this bucket. Table 3 shows the distribution of the labels for different decision moments.

Table 3. Distribution of target labels in the three decision moments

	Multi scale up required // Additional multi scale up	Routine incident // No additional multi scale up
Decision moment 1	2334	4381
Decision moment 2	277	1671
Decision moment 3	222	463

Secondly, new features are created based on the criteria found in the questionnaire. The attributes already in the data set are explored to find attributes describing these criteria, if possible. For example, for the criterion 'Incident type' the classification criteria are used. When no immediate matching is possible, new features are created. For example, for the criterion 'Own disciplines involved', a feature is created that counts the number of deployed vehicles at that moment.

Table 4. Summary of performance metrics for decision moment 1.

	Precision	Recall	F1 macro	F1 weighted
Decision tree	0.977	0.978	0.968	0.980
Random forest	0.975	0.969	0.972	0.975

Finally, the selected features are encoded to be given as input to the classification model. We use one-hot encoding for categorical features, while we need an ad-hoc encoding to represent the prefix. Since the prefixes in the different decision moments are not all the same length, we chose to use aggregation encoding to transform the *Prefix*. Aggregation encoding means that for each process activity, a feature is created, which represents the number of times this activity has appeared in the prefix. After the encoding step, there are 161 features in the train data of decision moment 1, 119 features in the second decision moment, and 92 features for the third decision moment.

As classifiers, we used the decision tree and the random forest, commonly used in decision-mining approaches. For the performance evaluation, we used the well-known metrics such as accuracy and F1 score.

6.1 Results

We used 70% of the samples for training and 30% for testing. Because the data is imbalanced it is chosen to use a stratified split. Besides, the samples are randomly distributed over the train and test set. Each decision moment is a separate data set. For each of these data sets, a decision tree model and random forest model are trained on the training data. The parameters of these models are optimized with a grid search, using 5-fold stratified cross-validation. With the grid search, all possible combinations of the provided parameter settings are fitted. For both models the class weight parameter 'balanced' is used to compensate for the class imbalance. The parameter 'balanced' adjusts the weights inversely proportional to the class frequencies. The other parameters optimized for both models are the maximum depth of the tree(s) and the minimal samples for each leaf. For the decision tree, the max depth is between 2 and 10, to ensure that it is possible to interpret the tree. The scoring used in the grid search is the weighted F1 score.

For each decision moment, we compared the performance of the classifiers and we assessed the importance of each feature to determine differences and similarities with the criteria defined by the human users. For the sake of space, below we only report the results obtained for the first decision moment. An overview of all performance metrics on the test set calculated for decision moment 1 is shown in Table 4. Both models achieved a good performance, with a weighted f1 score of 0.980 for the decision tree model and 0.975 for the random forest model.

The model explainability of the decision tree model and random forest model are examined next. The SHAP explainer is applied to both models. The summary plot shown in 3 shows the most important features for both models with their impact on multi scale up required. The upper two features are the same for both models. Overall, the decision tree seems to consider around seven features, while the random forest model considers many more. This is due to the bagging of the models in the random forest where other features are selected to build a tree each time. For all binary features, 1 is red and 0 is blue. So if there are only red values on the right side, the value 1 has a positive impact on multi classification and the other way around. For the continuous values dependency plots are shown.

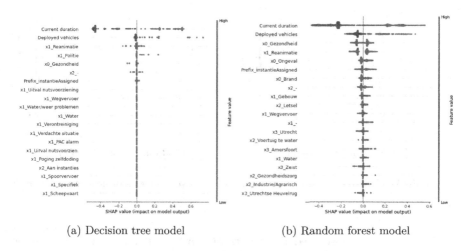

(a) Decision tree model   (b) Random forest model

Fig. 3. SHAP summary plots for Decision moment 1

In Fig. 4 and 5, the dependency plots for the features *Current duration* and *Deployed vehicles* can be found. In these plots the feature value is shown on the x-axis and the impact on the y-axis. For the *Current duration* both models have a similar plot, with positive SHAP values for short current duration and negative SHAP values for longer durations. For the *Deployed vehicles*, for the random forest model it can be seen that for more than zero vehicles the impact on the prediction for the multi-scale up required is negative or very low. For the decision tree model, this is true only between one and four deployed vehicles.

We analyzed in the same way the models extracted with the other buckets, and we compared the most important features in the different decision points with the questionnaire results and evaluated the results together with experts from VRU. For each feature, their meaning in multidisciplinary scale-up is of interest. In the following, for the sake of space, we only report few examples.

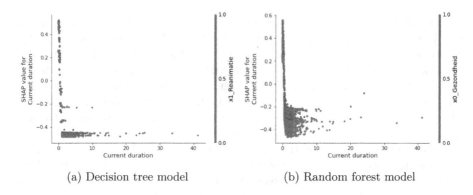

(a) Decision tree model   (b) Random forest model

Fig. 4. SHAP dependency plots for Current duration of decision moment 1

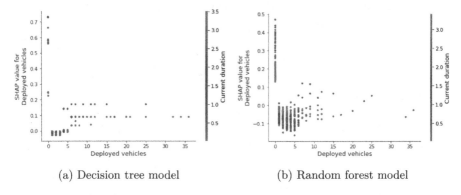

(a) Decision tree model   (b) Random forest model

Fig. 5. SHAP dependency plots for number of deployed vehicles of decision moment 1

The first feature discussed is the *Current duration* of an incident. This can be linked to the criteria 'Expected duration' and 'Duration of incident unknown' in the questionnaire. According to the model, a short current duration has a positive impact on scale-up in all decision points. From the perspective of the VRU, the suggestion is made, that these incidents have underlying reasons for scale-up soon. Determining these reasons requires further investigation.

The second feature is *Deployed vehicles* at the incident. This feature can be linked with the criteria 'Own disciplines involved' from the questionnaire. If there are more vehicles deployed by the incident, this results in more people that have to collaborate, which also require adequate coordination. The feature *Deployed vehicles* is mentioned as the second most important in all of the models and decision points. Later in the process model, the relevant number of deployed vehicles has a positive impact on predicting multidisciplinary scale-up increases. This is logical since if more vehicles are assigned to the incident over time it is reasonable that a scale-up is required.

Acknowledgement. The authors would like to thank the Veiligheidsregio Utrecht (VRU) for collaborating on this project. A special thanks to Michiel Rhoen and Arian van Donselaar for their support and providing invaluable insights into the problem.

References

1. Van der Aalst, W.: Process Mining?: Discovery, Conformance and Enhancement of Business Processes. Springer, Germany (2011). https://doi.org/10.1007/978-3-642-19345-3
2. Bennet, B.: Effective emergency management: a closer look at the incident command system. Prof. Saf. **56**(11), 28–37 (2011)
3. Di Ciccio, C., Marrella, A., Russo, A.: Knowledge-intensive processes: characteristics, requirements and analysis of contemporary approaches. J. Data Semant. **4**(1), 29–57 (2015)
4. van Duin, M., Wijkhuijs, V.: Instituut Fysieke Veiligheid, Boom bestuurskunde (2017)
5. van Duin, M., Wijkhuijs, V., Jong, W.: Instituut Fysieke Veiligheid, Boom bestuurskunde (2018)
6. Ernst, M.D., Cockrell, J., Griswold, W.G., Notkin, D.: Dynamically discovering likely program invariants to support program evolution. IEEE Trans. Softw. Eng. **27**(2), 99–123 (2001)
7. Fahland, D.: Extracting and pre-processing event logs (2021)
8. Herrera, M.P.R., Díaz, J.S.: Improving emergency response through business process, case management, and decision models. In: ISCRAM (2019)
9. Kushnareva, E., Rychkova, I., Le Grand, B.: Modeling business processes for automated crisis management support: lessons learned. In: 2015 IEEE 9th International Conference on Research Challenges in Information Science (RCIS), pp. 388–399. IEEE (2015)
10. Leemans, S.J.J., Fahland, D., van der Aalst, W.M.P.: Discovering block-structured process models from event logs - a constructive approach. In: Colom, J.-M., Desel, J. (eds.) PETRI NETS 2013. LNCS, vol. 7927, pp. 311–329. Springer, Heidelberg (2013). https://doi.org/10.1007/978-3-642-38697-8_17
11. Leewis, S., Smit, K., Zoet, M.: Putting decision mining into context: a literature study. In: Agrifoglio, R., Lamboglia, R., Mancini, D., Ricciardi, F. (eds.) Digital Business Transformation. LNISO, vol. 38, pp. 31–46. Springer, Cham (2020). https://doi.org/10.1007/978-3-030-47355-6_3
12. de Leoni, M., Dumas, M., García-Bañuelos, L.: Discovering branching conditions from business process execution logs. In: Cortellessa, V., Varró, D. (eds.) FASE 2013. LNCS, vol. 7793, pp. 114–129. Springer, Heidelberg (2013). https://doi.org/10.1007/978-3-642-37057-1_9
13. Mannhardt, F., de Leoni, M., Reijers, H.A., van der Aalst, W.M.P.: Decision mining revisited - discovering overlapping rules. In: Nurcan, S., Soffer, P., Bajec, M., Eder, J. (eds.) CAiSE 2016. LNCS, vol. 9694, pp. 377–392. Springer, Cham (2016). https://doi.org/10.1007/978-3-319-39696-5_23
14. Petrusel, R., Vanderfeesten, I., Dolean, C.C., Mican, D.: Making decision process knowledge explicit using the decision data model. In: Abramowicz, W. (ed.) BIS 2011. LNBIP, vol. 87, pp. 172–184. Springer, Heidelberg (2011). https://doi.org/10.1007/978-3-642-21863-7_15
15. Rozinat, A., der Aals, W.V.: Decision mining in business processes. BETA publicatie : working papers, Technische Universiteit Eindhoven (2006)

16. Rozinat, A., Van der Aalst, W.M.: Conformance checking of processes based on monitoring real behavior. Inf. Syst. **33**(1), 64–95 (2008)
17. Shahrah, A.Y., Al-Mashari, M.A.: Emergency response systems: research directions and current challenges. In: Proceedings of the Second International Conference on Internet of things, Data and Cloud Computing, pp. 1–6 (2017)
18. Slam, N., Wang, W., Xue, G., Wang, P.: A framework with reasoning capabilities for crisis response decision-support systems. Eng. Appl. Artif. Intell. **46**, 346–353 (2015)
19. Tax, N., Genga, L., Zannone, N.: On the use of hierarchical subtrace mining for efficient local process model mining. In: SIMPDA, pp. 8–22 (2017)
20. Teinemaa, I., Dumas, M., Rosa, M.L., Maggi, F.M.: Outcome-oriented predictive process monitoring: review and benchmark. ACM Trans. Knowl. Disc. Data (TKDD) **13**(2), 1–57 (2019)
21. Theeuwes, N., van Houtum, G., Zhang, Y.: Improving ambulance dispatching with machine learning and simulation. In: Dong, Y., Kourtellis, N., Hammer, B., Lozano, J. (eds.) Machine Learning and Knowledge Discovery in Databases. LNCS, pp. 302–318. Springer, Germany (2021)
22. Van Dongen, B., Carmona, J., Chatain, T.: Alignment-based metrics in conformance checking (summary). In: Fachgruppentreffen der GI-Fachgruppe Entwicklungsmethoden für Informationssysteme und deren Anwendung, pp. 87–90 (2016)
23. Weijters, A.J.M.M., van der Aalst, W.M., de Medeiros, A.K.A.: Process mining with the heuristicsminer algorithm (2006)
24. Wijkhuijs, V., van Duin, M.: Instituut Fysieke Veiligheid, Boom bestuurskunde (2020)

Job Shop Scheduling via Deep Reinforcement Learning: A Sequence to Sequence Approach

Giovanni Bonetta$^{(\boxtimes)}$ (ID), Davide Zago (ID), Rossella Cancelliere (ID), and Andrea Grosso (ID)

Department of Computer Science, University of Turin, 10149 Turin, Italy
{giovanni.bonetta,rossella.cancelliere,andrea.grosso}@unito.it,
zago@di.unito.it

Abstract. Job scheduling is a well-known Combinatorial Optimization problem with endless applications. Well planned schedules bring many benefits in the context of automated systems: among others, they limit production costs and waste. Nevertheless, the NP-hardness of this problem makes it essential to use heuristics whose design is difficult, requires specialized knowledge and often produces methods tailored to the specific task. This paper presents an original end-to-end Deep Reinforcement Learning approach to scheduling that automatically learns dispatching rules. Our technique is inspired by natural language encoder-decoder models for sequence processing and has never been used, to the best of our knowledge, for scheduling purposes. We applied and tested our method in particular to some benchmark instances of Job Shop Problem, but this technique is general enough to be potentially used to tackle other different optimal job scheduling tasks with minimal intervention. Results demonstrate that we outperform many classical approaches exploiting priority dispatching rules and show competitive results on state-of-the-art Deep Reinforcement Learning ones.

Keywords: Optimal Job Scheduling · Deep Reinforcement Learning · Combinatorial Optimization · Sequence to Sequence

1   Introduction

Job Shop Problem (JSP) is a well-known Combinatorial Optimization problem fundamental in various automated systems applications such as manufacturing, logistics, vehicle routing, telecommunication industry, etc. In short, some jobs with predefined processing constraints have to be assigned to a set of heterogeneous machines, to achieve the desired objective (e.g. minimizing the flowtime). Due to its NP-hardness, finding exact solutions to the JSP is often impractical (or impossible, in many real-world scenarios), but many tasks can be effectively addressed through heuristics [7,9] or approximate methods [11], that represent

---

G. Bonetta and D. Zago—Equal contribution.

the most suitable choice for large-scale problems, providing near optimal solutions with acceptable computational times.

Heuristic algorithms are classified as constructive or as local search methods. Constructive heuristics assemble the solution with an incremental process: at each step, the choice of the next element in the solution is made by examining some local information of the problem, and once one variable has been fixed it's not reconsidered. *Priority Dispatching Rules* (PDRs) [9] belong to the category of constructive approximate methods: each operation is allocated in a dispatching sequence following a monotonic utility measure.

The use of dispatching rules emerged very early in the scheduling area, and it is well established by now. Most dispatching rules are known to be less than a match for modern, sophisticated heuristic optimization techniques (e.g. simulated annealing, tabu search, etc.); despite this, they are still commonly used in many practical contexts because they are considered quick, flexible and adaptable to many situations. Besides, PDRs are widely used in real-world scheduling systems because they are intuitive and easy to implement. As a result, optimization literature is rich of PDR methods for the JSP [16], even if it is well known that designing an effective PDR is time-consuming and requires a substantial domain knowledge.

A possible solution is the automation of the process of designing dispatching rules: recent works on learning algorithms for Combinatorial Optimization (see [3] for a survey) show that Deep Reinforcement Learning (RL) could be an ideal technique for this purpose, and in particular that it can be considered a potential breakthrough in the construction of heuristic methods for the JSP [4]. Reinforcement Learning [18] is a subfield of Machine Learning (ML) that experienced a great development in recent years, mainly thanks to the contribution of Deep Learning.

The main idea of this paper is to treat the JSP as a sequence to sequence process: inspired by deep learning natural language models we propose a Deep Reinforcement Learning approach that, exploiting the encoder-decoder architecture typical of language, automatically learns robust dispatching rules. This leads us to consider PDRs as a reasonable match for deep RL-based optimization techniques that, it should be remembered, despite of the huge amount of works appearing on the subject, are still in their infancy.

Our method is able to learn dispatching rules with higher performance than traditional ones, e.g. *Shortest Processing Time* (SPT), *Most Work Remaining* (MWKR). On top of that, our approach shows competitive results against state-of-the-art Deep RL methods when tested on small and medium sized JSP benchmark instances. Besides, it shows a high degree of flexibility: *Flow Shop Problem* (FSP) instances can also be solved, and minimal modifications to the model would allow solving *Open Shop Problem* (OSP).

Since the model requires sequences as inputs and outputs we design an appropriate, yet compact and easily interpretable encoding for JSP instances and solutions. Besides, thanks to a tailored masking procedure, the model outputs a permutation of job operations (virtually a priority list) that respects precedence constraints and can be mapped to a *schedule*, i.e. the association of each operation to a specific starting time.

The rest of this paper is organized as follows: Sect. 2 contains an overview of related works concerning neural and Deep Reinforcement Learning methods for Combinatorial Optimization (CO). Section 3 provides the definition of Markov Decision Process (MDP) and the theoretical foundations of our model. Section 4 introduces the mathematical notation which formalizes the JSP. Section 5 describes our technique for sequence encoding, the neural architecture used and the proposed masking mechanism, the experimentation details and the results obtained.

2 Related Works

Before Deep RL gained the popularity it has today, many ML-based approaches have been applied to CO (see [17] for an in-depth overview), such as assignment problems, cutting stock and bin packing problems, knapsack problems, graph problems, shortest path problems, scheduling problems, vehicle routing problems and the Travelling Salesman Problem (TSP). In the last decade, Natural Language Processing research inspired the formulation of very effective models such as the *Pointer Networks* [21], a deep architecture which builds upon recurrent neural networks. These innovative models have the ability to tackle problems where the number of output tokens varies with the input, a feature that characterizes also many CO problems and, exploited for solving the TSP, showed interesting results and great potential.

One of the first attempts of applying the results of Vinyals et al. [21] is the work by Bello et al. [2], which successfully addresses the TSP and the Knapsack Problem (KP) in the context of Markov Decision Processes. It introduces active search, i.e. an RL-based technique that starting from a random (or pre-trained) policy iteratively optimizes the parameters on a single test instance. Deudon et al. [5] and Kool et al. [12] independently proposed a model inspired by the transformer architecture from Vaswani et al. [20] for solving the TSP. More specifically, the proposed architecture is made of an attention-based encoder in combination with a Pointer Network decoder.

The attempt to apply Deep RL to scheduling, and in particular to the JSP, is a phenomenon of growing research interest in recent years. We remand to Sect. 3 and Sect. 4 for all definitions concerning MDPs and the JSP. Waschneck et al., [22] present one of the first relevant works: in the context of MDPs each machine of the JSP is considered as an agent. The resulting multi-agent system is trained with Deep Q-Network (DQN) and, despite not showing higher performance with respect to other heuristics, this model obtains expert-level results. A similar multi-agent method is proposed by Liu et al. [14], where training is based on Deep Deterministic Policy Gradient (DDPG) algorithm. Their approach succeeds in reaching higher performance with respect to some dispatching rules.

The approach from Lin et al. [13] assigns a different dispatching rule to each machine. After the training, done using a multi-class DQN, their method performs better that individual dispatching rules, but is far from being optimal.

Another interesting approach focuses on the disjunctive graph representation of the JSP. Zhang et al. [24] use a Graph Neural Network to map the states into an embedding space, followed by a Multi-Layer Perceptron which provides a probability distribution over the possible actions. This method obtains competitive performance and can be easily scaled to larger instances. We chose to compare our proposed approach to this work since it is, at the best of our knowledge, the best performing Deep RL approach to the JSP.

Han and Yang's work [8] presents a technique which, differently from all other summarised here, utilizes a Convolutional Neural Network on images for encoding the state of the problem and operates as a state-action function approximator. The images, which are produced using the disjunctive graph, have three channels representing the features: processing time, current schedule, and machine availability. The action space corresponds to different dispatching rules, whereas the reward function highlights machine utilization.

3   Mathematical Foundations

RL substantially differs from other ML paradigms since it's concerned with how an agent learns to act in an environment: agents' behavior is optimized through a training phase, requiring the definition of a Markov Decision Process [1], focused on the maximization of a cumulative expected reward collected through a sequence of actions.

An MDP is a mathematical framework used to formalize a general decision making process involving a single agent acting in an environment. It is defined by a tuple $M = (S, A, R, T, \gamma, H)$ where:

- S - *state space.*
  It is the set of all the possible representations s of the environment and of the agent's internal state at a given time.
- A - *action space.*
  It is the set of all the possible actions a the agent can perform.
- R - *reward function* $R : S \times A \times S \to \mathbb{R}$.
  It is the reward given to the agent after doing action a in state s and landing in state s'.
- T - *transition function* $T(s'|s, a)$.
  It is the transition probability from state s to s' given that action a has been performed.
- γ - *discount factor.*
  It weights the rewards of future actions. $\gamma \in [0, 1]$.
- H - *time horizon.*
  It is the maximum number of transition that can occur before the decision process is halted.

The objective of RL is to maximize the expected return of the sequence of actions performed by the agent. Each action is sampled from a *stochastic policy* $\pi(a|s)$, with $a \in A$ and $s \in S$, i.e. a probability distribution over the set of actions given a particular state.

3.1   Policy Gradient Algorithms

Policy gradient (or *policy optimization*) methods [18] are widely used in Deep RL research and directly optimize the stochastic policy π_θ, which is approximated by a neural network with parameters θ.

By taking actions in the environment, the agent defines trajectories. A *trajectory* τ (alternatively *episode* or *rollout*) is a sequence of states and actions $(s_0, a_0, s_1, a_1, ..., s_{H-1}, a_{H-1}, s_H)$ and it has a return $R(\tau)$ associated to it:

$$R(\tau) = \sum_{t=0}^{H} R(s_t, a_t, s_{t+1}) \tag{1}$$

$R(\tau)$ is called *finite-horizon undiscounted return* since it's defined with horizon H. Moreover, the probability of a trajectory given the policy is:

$$P_\theta(\tau) = \rho(s_0) \prod_{t=0}^{H} T(s_{t+1}|s_t, a_t)\pi_\theta(a_t|s_t) \tag{2}$$

where $\rho(s_0)$ is the a priori probability of state s_0.

Given the parameterized stochastic policy π_θ, the learning objective is the maximization of the expected return w.r.t. a set of trajectories:

$$\max_\theta J(\pi_\theta), \text{ where } J(\pi_\theta) = \mathop{\mathbb{E}}_{\tau \sim \pi_\theta} [R(\tau)] \tag{3}$$

Considering a policy optimized with gradient ascent, the quantity $\nabla_\theta J(\pi_\theta)$ is called *policy gradient* and the following equation holds:

$$\nabla_\theta J(\pi_\theta) = \mathop{\mathbb{E}}_{\tau \sim \pi_\theta} \left[\sum_{t=0}^{H} \nabla_\theta \log \pi_\theta(a_t|s_t)R(\tau) \right] \tag{4}$$

This leads to the *REINFORCE* algorithm (Algorithm 1), also known as *Vanilla policy gradient*, for optimizing policies, first proposed by Williams in [23].

Algorithm 1: REINFORCE

Input: MDP $M = (S, A, T, R, \gamma, H)$
Output: policy π_{θ_k}
$\theta_0 \leftarrow$ INITIAL-PARAMETERS()
for $k \in (0, 1, 2, ...)$ **do**
    $\mathcal{D} \leftarrow$ COLLECT-TRAJECTORIES()
    $g_k \leftarrow \frac{1}{|\mathcal{D}|} \sum_{\tau \in \mathcal{D}} \sum_{t=0}^{H} \nabla_\theta \log \pi_\theta(a_t|s_t)R(\tau)$ $\triangleright$ *{policy gradient}
    $\theta_{k+1} \leftarrow \theta_k + \alpha g_k$ $\triangleright$ **{gradient ascent step}
return π_{θ_k}

Equation $*$ is the estimation of the policy gradient over the set of trajectories $\mathcal{D}$. Statement $**$ – i.e. the gradient ascent update rule—can be substituted with the update rule of a different optimization algorithm, e.g. Adam.

Unfortunately the *unbiased policy gradient* g_k suffers from high variance which hinders performance and learning stability. This can be addressed through the use of baselines, terms that only depend on the current state and are subtracted from the reward. Equation 5 is the policy gradient updated with a generic baseline term.

$$\nabla_\theta J(\pi_\theta) = \underset{\tau \sim \pi_\theta}{\mathbb{E}} \left[\sum_{t=0}^{H} \nabla_\theta \log \pi_\theta(a_t|s_t) \left(\sum_{t=0}^{H} R(s_t, a_t, s_{t+1}) - b(s_t) \right) \right] \quad (5)$$

4   The Job Shop Optimization Problem: Notation

Scheduling is a decision-making process consisting in the allocation of resources to tasks over a given time period, with the additional constraint of optimizing one (or more) objective functions. The JSP is one of the most studied scheduling problems, along with the Open Shop and the Flow Shop Problems. A $n \times m$ JSP instance is characterized by:

- n *jobs* J_i, with $i \in \{0, ..., n-1\}$, each one consisting of m operations (or tasks) O_{ij}, with $j \in \{0, ..., m-1\}$.
- m *machines* M_{ij}, with $j \in \{0, ..., m-1\}$. M_{ij} identifies the machine required to execute the j-th operation of job i.

We denote the execution time of an operation O_{ij} with p_{ij}; an operation execution cannot be interrupted and each operation of a given job must be executed on a different machine. A JSP solution is represented by a schedule. As an example, let us consider the JSP instance represented in Table 1. In this case there are three jobs J_i, with $i \in \{0, 1, 2\}$, and four operations O_{ij} for the i-th job, with $j \in \{0, 1, 2, 3\}$. Operation O_{ij} must be executed on machine $M_{ij} \in \{0, 1, 2, 3\}$ and has processing time p_{ij}.

Table 1. Example of a 3×4 JSP instance.

M_{ij}, p_{ij}	O_{*0}	O_{*1}	O_{*2}	O_{*3}
J_0	(0, 4)	(2, 2)	(1, 6)	(3, 2)
J_1	(0, 4)	(3, 5)	(2, 7)	(1, 8)
J_2	(2, 6)	(0, 4)	(1, 3)	(3, 1)

A useful tool for visualizing a schedule is Gantt charts [6]. Figure 1 represents the Gantt chart for a possible schedule of the JSP instance represented in Table 1.

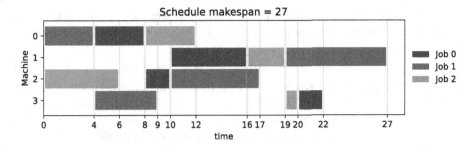

Fig. 1. One possible schedule for the JSP instance in Table 1.

The optimal solution of a JSP is the schedule that minimizes the makespan C_{max}, where $C_{max} = \max_i C_i$, and C_i is the completion time of the i-th job.

5 Our Sequence to Sequence Approach to the JSP

The main novelty we present is a sequence-based Deep RL approach applied to the JSP. Inspired by [2] and [12] we make use of a deep neural network used for NLG applications and we train it in a RL setting. Such model (see Fig. 2) combines a self-attention based encoder and a Pointer-Network decoder [21]. In order to apply it to the JSP, we formulate a sequence-based encoding of input and output, and design an appropriate masking mechanism to generate feasible solutions. Our code is available on Github[1].

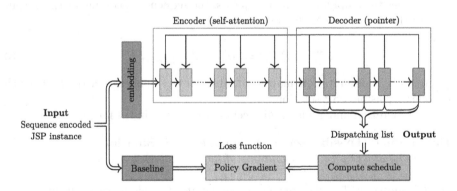

Fig. 2. Our encoder-decoder architecture for scheduling problems.

5.1 Sequence Encoding

The input (i.e. problem instance) and the model's output (i.e. solution) need to be encoded as sequences in order for the model to process them correctly.

---

[1] Github repository: https://github.com/dawoz/JSP-DeepRL-Seq2Seq.

We consider both the input and the output as sequences of operations and we define a 4-dimensional feature vector $\mathbf{o}_k$ for each operation O_{ij} as follows:

$$\mathbf{o}_k = [i \quad j \quad M_{ij} \quad p_{ij}] \qquad \text{with } k = m \cdot i + j \qquad (6)$$

where i is the index of the i-th job and j the index of its j-th operation. Consider a JSP instance S with n jobs J_i ($i \in \{0, ..., n-1\}$) and m operations O_{ij} for job J_i ($j \in \{0, ..., m-1\}$) with required machine $M_{ij} \in \{0, ..., m-1\}$ and execution time p_{ij}. S can be expressed with the following sequence encoding S^{seq}:

$$
S^{\text{seq}} =
\begin{bmatrix}
\mathbf{o}_0 \\
\mathbf{o}_1 \\
\vdots \\
\mathbf{o}_{m-1} \\
\mathbf{o}_m \\
\vdots \\
\mathbf{o}_{(m-1)(n-1)}
\end{bmatrix}
=
\begin{bmatrix}
\overset{i}{0} & \overset{j}{0} & \overset{M_{ij}}{M_{00}} & \overset{p_{ij}}{p_{00}} \\
0 & 1 & M_{01} & p_{01} \\
\vdots & \vdots & \vdots & \vdots \\
0 & m-1 & M_{0m-1} & p_{0m-1} \\
1 & 0 & M_{10} & p_{10} \\
\vdots & \vdots & \vdots & \vdots \\
n-1 & m-1 & M_{n-1m-1} & p_{n-1m-1}
\end{bmatrix}
\qquad (7)
$$

The matrix S^{seq} just defined determines which jobs/operations have to be handled: establishing an order in which to execute them allows to identify a schedule.

Besides, a correct encoding of the model's output sequence implies that if for job J_i operation O_{ij} must be executed before O_{ik}, then the vectors of the operations in the output sequence must occur in the same order (not consecutive, in general).

In order to comply with these requests, the sequence encoding for the output L^{seq} of the model has the following form:

$$L^{\text{seq}} = PS^{\text{seq}} = \left[\mathbf{o}'_0 \cdots \mathbf{o}'_{(n-1)(m-1)}\right]^T \qquad (8)$$

where P is a permutation matrix suitable for obtaining a matrix L^{seq} which encodes a feasible JSP solution.

The condition under which this occurs is explained in the following definition:

Definition 1 (Feasible sequence encoded JSP Solution).
Let O_{ij} and O_{ik} be the operations of the i-th job with $j < k$, and $\mathbf{o}'_s = [i \quad j \quad M_{ij} \quad p_{ij}]$, $\mathbf{o}'_r = [i \quad k \quad M_{ik} \quad p_{ik}]$.
The matrix L^{seq} is the sequence encoding of a feasible schedule iff the permutation P is such that $s < r$, for all s and r in $\{0, ..., (n-1)(m-1)\}$ (i.e. the order of operations for job i defined in S^{seq} is preserved).

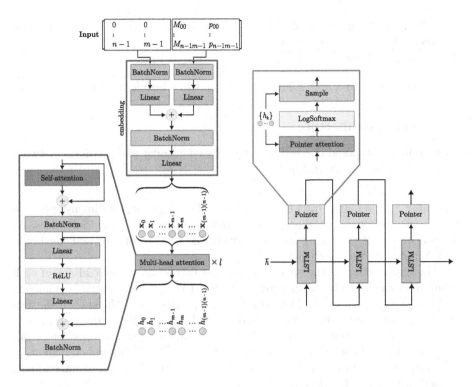

Fig. 3. *Left:* Encoder of our model. *Right:* Decoder with pointer mechanism.

As an example the sequence encodings of the JSP instance in Table 1 are the following:

$$
S^{\text{seq}} =
\begin{array}{c}
\overbrace{i \ j \ M_{ij} \ p_{ij}}
\end{array}
\begin{bmatrix}
0 & 0 & 0 & 4 \\
0 & 1 & 2 & 2 \\
0 & 2 & 1 & 6 \\
0 & 3 & 3 & 2 \\
1 & 0 & 0 & 4 \\
1 & 1 & 3 & 5 \\
1 & 2 & 2 & 7 \\
1 & 3 & 1 & 8 \\
2 & 0 & 2 & 6 \\
2 & 1 & 0 & 4 \\
2 & 2 & 1 & 3 \\
2 & 3 & 3 & 1
\end{bmatrix}
\qquad
L^{\text{seq}} =
\begin{array}{c}
\overbrace{i \ j \ M_{ij} \ p_{ij}}
\end{array}
\begin{bmatrix}
1 & 0 & 0 & 4 \\
0 & 0 & 0 & 4 \\
2 & 0 & 2 & 6 \\
1 & 1 & 3 & 5 \\
0 & 1 & 2 & 2 \\
2 & 1 & 0 & 4 \\
0 & 2 & 1 & 6 \\
2 & 2 & 1 & 3 \\
1 & 2 & 2 & 7 \\
2 & 3 & 3 & 1 \\
0 & 3 & 3 & 2 \\
1 & 3 & 1 & 8
\end{bmatrix}
$$

The output L^{seq} can be effectively interpreted as a dispatching list, which can be directly mapped to a schedule as follows:

1. Considering $\mathbf{o}'_p = [i \ l \ M_{il} \ p_{il}]$ (p-th row of L^{seq}), schedule operation O_{il} to the earliest time such that machine M_{il} is available and, if $l > 0$ (i.e. O_{il}

isn't the first operation of i-th job) the previous operation O_{il-1} has been executed.

2. Repeat for all the rows of L^{seq}.

Mapping L^{seq} to a JSP solution results in the schedule in Fig. 1.

5.2 Model Architecture

Our model is composed of a self-attention-based encoder and a Pointer Network used as decoder (shown in Fig. 3).

Encoder. Represented in (Fig. 3). The encoder's input is a 3-dimensional tensor $U \in \mathbb{R}^{N \times (nm) \times 4}$ that represents a batch of sequence-encoded instances. As defined in Sect. 4, n and m are respectively the number of jobs and machines, and N indicates the batch size.

The first portion of the encoder computes two separate embeddings of each input row, respectively for features (i, j) and (M_{ij}, p_{ij}), by batch-normalizing and projecting to the embedding dimension d_h. After that, the sum of the two vectors is batch-normalized and passed through a linear layer resulting in $X \in \mathbb{R}^{N \times (nm) \times d_h}$. X is then fed into l multi-head attention layers (we consider $l = 3$).

The output of the encoder is a tensor $H \in \mathbb{R}^{N \times (nm) \times d_h}$ of embeddings $h_k \in \mathbb{R}^{d_h}$, later used as input in the decoder. $\bar{h}$, the average of the these embeddings, is used to initialize the decoder.

Decoder. Represented in (Fig. 3), the decoder is a Pointer Network which generates the policy π_θ, a distribution of probability over the rows of the input S^{seq}, via the attention mechanism; during training, the next selected row $\mathbf{o}'_t$ is sampled from it. During evaluation instead, the row with highest probability is selected in a greedy fashion. π_θ is defined as follows:

$$\pi_\theta(\mathbf{o}'_t | \mathbf{o}'_0, ..., \mathbf{o}'_{t-1}, S^{\text{seq}}) = \text{softmax}\left(\text{mask}(u^t | \mathbf{o}'_0, ..., \mathbf{o}'_{t-1})\right) \quad (9)$$

where u^t is the score computed by the Pointer Network's attention mechanism over S^{seq} input rows. $\text{mask}(u^t | \mathbf{o}'_0, ..., \mathbf{o}'_{t-1})$ is a masking mechanism which depends on the sequence partially generated and enforces the constraint in Definition 1.

Masking. In order to implement the masking mechanism we use two boolean matrices M^{sched} and M^{mask} defined as follows:

Definition 2 (Boolean matrix M^{sched}). *Given the k-th instance in the batch and the j-th operation of the i-th job, the element M_{kp}^{sched} (which refers to $\mathbf{o}_p$, with $p = m \cdot i + j$) is true iff the j-th operation has already been scheduled.*

Definition 3 (Boolean matrix M^{mask}). *Given the k-th instance in the batch and the index l of an operation of the i-th job, the element M_{kp}^{mask} (which refers to $\mathbf{o}_p$, with $p = m \cdot i + l$) is true iff $l > j$, where j is the index of the next operation of the i-th job (i.e. scheduling the l-th operation would violate Definition 1).*

Given k-th instance in the batch and $\mathbf{o}_p$ feature vector of the operation scheduled at current time-step, we update M^{sched} and M^{mask} as follows.

$$M_{kp}^{\text{sched}} \leftarrow true, \qquad M_{kp+1}^{\text{mask}} \leftarrow false$$

At current step t, the resulting masking procedure of the score associated to input row index $p \in \{0, 1, ..., (m-1)(n-1)\}$ is the following:

$$\text{mask}(u_p^t | \mathbf{o}_0', ..., \mathbf{o}_{t-1}') = \begin{cases} -\infty, & \text{if } M_{kp}^{\text{sched}} \text{ OR } M_{kp}^{\text{mask}} \\ u_{p}^t, & \text{otherwise} \end{cases} \tag{10}$$

Masked scores result in a probability close to zero for operations that are already scheduled or cannot be scheduled. Figure 4 shows a possible generation procedure with the masking mechanism just described in order to solve the JSP instance represented in Table 2.

Table 2. Example of a 2×3 JSP instance.

M_{ij}, p_{ij}	O_{*0}	O_{*1}	O_{*2}
J_0	(1, 4)	(2, 7)	(0, 5)
J_1	(0, 7)	(1, 3)	(2, 7)

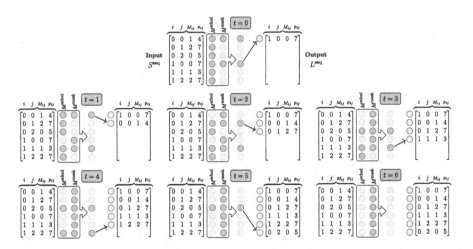

Fig. 4. Sequence generation with masking mechanism for the JSP. Light blue circles indicate masked rows and the arrows represent the agent's choices. (Color figure online)

Training Algorithm. The network is trained with REINFORCE [23] described in Subsect. 3.1 using the Adam optimizer. We use the following form of the policy gradient:

$$\nabla_\theta L(\pi_\theta) = \mathbb{E}\left[(C_{max}(L^{\text{seq}}) - b(S^{\text{seq}})) \nabla_\theta \log P_\theta(L^{\text{seq}}|S^{\text{seq}})\right] \quad (11)$$

where $P_\theta(L^{\text{seq}}|S^{\text{seq}}) = \prod_{t=0}^{nm-1} \pi_\theta(\mathbf{o}'_t|\mathbf{o}'_0, ..., \mathbf{o}'_{t-1}, S^{\text{seq}})$ is the probability of the solution L^{seq} and $b(S^{\text{seq}})$ is the greedy rollout baseline. After each epoch, the algorithm updates the baseline with the optimized policy's weights if the latter is statistically better. This is determined by evaluating both policies on a 10000 samples dataset and running a paired t-test with $\alpha = 0.05$ (see [12] for the detailed explanation). The periodic update ensures that the policy is always challenged by the best model, hence the reinforcement of actions is effective. From a RL perspective, $-C_{max}(L^{\text{seq}})$ is the reward of the solution—lower makespan implies higher reward. After training, the active search approach [2] is applied.

Solving Related Scheduling Problems. Our method represents a general approach to scheduling problems and, once trained on JSP instances, it can also solve the Flow Shop Problem. The Open Shop Problem can also be solved with a small modification of the masking mechanism. Since the order constraint between operations is dropped in the OSP, the feasible outputs of the model are all the permutations of the input sequence. This simplifies the masking mechanism, which can be done just by keeping track of the scheduled operations with matrix M^{sched}. In Fig. 5 we show three steps of the modified masking mechanism for solving the instance in Table 2, interpreted as an OSP.

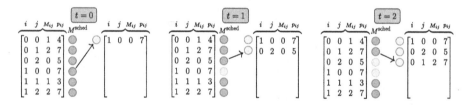

Fig. 5. Sequence generation with modified masking mechanism for the OSP.

5.3 Experiments and Results

In this section we present our experiments and results. We consider four JSP settings: 6×6, 10×10, 15×15 and 30×20. After hyperparameter tuning, we set the learning rate to 10^{-5} and gradient clipping to 0.5 in order to stabilize training. At each epoch, the model processes a dataset generated with the well-known Taillard's method [19]. Table 3 sums up training configurations for every experiment.

During training we note the average cost every 50 batches and the validation performance at the end of every epoch. Validation rollouts are done in a greedy fashion, i.e. by choosing actions with maximum likelihood. Training and validation curves are represented in Fig. 6.

Table 3. Training configurations for all the experiments. * Nvidia GPUs have been used.

Size	Epoch size	N° epochs	Batch size	GPU(s)*	Duration
6 × 6	640000	10	512	Titan RTX	30 m
10 × 10	640000	10	512	RTX A6000	1 h 30 m
15 × 15	160000	10	256	RTX A6000	1 h 30 m
30 × 20	16000	10	32	Titan RTX	1 h 45 m

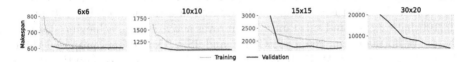

Fig. 6. Training and Validation curves for different JSPs.

Comparison with Concurrent Work. As already said in the Introduction, we compare our results with the work from Zhang et al. [24], and with a set of largely used dispatching rules: *Shortest Processing Time* (SPT), *Most Work Remaining* (MWKR), *Most Operations Remaining* (MOPNR), *minimum ratio of Flow Due Date to most work remaining* (FDD).

Table 4 shows the testing results obtained applying our technique on 100 instances generated by Zhang et al. with the Taillard's method.

We compare each solution with the optimal one obtained with Google OR-Tools' [15] solver; in the last column we report the percentage of instances for which OR-Tools returns optimal solutions in a limited computation time of 3600 s.

The column JSP settings shows the average makespan over the entire test dataset and the gap between $\overline{C}_{max}$ (the average makespan of heuristic solutions) and $\overline{C}^*_{max}$ (the average makespan of the optimal ones), defined as $\overline{C}_{max}/\overline{C}^*_{max} - 1$.

From Table 4 we can see that our model greatly outperforms the traditional dispatching rules even by a margin of 71% with respect to SPT. When compared

Table 4. Results over different JSP settings.

JSP settings		SPT	MWKR	FDD	MOPNR	Zhang [24]	Ours	Opt. Rate (%)
6 × 6	$\overline{C}_{max}$	691.95	656.95	604.64	630.19	574.09	**495.92**	100%
	Gap	42.0%	34.6%	24.0%	29.2%	17.7%	**1.7%**	
10 × 10	$\overline{C}_{max}$	1210.98	1151.41	1102.95	1101.08	988.58	**945.27**	100%
	Gap	50.0%	42.6%	36.6%	36.5%	22.3%	**16.9%**	
15 × 15	$\overline{C}_{max}$	1890.91	1812.13	1722.73	1693.33	**1504.79**	1535.14	99%
	Gap	59.2%	52.6%	45.1%	42.6%	**26.7%**	29.3%	
30 × 20	$\overline{C}_{max}$	3208.69	3080.11	2883.88	2809.62	**2508.27**	2683.05	12%
	Gap	65.3%	58.7%	48.6%	44.7%	**29.2%**	38.2%	

to [24] our model is superior in performance in the 6×6 and 10×10 cases, while having similar results in the 15×15 JSPs, and sligthly underperforming in the 30×20. Speculating about the drop in performance of our solution in the biggest settings (i.e. 30×20 JSPs) we think it could be due to the following reasons:

- Larger JSP instances are encoded by longer sequences: like traditional RNNs and transformers, our model tends to have a suboptimal representation of the input if the sequence is exceedingly long.
- As mentioned before, for execution time reasons we reduce the number of instances and examples in each batch: this implies a gradient estimate with higher variance, hence a potentially unstable and longer learning.

Improving Active Search Through Efficient Active Search. Efficient Active Search (EAS) is a technique introduced in a recent work by Hottung et al. [10] that extends and substantially improves active search, achieving state-of-the-art performance on the TSP, CVRP and JSP. The authors proposed three different techniques, EAS-Emb, EAS-Lay and EAS-Tab, all based on the idea of performing active search while adjusting only a small subset of model parameters. EAS-Emb achieves the best performance and works by keeping all model parameters frozen while optimizing the embeddings. As pointed out in [10], this technique can be applied in parallel to a batch of instances, greatly reducing the computing time. Here we present a preliminary attempt to extend our method applying EAS-Emb and we test it on the 10×10 JSP. Table 5 shows that our model greatly benefits from the use of EAS-Emb, although underperforming Hottung et al.'s approach.

Table 5. Efficient Active Search: comparison results

JSP settings		Hottung et al. [10]	Ours+EAS-Emb
10×10	$\overline{C}_{max}$	837.0	864.9
	Gap	3.7%	7.2%

6   Conclusions

In this work we designed a Sequence to Sequence model to tackle the JSP, a famous Combinatorial Optimization problem, and we demonstrated that it is possible to train such architecture with a simple yet effective RL algorithm. Our system automatically learns dispatching rules and relies on a specific masking mechanism in order to generate valid schedulings. Furthermore, it is easy to generalize this mechanism for the Flow Shop Problem and the Open Shop Problem with none or slight modifications. Our solution beats all the main traditional dispatching rules by great margins and achieve better or state of the art performance on small JSP instances.

For future works we plan to improve the performance of our method on larger JSP instances exploiting EAS-based approaches. Besides, although this work is mostly concerned with evaluating a Deep RL-based paradigm for combinatorial optimization, the idea of hybridizing these techniques with more classical heuristics remain viable. At last, one promising idea would be to improve our method using Graph Neural Networks as encoders. Graph-based models could produce more refined embeddings exploiting the disjunctive graph representation of scheduling problem instances.

Acknowledgements. The activity has been partially carried on in the context of the Visiting Professor Program of the Gruppo Nazionale per il Calcolo Scientifico (GNCS) of the Italian Istituto Nazionale di Alta Matematica (INdAM).

References

1. Bellman, R.: A Markovian decision process. J. Math. Mech. **6**(5), 679–684 (1957)
2. Bello, I., Pham, H., Le, Q.V., Norouzi, M., Bengio, S.: Neural combinatorial optimization with reinforcement learning. In: International Conference on Learning Representations (2017)
3. Bengio, Y., Lodi, A., Prouvost, A.: Machine learning for combinatorial optimization: a methodological tour d'horizon. Eur. J. Oper. Res. **290**(2), 405–421 (2021)
4. Cunha, B., Madureira, A.M., Fonseca, B., Coelho, D.: Deep reinforcement learning as a job shop scheduling solver: a literature review. In: Madureira, A.M., Abraham, A., Gandhi, N., Varela, M.L. (eds.) HIS 2018. AISC, vol. 923, pp. 350–359. Springer, Cham (2020). https://doi.org/10.1007/978-3-030-14347-3_34
5. Deudon, M., Cournut, P., Lacoste, A., Adulyasak, Y., Rousseau, L.-M.: Learning heuristics for the TSP by policy gradient. In: van Hoeve, W.-J. (ed.) CPAIOR 2018. LNCS, vol. 10848, pp. 170–181. Springer, Cham (2018). https://doi.org/10.1007/978-3-319-93031-2_12
6. Gantt, H.: A Graphical Daily Balance in Manufacture. ASME (1903)
7. Glover, F., Laguna, M.: Tabu Search. Springer, New York (1998). https://doi.org/10.1007/978-1-4615-6089-0
8. Han, B.A., Yang, J.J.: Research on adaptive job shop scheduling problems based on dueling double DQN. IEEE Access **8**, 186474–186495 (2020)
9. Haupt, R.: A survey of priority rule-based scheduling. Oper. Res. Spektrum **11**, 3–16 (1989). https://doi.org/10.1007/BF01721162
10. Hottung, A., Kwon, Y.D., Tierney, K.: Efficient active search for combinatorial optimization problems. In: International Conference on Learning Representations (2021)
11. Jansen, K., Mastrolilli, M., Solis-Oba, R.: Approximation algorithms for flexible job shop problems. In: Gonnet, G.H., Viola, A. (eds.) LATIN 2000. LNCS, vol. 1776, pp. 68–77. Springer, Heidelberg (2000). https://doi.org/10.1007/10719839_7
12. Kool, W., van Hoof, H., Welling, M.: Attention, learn to solve routing problems! In: International Conference on Learning Representations (2019)
13. Lin, C.C., Deng, D.J., Chih, Y.L., Chiu, H.T.: Smart manufacturing scheduling with edge computing using multiclass deep Q network. IEEE Trans. Ind. Inf. **15**(7), 4276–4284 (2019)
14. Liu, C.L., Chang, C.C., Tseng, C.J.: Actor-critic deep reinforcement learning for solving job shop scheduling problems. IEEE Access **8**, 71752–71762 (2020)

15. Perron, L.: Operations research and constraint programming at Google. In: Lee, J. (ed.) CP 2011. LNCS, vol. 6876, pp. 2–2. Springer, Heidelberg (2011). https://doi.org/10.1007/978-3-642-23786-7_2
16. Sels, V., Gheysen, N., Vanhoucke, M.: A comparison of priority rules for the job shop scheduling problem under different flow time-and tardiness-related objective functions. Int. J. Prod. Res. **50**(15), 4255–4270 (2012)
17. Smith, K.A.: Neural networks for combinatorial optimization: a review of more than a decade of research. INFORMS J. Comput. **11**(1), 15–34 (1999)
18. Sutton, R.S., Barto, A.G.: Reinforcement Learning: An Introduction. MIT Press, Cambridge (2018)
19. Taillard, E.: Benchmarks for basic scheduling problems. Eur. J. Oper. Res. **64**(2), 278–285 (1993)
20. Vaswani, A., et al.: Attention is all you need. In: Advances in Neural Information Processing Systems, vol. 30 (2017)
21. Vinyals, O., Fortunato, M., Jaitly, N.: Pointer networks. In: Advances in Neural Information Processing Systems, vol. 28 (2015)
22. Waschneck, B., et al.: Optimization of global production scheduling with deep reinforcement learning. Proc. CIRP **72**, 1264–1269 (2018)
23. Williams, R.J.: Simple statistical gradient-following algorithms for connectionist reinforcement learning. Mach. Learn. **8**(3), 229–256 (1992). https://doi.org/10.1007/BF00992696
24. Zhang, C., Song, W., Cao, Z., Zhang, J., Tan, P.S., Chi, X.: Learning to dispatch for job shop scheduling via deep reinforcement learning. In: Advances in Neural Information Processing Systems, vol. 33, pp. 1621–1632 (2020)

Generating a Graph Colouring Heuristic with Deep Q-Learning and Graph Neural Networks

George Watkins$^{(\boxtimes)}$, Giovanni Montana, and Juergen Branke

University of Warwick, Coventry, UK
{george.watkins,g.montana}@warwick.ac.uk, juergen.branke@wbs.ac.uk

Abstract. The graph colouring problem consists of assigning labels, or colours, to the vertices of a graph such that no two adjacent vertices share the same colour. In this work we investigate whether deep reinforcement learning can be used to discover a competitive construction heuristic for graph colouring. Our proposed approach, ReLCol, uses deep Q-learning together with a graph neural network for feature extraction, and employs a novel way of parameterising the graph that results in improved performance. Using standard benchmark graphs with varied topologies, we empirically evaluate the benefits and limitations of the heuristic learned by ReLCol relative to existing construction algorithms, and demonstrate that reinforcement learning is a promising direction for further research on the graph colouring problem.

Keywords: Graph Colouring · Deep Reinforcement Learning · Graph Neural Networks

1 Introduction

The Graph Colouring Problem (GCP) is among the most well-known and widely studied problems in graph theory [12]. Given a graph G, a solution to GCP is an assignment of colours to vertices such that adjacent vertices have different colours; the objective is to find an assignment that uses the minimum number of colours. This value is called the *chromatic number* of G, and denoted $\chi(G)$. GCP is one of the most important and relevant problems in discrete mathematics, with wide-ranging applications from trivial tasks like sudoku through to vital logistical challenges like scheduling and frequency assignment [1]. Given that GCP has been proven to be NP-Complete for general graphs [15], no method currently exists that can optimally colour any graph in polynomial time. Indeed it is hard to find even approximate solutions to GCP efficiently [28] and currently no algorithm with reasonable performance guarantees exists [22].

Many existing methods for GCP fall into the category of *construction heuristics*, which build a solution incrementally. Designing an effective construction heuristic is challenging and time-consuming and thus there has been a lot of

M. Sellmann and K. Tierney (Eds.): LION 2023, LNCS 14286, pp. 491–505, 2023.
https://doi.org/10.1007/978-3-031-44505-7_33

interest in ways to generate heuristics *automatically*. This has previously been very successful in, for example, job shop scheduling [8]. Among the simplest construction methods for GCP are *greedy algorithms* [25], in which vertices are selected one by one and assigned the 'lowest' permissible colour based on some pre-defined ordering of colours.

In this work we investigate the use of reinforcement learning (RL) to learn a greedy construction heuristic for GCP by framing the selection of vertices as a sequential decision-making problem. Our proposed algorithm, ReLCol, uses deep Q-learning (DQN) [30] together with a graph neural network (GNN) [5,33] to learn a policy that selects the vertices for our greedy algorithm. Using existing benchmark graphs, we compare the performance of the ReLCol heuristic against several existing greedy algorithms, notably Largest First, Smallest Last and DSATUR. Our results indicate that the solutions generated by our heuristic are competitive with, and in some cases better than, these methods. As part of ReLCol, we also present an alternative way of parameterising the graph within the GNN, and show that our approach significantly improves performance compared to the standard representation.

2   Related Work

Graph Colouring. Methods for GCP, as with other combinatorial optimisation (CO) problems, can be separated into exact solvers and heuristic methods. Exact solvers must process an exponentially large number of solutions to guarantee optimality; as such, they quickly become computationally intractable as the size of the problem grows [26]. Indeed exact algorithms are generally not able to solve GCP in reasonable time when the number of vertices exceeds 100 [31].

When assurances of optimality are not required, heuristic methods offer a compromise between good-quality solutions and reasonable computation time. Heuristics may in some cases produce optimal solutions, but offer no guarantees for general graphs. Considering their simplicity, greedy algorithms are very effective: even ordering the vertices at random can yield a good solution. And crucially, for every graph there exists a vertex sequence such that greedily colouring the vertices in that order will yield an optimal colouring [25].

Largest-First (LF), Smallest-Last (SL) and DSATUR [9] are the three most popular such algorithms [20], among which DSATUR has become the de facto standard for GCP [32]. As such, we have chosen these three heuristics as the basis for our comparisons. Both LF and SL are *static* methods, meaning the vertex order they yield is fixed at the outset. LF chooses the vertices in decreasing order by degree; SL also uses degrees, but selects the vertex v with smallest degree to go last, and then repeats this process with the vertex v (and all its incident edges) removed. Conversely, DSATUR is a *dynamic* algorithm: at a given moment the choice of vertex depends on the previously coloured vertices. DSATUR selects the vertex with maximum *saturation*, where saturation is the number of distinct colours assigned to its neighbours. Similarly, the Recursive

Largest First algorithm [23] is dynamic. At each step it finds a maximal independent set and assigns the same colour to the constituent vertices. The coloured vertices are then removed from the graph and the process repeats.

Improvement algorithms take a different approach: given a (possibly invalid) colour assignment, these methods use local search to make small adjustments in an effort to improve the colouring, either by reducing the number of colours used or eliminating conflicts between adjacent vertices. Examples include TabuCol [17], simulated annealing [2] and evolutionary algorithms [14].

Machine Learning Methods for CO Problems. Given how difficult it is to solve CO problems exactly, and the reliance on heuristics for computationally tractable methods, machine learning appears to be a natural candidate for addressing problems like GCP. Indeed there are many examples of methods for CO problems that use RL [4,19,29] or other machine learning techniques [6,35].

For GCP, supervised learning has been used to predict the chromatic number of a graph [24] but this is dependent on having a labelled training dataset. Given the computational challenges inherent in finding exact solutions, this imposes limitations on the graphs that can be used for training. Conversely, RL does not require a labelled dataset for training. In [41], RL is used to support the local search component of a hybrid method for the related k-GCP problem by learning the probabilities with which each vertex should be assigned to each colour. While iteratively solving k-GCP for decreasing values of k is a valid (and reasonably common) method for solving GCP [14,27], it is inefficient.

On the other hand, [18] addresses GCP directly with a method inspired by the success of AlphaGo Zero [34]. In contrast to greedy algorithms, this approach uses a pre-determined vertex order and learns the mechanism for deciding the *colours*. During training they use a computationally demanding Monte Carlo Tree Search using 300 GPUs; due to the computational overhead and lack of available code we were unable to include this algorithm in our study.

Our proposed algorithm is most closely related to [16], in which the authors present a greedy construction heuristic that uses RL with an attention mechanism [37] to select the vertices. There are, however, several key differences between the two methods. Their approach uses the REINFORCE algorithm [40] whereas we choose DQN [30] because the action space is discrete; they incorporate spatial and temporal locality biases; and finally, we use a novel state parameterisation, which we show improves the performance of our algorithm.

3 Problem Definition

A k-*colouring* of a graph $G = (V, E)$ is a partition of the vertices V into k disjoint subsets such that, for any edge $(u, v) \in E$, the vertices u and v are in different subsets. The subsets are typically referred to as *colours*. GCP then consists of identifying, for a given graph G, the minimum number of colours for which a k-colouring exists and the corresponding colour assignment. This number is known as the *chromatic number* of G, denoted $\chi(G)$.

Given any graph, a greedy construction heuristic determines the order in which vertices are to be coloured, sequentially assigning to them the lowest permissible colour according to some pre-defined ordering of colours. In this work we address the problem of automatically deriving a greedy construction heuristic that colours general graphs using as few colours as possible.

4 Preliminaries

Markov Decision Processes. A Markov Decision Process (MDP) is a discrete-time stochastic process for modelling the decisions taken by an agent in an environment. An MDP is specified by the tuple $(\mathcal{S}, \mathcal{A}, \mathcal{P}, \mathcal{R}, \gamma)$, where $\mathcal{S}$ and $\mathcal{A}$ represent the state and action spaces; $\mathcal{P}$ describes the environment's transition dynamics; $\mathcal{R}$ is the reward function; and γ is the discount factor. The goal of reinforcement learning is to learn a decision policy $\pi : \mathcal{S} \to \mathcal{A}$ that maximises the expected sum of discounted rewards, $\mathbb{E}\left[\sum_{t=1}^{\infty} \gamma^t R_t\right]$.

Deep Q-Learning. Q-learning [38] is a model-free RL algorithm that learns $Q^*(s, a)$, the value of taking action a in a state s and subsequently behaving optimally. Known as the *optimal action value function*, $Q^*(s, a)$ is defined as

$$Q^*(s, a) = \mathbb{E}\left[\sum_{i=1}^{\infty} \gamma^i R_{t+i} \middle| S_t = s, A_t = a\right] \quad (1)$$

where S_t, A_t and R_t are random variables representing respectively the state, action and reward at timestep t. *Deep* Q-learning (DQN) [30] employs a Q-network parameterised by weights θ to approximate $Q^*(s, a)$. Actions are chosen greedily with respect to their values with probability $1 - \epsilon$, and a random action is taken otherwise to facilitate exploration. Transitions (s, a, r, s') - respectively the state, action, reward and next state - are added to a buffer and the Q-network is trained by randomly sampling transitions, backpropagating the loss

$$L(\theta) = \left(\left[r + \gamma \max_{a'} Q_{\hat{\theta}}(s', a')\right] - Q_\theta(s, a)\right)^2 \quad (2)$$

and updating the weights using stochastic gradient descent. Here $Q_\theta(s, a)$ and $Q_{\hat{\theta}}(s, a)$ are estimates of the value of state-action pair (s, a) using the Q-network and a *target network* respectively. The target network is a copy of the Q-network, with weights that are updated via periodic soft updates, $\hat{\theta} \leftarrow \tau\theta + (1 - \tau)\hat{\theta}$. Using the target network in the loss helps to stabilise learning [30].

Graph Neural Networks. Graph neural networks (GNNs) [5,33] support learning over graph-structured data. GNNs consist of blocks; the most general GNN block takes a graph G with vertex-, edge- and graph-level features, and outputs a new graph G' with the same topology as G but with the features replaced by vertex-, edge- and graph-level embeddings [5]. The embeddings are generated via a message-passing and aggregation mechanism whereby information flows between pairs of neighbouring vertices. Stacking multiple GNN blocks allows for more complex dependencies to be captured. The steps within a single GNN block are demonstrated in Fig. 1; in our method we do not use graph-level features so for simplicity these have been omitted.

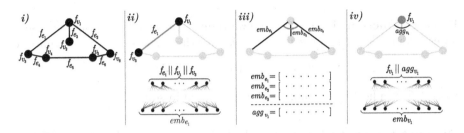

Fig. 1. Demonstration of how a GNN block generates edge and vertex embeddings. Blue indicates the element that is updated in that step. *i)* (Color figure online) The input graph with edge and vertex features; *ii)* For each edge e, concatenate the features of e with the features of the vertices it connects, and pass the resulting vector through a small neural network to generate the edge embedding emb_e; *iii)* For each vertex v, aggregate the embeddings of the incident edges using an elementwise operation like *sum* or *max* to generate the edge aggregation agg_v; *iv)* For each vertex v, concatenate the features of v with the associated edge aggregation agg_v, and pass the resulting vector through a small network to generate the vertex embedding emb_v. Blocks can be stacked by repeating this process with the previous block's edge and vertex embeddings used as the features.

5   Methodology

5.1   Graph Colouring as a Markov Decision Process

States. A state $s \in S$ for graph $G = (V, E)$ is a vertex partition of V into subsets P_i, $i \in \{-1, 0, 1, 2, ...\}$. For $i \neq -1$, the partition P_i contains the vertices currently assigned colour i, and P_{-1} represents the set of currently un-coloured vertices. States in which $P_{-1} = \emptyset$ are terminal. Our method for parameterising the state, which results in improved performance, is described in Sect. 5.2.

Actions. An action $a \in \mathcal{A}$ is an un-coloured vertex (i.e. $a \in P_{-1}$) indicating the next vertex to be coloured. The complete mechanism by which ReLCol chooses actions is described in Sect. 5.4.

Transition Function. Given an action a, the transition function $\mathcal{P} : \mathcal{S} \times \mathcal{A} \to \mathcal{S}$ updates the state s of the environment to s' by assigning the lowest permissible colour to vertex a. Note that choosing colours in this way does not preclude finding an optimal colouring [25] as every graph admits a sequence that will yield an optimal colouring. The transition function $\mathcal{P}$ is deterministic: given a state s and an action a, there is no uncertainty in the next state s'.

Reward Function. For GCP, the reward function should encourage the use of fewer colours. As such our reward function for the transition (s, s') is defined as

$$\mathcal{R}(s, s') = -1(C(s') - C(s))$$

where $C(s)$ indicates the number of colours used in state s.

Discount Factor. GCP is an episodic task, with an episode corresponding to colouring a single graph G. Given that each episode is guaranteed to terminate after n steps, where n is the number of vertices in G, we set $\gamma = 1$. Using $\gamma < 1$ would bias the heuristic towards deferring the introduction of new colours, which may be undesirable.

5.2 Parameterising the State

Recall that for the graph $G = (V, E)$, a state $s \in \mathcal{S}$ is a partition of V into subsets P_i, $i \in \{-1, 0, 1, 2, ...\}$. We represent the state using a *state graph* $G_s = (V, E_s, F_s^v, F_s^e)$: respectively the vertices and edges of G_s, together with the associated vertex and edge features.

State Graph Vertices. Note that the vertices in G_s are the same as the vertices in the original graph G. Then, given a state s, the feature f_s^v of vertex v is a 2-tuple containing: i) A vertex name $\in \{0, 1, 2, ..., n-1\}$ and ii) The current vertex colour $c_v \in \{-1, 0, 1, 2, ...\}$ (where $c_v = -1$ if and only if v has not yet been assigned a colour).

State Graph Edges. In the standard GNN implementation, messages are only passed between vertices that are joined by an edge. In our implementation we choose to represent the state as a complete graph on V to allow information to flow between all pairs of vertices. We use a binary edge feature f_s^e to indicate whether the corresponding edge was in the original graph G:

$$f_s^e = \begin{cases} -1 & \text{if } e = (v_i, v_j) \in E \\ 0 & \text{otherwise} \end{cases}$$

Our state parameterisation, which is the input to the Q-network, allows messages to be passed between all pairs of vertices, including those that are not connected; in Sect. 6.4 we show that this representation results in improved performance.

5.3 Q-Network Architecture

Our Q-network is composed of several stacked GNN blocks followed by a feed-forward neural network of fully connected layers with ReLU activations. Within the aggregation steps in the GNN we employ an adaptation of Principal Neighbourhood Aggregation [10] which has been shown to mitigate information loss. The GNN takes the state graph G_s as input, and returns a set of vertex embeddings. For each vertex v, the corresponding embedding is passed through the fully connected layers to obtain the value for the action of choosing v next.

5.4 Selecting Actions

In general actions are selected using an ϵ-greedy policy with respect to the vertices' values. However using the Q-network is (relatively) computationally expensive. As such, where it has no negative effect - in terms of the ultimate number of colours used - we employ alternative mechanisms to select vertices.

First Vertex Rule. The first vertex to be coloured is selected at random. By Proposition 1, this does not prevent an optimal colouring being found.

Proposition 1. *An optimal colouring remains possible regardless of the first vertex to be coloured.*

Proof. Let $G = (V, E)$ be a graph with $\chi(G) = k^*$, and $\mathcal{P}^* = P_0^*, P_1^*, ..., P_{k^*-1}^*$ an optimal colouring of G, where the vertices in P_i^* are all assigned colour i. Suppose vertex v is the first to be selected, and j is the colour of v in $\mathcal{P}^*$ (i.e. $v \in P_j^*$, where $0 \leq j \leq k^* - 1$). Simply swap the labels P_0^* and P_j^* so that the colour assigned to v is the 'first' colour. Now, using this new partition, we can use the construction described in [25] to generate an optimal colouring. ∎

Isolated Vertices Rule. We define a vertex to be *isolated* if all of its neighbours have been coloured. By Proposition 2, we can immediately colour any such vertices without affecting the number of colours required.

Proposition 2. *Immediately colouring isolated vertices has no effect on the number of colours required to colour the graph.*

Proof. Let $G = (V, E)$ be a graph with $\chi(G) = k^*$. Suppose also that $\mathcal{P} = P_{-1}, P_0, P_1, ..., P_{k^*-1}$ is a partial colouring of G (with P_{-1} the non-empty set of un-coloured vertices). Let $v \in P_{-1}$ be an un-coloured, isolated vertex (i.e. it has no neighbours in P_{-1}). No matter when v is selected, its colour will be the first that is different from all its neighbours. Also, given that v has no un-coloured neighbours, it has no influence on the colours assigned to subsequent vertices. Therefore v can be chosen at any moment (including immediately) without affecting the ultimate number of colours used. ∎

5.5   The ReLCol Algorithm

Figure 2 demonstrates how the state graph is constructed, and how it evolves as the graph is coloured. The full ReLCol algorithm is presented in Algorithm 1.

6   Experimental Results

In this section we evaluate the performance of the heuristic learned by ReLCol against existing algorithms. Because the learning process is inherently stochastic, we generate 12 ReLCol heuristics, each using a different random seed. In our experiments we apply all 12 to each graph and report the average number of colours required. We note that although ReLCol refers to the generating algorithm, for brevity we will also refer to our learned heuristics as ReLCol.

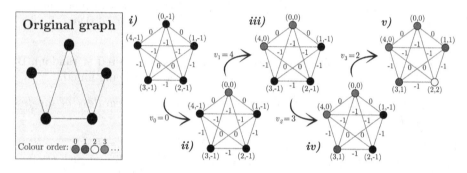

Fig. 2. Example of graph parameterisation and colouring rules, where $v_t = j$ indicates the vertex with name j is selected at step t. *i)* Initial parameterisation of the original graph; *ii)* First vertex $v_0 = 0$ is selected at random and assigned colour 0; *iii)* Vertex $v_1 = 4$ can also be assigned colour 0; *iv)* Vertex $v_2 = 3$ cannot be assigned colour 0 so takes colour 1; *v)* Vertex $v_3 = 2$ cannot be assigned colour 0 or 1 so takes colour 2, leaving vertex 1 isolated; vertex 1 cannot be assigned colour 0 so takes colour 1.

Architecture and Hyperparameters. Our Q-network consists of 5 GNN blocks and 3 fully connected layers with weights initialised at random. Our GNN blocks use only edge and vertex features; we experimented with including global features but found no evidence of performance improvement. The vertex and edge embeddings, as well as the hidden layers in all fully connected neural networks, have 64 dimensions. We use the Adam optimiser [21] with learning rate 0.001 and batch size 64, $\tau = 0.001$, and an ϵ-greedy policy for exploration, where ϵ decays exponentially from 0.9 to 0.01 through 25000 episodes of training.

Training Data. We have constructed a dataset of 1000 training graphs of size $n \in [15, 50]$, composed of 7 different graph types: Leighton graphs [23], Queen graphs [13], Erdos-Renyi graphs [11], Watts-Strogatz graphs [39], Barabasi-Albert graphs [3], Gaussian Random Partition graphs [7] and graphs generated by our own method which constructs graphs with known upper bound on the chromatic number. Each training graph was constructed by choosing its type and size uniformly at random. Where the generating process of the chosen graph type has its own parameters these too were chosen at random to ensure as much diversity as possible amongst the training graphs. Our datasets, together with our code, are available on GitHub[1].

6.1   Comparison with Existing Algorithms

We first compare ReLCol to existing construction algorithms, including Largest First, Smallest Last, DSATUR and Random (which selects the vertices in a random order). We also compare to the similar RL-based method presented in Gianinazzi et al. [16]. Using their implementation we generate 12 heuristics with different random seeds and report the average result. Note that in their paper the authors present both a deterministic and a stochastic heuristic, with the stochastic version generated by taking a softmax over the action weights. At test time their stochastic heuristic is run 100 times and the best colouring is returned. Given that with enough attempts even a random algorithm will eventually find an optimal solution, we consider heuristics that return a colouring in a single pass to be more interesting. As such we consider only the deterministic versions of the Gianinazzi et al. algorithm and ReLCol.

Algorithm 1 ReLCol Algorithm

Initialise Q-network Q_θ; target network $Q_{\hat\theta}$; replay buffer $\mathcal{B}$; and training dataset $\mathcal{D}$
set $\epsilon = 0.9$
for episode in $1, \ldots, E$ **do**
  Sample a graph G from $\mathcal{D}$ and construct state graph G_s
  set $n_c = 0$ ▷ n_c = no. of coloured vertices
  Colour a random vertex, increment n_c and update G_s ▷ First vertex rule
  Colour isolated vertices, increment n_c and update G_s ▷ Isolated vertices rule
  while $n_c < n$ **do**   ▷ n = no. of vertices in G
    let $G_s' = G_s$, $n_c' = n_c$
    Generate embeddings for uncoloured vertices using GNN component of Q_θ
    Pass each vertex embedding through dense layers of Q_θ to get action values
    Select vertex a using ϵ-greedy policy
    Colour a, increment n_c' and update G_s'
    Colour isolated vertices, increment n_c' and update G_s'
    Calculate reward $r = -1(n_c' - n_c)$
    Add transition (G_s, a, r, G_s', d) to $\mathcal{B}$ ▷ $d = (n_c' = n)$ is the 'done' flag
    if $\mathcal{B}$ contains more than N transitions **then**   ▷ Perform learning step
      Sample a batch B of N transitions from $\mathcal{B}$.
      for $(G_{s\,i}, a_i, r_i, G_{s\,i}', d_i)$ in B **do**
        let $y_i = r_i + (1 - d_i) \max_{a' \in \mathcal{A}} Q_{\hat\theta}(G_{s\,i}', a')$
      let $\mathcal{L} = \frac{1}{N} \sum_{i=1}^{N} (Q_\theta(G_{s\,i}, a_i) - y_i)^2$ ▷ Calculate the loss
      Update Q_θ using gradient descent on $\mathcal{L}$
    $G_s \leftarrow G_s'$; $n_c \leftarrow n_c'$
    every U steps
      $\hat\theta \leftarrow \tau\theta + (1 - \tau)\hat\theta$ ▷ Soft update of target network
  Decay ϵ

---

[1] https://github.com/gpdwatkins/graph_colouring_with_RL.

Each heuristic is applied to the benchmark graphs used in [24] and [16], which represent a subset of the graphs specified in the *COLOR02: Graph Colouring and its Generalizations* series[2]. For these graphs the chromatic number is known; as such we report the *excess* number of colours used by an algorithm (i.e. 0 would mean the algorithm has found an optimal colouring for a graph). The results are summarised in Table 1. On average over all the graphs DSATUR was the best performing algorithm, using 1.2 excess colours, closely followed by our heuristic with 1.35 excess colours. The other tested algorithms perform significantly worse on these graphs, even slightly worse than ordering the vertices at random. The test set contains a mix of easier graphs - all algorithms manage to find the chromatic number for huck - as well as harder ones - even the best algorithm uses four excess colours on queen13_13. DSATUR and ReLCol each outperform all other methods on 4 of the graphs.

Table 1. Comparison of ReLCol with other construction algorithms on graphs from the COLOR02 benchmark dataset. Values indicate how many more colours are required than the chromatic number, χ. For each graph, Random is run 100 times and the average and standard error are reported. For Gianinazzi et al. and ReLCol, the 12 heuristics are run and the average and standard error are reported. A bold number indicates that an algorithm has found the unique best colouring amongst the algorithms.

Graph instance	n	χ	Random	LF	SL	DSATUR	Reinforce	ReLCol
queen5_5	25	5	$2.3^{\pm 0.1}$	2	3	**0**	$2.1^{\pm 0.3}$	$0.2^{\pm 0.11}$
queen6_6	36	7	$2.4^{\pm 0.06}$	2	4	2	$3.2^{\pm 0.16}$	$1^{\pm 0}$
myciel5	47	6	$0.1^{\pm 0.03}$	0	0	0	$0.1^{\pm 0.08}$	$0^{\pm 0}$
queen7_7	49	7	$4^{\pm 0.06}$	5	3	4	$3.3^{\pm 0.32}$	$\mathbf{2.2^{\pm 0.16}}$
queen8_8	64	9	$3.4^{\pm 0.07}$	4	5	3	$3.3^{\pm 0.24}$	$\mathbf{2.1^{\pm 0.14}}$
1-Insertions_4	67	4	$1.2^{\pm 0.04}$	1	1	1	$1^{\pm 0}$	$1^{\pm 0}$
huck	74	11	$0^{\pm 0}$	0	0	0	$0^{\pm 0}$	$0^{\pm 0}$
jean	80	10	$0.3^{\pm 0.05}$	0	0	0	$0^{\pm 0}$	$0.3^{\pm 0.12}$
queen9_9	81	10	$3.9^{\pm 0.07}$	5	5	3	$5^{\pm 0}$	$\mathbf{2.7^{\pm 0.25}}$
david	87	11	$0.7^{\pm 0.07}$	0	0	0	$0.4^{\pm 0.14}$	$1.2^{\pm 0.33}$
mug88_1	88	4	$0.1^{\pm 0.02}$	0	0	0	$0^{\pm 0}$	$0^{\pm 0}$
myciel6	95	7	$0.3^{\pm 0.05}$	0	0	0	$0.8^{\pm 0.17}$	$0^{\pm 0}$
queen8_12	96	12	$3.3^{\pm 0.06}$	3	3	**2**	$4^{\pm 0}$	$2.6^{\pm 0.18}$
games120	120	9	$0^{\pm 0.01}$	0	0	0	$0^{\pm 0}$	$0^{\pm 0}$
queen11_11	121	11	$5.9^{\pm 0.07}$	6	6	4	$6^{\pm 0}$	$5.4^{\pm 0.25}$
anna	138	11	$0.2^{\pm 0.04}$	0	0	0	$0^{\pm 0}$	$0.4^{\pm 0.18}$
2-Insertions_4	149	4	$1.5^{\pm 0.05}$	1	1	1	$1^{\pm 0}$	$1^{\pm 0}$
queen13_13	169	13	$6.5^{\pm 0.07}$	10	9	**4**	$8^{\pm 0}$	$6.3^{\pm 0.24}$
myciel7	191	8	$0.6^{\pm 0.06}$	0	0	0	$0.3^{\pm 0.14}$	$0.2^{\pm 0.11}$
homer	561	13	$1^{\pm 0.07}$	0	0	0	$0^{\pm 0}$	$0.6^{\pm 0.22}$
Average			$1.9^{\pm 0.01}$	1.95	2	1.2	$1.92^{\pm 0.05}$	$1.35^{\pm 0.05}$

---

[2] https://mat.tepper.cmu.edu/COLOR02/.

6.2 A Class of Graphs on Which ReLCol Outperforms DSATUR

Although the previous results suggest that ReLCol does not outperform DSATUR on general graphs, there do exist classes of graphs on which DSATUR is known to perform poorly. One such class is presented in [36]; these graphs, which we refer to as *Spinrad graphs*, are constructed as follows:

1. Fix the number of vertices n such that $n \pmod 7 = 3$, and let $m = \frac{n+4}{7}$.
2. Partition the vertices into 5 disjoint sets as follows:

$$A = \{a_1, a_2, \cdots, a_{m-2}\} \quad B = \{b_1, b_2, \cdots, b_{m-1}\} \quad C = \{c_2, c_3, \cdots, c_m\}$$
$$B' = \{b'_1, b'_2, \cdots, b'_{2m}\} \quad C' = \{c'_1, c'_2, \cdots, c'_{2m}\}$$

3. Add the following sets of edges:

$$E^B_A = \{(a_i, b_j) : i \neq j\} \quad E^C_A = \{(a_i, c_j) : i < j\} \quad E^C_B = \{(b_{i-1}, c_i) : 2 < i < m\}$$

Plus:
  - $\forall b \in B$, add edges to vertices in B' such that the degree of b is $2m$.
  - $\forall c \in C$, add edges to vertices in C' such that the degree of c is $2m$.

An example of such a graph with $m = 4$ is shown in Fig. 3. Note that some of the vertices in B' and C' may be disconnected; they exist simply to ensure that the vertices in B and C all have degree $2m$.

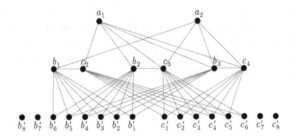

Fig. 3. Example of a Spinrad graph on 24 vertices, generated with $m = 4$.

The vertices can be partitioned into 3 disjoint sets $A \cup B' \cup C'$, B and C. Given that there are no edges between pairs of vertices in the same set, the chromatic number of the graph is 3. However the DSATUR algorithm assigns the same colour to vertices a_i, b_i and c_i, meaning it uses m colours for the whole graph. A proof of this is provided in [36].

Figure 4 shows that ReLCol vastly outperforms DSATUR on Spinrad graphs, in most cases identifying an optimal colouring using the minimum number of colours. This indicates that despite their similar performance on general graphs, ReLCol has learned a heuristic that is selecting vertices differently to DSATUR.

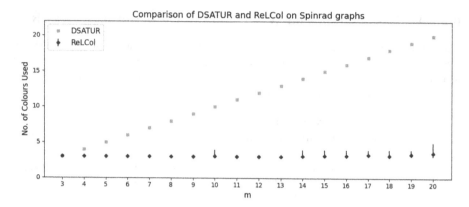

Fig. 4. ReLCol outperforms DSATUR on Spinrad graphs. Error bars show the maximum and minimum number of colours used by the 12 ReLCol-generated heuristics.

6.3   Scalability of ReLCol

While we have shown that ReLCol is competitive with existing construction heuristics, and can outperform them on certain graph classes, our results suggest that the ability of ReLCol to colour general graphs effectively may reduce when the test graphs are significantly larger than those used for training.

This can be observed in Fig. 5, which compares the performance of DSATUR, ReLCol, and Random on graphs of particular sizes generated using the same process as our training dataset. The degradation in performance could be a result of the nature of the training dataset, whose constituent graphs have no more than 50 vertices: for graphs of this size DSATUR and ReLCol seem to achieve comparable results, and much better than Random, but the performance of ReLCol moves away from DSATUR towards Random as the graphs grow in size.

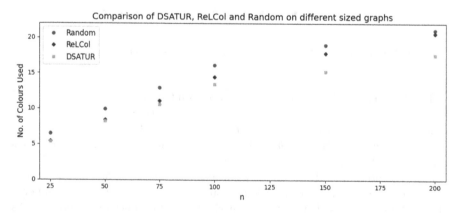

Fig. 5. As the graph size increases, the performance of ReLCol moves from being similar to DSATUR towards Random. This suggests that there are limitations to how well ReLCol generalises to graphs larger than those seen during training.

6.4 Representing the State as a Complete Graph

To demonstrate the benefit of the proposed complete graph representation, we compare ReLCol with a version that preserves the topology of the original graph, meaning that within the GNN, messages are only passed between pairs of adjacent vertices. Figure 6 compares the number of colours used by each version when applied to a validation dataset periodically during training. The validation dataset is composed of 100 graphs generated by the same mechanism as the training dataset. The complete graph representation clearly leads to faster learning and significantly better final performance.

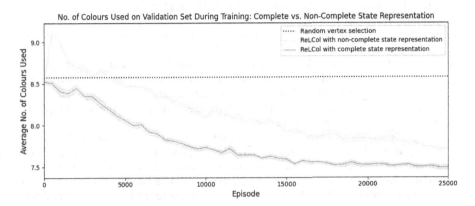

Fig. 6. Our complete graph representation results in faster learning and better final performance compared to the standard GNN representation.

7 Conclusions

We have proposed ReLCol, a reinforcement learning algorithm based on graph neural networks that is able to learn a greedy construction heuristic for GCP. The ReLCol heuristic is competitive with DSATUR, a leading greedy algorithm from the literature, and better than several other comparable methods. We have demonstrated that part of this success is due to a novel (to the best of our knowledge) complete graph representation of the graph within the GNN. Since our complete graph representation seems to perform much better than the standard GNN representation, we intend to investigate its effect in further RL tasks with graph-structured data. We also plan to incorporate techniques for generalisability from the machine learning literature to improve the performance of the ReLCol heuristic on graphs much larger than the training set.

An advantage of automatically generated heuristics is that they can be tuned to specific classes of problem instances by amending the training data, so exploring the potential of ReLCol to learn an algorithm tailored to specific graph types would be an interesting direction. Finally, given that the ReLCol heuristic appears to work quite differently from DSATUR, further analysis of how it selects vertices may yield insights into previously unknown methods for GCP.

Acknowledgements. G. Watkins acknowledges support from EPSRC under grant EP/L015374/1.
G. Montana acknowledges support from EPSRC under grant EP/V024868/1.
We thank L. Gianinazzi for sharing the code for the method presented in [16].

References

1. Ahmed, S.: Applications of graph coloring in modern computer science. Int. J. Comput. Inf. Technol. **3**(2), 1–7 (2012)
2. Aragon, C.R., Johnson, D., McGeoch, L., Schevon, C.: Optimization by simulated annealing: an experimental evaluation; part II, graph coloring and number partitioning. Oper. Res. **39**(3), 378–406 (1991)
3. Barabási, A.L., Albert, R.: Emergence of scaling in random networks. Science **286**(5439), 509–512 (1999)
4. Barrett, T., Clements, W., Foerster, J., Lvovsky, A.: Exploratory combinatorial optimization with reinforcement learning. In: Proceedings of the AAAI Conference on Artificial Intelligence, vol. 34, pp. 3243–3250 (2020)
5. Battaglia, P.W., et al.: Relational inductive biases, deep learning, and graph networks. arXiv:1806.01261 (2018)
6. Bengio, Y., Lodi, A., Prouvost, A.: Machine learning for combinatorial optimization: a methodological tour d'Horizon. Eur. J. Oper. Res. **290**(2), 405–421 (2021)
7. Brandes, U., Gaertler, M., Wagner, D.: Experiments on graph clustering algorithms. In: Di Battista, G., Zwick, U. (eds.) ESA 2003. LNCS, vol. 2832, pp. 568–579. Springer, Heidelberg (2003). https://doi.org/10.1007/978-3-540-39658-1_52
8. Branke, J., Nguyen, S., Pickardt, C.W., Zhang, M.: Automated design of production scheduling heuristics: a review. IEEE Trans. Evol. Comput. **20**(1), 110–124 (2015)
9. Brélaz, D.: New methods to color the vertices of a graph. Commun. ACM **22**(4), 251–256 (1979)
10. Corso, G., Cavalleri, L., Beaini, D., Liò, P., Veličković, P.: Principal neighbourhood aggregation for graph nets. In: Advances in Neural Information Processing Systems, vol. 33, pp. 13260–13271 (2020)
11. Erdős, P., Rényi, A.: On random graphs I. Publicationes Math. **6**(1), 290–297 (1959)
12. Formanowicz, P., Tanaś, K.: A survey of graph coloring - its types, methods and applications. Found. Comput. Decis. Sci. **37**(3), 223–238 (2012)
13. Fricke, G., et al.: Combinatorial problems on chessboards: a brief survey. In: Quadrennial International Conference on the Theory and Applications of Graphs, vol. 1, pp. 507–528 (1995)
14. Galinier, P., Hao, J.K.: Hybrid evolutionary algorithms for graph coloring. J. Comb. Optim. **3**(4), 379–397 (1999)
15. Garey, M.R., Johnson, D.S.: Computers and intractability, vol. 174. Freeman, San Francisco (1979)
16. Gianinazzi, L., Fries, M., Dryden, N., Ben-Nun, T., Besta, M., Hoefler, T.: Learning combinatorial node labeling algorithms. arXiv preprint arXiv:2106.03594 (2021)
17. Hertz, A., de Werra, D.: Using tabu search techniques for graph coloring. Computing **39**(4), 345–351 (1987)

18. Huang, J., Patwary, M., Diamos, G.: Coloring big graphs with alphagozero. arXiv preprint arXiv:1902.10162 (2019)
19. Ireland, D., Montana, G.: Lense: Learning to navigate subgraph embeddings for large-scale combinatorial optimisation. In: International Conference on Machine Learning (2022)
20. Janczewski, R., Kubale, M., Manuszewski, K., Piwakowski, K.: The smallest hard-to-color graph for algorithm DSATUR. Discret. Math. **236**(1–3), 151–165 (2001)
21. Kingma, D.P., Ba, J.: Adam: a method for stochastic optimization. arXiv preprint arXiv:1412.6980 (2014)
22. Korte, B., Vygen, J.: Combinatorial Optimization. AC, vol. 21. Springer, Heidelberg (2018). https://doi.org/10.1007/978-3-662-56039-6
23. Leighton, F.T.: A graph coloring algorithm for large scheduling problems. J. Res. Natl. Bur. Stand. **84**(6), 489–506 (1979)
24. Lemos, H., Prates, M., Avelar, P., Lamb, L.: Graph colouring meets deep learning: effective graph neural network models for combinatorial problems. In: International Conference on Tools with Artificial Intelligence, pp. 879–885. IEEE (2019)
25. Lewis, R.M.R.: Guide to Graph Colouring. TCS, Springer, Cham (2021). https://doi.org/10.1007/978-3-030-81054-2
26. de Lima, A.M., Carmo, R.: Exact algorithms for the graph coloring problem. Revista de Informática Teórica e Aplicada **25**(4), 57–73 (2018)
27. Lü, Z., Hao, J.K.: A memetic algorithm for graph coloring. Eur. J. Oper. Res. **203**(1), 241–250 (2010)
28. Lund, C., Yannakakis, M.: On the hardness of approximating minimization problems. J. ACM (JACM) **41**(5), 960–981 (1994)
29. Mazyavkina, N., Sviridov, S., Ivanov, S., Burnaev, E.: Reinforcement learning for combinatorial optimization: a survey. Comput. Oper. Res. **134**, 105400 (2021)
30. Mnih, V., et al.: Playing atari with deep reinforcement learning. arXiv:1312.5602 (2013)
31. Moalic, L., Gondran, A.: Variations on memetic algorithms for graph coloring problems. J. Heuristics **24**(1), 1–24 (2018)
32. Sager, T.J., Lin, S.J.: A pruning procedure for exact graph coloring. ORSA J. Comput. **3**(3), 226–230 (1991)
33. Scarselli, F., Gori, M., Tsoi, A.C., Hagenbuchner, M., Monfardini, G.: The graph neural network model. IEEE Trans. Neural Netw. **20**(1), 61–80 (2008)
34. Silver, D., et al.: Mastering the game of go without human knowledge. Nature **550**(7676), 354–359 (2017)
35. Smith, K.A.: Neural networks for combinatorial optimization: a review of more than a decade of research. INFORMS J. Comput. **11**(1), 15–34 (1999)
36. Spinrad, J.P., Vijayan, G.: Worst case analysis of a graph coloring algorithm. Discret. Appl. Math. **12**(1), 89–92 (1985)
37. Vaswani, A., et al.: Attention is all you need. In: Advances in Neural Information Processing Systems, vol. 30 (2017)
38. Watkins, C.J., Dayan, P.: Q-learning. Mach. Learn. **8**(3), 279–292 (1992)
39. Watts, D.J., Strogatz, S.H.: Collective dynamics of 'small-world' networks. Nature **393**(6684), 440–442 (1998)
40. Williams, R.J.: Simple statistical gradient-following algorithms for connectionist reinforcement learning. Mach. Learn. **8**(3), 229–256 (1992)
41. Zhou, Y., Hao, J.K., Duval, B.: Reinforcement learning based local search for grouping problems: a case study on graph coloring. Expert Syst. Appl. **64**, 412–422 (2016)

Multi-task Predict-then-Optimize

Bo Tang[iD] and Elias B. Khalil[(✉)][iD]

SCALE AI Research Chair in Data-Driven Algorithms for Modern Supply Chains,
Department of Mechanical and Industrial Engineering, University of Toronto,
Toronto, Canada
{botang,khalil}@mie.utoronto.ca

Abstract. The predict-then-optimize framework arises in a wide variety of applications where the unknown cost coefficients of an optimization problem are first predicted based on contextual features and then
used to solve the problem. In this work, we extend the predict-then-
optimize framework to a multi-task setting: contextual features must be
used to predict cost coefficients of multiple optimization problems, possibly with different feasible regions, simultaneously. For instance, in a vehicle dispatch/routing application, features such as time-of-day, traffic, and
weather must be used to predict travel times on the edges of a road network for multiple traveling salesperson problems that span different target
locations and multiple $s - t$ shortest path problems with different source-
target pairs. We propose a set of methods for this setting, with the most
sophisticated one drawing on advances in multi-task deep learning that
enable information sharing between tasks for improved learning, particularly in the small-data regime. Our experiments demonstrate that multi-
task predict-then-optimize methods provide good tradeoffs in performance
among different tasks, particularly with less training data and more tasks.

Keywords: multi-task learning · predict-then-optimize · data-driven
optimization · machine learning

1 Introduction

The predict-then-optimize framework, in which the unknown coefficients for an
optimization problem are predicted and then used to solve the problem, is emerging as a useful framework in some applications. For instance, in vehicle routing
and job scheduling, we often require optimization where the model's cost coefficients, e.g., travel time and execution time, are unknown but predictable at decision time. In the conventional two-stage method, a learning model is first trained
to predict cost coefficients, after which a solver separately optimizes accordingly.
However, end-to-end approaches that learn predictive models that minimize the
decision error directly have recently gained interest due to some improvements
in experimental performance [4,10,12,15,22,23]. Although there has been some
recent work in predicting elements of the constraint matrix in a linear programming setting [13,17], our focus here is on the predominant line of research that
has focused on unknown cost coefficients [1,5,8,10,12,14,22,23,30,32].

M. Sellmann and K. Tierney (Eds.): LION 2023, LNCS 14286, pp. 506–522, 2023.
https://doi.org/10.1007/978-3-031-44505-7_34

Previous work on predict-then-optimize has focused on learning the cost coefficients for a single optimization task. However, it is natural to consider the setting where multiple related tasks can share information and representations. For example, a vehicle routing application requires predicting travel times on the edges of a road network for multiple traveling salesperson problems (TSPs) that span different target locations and multiple shortest path problems with different source-target pairs. These travel time predictions should be based on the *same contextual information*, e.g., if the tasks are to be executed at the same time-of-day, then the travel times that should be predicted for the different tasks depend on the same features. Another case is in package delivery, where distributing packages from one depot to multiple depots results in independent delivery tasks that nonetheless share travel time predictions since they use the same road network. To that end, we introduce multi-task end-to-end predict-then-optimize, which simultaneously solves multiple optimization problems with a loss function that relates to the decision errors of all such problems.

Multi-task learning has been successfully applied to natural language processing, computer vision, and recommendation systems. However, its applicability to the predict-then-optimize paradigm is yet to be explored. Predict-then-optimize with multi-task learning is attractive because of the ability to improve model performance in the small-data regime. Machine learning, especially with deep neural networks, is data-intensive and prone to overfitting, which might limit applicability to the predict-then-optimize paradigm. The need to simultaneously minimize the losses of different tasks helps reduce overfitting and improve generalization. Multi-task learning combines the data of all tasks, which increases the overall training data size and alleviates task-specific noise.

To the best of our knowledge, we introduce multi-task learning for end-to-end predict-then-optimize for the first time. We motivate and formalize this problem before proposing a set of methods for combining the different task losses. Our experiments show that multi-task end-to-end predict-then-optimize provides performance benefits and good tradeoffs in performance among different tasks, especially with less training data and more tasks, as compared to applying standard single-task methods independently for each task. As an additional contribution, we distinguish end-to-end predict-then-optimize approaches that learn from observed costs (the usual setting of Elmachtoub and Grigas [12]) and those that learn directly from (optimal) solutions without the objective function costs themselves. This extends our framework to applications where there are no labeled coefficients in the training data, e.g., the Amazon Last Mile Routing Challenge [33]. The open-source code is available[1].

2   Related Work

2.1   Differentiable Optimization

The key component of gradient-based end-to-end predict-then-optimize is differentiable optimization, which allows the backpropagation algorithm to update

---

[1] https://github.com/khalil-research/Multi-Task_Predict-then-Optimize.

model parameters from the decision made by the optimizer. Based on the KKT conditions, Amos and Kolter [3] introduced OptNet, a differentiable optimizer to make quadratic optimization problems learnable. With OptNet, Donti et al. [10] investigated a learning framework for quadratic programming; Wilder et al. [32] then added a small quadratic regularization to the linear objective function for linear programming; Ferber et al. [14] extended the method to the discrete model with the cutting plane; Mandi and Guns [22] adopted log-barrier regularization instead of a quadratic one. Besides OptNet, Agrawal et al. [1] leveraged KKT conditions to differentiate conic programming.

Except for the above approaches with KKT, an alternative methodology is to design gradient approximations to avoid ill-defined gradients from predicted costs to optimal solutions. Elmachtoub and Grigas [12] proposed a convex surrogate loss. Vlastelica et al. [30] developed a differentiable optimizer through implicit interpolation. Berthet et al. [5] demonstrated a method with stochastic perturbation to smoothen the loss function and further constructed the Fenchel-Young loss. Dalle et al. [8] extended the perturbation approach to the multiplicative perturbation and the Frank-Wolfe regularization. Mulamba et al. [24] studied a solver-free contrastive loss. Moreover, Shah et al. [26] provided an alternate paradigm that additionally trains a model to predict decision errors to replace the solver.

2.2 Multi-task Learning

Multi-task learning, first proposed by Caruana [6], aims to learn multiple tasks together with joint losses. In short, a model with multiple loss functions is defined as multi-task learning. Much research has focused on the (neural network) model architecture: the most basic model is a shared-bottom model ([6]), including shared hidden layers at the bottom and task-specific layers at the top. Besides such hard parameter sharing schemes, there is also soft sharing so that each task keeps its own parameters. Duong et al. [11] added l_2 norm regularization to encourage similar parameters between tasks. Furthermore, neural networks with different experts and some gates [21,27,29] were designed to fuse information adaptively.

Another crucial issue is resolving the unbalanced (in magnitude) and conflicting gradients from different tasks. There are weighting approaches, such as UW [18], GradNorm [7], and DWA [20], that have been proposed to adjust the weighting of different losses. Other methods, such as PCGrad [34], GradVec [31], and CAGrad [19], were designed to alter the direction of the overall gradient to avoid conflicts and accelerate convergence.

3   Building Blocks

3.1   Optimization Problem

For multi-task end-to-end predict-then-optimize, each task t is a separate (integer) linear optimization problem, defined as follows:

$$\min_{w^t} \quad c^{t^T} w^t$$

$$\text{s.t.} \quad A^t w^t \le b^t$$

$$w^t \ge 0$$

Some w_i^t are integer.

The decision variables are w^t, the constraint coefficients are A^t, the right-hand sides of the constraints are b^t, and the unknown cost coefficients are c^t. Given a cost c^t, $w_{c^t}^{t*}$ is the corresponding optimal solution, and $z_{c^t}^t$ is the optimal value. Additionally, we define $S^t = \{A^t w^t \le b^t, w^t \ge 0, ...\}$ as the feasible region of decision variables w^t.

3.2 Gradient-Based Learning

End-to-end predict-then-optimize aims to minimize a decision loss directly. As a supervised learning problem, it requires a labeled dataset $\mathcal{D}$ consisting of features x and labels c or w_c^*. We will further discuss the difference between cost labels c and solution labels w_c^* in Sect. 4.2. As shown in Fig. 1, multi-task end-to-end predict-then-optimize predicts the unknown costs for multiple optimization problems and then solves these optimization problems with the predicted costs. The critical component is a differentiable optimization embedded into a differentiable predictive model. However, the only learnable part is the prediction model $g(x; \theta)$, since there are no parameters to update in the solver and loss function.

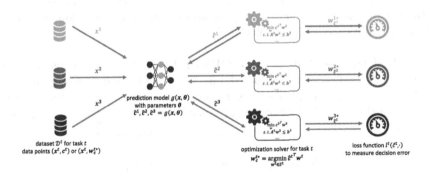

Fig. 1. Illustration of end-to-end multi-task end-to-end predict-then-optimize: labeled datasets $\mathcal{D}^1, \mathcal{D}^2, ..., \mathcal{D}^t$ are used to fit a machine learning predictor $g(x; \theta)$ that predicts costs $\hat{c}^t$ for each task t. The loss function $l^t(\hat{c}^t, \cdot)$ to be minimized measures decision error instead of prediction error.

3.3 Decision Losses

Loss functions for end-to-end predict-then-optimize aim to measure the error in decision-making. For instance, regret is defined as the difference in objective values due to an optimal solution that is based on the true costs and another that is based on the predicted costs:

$$l_{\text{Regret}}(\hat{c}, c) = c^T w_{\hat{c}}^* - z_c^*$$

However, with a linear objective function, regret does not provide useful gradients for learning [12]. Besides regret, decision error can also be defined as the difference between the true solution and its prediction, such as using the Hamming distance [30] or the squared error of a solution to an optimal one [5]. Because the function from costs c to optimal solutions w_c^* is piecewise constant, a solver with any of the aforementioned losses has no nonzero gradients to update the model parameter θ. Thus, the state-of-art methods, namely Smart Predict-then-Optimize (SPO+) [12] and Perturbed Fenchel-Young Loss (PFYL), both design surrogate decision losses which allow for a nonzero approximate gradient (or subgradient), $\frac{\partial l(\cdot)}{\partial \hat{c}}$.

Smart Predict-then-Optimize Loss (SPO+). SPO+ loss [12] is a differentiable convex upper bound on the regret:

$$l_{\text{SPO+}}(\hat{c}, c) = -\min_{w \in S}\{(2\hat{c} - c)^T w\} + 2\hat{c}^T w_c^* - z_c^*.$$

One proposed subgradient for this loss writes as follows:

$$2(w_c^* - w_{2\hat{c}-c}^*) \in \frac{\partial l_{\text{SPO+}}(\hat{c}, c)}{\partial \hat{c}}.$$

In theory, the SPO+ framework can be applied to any problem with a linear objective. Elmachtoub and Grigas [12] have conducted experiments on shortest path as a representative linear program and portfolio optimization as a representative quadratically-constrained problem.

Perturbed Fenchel-Young Loss (PFYL). PFYL [5] leverages Fenchel duality, where Berthet et al. [5] discussed only the case of linear programming. The predicted costs are sampled with Gaussian perturbation ξ, and the expected function of the perturbed minimizer is defined as $F(c) = \mathbb{E}_\xi[\min_{w \in S}\{(c + \sigma\xi)^T w\}]$. With $\Omega(w_c^*)$, the dual of $F(c)$, the Fenchel-Young loss reads:

$$l_{\text{FY}}(\hat{c}, w_c^*) = \hat{c}^T w_c^* - F(\hat{c}) - \Omega(w_c^*).$$

Then, we can estimate the gradients by M samples Monte Carlo:

$$\frac{\partial l_{\text{FY}}(\hat{c}, w_c^*)}{\partial \hat{c}} \approx w_c^* - \frac{1}{M} \sum_m^M \operatorname*{argmin}_{w \in S}\{(\hat{c} + \sigma\xi_m)^T w\}$$

3.4   Multi-task Loss Weighting Strategies

The general idea of multi-task learning is that multiple tasks are solved simultaneously by the same predictive model. It is critical for a multi-task neural

Table 1. Losses of Various Training Strategies

	Strategy	Losses
Single-Task	mse	$l_{\mathrm{MSE}}(c, \hat{c})$
	separated	Separate $l_{\mathrm{Decision}}(\hat{c}^t, \cdot)$ for each task t
	separated+mse	Separate $l_{\mathrm{Decision}}(\hat{c}^t, \cdot) + l_{\mathrm{MSE}}(c^t, \hat{c}^t)$ for each task t
Multi-Task	comb	$\sum_t^T l_{\mathrm{Decision}}^t(\hat{c}^t, \cdot)$
	comb+mse	$\sum_t^T l_{\mathrm{Decision}}^t(\hat{c}^t, \cdot) + \sum_t^T l_{\mathrm{MSE}}^t(c^t, \hat{c}^t)$
	gradnorm	$\sum_t^T u_{\mathrm{Ada}}^t l_{\mathrm{Decision}}^t(\hat{c}^t, \cdot)$
	gradnorm+mse	$\sum_t^T u_{\mathrm{Ada}_1}^t l_{\mathrm{Decision}}^t(\hat{c}^t, \cdot) + \sum_t^T u_{\mathrm{Ada}_2}^t l_{\mathrm{MSE}}^t(c^t, \hat{c}^t)$

network, one such flexible class of models, to balance losses among tasks with the loss weights u_t for task t. The weighting approaches we evaluated include a uniform combination (all $u^t = 1$) and GradNorm [7], an adaptive loss weighting approach. The latter provides adaptive per-task weights u_{Ada}^t that are dynamically adjusted during training in order to keep the scale of the gradients similar. In this work, we set the GradNorm hyperparameters of "restoring force" to 0.1 and the learning rate of loss weights to 0.005. Further tuning is possible but was not needed for our experiments.

All the training strategies we have explored in this paper, including baseline approaches, are summarized in Table 1. Let T be the number of tasks, and cost coefficient prediction for task t be $\hat{c}^t$. "mse" is the usual two-stage baseline of training a regression model that minimizes cost coefficients mean-squared error $l_{\mathrm{MSE}} = \frac{1}{n} \sum_i^n \|\hat{c}_i - c_i\|^2$ only without regard to the decision. "separated" trains one model per task, minimizing, for each task, a decision-based loss such as SPO+ or PFYL from Sect. 3.3. "comb" simply sums up the per-task decision losses, whereas "gradnorm" does so in a weighted adaptive way. For any of these methods, whenever "+mse" is appended to the method name, a variant of the method is obtained that combines additional mean-squared error l_{MSE} in the cost predictions with the decision loss. Such a regularizer is known to be useful in practice, even with the primary evaluation metric of a trained model being its decision regret [12]. Although we refer to "separated+mse" as a single-task method, it can also be considered as a multi-task learning method in a broad sense because of the inclusion of two losses.

4 Learning Architectures

4.1 Shared Learnable Layers

The model class we will explore is deep neural networks. Besides their capacity to represent complex functions from labeled data, neural networks have a compositional structure that makes them particularly well-suited for multi-task learning. A multi-task neural network shares hidden layers across all tasks and keeps specific layers for each task. Figure 2 illustrates that the sharing part of multi-task end-to-end predict-then-optimize depends on the consistency of the predicted coefficients, which we will define next. At a high level, Fig. 2 distinguishes two settings. On the left, the different tasks use the exact same predicted cost vector. On the right, each task could have a different cost vector. In both settings, the predictions are based on the same input feature vector.

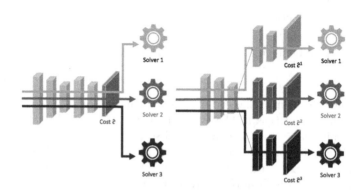

Fig. 2. Illustration of two types of multi-task end-to-end predict-then-optimize: On the left, all optimization tasks require the same prediction as cost coefficients. On the right, different tasks share some layers as feature embeddings and make different coefficient predictions.

Shared Predicted Coefficients (Single-Cost). In this setting (left of Fig. 2), which we will refer to as *single-cost*, the same cost coefficients are shared among all tasks. For example, multiple navigation tasks on a single map are shortest-path problems with different source-target pairs that share the same distance matrix (i.e., costs). In this case, the cost coefficients c^t for task t are equal to or a subset of the shared costs c. Thus, the prediction model is defined as

$$\hat{c} = g(x; \theta_{\text{Shared}}),$$

which is the same as a single-task model, and the multiple tasks combine their losses,

$$\sum_t u^t l^t_{\text{Decision}}(\hat{c}, \cdot),$$

based on the shared prediction, $\hat{c}$. Therefore, as Algorithm 1 shows, all learnable layers are shared. In addition, the baseline methods we referred to as "separated" and "separated+mse" are not practical in this same-costs setting as they, inconsistently, produce different cost predictions for each task, even when that is not required. Nonetheless, an experimental assessment of their performance will be conducted in order to contrast it with using multi-task learning.

Algorithm 1. Single-Cost Multi-Task Gradient Descent

Require: coefficient matrix $A^1, A^2, ...$; right-hand side $b^1, b^2, ...$; training data $\mathcal{D}$
1: Initialize predictor parameters θ_{Shared} for predictor $g(x; \theta_{\text{Shared}})$
2: **for** epochs **do**
3:     **for** each batch of training data (x, c) or $(x, w_c^{1*}, w_c^{2*}, ...)$ **do**
4:        Forward pass to predict cost $\hat{c} := g(x; \theta_{\text{Shared}})$
5:        Forward pass to solve optimal solution $w_{\hat{c}}^{t*} := \text{argmin}_{w^t \in S^t} \hat{c}^T w^t$ per task
6:        Forward pass to sum weighted decision losses $l(\hat{c}, \cdot) := \sum_t u^t l^t(\hat{c}, \cdot)$
7:        Backward pass from loss $l(\hat{c}, \cdot)$ to update parameters θ_{Shared}
8:     **end for**
9: **end for**

Shared Features Embeddings (Multi-cost). In many applications of predict-then-optimize, the optimization problem requires cost coefficients that are specific and heterogeneous to each task, but that can be inferred from homogeneous contextual features. For instance, in a vehicle routing application, features such as time of day and weather predict travel time in different regions. Compared to the single-cost setting we just introduced, the *multi-cost* predictor here has the form

$$\hat{c}^t = g(x; \theta_{\text{Shared}}; \theta_t).$$

Per-task predictions are made by leveraging the same information embedding in the layers of the neural network that are shared across the tasks (see right of Fig. 2). Thus, the corresponding loss function is

$$\sum_t u^t l^t_{\text{Decision}}(\hat{c}^t, \cdot),$$

and Algorithm 2 updates the parameters of the predictor.

Algorithm 2. Multi-Cost Multi-Task Gradient Descent

Require: coefficient matrix $A^1, A^2, ...$; right-hand side $b^1, b^2, ...$; training data $\mathcal{D}^1, \mathcal{D}^2, ...$
1: Initialize predictor parameters $\theta_{\text{Shared}}, \theta_1, \theta_2, ...$ for predictor $g(x; \theta_{\text{Shared}}; \theta_t)$
2: **for** epochs **do**
3:     **for** each batch of training data $(x^1, x^2, ..., c^1, c^2, ...)$ or $(x^1, x^2, ..., w_{c1}^{1*}, w_{c2}^{2*}, ...)$ **do**
4:        Forward pass to predict cost $\hat{c}^t := g(x^t; \theta_{\text{Shared}}; \theta_t)$ per task
5:        Forward pass to solve optimal solution $w_{\hat{c}^t}^{t*} := \text{argmin}_{w^t \in S^t} \hat{c}^{t T} w^t$ per task
6:        Forward pass to sum weighted decision losses $l(\hat{c}^1, \hat{c}^2, ..., \cdot) := \sum_t u^t l^t(\hat{c}, \cdot)$
7:        Backward pass from loss $l(\hat{c}^1, \hat{c}^2, ..., \cdot)$ to update parameters $\theta_{\text{Shared}}, \theta_1, \theta_2, ...$
8:     **end for**
9: **end for**

4.2   Label Accessibility and Learning Paradigms

We distinguish two learning paradigms that require different kinds of labels (cost coefficients c or optimal solutions w_c^*) in the training data: **learning from costs** and **learning from (optimal) solutions**. This distinction is based on the availability of labeled cost coefficients c. Thus, SPO+ is learning from costs because the calculation of SPO+ loss involves true cost coefficients c, whereas PFYL is learning from solutions that do not require access to c; see Sect. 3.3.

The need for the true cost coefficients as labels in the training data is a key distinguishing factor because these cost coefficients provide additional information that can be used to train the model, but they may be absent in the data. Deriving optimal solutions from the cost coefficients is trivial, but the opposite is intricate as it requires some form of inverse optimization [2]. The ability to directly learn from solutions extends the applicability of end-to-end predict-then-optimize beyond what a two-stage approach, which is based on regressing on the cost coefficients, can do. Indeed, the recent MIT-Amazon Last Mile Routing Challenge [33] is one such example in which good TSP solutions are observed on historical package delivery instances, but the corresponding edge costs are unobserved. Those good solutions are based on experienced drivers' tacit knowledge.

5   Experiments

In this section, we present experimental results for multi-task end-to-end predict-then-optimize. In our experiments, we evaluate decision performance using regret, and we use mean-squared error (MSE) to measure the prediction of cost coefficients $\hat{c}$. We use SPO+ and PFYL as typical methods for learning from costs and learning from solutions and adopt various multi-task learning strategies discussed in Sect. 3.4, as well as two-stage and single-task baselines. Our experiments are conducted on two datasets, including graph routing on PyEPO TSP dataset[2] [25], and adjusted Warcraft terrain[3] [30] to learn single-cost decisions and multi-cost decisions. We also vary the amount of training data size and the number of tasks.

All the numerical experiments were conducted in Python v3.7.9 with two Intel E5-2683 v4 Broadwell CPUs, two NVIDIA P100 Pascal GPUs, and 8GB memory. Specifically, we used PyTorch [25] v1.10.0 for the prediction model and Gurobi [16] v9.1.2 for the optimization solver, and PyEPO [28] v0.2.0 for SPO+ and PFYL autograd functions.

5.1   Benchmark Datasets and Neural Network Architecture

Graph Routing with Multiple Tasks. We used the traveling salesperson problem dataset generated from PyEPO [28], which uses the Euclidean distance among nodes plus polynomial function $f(x_i) = (\frac{1}{\sqrt{p}}(\mathcal{B}x_i)_j + 3)^4$ (where $\mathcal{B}$ is a random matrix) with random noise perturbations $f(x_i) \cdot \epsilon$ to map the features x into a symmetric distance matrix of a complete graph. We discuss both learning from costs and learning from solutions. In this experiment, the number of features p is 10, the number of nodes m is 30, the polynomial degree of function $f(x_i)$ is 4, and the noise ϵ comes from $U(0.5, 1.5)$. We sample $15 - 22$ nodes as target

---

[2] https://khalil-research.github.io/PyEPO.
[3] https://drive.google.com/file/d/1lYPT7dEHtH0LaIFjigLOOkxUq4yi94fy.

locations for multiple traveling salesperson problems (TSPs) and 54 undirected edges for multiple shortest paths (SPs) with different source-target pairs. Thus, all TSP and SP tasks share the same cost coefficients.

Since the multiple routing tasks require consistent cost coefficients, the model $g(x; \theta)$ makes one prediction of the costs that is used for all of the tasks. The architecture of the regression network is one fully-connected layer with a softplus activation to prevent negative cost predictions, and all tasks share the learnable layer. For the hyperparameters, the learning rate is 0.1, the batch size is 32, and the max training iterations is 30000 with 5 patience early stopping. For PFYL, the number of samples M is 1, and the perturbation temperature σ is 1.0. We formulate SP as a network flow problem and use the Dantzig-Fulkerson-Johnson (DFJ) formulation [9] to solve TSP.

Warcraft Shortest Path with Various Species. The Warcraft map shortest path dataset [30] allows for the learning of the shortest path from RGB terrain images, and we use 96×96 RGB images for 12×12 grid networks and sample 3 small subsets from 10000 original training data points for use in training. As shown in Fig. 3, we modify the cost coefficients for different species (human, naga, dwarf) and assume that the cost coefficients are not accessible in the data. This means there are three separate datasets of feature-solution pairs, which require learning from solutions using the PFYL method. Similar to SP tasks in Graph Routing, the shortest path optimization model is a linear program.

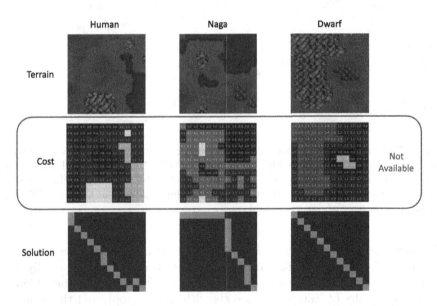

Fig. 3. Multiple datasets of Warcraft terrain images for different species, where labeled cost coefficients are unavailable.

Since the multiple Warcraft shortest paths tasks require us to predict cost coefficients for different species, the prediction model should incorporate task-specific layers. Following Vlastelica et al. [30], we train a truncated ResNet18 (first five layers) for 50 epochs with batches of size 70, and learning rate 0.0005 decaying at the epochs 30 and 40. The first three layers are the shared-bottom. The number of samples M is 1, and the perturbation temperature σ is 1.0.

5.2   Performance Advantage of Multi-task Learning

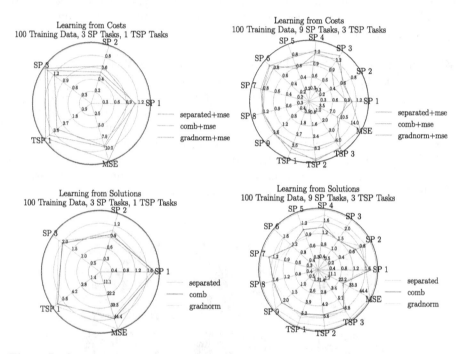

Fig. 4. Performance Radar Plot for Graph Routing: Average performance for different tasks on the test set, trained with SPO+ (Top) and PFYL (Bottom), and 100 training data points, including regrets and cost MSE, lower is better. SP i is the regret for shortest path task i, TSP i is the regret for traveling salesperson task i, MSE is the mean squared error of cost coefficients.

Experimental results on multiple routing tasks on a graph, as shown in Fig. 4, demonstrate that multi-task end-to-end predict-then-optimize, especially with GradNorm, has a performance advantage over single-task approaches. Figure 4 shows the results of learning from costs with SPO+ (top), and the results of learning from solutions with PFYL (bottom). In these "radar plots", the per-task regret on unseen test data is shown along each dimension (lower is better). It can be seen that the innermost method in these figures (best seen in color) is the red one, "gradnorm+mse".

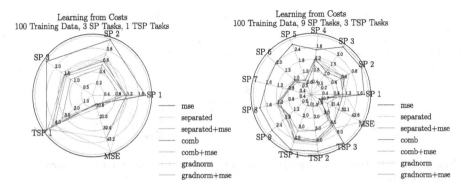

Fig. 5. Learning from Costs for Graph Routing: Average performance for different tasks on the test set, trained with SPO+ and 100 training data points for more methods, lower is better. SP i is the regret for shortest path task i, TSP i is the regret for traveling salesperson task i, MSE is the mean squared error of cost coefficients.

More experiments are shown in Fig. 5. The investigation includes the two-stage method, single-task, and multi-task with and without cost MSE as regularization; these two plots include a superset of the methods in Fig. 4. Despite achieving a lower MSE, the two-stage approaches exhibit significantly worse regret than end-to-end learning. Additionally, adding an MSE regularizer on the cost coefficients consistently improves end-to-end learning approaches. Thus, we always include the additional cost of MSE regularizer when learning from costs. However, since labeled costs are absent when learning directly from solutions, PFYL cannot add cost MSE ("+mse") as regularization and cannot be compared with the two-stage method (which requires cost labels.)

5.3 Efficiency Benefit of Multi-task Learning

Figure 6 shows the training time for SPO+ and PFYL models when using early stopping when five consecutive epochs exhibit non-improving loss values on held-out validation data, a standard trick in neural network training. For "separated" and "separated+mse", the training time is the sum of each individual model. We can see that the use of GradNorm to adjust the weights dynamically allows for efficient model training as faster convergence is achieved. The "separated+mse" baseline typically requires more time to converge, but also converges to worse models in terms of regret, as seen in the previous paragraph. Furthermore, "comb" and "comb+mse" usually require more time to converge, which highlights the importance of an effective weighting scheme for multi-tasks approaches.

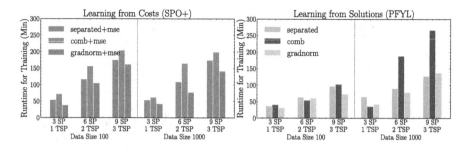

Fig. 6. Training time for Graph Routing: The elapsed time of training to convergence at different settings.

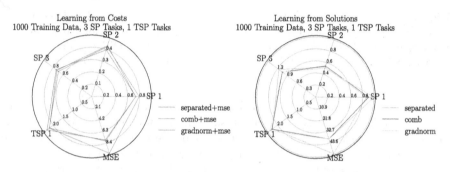

Fig. 7. More Training Data for Graph Routing: Average performance for different graph routing tasks on the test set for SPO+ (left) and PFYL (right), trained with PFYL and 1000 training data points, including regrets and cost MSE, lower is better. SP i is the regret for shortest path task i, TSP i is the regret for traveling salesperson task i, MSE is the mean squared error of cost coefficients.

5.4   Learning Under Data Scarcity

In this section, we claim that the multi-task end-to-end predict-then-optimize framework is particularly effective in the small-data regime. Compared to Figs. 4, we find that multi-task learning for graph routing loses its advantage with more training data (Fig. 7). In Warcraft shortest path problem, Fig. 8 shows that the performance of the separated single-task model gradually improves and may even surpass multi-task learning as the amount of training data increases. These figures show that multi-task end-to-end predict-then-optimize can effectively leverage information from related datasets when the size of the individual dataset is limited. Therefore, multi-task learning is a reliable option under data scarcity.

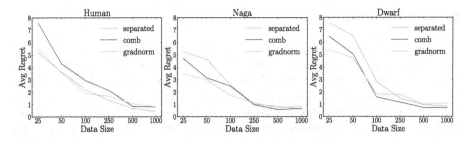

Fig. 8. Performance on Warcraft Shortest Path: Average regrets of different strategies, trained with PFYL, decreases as the amount of training data increases; lower is better.

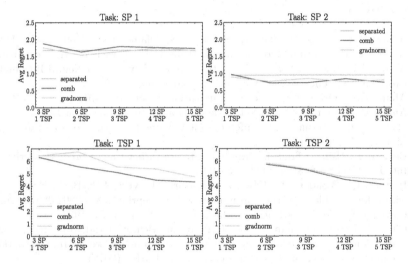

Fig. 9. Performance on Graph Routing: Average regrets of a different strategy, trained with PFYL and 100 training data, decreases as the amount of tasks increases, lower is better.

5.5   Learning Under Task Redundancy

Figure 9 indicates that increasing the number of related tasks improves the model performance, especially for complicated tasks such as TSP. This performance improvement can be attributed to the positive interaction between the losses of the related tasks. This finding suggests the potential for using auxiliary tasks to enhance model performance.

6 Conclusion

We extend the end-to-end predict-then-optimize framework to multi-task learning, which jointly minimizes decision error for related optimization tasks. Our results demonstrate the benefits of this approach, including an improved performance with less training data points and the ability to handle multiple tasks simultaneously. Future work in this area could include the application of this method to real-world problems, as well as further exploration of techniques for multi-task learning, such as current and novel multi-task neural network architectures and gradient calibration methods.

References

1. Agrawal, A., Amos, B., Barratt, S., Boyd, S., Diamond, S., Kolter, J.Z.: Differentiable convex optimization layers. In: Wallach, H., Larochelle, H., Beygelzimer, A., d' Alché-Buc, F., Fox, E., Garnett, R. (eds.) Advances in Neural Information Processing Systems, vol. 32, Curran Associates, Inc. (2019)
2. Ahuja, R.K., Orlin, J.B.: Inverse optimization. Oper. Res. **49**(5), 771–783 (2001)
3. Amos, B., Kolter, J.Z.: OptNet: differentiable optimization as a layer in neural networks. In: International Conference on Machine Learning, pp. 136–145, PMLR (2017)
4. Bengio, Y.: Using a financial training criterion rather than a prediction criterion. Int. J. Neural Syst. **8**(04), 433–443 (1997)
5. Berthet, Q., Blondel, M., Teboul, O., Cuturi, M., Vert, J.P., Bach, F.: Learning with differentiable perturbed optimizers. arXiv preprint arXiv:2002.08676 (2020)
6. Caruana, R.: Multitask learning. Mach. Learn. **28**(1), 41–75 (1997)
7. Chen, Z., Badrinarayanan, V., Lee, C.Y., Rabinovich, A.: GradNorm: gradient normalization for adaptive loss balancing in deep multitask networks. In: International Conference on Machine Learning, pp. 794–803, PMLR (2018)
8. Dalle, G., Baty, L., Bouvier, L., Parmentier, A.: Learning with combinatorial optimization layers: a probabilistic approach. arXiv preprint arXiv:2207.13513 (2022)
9. Dantzig, G., Fulkerson, R., Johnson, S.: Solution of a large-scale traveling-salesman problem. J. Oper. Res. Soc. Am. **2**(4), 393–410 (1954)
10. Donti, P.L., Amos, B., Kolter, J.Z.: Task-based end-to-end model learning in stochastic optimization. In: Advances in Neural Information Processing Systems (2017)
11. Duong, L., Cohn, T., Bird, S., Cook, P.: Low resource dependency parsing: cross-lingual parameter sharing in a neural network parser. In: Proceedings of the 53rd Annual Meeting of the Association for Computational Linguistics and the 7th International Joint Conference on Natural Language Processing (vol. 2: short papers), pp. 845–850 (2015)
12. Elmachtoub, A.N., Grigas, P.: Smart "predict, then optimize". Manage. Sci. **68**, 9–26 (2021)
13. Estes, A.S., Richard, J.P.P.: Smart predict-then-optimize for two-stage linear programs with side information. INFORMS Journal on Optimization (2023)
14. Ferber, A., Wilder, B., Dilkina, B., Tambe, M.: MIPaaL: mixed integer program as a layer. In: Proceedings of the AAAI Conference on Artificial Intelligence, vol. 34, pp. 1504–1511 (2020)

15. Ford, B., Nguyen, T., Tambe, M., Sintov, N., Fave, F.D.: Beware the soothsayer: from attack prediction accuracy to predictive reliability in security games. In: Khouzani, M.H.R., Panaousis, E., Theodorakopoulos, G. (eds.) GameSec 2015. LNCS, vol. 9406, pp. 35–56. Springer, Cham (2015). https://doi.org/10.1007/978-3-319-25594-1_3

16. Gurobi Optimization, LLC: Gurobi Optimizer Reference Manual (2021). https://www.gurobi.com/

17. Hu, X., Lee, J.C., Lee, J.H.: Predict+ optimize for packing and covering LPS with unknown parameters in constraints. In: Proceedings of the AAAI Conference on Artificial Intelligence (2023)

18. Kendall, A., Gal, Y., Cipolla, R.: Multi-task learning using uncertainty to weigh losses for scene geometry and semantics. In: Proceedings of the IEEE Conference on Computer Vision and Pattern Recognition, pp. 7482–7491 (2018)

19. Liu, B., Liu, X., Jin, X., Stone, P., Liu, Q.: Conflict-averse gradient descent for multi-task learning. In: Advances in Neural Information Processing Systems, vol. 34, pp. 18878–18890 (2021)

20. Liu, S., Johns, E., Davison, A.J.: End-to-end multi-task learning with attention. In: Proceedings of the IEEE/CVF Conference on Computer Vision and Pattern Recognition, pp. 1871–1880 (2019)

21. Ma, J., Zhao, Z., Yi, X., Chen, J., Hong, L., Chi, E.H.: Modeling task relationships in multi-task learning with multi-gate mixture-of-experts. In: Proceedings of the 24th ACM SIGKDD International Conference on Knowledge Discovery & Data Mining, pp. 1930–1939 (2018)

22. Mandi, J., Guns, T.: Interior point solving for LP-based prediction+optimisation. In: Larochelle, H., Ranzato, M., Hadsell, R., Balcan, M.F., Lin, H. (eds.) Advances in Neural Information Processing Systems, vol. 33, pp. 7272–7282, Curran Associates, Inc. (2020)

23. Mandi, J., Stuckey, P.J., Guns, T.: Smart predict-and-optimize for hard combinatorial optimization problems. In: Proceedings of the AAAI Conference on Artificial Intelligence, vol. 34, pp. 1603–1610 (2020)

24. Mulamba, M., Mandi, J., Diligenti, M., Lombardi, M., Bucarey, V., Guns, T.: Contrastive losses and solution caching for predict-and-optimize. arXiv preprint arXiv:2011.05354 (2020)

25. Paszke, A., et al.: PyTorch: an imperative style, high-performance deep learning library. In: Advances in Neural Information Processing Systems, vol. 32 (2019)

26. Shah, S., Wang, K., Wilder, B., Perrault, A., Tambe, M.: Decision-focused learning without decision-making: learning locally optimized decision losses. In: Advances in Neural Information Processing Systems (2022)

27. Shazeer, N., et al.: Outrageously large neural networks: the sparsely-gated mixture-of-experts layer. arXiv preprint arXiv:1701.06538 (2017)

28. Tang, B., Khalil, E.B.: PyEPO: a PyTorch-based end-to-end predict-then-optimize library for linear and integer programming. Mathematical Programming Computation (2022 in submission)

29. Tang, H., Liu, J., Zhao, M., Gong, X.: Progressive layered extraction (PLE): a novel multi-task learning (MTL) model for personalized recommendations. In: Fourteenth ACM Conference on Recommender Systems, pp. 269–278 (2020)

30. Vlastelica, M., Paulus, A., Musil, V., Martius, G., Rolínek, M.: Differentiation of blackbox combinatorial solvers. arXiv preprint arXiv:1912.02175 (2019)

31. Wang, Z., Tsvetkov, Y., Firat, O., Cao, Y.: Gradient vaccine: investigating and improving multi-task optimization in massively multilingual models. arXiv preprint arXiv:2010.05874 (2020)

32. Wilder, B., Dilkina, B., Tambe, M.: Melding the data-decisions pipeline: decision-focused learning for combinatorial optimization. In: Proceedings of the AAAI Conference on Artificial Intelligence, vol. 33, pp. 1658–1665 (2019)
33. Winkenbach, M., Parks, S., Noszek, J.: Technical proceedings of the amazon last mile routing research challenge (2021)
34. Yu, T., Kumar, S., Gupta, A., Levine, S., Hausman, K., Finn, C.: Gradient surgery for multi-task learning. In: Advances in Neural Information Processing Systems, vol. 33, pp. 5824–5836 (2020)

Integrating Hyperparameter Search into Model-Free AutoML with Context-Free Grammars

Hernán Ceferino Vázquez$^{(\boxtimes)}$, Jorge Sanchez, and Rafael Carrascosa

MercadoLibre Inc., Buenos Aires, Argentina
{hernan.vazquez,jorge.sanchez,rafael.carrascosa}@mercadolibre.com

Abstract. Automated Machine Learning (AutoML) has become increasingly popular in recent years due to its ability to reduce the amount of time and expertise required to design and develop machine learning systems. This is very important for the practice of machine learning, as it allows building strong baselines quickly, improving the efficiency of the data scientists, and reducing the time to production. However, despite the advantages of AutoML, it faces several challenges, such as defining the solutions space and exploring it efficiently. Recently, some approaches have been shown to be able to do it using tree-based search algorithms and context-free grammars. In particular, GramML presents a model-free reinforcement learning approach that leverages pipeline configuration grammars and operates using Monte Carlo tree search. However, one of the limitations of GramML is that it uses default hyperparameters, limiting the search problem to finding optimal pipeline structures for the available data preprocessors and models. In this work, we propose an extension to GramML that supports larger search spaces including hyperparameter search. We evaluated the approach using an OpenML benchmark and found significant improvements compared to other state-of-the-art techniques.

Keywords: Automated Machine Learning · Hyperparameter Optimization · Context-free Grammars · Monte Carlo Tree Search · Reinforcement Learning · Model-Free

1 Introduction

The practice of Machine Learning (ML) involves making decisions to find the best solution for a problem among a myriad of candidates. Machine learning practitioners often find themselves in a process of trial and error in order to determine optimal processing pipelines (data preprocessing, model architectures, etc.) and their configurations (hyperparameters) [4]. These decisions go hand in hand with the many constraints imposed by the actual application domain and have a great impact on the final system performance. These constraints can be either intrinsic, such as working with heterogeneous data, extrinsic, such as requiring using limited computational resources and/or short development cycles, or both. Solving real-world problems by means of ML is a time-consuming and difficult process [5].

In this context, the development of methods and techniques that allow for the automated generation of machine learning solutions [30] is a topic of great practical importance, as they help reduce the time and expertise required for the design and evaluation of a broad range of ML-based products and applications.

Besides the benefits it brings to the practice of machine learning, AutoML poses several challenges. In our case, these challenges are how to define the solution space (the set of possible pipeline configurations, i.e. pipeline structure and its hyperparameters) in a way that is both comprehensive and computationally feasible, and how to explore it efficiently both in terms of time and resources.

Many techniques have been developed over the years to address these problems, most of which rely on parametric models to guide the exploration of the search space. However, such model-based solutions require a characterization of configuration space and its possible solutions via meta-features, which are not always easy to define and evolve over time, adding an extra (meta) learning step as well as an additional layer of hyperparameters (those of the meta-model) [28]. An alternative to this family of models corresponds to the model-free counterpart. These approaches rely on different strategies to guide the search stage, in which evolutionary [14,23] and tree-based techniques [21,24] appear as the most prominent ones.

A particularly interesting approach consists of using context-free grammars for the search space specification and an efficient search strategy over the production tree [32] derived from it [10,21,31]. A specific algorithm that has been used in this case is Monte Carlo Tree Search (MCTS), which allows for a more efficient exploration of the search space [24]. In particular, [31] proposes GramML, a simple yet effective model-free reinforcement learning approach based on an adaptation of the MCTS for trees and context-free grammars that show superior performance compared to the state-of-the-art. However, one of the limitations of their model is that it disregards hyperparameters, leaving them to their default values. This effectively reduces the search space pipeline architectures consisting of data preprocessors and generic model configurations.

In this work, we propose an extension to GramML that supports larger search spaces by incorporating hyperparameters into the model. We extend the basic MCTS algorithm to account for this increase in complexity, by incorporating a pruning strategy and the use of non-parametric selection policies. In addition, we run experiments to evaluate the performance of our approach and compare it with state-of-the-art techniques. The results show that our method significantly improves performance and demonstrates the effectiveness of incorporating hyperparameter search into grammar-based AutoML.

The rest of the paper is structured as follows. Section 2 discusses the related work. Section 3 explains the main concepts used in this work. Section 4 presents the core of our approach. Section 5 describes the experiments we conducted and highlights the main results. Finally, Sect. 6 concludes and outlines our future work.

2   Related Work

This section provides a succinct overview of prior studies related to AutoML. The main representative for solving this class of problems is AutoSklearn [15], which is based on the Sequential Model-Based Algorithm Configuration (SMAC) algorithm [18] and incorporates Bayesian optimization (BO) and surrogate models to determine optimal machine learning pipelines. In AutoSklearn hyperparameters and pipeline selection problem are tackled as a single, structured, joint optimization problem named CASH (Combined Algorithm Selection and Hyperparameter optimization [27]). Recent research has produced model-based approaches that exhibit comparable performance to AutoSklearn. For instance, MOSAIC [24] leverages a surrogate model formulation in conjunction with a Monte Carlo Tree Search (MCTS) to search the hypothesis space. MOSAIC tackle both pipeline selection and hyperparameter optimization problems using a coupled hybrid strategy: MCTS is used for the first and Bayesian optimization for the latter, where the coupling is enforced by using a surrogate model. AlphaD3M [11] also builds on MCTS and uses a neural network as surrogate model, considering only partial pipelines rather than whole configurations. The difference with MOSAIC is that the pipeline selection is decoupled from the hyperparameter search, it first looks for the best pipelines and then performs hyperparameter optimization of the best pipelines using SMAC. [10] studied the use of AlphaD3M in combination with a grammar, relying on pre-trained models for guidance during the search. Another approach, PIPER [21], utilizes a context-free grammar and a greedy heuristic, demonstrating competitive results relative to other techniques. In PIPER, the best pipeline structure is found and then its hyperparameters are searched with a standard CASH optimizer. Other approaches include the use of genetic programming [23,25] and hierarchical planning [19,22] techniques.

3   Background

A formal grammar is a collection of symbols and grammar rules that describe how strings are formed in a specific formal language [26]. There are various types of grammar, including context-free grammars [8]. In context-free grammars, the grammar rules consist of two components: the left-hand side of a grammar rule is always a single non-terminal symbol. These symbols never appear in the strings belonging to the language. The right-hand side contains a combination of terminal and non-terminal symbols that are equivalent to the left-hand side. Therefore, a non-terminal symbol can be substituted with any of the options specified on the right-hand side.

Once specified, the grammar provide a set of rules by which non-terminal symbols can be substituted to form *strings*. These strings belong to the language represented by the grammar. This generation process (production task) is performed by starting from an initial non-terminal symbol and replacing it recursively using the grammar rules available. An important property, in this case, is that the set of rules can be expressed in the form of a tree and the generation of any given production can be seen as a path from the root to a particular leaf of this tree. The generation process can be thus seen as a search problem over the tree induced by the grammar.

In the context of the AutoML problem, the grammar is used to encode the different components and their combination within a desired pipeline. For instance, we can set a grammar that accounts for different data preprocessing strategies, like *PREPROCESSING := "NumericalImputation" or SCALING := "MinMaxScaler" or "StandardScaler"* or the type of weight regularization used during learning, as *PENALTY := "l1" or "l2"*. Finding optimal ML pipelines corresponds to searching for a path from the root (the initial symbol of the grammar) to a leaf of the tree (the last rule evaluated) that performs best in the context of a given ML problem. Interestingly, we could also encode hyperparameter decisions within the same grammar, *e.g.* by listing the possible values (or ranges) that a particular parameter can take. By doing so, we grow the tree size (configuration space of all possible ML pipelines and their configurations) exponentially.

From the above, a key component of grammar-based approaches to AutoML is the search algorithm, as it has to be efficient and scale gracefully with the size of the tree. In the literature, one of the most reliable alternatives is the Monte Carlo Tree Search (MCTS) algorithm [20]. MCTS is a search algorithm that seeks to expand the search tree by using random sampling heuristics. It consists of four steps: *selection, expansion, simulation* and *backpropagation*. During the selection step, the algorithm navigates the search tree to reach a node that has not yet been explored. Selection always starts from the root node and, at each level, selects the next node to visit according to a given *selection policy*. The expansion, simulation, and backpropagation steps ensure that the tree is expanded according to the rules available at each node and, once a terminal node is reached, that the information about the reliability of the results obtained at this terminal state is propagated back to all the nodes along the path connecting this node to the root. As a reference, Algorithm 1 illustrates the general form of the MCTS algorithm.

Algorithm 1. General MCTS approach

1: **function** SEARCH(s_0)
2:     create root node a_0 with state s_0
3:     **while** within budget **do**
4:         $a_i \leftarrow$ Selection($a_0, SelectionPolicy$)
5:         $a_i \leftarrow$ Expansion(a_i)
6:         $\Delta \leftarrow$ Simulation($s(a_i)$)
7:         Backpropagation(a_i, Δ)
    return BestChild($a_0, SelectionPolicy$)

In general, MCTS is a powerful method that allows to explore the search tree in a systematic and efficient way and can be used for AutoML. However, balancing the exploration and exploitation trade-off depends on find best selection policy to use. Furthermore, when the tree grows up exponentially as in the case of hyper-parameter search problem, finding good solutions in an acceptable time can be challenging.

4 Hyperparameter Search in Grammar-Based AutoML

In grammar-based AutoML, the objective is to generate pipeline configurations using the available production rules that encode the choices for a particular nonterminal symbol. In our case, we extend the GramML[1] formulation of [31] to account for the hyperparameters associated with the different ML components that compose an ML pipeline. For continuous numerical hyperparameters, we use the same sampling intervals as AutoSklearn for a fairer comparison. In the case of categorical or discrete parameters, we incorporate their values directly into the grammar.

Using MCTS the grammar is traversed generating productions and looking for the best pipeline configurations. The algorithm can be stopped at any point, and the current best pipeline configuration is returned. The longer the algorithm runs, the more productions will be generated and evaluated, and the more likely that the recommended pipeline is indeed the optimal one. The process of generating grammar productions and evaluating pipeline configurations consists of the traditional four steps of MCTS; however, it has some considerations given the nature of the problem. In our setting, the only valid configurations are leaf nodes, as they contain the complete configuration object. Any attempt to execute a partial configuration would result in an error. Additionally, each iteration of the algorithm (commonly referred to as an episode in RL) ends in a leaf node or terminal state. These differences lead to some changes in the original algorithm.

Algorithm 2 illustrates an adaptation of the MCTS algorithm for AutoML. The SEARCH function is called by the main program until a budget is exhausted. The budget can be specified in terms of the number of iterations or running time. Each call to the SEARCH function returns a pipeline configuration. This function is executed until either a pipeline configuration is found or the budget is exhausted.

In each step of the algorithm, if the selected node is not terminal, the search function performs the expansion and simulation steps. In the expansion step, the tree is navigated from the selected node to find a node that has not been expanded. Once found, it is expanded, and the algorithm continues. In the simulation step, the algorithm randomly selects a path from the selected node to a leaf. Upon reaching a leaf, a reward function is computed and returned. In the case of the selected node being terminal, this reward is used for backpropagation. Finally, the selected leaf is pruned from the tree. If all leaves of a branch are pruned, then the branch is also pruned.

---

[1] The Extended GramML source code is available at: github.com/mercadolibre/fury_gramml-with-hyperparams-search.

Algorithm 2. MCTS for AutoML

1: **function** SEARCH(s_0)
2:     $a_i \leftarrow a_0$
3:     **while** used budget < budget & a_i is not Terminal **do**
4:         $a_i \leftarrow$ Selection($a_0, SelectionPolicy$)
5:         **if** a_i is not Terminal **then**
6:             $a_i \leftarrow$ Expansion(a_i)
7:             $\Delta \leftarrow$ Simulation($s(a_i)$)
8:         **else**
9:             $\Delta \leftarrow$ GetReward($s(a_i)$)
10:        Backpropagation(a_i, Δ)
11:        $p_i^* \leftarrow$ BestLeaf($a_0, SelectionPolicy$)
12:        PruneTree(p_i^*)
13:        **return** p_i^*

Selection and backpropagation steps depend on the actual selection policy. Selection policies are important as they guide the search, regulating the trade-off between exploration and exploitation. In what follows, we present three different alternatives to this combination that appear as particularly suited to our problem.

Upper Confidence Bound Applied to Trees (UTC). UTC [20] is a standard algorithm that seeks to establish a balance between exploration and exploitation to find the best option among a set of candidates. It uses a constant C as the only parameter that regulates this trade-off. UCT is based on the computation of two statistics, a cumulative sum of the rewards obtained and the number of visits for each node. base on these values, we compute a score for each node as:

$$v(a_i) = \frac{reward(a_i)}{visits(a_i)} + C\sqrt{\frac{\log\left(visits(a_i)\right)}{visits(parent(a_i)))}}, \tag{1}$$

The intuition behind this expression is constructing confidence intervals around the estimated rewards of each node. The UCT formula essentially selects the node with the highest upper confidence bound (UCB). By selecting nodes based on this UCB, the algorithm ensures fair exploration of all nodes, avoiding local optima, and achieving a balance between exploiting known rewards and exploring potentially better options. The parameter C controls how narrow or wider the confidence bound will be and with this balance the tradeoff between exploitation (first term) and exploration (second term).

The selection policy corresponds to choosing the node with the highest score. Algorithm 3 shows the process by which rewards and visits are backpropagated from a leaf up to the root.

Bootstrap Thompson Sampling (BTS). In Thompson Sampling (TS) the basic idea is to select actions stochastically [1]. TS requires being able to compute the

Algorithm 3. Backpropagation for UCT

1: **function** BACKPROPAGATION(a, Δ)
2:     $a_i \leftarrow a$ ▷ a is a leaf node
3:     **while** a_i is not Null **do**
4:         $reward_i \leftarrow reward_i + \Delta$
5:         $visits_i \leftarrow visits_i + 1$
6:         $a_i \leftarrow parent(a_i)$ ▷ if a_i is the root node parent is Null

exact posterior [7]. In cases where this is not possible, the posterior distribution can be approximated using bootstrapping [13]. This method is known as Bootstrap Thompson Sampling (BTS) [12]. In BTS the bootstrap distribution is a collection of bootstrap replicates $j \in \{1, ..., J\}$, where J is a tunable hyperparameter that balances the trade-off between exploration and exploitation. A smaller J value results in a more greedy approach, while a larger J increases exploration, but has a higher computational cost [12]. Each replicate j has some parameters θ, which are used to estimate j's expected utility given some prior distribution $P(\theta)$. At decision time, the bootstrap distribution for each node is sampled, and the child with the highest expected utility is selected. During backpropagation, the distribution parameters are updated by simulating a coin flip for each replicate j. In our scenario, we use a Normal distribution as prior and two parameters α y β (by consistency to other works [17]). If the coin flip comes up heads, the α and β parameters for j are re-weighted by adding the observed reward to α_j and the value 1 to β_j. Algorithm 4 illustrates the backpropagation step for the BTS strategy.

Algorithm 4. Backpropagation for BTS

1: **function** BACKPROPAGATION(a, Δ)
2:     $a_i \leftarrow a$ ▷ a is a leaf node
3:     **while** a_i is not Null **do**
4:         **for** $j \in J$ **do**
5:             sample d_j from Bernoulli($1/2$)
6:             **if** $d_j = 1$ **then**
7:                 $\alpha_{a_i} \leftarrow \alpha_{a_i} + \Delta$
8:                 $\beta_{a_i} \leftarrow \beta_{a_i} + 1$
9:         $a_i \leftarrow parent(a_i)$ ▷ if a_i is the root node parent is Null

Regarding the selection policy, we estimate the value of the nodes as α_{a_i}/β_{a_i}, which represents the largest point estimate of our prior (the mean). Afterwards, we select a child with a probability proportional to that value.

Tree Parzen Estimator (TPE). TPE is a widely used decision algorithm in hyperparameter optimization [2] that can be easily adapted to work on trees. TPE defines two densities, $l(a_i)$ and $g(a_i)$, where $l(a_i)$ is computed by using the observations $\{a_i\}$ such that corresponding reward $y = reward(a_i)$ is less than a threshold y^*, while $g(a_i)$ is computed using the remaining observations. The value of y^* is chosen as the γ-quantile of the observed y values.

To compute the functions $g(a_i)$ and $l(a_i)$, it is necessary to keep track of the rewards obtained from each node instead of accumulating them as in the case of UCT. Algorithm 5 shows the overall process.

Algorithm 5. Backpropagation for TPE

1: **function** BACKPROPAGATION(a, Δ)
2:     $a_i \leftarrow a$ ▷ a is a leaf node
3:     **while** a_i is not Null **do**
4:         $[reward(a_i)] \leftarrow [reward(a_i)] + [\Delta]$
5:         $a_i \leftarrow parent(a_i)$ ▷ if a_i is the root node parent is Null

---

The tree-structured induced by the partitions used to compute l and g makes it easy to evaluate candidates based on the ratio $g(a_i)/l(a_i)$, similar to the selection policy used in BTS.

5   Experiments

To evaluate our work, we conducted experiments in two parts. In the first part, we performed an ablation study of each of the non-parametric selection policies proposed in the context of the proposed MCTS algorithm. In the second part, we compare our approach with other from the state-of-the-art.

5.1   Experimental Setup

The complete grammar we use in our experiments is based on the same set of components as in AutoSklearn. We employ the same hyperparameter ranges and functions as AutoSklearn when sampling and incorporates them into the grammar. For our experiments, we sample 3 values of each hyperparameter that, added to the pipeline options, extend the space to more than 183 billions possible combinations. This has to be compared with the $\sim$24K configurations of the original GramML method [31].

For evaluation, we use the OpenML-CC18 benchmark [29], a collection of 72 binary and multi-class classification datasets carefully curated from the thousands of datasets on OpenML [3]. We use standard train-test splits, as provided by the benchmark suite. Also, to calculate validation scores (e.g. rewards), we use AutoSklearn's default strategy which consists on using 67% of the data for training and 33% for validation. All experiments were performed on an Amazon EC2 R5 spot instance (8 vCPU and 64 GB of RAM).

5.2   Ablation Study

To conduct the ablation study, four tasks with low computational cost were selected from the OpenML-CC18 benchmark[2]. For each task, we run 100 iterations of the algorithm for each policy configuration and report the mean and standard deviation for the following metrics: time per iteration (Time Iter), number of actions per iteration (Act/Iter), simulations' repetition rate (Rep Ratio), time spent by the first iteration (Time 1st), the number of actions for the first iteration (1st Act), and the total time (Tot Time) and total number of actions (Tot Act) at the end of the 100 iterations. Time measurements only consider the time of the algorithm and do not take into account the fitting time. The simulations' repetition ratio, *i.e.* the number of times the algorithm repeats the same paths during simulation, can be seen as a measure of the exploration efficiency of the algorithm (the lower this value, the more efficient the exploration of the search space).

Tables 1, 2 and 3 show results for the UTC, BTS and TPE strategies, respectively. All three methods depend on a single hyperparameter to control the overall behavior of the search algorithm, *i.e.* the exploration-exploitation trade-off. An immediate effect of varying such parameters can be observed on the time it takes for the algorithm to complete an iteration. This has to do with the iteration ending when the algorithm reaches a leaf. If we favor exploration (i.e. increase C and J for UCT and BTS, or decrease γ for TPE), the algorithm explores the tree in width, performs more *total actions*, and takes longer to reach a leaf. On the other hand, by favoring exploitation, the algorithm performs fewer actions and reaches the terminal nodes faster. The number of actions per iteration and the time per iteration are strongly correlated.

Comparing the different strategies, we observe that UCT with $C = 0.7$ is the most efficient alternative (in terms of the repetition ratio), while UCT with $C = 0$ is the fastest, with the later being also the less efficient in terms of exploration.

Finally, we consider the best version of each strategy prioritizing exploration efficiency (low iteration rate) trying to keep the time per iteration limited to $\sim$0.1. We select UCT with $C = 0.7$, BTS with $J = 1$, and TPE with $\gamma = 0.85$ and name them GramML$_{\text{UCT}}^{++}$, GramML$_{\text{BTS}}^{++}$, and GramML$_{\text{TPE}}^{++}$, respectively.

5.3   Comparison with Other Techniques

In this section, we compare our methods with others from the literature. For each task, each method is run for an hour (time budget). We report performance on the test set proposed in the benchmark suite. We compare the different variations of our approach to AutoSklearn [15] and MOSAIC [24] as they allow us to rely on the exact same set of basic ML components. In all cases, we set a maximum fitting time of 300 s, as the time limit for a single call to the machine learning model. If the algorithm runs beyond this time, the fitting is terminated. In our

---

[2] These tasks, identified by task IDs 11, 49, 146819, and 10093.

Table 1. Ablation results for the UCT strategy for different values of the parameter C.

	$C = 0$	$C = 0.1$	$C = 0.7$	$C = 1$
Time Iter	0,04 (0,01)	0,05 (0,01)	0,10 (0,02)	0,14 (0,04)
Time 1st	0,76 (0,31)	1,65 (0,49)	7,16 (2,50)	10,64 (3,99)
Tot Time	3,94 (0,64)	4,90 (0,54)	9,71 (2,14)	14,21 (3,84)
Act/Iter	5,8 (0,3)	7,1 (0,9)	19,4 (4,7)	28,1 (7,7)
1st Act	165,5 (52,7)	324,2 (95,9)	1535,0 (545,8)	2201,2 (810,5)
Tot Act	682,5 (35,4)	817,2 (95,7)	2047,0 (475,9)	2914,0 (776,5)
Rep Ratio	0,65 (0,04)	0,58 (0,07)	0,40 (0,08)	0,45 (0,10)

Table 2. Ablation results for the BTS strategy for different values of the parameter J.

	$J = 1$	$J = 10$	$J = 100$	$J = 1000$
Time Iter	0,15 (0,05)	0,15 (0,06)	0,33 (0,23)	0,28 (0,12)
Time 1st	7,13 (2,19)	7,42 (3,86)	18,90 (15,42)	14,56 (7,05)
Tot Time	14,72 (4,51)	14,98 (6,27)	33,21 (23,20)	28,18 (12,40)
Act/Iter	24,6 (7,8)	24,3 (11,5)	46,0 (30,0)	36,6 (15,4)
1st Act	1307,5 (398,8)	1333,2 (713,6)	2744,0 (2075,2)	2027,0 (927,2)
Tot Act	2561,2 (783,8)	2531,5 (1151,3)	4706,7 (3008,7)	3760,2 (1546,0)
Rep Ratio	0,50 (0,13)	0,51 (0,10)	0,54 (0,09)	0,59 (0,04)

Table 3. Ablation results for the TPE strategy for different values of the parameter γ.

	$\gamma = 50$	$\gamma = 65$	$\gamma = 75$	$\gamma = 85$
Time Iter	0,33 (0,16)	0,19 (0,03)	0,18 (0,06)	0,12 (0,02)
Time 1st	11,16 (7,55)	5,93 (1,47)	4,02 (2,50)	1,71 (0,57)
Tot Time	33,31 (15,82)	18,92 (3,35)	17,86 (6,05)	12,33 (1,76)
Act/Iter	15,0 (5,5)	9,1 (1,3)	7,8 (1,8)	5,9 (0,2)
1st Act	740,2 (382,0)	434,7 (81,8)	286,5 (128,4)	143,0 (25,7)
Tot Act	1605,5 (555,8)	1018,7 (129,8)	882,5 (187,2)	697,0 (23,0)
Rep Ratio	0,47 (0,06)	0,59 (0,06)	0,60 (0,05)	0,68 (0,01)

case, we return a reward of zero for methods exceeding this time limit. For each task, we ranked the performance of all systems and reported the average ranking (lower is better) and average performance score (higher is better) calculated as 1 - regret (the difference between maximum testing performance found so far by each method and the true maximum) at each time step.

Comparative results are shown in Fig. 1. Figure 1a shows the average rank across time. A marked difference of GramML$_{\text{BTS}}^{++}$ variant can be seen over

MOSAIC, AutoSklearn and GramML. In particular, $GramML^{++}_{BTS}$ have the best ranking over time, although it is surpassed in the first few minutes by MOSAIC and GramML. There also seems to be a slight decreasing trend for $GramML^{++}_{BTS}$ which would seem to indicate that the results could improve with more time.

In addition, Fig. 1b shows the average score across time. $GramML^{++}$ variants have a visible difference from MOSAIC, AutoSKLearn and GramML. The best average score is also achieved by $GramML^{++}_{BTS}$. An interesting observation is that the previous version of GramML was eventually slightly outperformed in terms of average score by AutoSKLearn. One reason for this was that GramML did not utilize hyperparameters. Conversely, the variants of $GramML^{++}$ demonstrate a significant improvement over their predecessor and AutoSKLearn. It should be noted that the average regret measure may not be very representative if the accuracy varies greatly between tasks. However, we believe that the graph provides important information together with the average ranking for the comparison between techniques in the OpenML-CC18 benchmark.

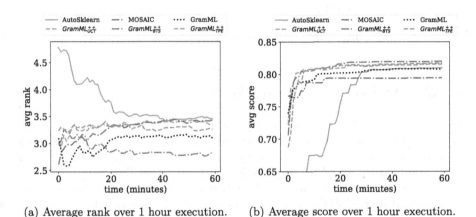

(a) Average rank over 1 hour execution.   (b) Average score over 1 hour execution.

Fig. 1. Results of the approaches on OpenML-CC18 benchmark.

In order to check the presence of statistically significant differences in the average rank distributions, Fig. 2 shows the results using critical difference (CD) diagrams [9] at 15, 30, 45 and 60 min. We use a non-parametric Friedman test at $p < 0.05$ and a Nemenyi post-hoc test to find which pairs differ [16]. If the difference is significant, the vertical lines for the different approaches appear as well separated. If the difference is not significant, they are joined by a thick horizontal line.

The diagrams show statistically significant differences at 15 min between AutoSKLearn and the other variants. At 30 min both $GramML^{++}_{BTS}$ and GramML have significant difference with MOSAIC and AutoSKLearn. At minute 45, the difference between $GramML^{++}_{BTS}$ and the other techniques remains, but there is no significant difference between the previous version of GramML and

GramML$^{++}_{TPE}$. Finally, at 60 min, GramML^{++}BTS showed a significant difference compared to MOSAIC and AutoSKLearn, while GramML^{++}TPE showed a significant difference only compared to MOSAIC. However, GramML$^{++}_{UCT}$ did not show a statistically significant difference compared to either MOSAIC or AutoSKLearn.

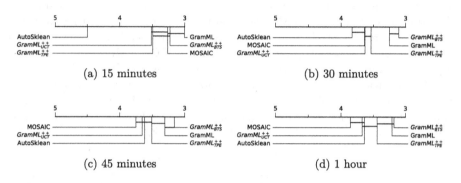

Fig. 2. CD plots with Nimenyi post-hoc test

6   Conclusions and Future Work

This article presents a model-free approach for grammar-based AutoML that integrates hyperparameter search as an extension to GramML [31]. The approach involves two steps: (1) incorporating hyperparameter values into the grammar using grammar rules and (2) modifying the Monte Carlo Tree Search (MCTS) algorithm to improve the collection of the best pipeline configuration, implement pruning, and support different selection policies and backpropagation functions. The evaluation involved an ablation study to assess the efficiency of each selection policy and a comparison with state-of-the-art techniques, which showed significant improvements, particularly with a variant that uses bootstrapped Thompson Sampling [12]. Consequently, this work demonstrates the effectiveness of incorporating hyperparameter search into grammar-based AutoML, and provides a promising approach for addressing the challenges of larger search spaces.

Furthermore, this work opens up several avenues for future research. In model-based AutoML, meta-learning can improve exploration efficiency [28]. However, the application of meta-learning in model-free AutoML based on grammars remains an open question. Additionally, incorporating resource information into the AutoML objective function is also an important direction to follow [30]. Finally, the speed of the search algorithms is an important factor that can be improved through horizontal scalability [6]. Therefore, future research should aim to parallelize the algorithm for more efficient resource utilization.

References

1. Bai, A., Wu, F., Chen, X.: Bayesian mixture modelling and inference based Thompson sampling in Monte-Carlo tree search. In: Advances in Neural Information Processing Systems, vol. 26 (2013)
2. Bergstra, J., Komer, B., Eliasmith, C., Yamins, D., Cox, D.D.: Hyperopt: a python library for model selection and hyperparameter optimization. Comput. Sci. Discov. **8**(1), 014008 (2015)
3. Bischl, B., et al.: OpenML benchmarking suites. arXiv:1708.03731v2 [stat.ML] (2019)
4. Bishop, C.M., Nasrabadi, N.M.: Pattern Recognition and Machine Learning, vol. 4. Springer, Heidelberg (2006)
5. Bouneffouf, D., et al.: Survey on automated end-to-end data science? In: Proceedings of the International Joint Conference on Neural Networks (2020). https://www.scopus.com, Cited By: 2
6. Bourki, A., et al.: Scalability and parallelization of Monte-Carlo tree search. In: van den Herik, H.J., Iida, H., Plaat, A. (eds.) CG 2010. LNCS, vol. 6515, pp. 48–58. Springer, Heidelberg (2011). https://doi.org/10.1007/978-3-642-17928-0_5
7. Chapelle, O., Li, L.: An empirical evaluation of Thompson sampling. In: Advances in Neural Information Processing Systems, vol. 24 (2011)
8. Chomsky, N.: Syntactic structures. In: Syntactic Structures. De Gruyter Mouton (2009)
9. Demšar, J.: Statistical comparisons of classifiers over multiple data sets. J. Mach. Learn. Res. **7**, 1–30 (2006)
10. Drori, I., et al.: Automatic machine learning by pipeline synthesis using model-based reinforcement learning and a grammar. arXiv preprint arXiv:1905.10345 (2019)
11. Drori, I., et al.: AlphaD3M: machine learning pipeline synthesis. arXiv preprint arXiv:2111.02508 (2021)
12. Eckles, D., Kaptein, M.: Thompson sampling with the online bootstrap. arXiv preprint arXiv:1410.4009 (2014)
13. Efron, B.: Bayesian inference and the parametric bootstrap. Ann. Appl. Stat. **6**(4), 1971 (2012)
14. Evans, B., Xue, B., Zhang, M.: An adaptive and near parameter-free evolutionary computation approach towards true automation in AutoML. In: 2020 IEEE Congress on Evolutionary Computation (CEC), pp. 1–8. IEEE (2020)
15. Feurer, M., Klein, A., Eggensperger, K., Springenberg, J., Blum, M., Hutter, F.: Efficient and robust automated machine learning. In: Advances in Neural Information Processing Systems, vol. 28 (2015)
16. Gijsbers, P., et al.: AMLB: an AutoML benchmark. arXiv preprint arXiv:2207.12560 (2022)
17. Hayes, C.F., Reymond, M., Roijers, D.M., Howley, E., Mannion, P.: Distributional Monte Carlo tree search for risk-aware and multi-objective reinforcement learning. In: Proceedings of the 20th International Conference on Autonomous Agents and Multiagent Systems, pp. 1530–1532 (2021)
18. Hutter, F., Hoos, H.H., Leyton-Brown, K.: Sequential model-based optimization for general algorithm configuration. In: Coello, C.A.C. (ed.) LION 2011. LNCS, vol. 6683, pp. 507–523. Springer, Heidelberg (2011). https://doi.org/10.1007/978-3-642-25566-3_40

19. Katz, M., Ram, P., Sohrabi, S., Udrea, O.: Exploring context-free languages via planning: the case for automating machine learning. In: Proceedings of the International Conference on Automated Planning and Scheduling, vol. 30, pp. 403–411 (2020)
20. Kocsis, L., Szepesvári, C.: Bandit based Monte-Carlo planning. In: Fürnkranz, J., Scheffer, T., Spiliopoulou, M. (eds.) ECML 2006. LNCS (LNAI), vol. 4212, pp. 282–293. Springer, Heidelberg (2006). https://doi.org/10.1007/11871842_29
21. Marinescu, R., et al.: Searching for machine learning pipelines using a context-free grammar. In: Proceedings of the AAAI Conference on Artificial Intelligence, vol. 35, pp. 8902–8911 (2021)
22. Mohr, F., Wever, M., Hüllermeier, E.: ML-plan: automated machine learning via hierarchical planning. Mach. Learn. **107**(8), 1495–1515 (2018)
23. Olson, R.S., Moore, J.H.: TPOT: a tree-based pipeline optimization tool for automating machine learning. In: Workshop on Automatic Machine Learning, pp. 66–74. PMLR (2016)
24. Rakotoarison, H., Sebag, M.: AutoML with Monte Carlo tree search. In: Workshop AutoML 2018@ ICML/IJCAI-ECAI (2018)
25. de Sá, A.G.C., Pinto, W.J.G.S., Oliveira, L.O.V.B., Pappa, G.L.: RECIPE: a grammar-based framework for automatically evolving classification pipelines. In: McDermott, J., Castelli, M., Sekanina, L., Haasdijk, E., García-Sánchez, P. (eds.) EuroGP 2017. LNCS, vol. 10196, pp. 246–261. Springer, Cham (2017). https://doi.org/10.1007/978-3-319-55696-3_16
26. Segovia-Aguas, J., Jiménez, S., Jonsson, A.: Generating context-free grammars using classical planning. In: Proceedings of the Twenty-Sixth International Joint Conference on Artificial Intelligence (IJCAI 2017), Melbourne, Australia, 19–25 August 2017, pp. 4391–7. IJCAI (2017)
27. Thornton, C., Hutter, F., Hoos, H.H., Leyton-Brown, K.: Auto-WEKA: combined selection and hyperparameter optimization of classification algorithms. In: Proceedings of the 19th ACM SIGKDD International Conference on Knowledge Discovery and Data Mining, pp. 847–855 (2013)
28. Vanschoren, J.: Meta-learning. In: Automated Machine Learning: Methods, Systems, Challenges, pp. 35–61 (2019)
29. Vanschoren, J., Van Rijn, J.N., Bischl, B., Torgo, L.: OpenML: networked science in machine learning. ACM SIGKDD Explor. Newsl. **15**(2), 49–60 (2014)
30. Vazquez, H.C.: A general recipe for automated machine learning in practice. In: Bicharra Garcia, A.C., Ferro, M., Rodríguez Ribón, J.C. (eds.) IBERAMIA 2022. LNCS, vol. 13788, pp. 243–254. Springer, Cham (2022). https://doi.org/10.1007/978-3-031-22419-5_21
31. Vazquez, H.C., Sánchez, J., Carrascosa, R.: GramML: exploring context-free grammars with model-free reinforcement learning. In: Sixth Workshop on Meta-Learning at the Conference on Neural Information Processing Systems (2022). https://openreview.net/forum?id=OpdayUqlTG
32. Waddle, V.E.: Production trees: a compact representation of parsed programs. ACM Trans. Program. Lang. Syst. (TOPLAS) **12**(1), 61–83 (1990)

Improving Subtour Elimination Constraint Generation in Branch-and-Cut Algorithms for the TSP with Machine Learning

Thi Quynh Trang Vo[1], Mourad Baiou[1], Viet Hung Nguyen[1(✉)],
and Paul Weng[2]

[1] INP Clermont Auvergne, Univ Clermont Auvergne, Mines Saint-Etienne, CNRS,
UMR 6158 LIMOS, 1 Rue de la Chebarde, Aubiere Cedex, France
{thi_quynh_trang.vo,mourad.baiou,viet_hung.nguyen}@uca.fr
[2] Shanghai Jiao Tong University, Shanghai, China
paul.weng@sjtu.edu.cn

Abstract. Branch-and-Cut is a widely-used method for solving integer programming problems exactly. In recent years, researchers have been exploring ways to use Machine Learning to improve the decision-making process of Branch-and-Cut algorithms. While much of this research focuses on selecting nodes, variables, and cuts [10, 12, 27], less attention has been paid to designing efficient cut generation strategies in Branch-and-Cut algorithms, despite its large impact on the algorithm performance. In this paper, we focus on improving the generation of subtour elimination constraints, a core and compulsory class of cuts in Branch-and-Cut algorithms devoted to solving the Traveling Salesman Problem, which is one of the most studied combinatorial optimization problems. Our approach takes advantage of Machine Learning to address two questions before executing the separation routine to find cuts at a node of the search tree: 1) Do violated subtour elimination constraints exist? 2) If yes, is it worth generating them? We consider the former as a binary classification problem and adopt a Graph Neural Network as a classifier. By formulating subtour elimination constraint generation as a Markov decision problem, the latter can be handled through an agent trained by reinforcement learning. Our method can leverage the underlying graph structure of fractional solutions in the search tree to enhance its decision-making. Furthermore, once trained, the proposed Machine Learning model can be applied to any graph of any size (in terms of the number of vertices and edges). Numerical results show that our approach can significantly accelerate the performance of subtour elimination constraints in Branch-and-Cut algorithms for the Traveling Salesman Problem.

Keywords: Traveling Salesman Problem · Subtour elimination constraints · Branch-and-Cut · Cut generation · Machine Learning

1 Introduction

Branch-and-Cut (B&C) is a popular method for solving integer programming (IP) problems exactly. B&C is the combination of two methods: branch-and-

M. Sellmann and K. Tierney (Eds.): LION 2023, LNCS 14286, pp. 537–551, 2023.
https://doi.org/10.1007/978-3-031-44505-7_36

bound and cutting-plane. While branch-and-bound breaks down the problem into subproblems by a divide-and-conquer strategy, the cutting-plane method tightens these subproblems by adding valid inequalities. B&C contains a sequence of decision problems such as variable selection, node selection, and cut generation. Consequently, its performance heavily depends on decision-making strategies.

One of the critical components of B&C is the cutting-plane method that strengthens linear programming (LP) relaxations (subproblems) of the IP problem by adding valid inequalities (a.k.a. *cuts*). More precisely, given a solution x^* obtained by solving some LP relaxations of the IP problem, we solve a separation problem, which either asserts the feasibility of x^* or generates a cut violated by x^*. Adding cuts can remove a large portion of the infeasible region and improve the performance. In general, cuts are categorized into general-purpose cuts obtained by the variable's integrality conditions and combinatorial cuts arising from the underlying combinatorial structure of the problem.

Generating cuts within B&C is a delicate task [9]. One of the design challenges of using cuts is balancing the separation routine's computational cost and the benefits of generated cuts. Generating cuts in a naive way can reduce the branch-and-bound tree's size but potentially increase the overall computing time due to the time spent executing the separation routine and solving the LP relaxations in the search tree. Thus, learning a deft policy for cut generation is crucial. In spite of its importance, cut generation is less studied than other related decision-making problems in B&C. To the best of our knowledge, only a few simple heuristics [5,21] have been proposed for cut generation, and concrete work has yet to be investigated to learn a cut generation policy.

In this paper, we focus on the generation of subtour elimination constraints (SECs)—a core class of cuts—for the Traveling Salesman Problem (TSP) in B&C. SECs were proposed by Dantzig, Fulkerson, and Johnson [8] to ensure the biconnectivity of solutions. They are well-known facet-defining inequalities for the TSP polytope. Due to their exponential number, SECs are usually served as cuts in the course of B&C. The separation problem of SECs is solvable in polynomial time [21] by using the Gomory-Hu procedure [11] to find a minimum cut in a graph. Although adding SECs is able to decrease the number of branching nodes, generating all possible SECs can decelerate the B&C performance, as the separation procedure of SECs is computationally expensive, especially for large-sized instances.

To improve SEC generation in B&C for the TSP, we propose an approach based on Machine Learning (ML) to handle two questions before executing the separation routine at a node of the branch-and-bound tree: 1) Do violated SECs exist? 2) If yes, is it worth generating them? The first question is to avoid solving redundant separation problems that do not provide any SEC. We treat this question as a binary classification problem and train a Graph Neural Network (GNN) in a supervised fashion. The second one is to predict the benefit of generating SECs compared to branching. To this end, we formulate the sequential decision-making process of SEC generation as a Markov decision problem and train a policy by reinforcement learning (RL). Our GNN-RL framework can leverage the underlying graph structure of fractional solutions to predict the SEC existence and capture the context of nodes in the search tree to make SEC generation decisions. Furthermore, it offers flexibility over instance size, namely that our model can be used for any instance (of arbitrary size) while being only trained with fixed-size graphs. Experimental results show that our trained policy for SEC generation significantly accelerates the B&C performance to solve the TSP, even on instances of different sizes from its training counterparts.

2 Related Work

Most approaches in the literature for cut generation exist in heuristic forms. Padberg and Rinaldi, in their research on B&C for large-scale TSP [20], empirically discovered the tailing-off phenomenon of cuts [20, Section 4.3], which shows the cut generator's inability to produce cuts that can assist the optimal LP solution to escape the corner of the polytope where it is "trapped". To deal with the tailing-off, the authors proposed to stop generating cuts if the objective value of the relaxed LP does not improve sufficiently within a given window and switch to branching. Another approach to control cut generation introduced by Balas et al. [5] is generating cuts at every k nodes of the search tree. The number k, named "skip factor" in [5], determines the frequency of generating cuts. It can be chosen either as a fixed constant or as an adaptive value varying throughout the search tree. Another commonly used strategy is the so-called cut-and-branch which only generates cuts at the root node of the search tree. Overall, despite its importance, the question of the branching versus cutting decision has yet to receive the attention it deserves.

In contrast, a closely-related problem to cut generation, *cut selection*, has been studied extensively in the literature. While cut generation decides whether to launch separation processes to generate cuts, cut selection requires selecting cuts from a candidate set obtained by solving separation problems. Cut selection is usually considered for general-purpose cuts whose separation procedure is computationally cheap and provides many cuts. Due to its definition, cut selection can be viewed as a ranking problem where cuts are sorted and chosen based on some criteria. This point of view opened up many different approaches based on many measurements of the cut quality. Among the most popular scores are efficacy [5], objective parallelism [1], and integral support [28], to name a few. Another research line on cut selection is to use ML to learn the ranking of cuts. Most works of this approach fall into two categories: supervised learning and RL. In the former, cuts are scored (or labeled) by an expert, and a cut ranking function (usually a neural network) is trained to be able to choose the best ones [14]. For the latter, one can formulate the problem of sequentially selecting cuts as a Markov decision process. An agent can then be trained to either directly optimize the objective value (RL) [27] or mimic a look-ahead expert (imitation learning) [22].

In recent years, using ML to enhance fundamental decisions in branch-and-bound is an active research domain; we refer to [6] for a summary of this line of work and to [4] for a more general discussion focused on routing problems. Specific examples contain learning to branch [2,10,16], learning to select nodes [12], and learning to run primal heuristics [7,17]. Similar to cut selection, these problems can be reformulated as ranking [10,14,16], regression [2], or classification problems [17], and can then be treated correspondingly. Most of these reformulations are possible due to the existence of an expensive expert (for example, the strong branching expert for variable selection), which can be used to calculate the score, label the instances, or act as an agent to be mimicked. In the case of cut generation, such an expert is too expensive to obtain. To the best of our knowledge, our paper is the first work to build an ML framework for cut generation.

3   SEC Generation in B&C for the TSP

3.1   IP Formulation

Given an undirected graph $G = (V, E)$ with a cost vector $\boldsymbol{c} = (c_e)_{e \in E}$ associated with E, the TSP seeks a Hamiltonian cycle (a.k.a. tour) that minimizes the total edge cost. For all edges $e \in E$, we denote by x_e a binary variable such that $x_e = 1$ if edge e occurs in the tour and $x_e = 0$ otherwise. We denote by $\delta(S)$ the set of edges that have exactly one end-vertex in $S \subset V$; $\delta(\{v\})$ is abbreviated as $\delta(v)$ for $v \in V$. Let $x(F) = \sum_{e \in F} x_e$ for $F \subseteq E$, the TSP can be formulated as an integer program as follows:

$$\min \boldsymbol{c}^T \boldsymbol{x} \tag{1a}$$
$$\text{s.t. } x(\delta(v)) = 2 \qquad \forall\, v \in V \tag{1b}$$
$$x(\delta(S)) \geq 2 \qquad \forall\, \emptyset \neq S \subset V \tag{1c}$$
$$x_e \in \{0, 1\} \qquad \forall\, e \in E \tag{1d}$$

where $\boldsymbol{x} = (x_e)_{e \in E}$. The objective function (1a) represents the total cost of edges selected in the tour. Constraints (1b) are *degree constraints* assuring that each vertex in the tour is the end-vertex of precisely two edges. Constraints (1c) are *subtour elimination constraints*, which guarantee the non-existence of cycles that visit only a proper subset of V. Finally, (1d) are *integrality constraints*.

Note that this formulation, introduced by Dantzig, Fulkerson, and Johnson [8], is widely used in most B&C algorithms for the TSP.

3.2   B&C Framework for the TSP

One of the most successful approaches for exactly solving the TSP is B&C. Intuitively, B&C starts by solving a relaxation of the TSP where all SECs are omitted and the integrality constraints are relaxed to $x_e \in [0, 1] \ \forall e \in E$. At each node of the branch-and-bound tree, the LP relaxation is solved, and SECs violated by the optimal LP solution are generated as cuts through the separation routine. This principle of generating SECs is used in most B&C algorithms for the TSP, including *Concorde*—the acknowledged best exact algorithm for the TSP [3].

We denote (α, β) an inequality $\alpha^T x \leq \beta$, $\mathcal{C}$ a set of valid inequalities for the TSP and $\langle F_0, F_1 \rangle$ an ordered pair of disjoint edge sets. Let $LP(\mathcal{C}, F_0, F_1)$ be the following LP problem:

$$\min c^T x$$
$$\text{s.t } x(\delta(v)) = 2 \qquad \forall v \in V$$
$$\alpha^T x \leq \beta \qquad \forall (\alpha, \beta) \in \mathcal{C}$$
$$x_e = 0 \qquad \forall e \in F_0$$
$$x_e = 1 \qquad \forall e \in F_1$$
$$x_e \in [0, 1] \qquad \forall e \in E.$$

A basic B&C framework based on SECs is sketched as follows:

1. *Initialization.* Set $\mathcal{S} = \{\langle F_0 = \emptyset, F_1 = \emptyset \rangle\}$, $\mathcal{C} = \emptyset, \overline{x} = $ NULL and $UB = +\infty$.
2. *Node selection.* If $\mathcal{S} = \emptyset$, return $\overline{x}$ and terminate. Otherwise, select and remove an ordered pair $\langle F_0, F_1 \rangle$ from $\mathcal{S}$.
3. Solve $LP(\mathcal{C}, F_0, F_1)$. If the problem is infeasible, go to step 2; otherwise, let x^* be its optimal solution. If $c^T x^* \geq UB$, go to step 2.
4. *SEC verification.* If x^* is integer, verify SECs with x^*. If x^* satisfies all SECs, replace $\overline{x}$ by x^*, UB by $c^T x^*$ and go to step 2. Otherwise, add violated SECs to $\mathcal{C}$ and go to step 3.
5. *Branching versus cut generation.* Should SECs be generated? If yes, go to step 6, else go to step 7.
6. *Cut generation.* Solve the separation problem. If violated SECs are found, add them to $\mathcal{C}$ and go to step 3.
7. *Branching.* Pick an edge e such that $0 < x_e^* < 1$. Add $\langle F_0 \cup \{e\}, F_1 \rangle$, $\langle F_0, F_1 \cup \{e\} \rangle$ to $\mathcal{S}$ and go to step 2.

When the algorithm terminates, $\overline{x}$ is an optimal solution of the TSP. Notice that the basic B&C framework stated above simply contains fundamental steps, but it could be easily extended with additional techniques, such as the use of other valid inequality classes for cut generation, branching strategies, and primal heuristics. Any improvement for this basic B&C framework will also be valid for the extensions.

Separation Routine for SECs. We now describe an exact separation algorithm to find violated SECs in polynomial time, proposed by Crowder and Padberg [21]. The input of the separation algorithm is the optimal solution x^* of the current LP relaxation. We then construct from x^* the so-called *support graph* $G_{x^*} = (V, E_{x^*})$ where $E_{x^*} = \{e \in E \mid x_e^* > 0\}$. For each edge e in E_{x^*}, we set x_e^* as its capacity. Due to the construction of G_{x^*}, the value $x(\delta(S))$ for $S \subset V$ is precisely the capacity of the cut $(S, V \setminus S)$ in G_{x^*}. Therefore, an SEC violated by x^* is equivalent to a cut with a capacity smaller than 2 in G_{x^*}. Such a cut can be found by using the Gomory-Hu procedure [11] with $|V| - 1$ maximum flow computations. Thus, it is computationally expensive, especially for instances with large-sized graphs.

Note that when executing the separation routine for SECs, one can either build the Gomory-Hu tree completely and get all violated SECs from the tree or terminate the process as soon as a violated SEC is found. Our experimental results show that the former is more efficient than the latter in terms of overall solving time. Hence, in our implementation, we generate all violated SECs from the Gomory-Hu tree each time the separation routine is called.

3.3   SEC Generation Problem

One of the primary decisions to make in B&C for the TSP is to decide whether to generate SECs or to branch in Step 5, which has a tremendous impact on the B&C performance. On the one hand, generating SECs can help tighten the LP relaxations, reduce the number of nodes in the branch-and-bound tree, and significantly improve computing time. On the other hand, SEC generation can also worsen the B&C performance. One reason is the computational cost of the SEC separation routine, which can be time-consuming when the instance size is large. Furthermore, not all separation processes can produce violated SECs, and thus launching the separation routine when the optimal LP solution satisfies all SECs is wasteful. Another reason is that generating SECs is useless at some nodes of the search tree where additional SECs may not provide new information to improve the LP relaxation.

To illustrate the impact of SEC generation on the B&C performance, we consider the following experimental example. We solve the TSP on the instance rat195 from TSPLIB [25] by the commercial solver CPLEX 12.10 with three different SEC generation strategies in Step 5 of the basic B&C framework. In the first strategy (*No cut*), we do not generate any SECs; in the second one (*Every node*), SECs are generated at every node of the search tree. The last strategy, *Sample cut*, solves the separation problem exactly 100 times: at each node of the search tree, we will perform the separation routine with the probability 1/2 and stop doing so after solving the 100th (separation) problem. The CPU time limit is set to 3600 s. Table 1 shows the results of the strategies. *Sample 1* and *Sample 2* are two different runs of the strategy *Sample cut*. Column *"CPU time"* gives the running time in seconds of B&C, in which the time spent by the separation routine is shown in column *"Separation time"*. Column *"Nodes"* reports the number of nodes in the search tree, and column *"Cuts"* indicates the number of generated SECs. Columns *"Separations"* and *"Separations with cuts"* give the number of separation routine executions and the number of executions that can obtain violated SECs, respectively.

Table 1 shows that the SEC generation strategies may significantly affect the algorithm performance. Obviously, generating SECs is crucial, as B&C cannot solve the instance to optimality without it under the given CPU time budget. In addition, adding SECs substantially reduced the search tree size. However,

Table 1. The results of the SEC generation strategies on the instance rat195. The asterisk in the "CPU time" column indicates strategies that fail to solve the TSP within the time limit.

Strategy	CPU time	Separation time	Nodes	Cuts	Separations	Separations with cuts
No cut	3601.8*	0	1506514	0	0	0
Every node	1365.7	1340.5	4105	1116	2992	134
Sample 1	65.5	48.3	3834	359	100	21
Sample 2	114.5	39.5	10543	727	100	43

solving SEC separation problems might take a major portion of computing time, and only a few separation executions obtained violated SECs. For example, with the strategy generating SECs at every tree node, B&C spent 98% of the CPU time to execute separation routines, but only 134 out of 2992 executions yielded violated SECs. Table 1 also indicates that the effectiveness of the strategies relies not only on the number of solved separation problems but also on specific nodes where violated SECs are generated. Indeed, although the number of times the separation problem is solved is the same, the difference in nodes generating SECs makes the strategy *Sample 1* outperform *Sample 2*.

Motivated by this issue, in this paper, we study the SEC generation problem stated as follows: *"Given a fractional solution at a node of the search tree, decide whether to generate SECs or to branch"*.

4   The GNN-RL Framework for SEC Generation

In this section, we describe our GNN-RL framework to learn an SEC generation strategy in B&C for the TSP. Our GNN-RL contains two separate components: a cut detector (i.e., a GNN) to predict the existence of violated SECs and a cut evaluator (i.e., a Q-value function) to decide whether to generate SECs or to branch when the GNN has predicted the existence of SECs.

Figure 1 provides the flowchart of GNN-RL at a node of the search tree. After obtaining an optimal solution to the LP relaxation at the node, the cut detector predicts whether the solution violates SECs. If it predicts that no SEC is violated, we skip to the branching step. Otherwise, the cut evaluator will assess the effectiveness of additional SECs to select the next action to perform.

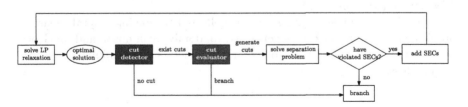

Fig. 1. The flowchart of GNN-RL

4.1   Cut Detector

Given a fractional solution, the cut detector predicts whether there exists any violated SEC. Therefore, one can view the cut detector as a binary classifier that takes a fractional solution x as input and returns:

$$y = \begin{cases} 1 & \text{if there exists any SEC violated by } x, \\ 0 & \text{otherwise.} \end{cases}$$

We adopt a GNN [26], a message-passing (MP) neural network, for this classification task to take into account the underlying graph structure of fractional solutions, which is critical for the separation problem. Furthermore, GNN possesses many properties which make it a natural choice for graph-related ML problems, such as permutation invariance or independence from the instance size.

We parameterize the cut detector as follows. Given a fractional solution x, we construct from x its support graph $G_x = (V, E_x)$ where the capacity w_e of edge e is x_e. For each node $i \in V$, we define the node feature as its degree d_i in G_x, and embed d_i to a h-dimensional vector by a multi-layer perceptron (MLP):

$$h_i^{(0)} = W^{(0)} d_i + b^{(0)}$$

where $W^{(0)} \in \mathbb{R}^{h \times 1}$ and $b^{(0)} \in \mathbb{R}^h$. To update the node embeddings, we use two MP layers [19]:

$$h_i^{(l)} = \text{ReLU} \left(W_1^{(l)} h_i^{(l-1)} + W_2^{(l)} \sum_{j \in N(i)} w_{(i,j)} \cdot h_j^{(l-1)} \right)$$

where $h_i^{(l-1)}$ is the representation of node i in layer $l - 1$, $\mathcal{N}(i)$ is the set of i's neighbors in G_x, $W_1^{(l)}$ and $W_2^{(l)}$ are weight matrices in the l-th layer, and $w_{(i,j)}$ is the capacity associated with edge (i, j). To obtain a representation of the entire graph G_x, we apply a min-cut pooling [29] layer to assign nodes into two clusters, compute the element-wise addition of all node vectors in each cluster and concatenate the two cluster vectors. Finally, an MLP with a softmax activation function is used to predict the probability of the solution's labels, i.e., $P(y = 0|x)$ and $P(y = 1|x)$.

Our training dataset $\{(x^i, y^i)\}_{i=1}^N$ is collected by solving several separation problems on training instances generated randomly by Johnson and McGeoch's generator [15]. We train the cut detector's parameters Θ_G to minimize the cross-entropy loss:

$$L(\Theta_G) = - \sum_{i=1}^N \left(y^i \cdot \log P_{\Theta_G}(y^i = 1|x^i) + (1 - y^i) \cdot \log(1 - P_{\Theta_G}(y^i = 0|x^i)) \right).$$

4.2   Cut Evaluator

We now formulate SEC generation as a Markov decision process (MDP) [23]. Considering the IP solver as the environment and the cut evaluator as the agent, we define the state space, action space, and transition and reward functions of the MDP as follows.

State Space. At iteration t, a state s_t contains the information about the TSP instance and the current search tree, which comprises branching decisions, the lower and upper bounds, the LP relaxations at nodes with SECs added so far, and the considered node. A terminal state is achieved when the instance is solved to optimality.

Due to the search tree complexity, we represent a state s_t as a collection of the following elements:

- *An optimal solution x_t to the LP relaxation of the considered node.* It is used to provide information about the separation problem for the agent. We represent this solution as its corresponding support graph $G_{x_t} = (V, E_{x_t})$ with edge capacities x_t.
- *The TSP instance.* Encoding the TSP instance in the state representation is essential, as SEC generation is naturally instance-dependent. Recall that the TSP instance is an undirected graph $G = (V, E)$ with edge costs c. We define the node features as the node degrees in G. The edge features contain the edge costs and information about the variables representing edges at the considered node (values in the optimal LP solution, lower and upper bounds), which is used to encode the context of the considered node in the search tree.
- *Features of the search tree.* To enrich the information about the search tree in the state representation, we design 11 tree features based on our experimental observations and inspired by hand-crafted input features for branching variable selection proposed in [30]. The features are shown in Table 2. The top four features correspond to the incumbent existence, the IP relative gap (i.e., $|L - U|/U$ where L, U are respectively the lower and upper bounds), and the portions of processed and unprocessed nodes, which help to capture the state of the search tree. The remaining features are extracted at the considered node to describe its context through depth, objective value, optimal solution, and fixed variables. Each feature is normalized to the range $[0, 1]$.

Action Space. Given a non-terminal state s_t, the RL agent selects an action a_t from the action space $\mathcal{A} = \{generate\ SECs,\ branch\}$.

Transition. After selecting an action a_t, the new state s_{t+1} is determined as follows. If a_t is to branch, the solver selects a branching variable to create two child nodes, picks the next node to explore, and solves the corresponding LP relaxation to get an optimal solution x_{t+1}. Otherwise, if a_t is to generate SECs,

Table 2. The features extracted from the search tree

Feature group	Feature	Description	Ref.
Tree (4)	has_incumbent	1 if an integer feasible solution is found and 0 otherwise	
	IP_rel_gap	(upper bound - lower bound)/upper bound	[30]
	processed_nodes	the number of processed nodes/the total nodes in the current search tree	[30]
	unprocessed_nodes	the number of unprocessed nodes/the total nodes in the current search tree	[30]
Node (7)	node_depth	max(1, the node depth/$|V|$)	[30]
	obj_quality	objective value/upper bound	
	vars_1	the number of variables equal to 1 in the solution/$|V|$	
	fixed_vars	the number of fixed variables/$|E|$	
	unfixed_vars	the number of unfixed variables/$|E|$	
	vars_fixed_1	the number of variables fixed to 1/$|V|$	
	vars_fixed_0	the number of variables fixed to 0/$|E|$	

the solver launches the separation routine to yield SECs violated by x_t, adds them to the formulation, and solves the LP relaxation again with the new cuts to obtain x_{t+1}. If no cut is found, the next state s_{t+1} is determined in the same way as when performing the branching action.

Reward Function. Since we want to solve the instance as fast as possible, we consider the reduction of the IP relative gap to define the reward function. The faster the IP relative gap drops, the faster the instance is solved. Formally, let γ_t be the IP relative gap at iteration t, the reward at iteration t is defined as

$$r_t = r(s_t, a_t, s_{t+1}) = \gamma_t - \gamma_{t+1}. \tag{3}$$

An issue of this reward function is its sparsity, namely that most rewards are 0; thus, it rarely gives feedback to guide the agent. To deal with this issue, we add additional rewards, a.k.a *reward shaping*, to provide more frequent feedback for training the agent. In particular, to encourage the solver to terminate as soon as possible, we set penalties for each additional iteration, and each solved redundant separation problem in the cases where the cut detector predicts incorrectly. Moreover, we also give a bonus for each SEC found by the separation routine. Details of the additional rewards are shown in Table 3.

Policy Parametrization. We parameterize the cut evaluator (i.e., a Q-value function) as a neural network consisting of two parts: one to embed a state into a vector and one to approximate the Q-value of actions. In the first part, we use three separate models to encode the state components, i.e., a GNN for the optimal LP solution, another GNN for the TSP instance, and an MLP for the tree features. The state embedding is the concatenation of the outputs of

Table 3. The additional rewards for SEC generation

Additional reward	Value
Penalty for each additional iteration	−0.01
Penalty for solving redundant separation problem	−0.10
Bonus for an SEC	0.01

these three models. We then pass this embedding to a 3-layer perceptron to get the Q-value approximation of actions. Figure 2 illustrates the cut evaluator architecture.

Training. To train the cut evaluator, we use the Deep Q-Network algorithm [18], i.e., the parameters of the cut evaluator are updated to minimize an L2 loss defined with a target network using data sampled from a replay buffer filled with transitions generated during online interactions with the environment. For simplicity, an ϵ-greedy policy is used for exploration.

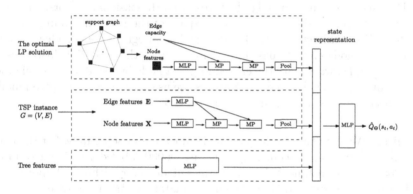

Fig. 2. The neural network architecture for the cut evaluator

5   Experiments

In this section, we demonstrate the effectiveness of the GNN-RL framework in controlling SEC generation for the TSP in B&C algorithms.

5.1   Setup

All experiments are conducted on a computing server with AMD EPYC 7742 64-core CPU, 504 GB of RAM, and an RTX A4000 GPU card with 48 GB graphic memory.

B&C Solver. We use the commercial solver CPLEX 12.10 as a backend solver, and CPLEX UserCutCallback to generate SECs in the tree nodes and integrate our method into the solver. We keep the CPLEX's default settings which are expertly tuned. However, to focus on evaluating the benefit of SECs, we switch off the CPLEX's cuts. The solver time limit is 3600 s per instance.

Benchmarks. We train and evaluate our method on random TSP instances generated following Johnson and McGeoch's generator used for DIMACS TSP Challenge [15]. These instances are complete graphs. In particular, we train on 200 instances with graphs of 200 vertices and evaluate on three instance groups: instances with graphs of 200 (small), 300 (medium), and 500 vertices (large) and 100 instances per group. Furthermore, we also assess the proposed method on 29 instances with graphs of 200 to 1000 vertices from TSPLIB [25], a well-known library of sample instances for the TSP.

Neural Network Architecture. We describe here the model architectures for encoding the state components. For the optimal LP solution, we use a GNN with the same architecture as the cut detector without the last MLP layer. For the TSP instance, since the edge features are 4-dimensional vectors, we use an MLP layer to embed them into the same space of the node embeddings and the modified GIN architecture introduced in [13] to integrate the edge features into updating the node embeddings. Furthermore, since the used TSP instances are complete graphs, we update the embedding of a node by its 10 nearest neighbors in terms of edge costs. For the tree features, we use a two-layer perceptron model. For all architectures, the feature dimension is 64.

Training. We train the cut detector and cut evaluator separately. For the training phase of the cut detector, we generate 96000 labeled fractional solutions from 200 random instances. We train the cut detector within 100 epochs, and the learning rate is 0.0001. For the cut evaluator, we train the Q-learning network on the 200 training instances with one million steps using the package *stable-baselines3* [24].

Baselines. We compare the performance of GNN-RL with the fixed and automatic strategies proposed in [5], which generate cuts for every k nodes. For the fixed strategies, we use $k = 1$ (FS-1) as the default strategy and $k = 8$ (FS-8), which gave the best results in [5] and is one of the best skip factors in our own experiments. For the automatic strategy (AS) where the skip factor is chosen based on the instance to be solved, k is computed as follows:

$$k = \min\left\{ \text{KMAX}, \left\lceil \frac{f}{cd\log_{10} p} \right\rceil \right\}$$

where f is the number of cuts generated at the root node, d is the average distance cutoff of these cuts (a distance cutoff of a cut is the Euclidean distance between the optimal solution and the cut [5]), p is the number of variables in the formulation, and KMAX,c are constants. In our implementation, we set KMAX $= 32$ as in [5] and $c = 100$ since the average distance cutoff of SECs is small. Moreover, we experimentally observe that SECs are very efficient at the root node and inefficient at the late stage of the computation. Thus, we always generate SECs at the root node and stop generating SECs when the IP relative gap is less than 1%, regardless of the strategies used.

5.2 Results

Table 4 shows the results of GNN-RL and the three baselines on both the random and TSPLIB instances. For each instance group, we report the number of instances that can be solved to optimality within the CPU time limit over the total instances (column *"Solved"*), the average CPU time in seconds (including also the running times of instances that cannot be solved to optimality within the CPU time limit) (column *"Time"*), the average number of nodes in the search tree (column *"Nodes"*), and the average number of generated SECs (column *"Cuts"*). Recall that our goal in this paper is to accelerate the B&C algorithm; thus, the main criterion for comparison is the CPU running time.

Table 4. The numerical results of the SEC generation strategies

	Strategy	Solved	CPU Time	Nodes	Cuts
SMALL	FS-1	100/100	109.9	**2769.0**	506.9
	FS-8	100/100	56.8	3090.1	493.8
	AS	100/100	48.1	3521.1	439.3
	GNN-RL	100/100	**34.4**	3185.7	**423.7**
MEDIUM	FS-1	96/100	511.1	**11969.1**	956.8
	FS-8	98/100	424.6	15983.4	970.2
	AS	96/100	441.0	26759.5	861.7
	GNN-RL	**99/100**	**288.5**	17390.6	**726.1**
LARGE	FS-1	32/100	2998.0	**37698.9**	2330.7
	FS-8	35/100	2916.4	55882.8	2425.0
	AS	33/100	2922.4	71455.1	2235.9
	GNN-RL	**37/100**	**2889.7**	72160.1	**1965.9**
TSPLIB	FS-1	**15/29**	2062.3	**15114.4**	2412.9
	FS-8	14/29	2056.7	19797.6	2694.7
	AS	13/29	2087.7	23202.5	2967.0
	GNN-RL	**15/29**	**1890.1**	30995.7	2622.4

As shown in Table 4, our method outperforms all the baselines on all instance groups. Indeed, GNN-RL solves more instances to optimality within a smaller average CPU time. Compared to FS-8, GNN-RL is faster by 5% on average over all random instances, i.e., is 39%, 32%, and 1% faster for small, medium, and large instances, respectively. For the TSPLIB instances, GNN-RL is faster by 9%, 8%, and 8% compared to AS, FS-8, and FS-1.

As predicted, FS-1 has the smallest tree size on average over all instances, but its running time is the highest due to the extra time spent on generating SECs. On the other hand, too few cuts might be detrimental to the B&C performance. It can be seen in the comparison between FS-8 and AS strategies on large and medium instances. Indeed, AS requires more computing time than FS-8 despite generating fewer SECs. The numerical results give evidence that GNN-RL can balance the separation cost and the benefit of generated SECs.

6   Conclusion

In this paper, we proposed a GNN-RL framework, the first ML-based approach, to improve SEC generation in B&C for the TSP. Experimental results showed that the policy learned by GNN-RL outperformed the previously proposed heuristics for cut generation. Most importantly, the GNN-RL architecture allows the trained policy to generalize to unseen instances with larger-sized graphs. Our future work will extend this framework to other valid inequality classes for the TSP, such as the comb and 2-matching inequalities. We will also integrate GNN-RL into B&C algorithms for other combinatorial optimization problems, for example, the max-cut and vehicle routing problems.

Acknowledgements. This work has been supported in part by the program of National Natural Science Foundation of China (No. 62176154).

References

1. Achterberg, T.: Constraint integer programming. Ph.D. thesis (2007)
2. Alvarez, A.M., Louveaux, Q., Wehenkel, L.: A machine learning-based approximation of strong branching. INFORMS J. Comput. **29**(1), 185–195 (2017)
3. Applegate, D., Bixby, R., Chvatal, V., Cook, W.: Concorde TSP solver (2006)
4. Bai, R., et al.: Analytics and machine learning in vehicle routing research. Int. J. Prod. Res. **31**, 4–30 (2021)
5. Balas, E., Ceria, S., Cornuéjols, G.: Mixed 0–1 programming by lift-and-project in a branch-and-cut framework (1996)
6. Bengio, Y., Lodi, A., Prouvost, A.: Machine learning for combinatorial optimization: a methodological tour d'horizon. Eur. J. Oper. Res. **290**, 405–421 (2021)
7. Chmiela, A., Khalil, E., Gleixner, A., Lodi, A., Pokutta, S.: Learning to schedule heuristics in branch and bound. In: NeurIPS, vol. 34, pp. 24235–24246 (2021)
8. Dantzig, G., Fulkerson, R., Johnson, S.: Solution of a large-scale traveling-salesman problem. J. Oper. Res. Soc. Am. **2**, 393–410 (1954)
9. Dey, S.S., Molinaro, M.: Theoretical challenges towards cutting-plane selection. Math. Program. **170**, 237–266 (2018)

10. Gasse, M., Chételat, D., Ferroni, N., Charlin, L., Lodi, A.: Exact combinatorial optimization with graph convolutional neural networks. In: Advances in Neural Information Processing Systems (2019)
11. Gomory, R.E., Hu, T.C.: Multi-terminal network flows. J. Soc. Ind. Appl. Math. 9(4), 551–570 (1961)
12. He, H., Daume, H., III., Eisner, J.M.: Learning to search in branch and bound algorithms. In: Advances in Neural Information Processing Systems (2014)
13. Hu, W., et al.: Strategies for pre-training graph neural networks. arXiv preprint arXiv:1905.12265 (2019)
14. Huang, Z., et al.: Learning to select cuts for efficient mixed-integer programming. Pattern Recogn. **123**, 108353 (2022)
15. Johnson, D.S., McGeoch, L.A.: Benchmark code and instance generation codes (2002). https://dimacs.rutgers.edu/archive/Challenges/TSP/download.html
16. Khalil, E., Bodic, P.L., Song, L., Nemhauser, G., Dilkina, B.: Learning to branch in mixed integer programming. In: AAAI (2016)
17. Khalil, E.B., Dilkina, B., Nemhauser, G.L., Ahmed, S., Shao, Y.: Learning to run heuristics in tree search. In: IJCAI, pp. 659–666 (2017)
18. Mnih, V.: Human-level control through deep reinforcement learning. Nature **518**, 529–533 (2015)
19. Morris, C., et al.: Weisfeiler and leman go neural: higher-order graph neural networks. In: AAAI, vol. 33, pp. 4602–4609 (2019)
20. Padberg, M., Rinaldi, G.: Branch-and-cut algorithm for the resolution of large-scale symmetric traveling salesman problems. SIAM Rev. **33**, 60–100 (1991)
21. Padberg, M.W., Hong, S.: On the symmetric travelling salesman problem: a computational study. In: Padberg, M.W. (ed.) Combinatorial Optimization Mathematical Programming Studies, vol. 12. Springer, Heidelberg (1980). https://doi.org/10.1007/BFb0120888
22. Paulus, M.B., Zarpellon, G., Krause, A., Charlin, L., Maddison, C.: Learning to cut by looking ahead: cutting plane selection via imitation learning. In: ICML (2022)
23. Puterman, M.L.: Markov Decision Processes: Discrete Stochastic Dynamic Programming. Wiley, Hoboken (1994)
24. Raffin, A., Hill, A., Gleave, A., Kanervisto, A., Ernestus, M., Dormann, N.: Stable-baselines3: reliable reinforcement learning implementations. J. Mach. Learn. Res. **22**(268), 1–8 (2021)
25. Reinelt, G.: Tspliba traveling salesman problem library. ORSA J. Comput. **3**(4), 376–384 (1991)
26. Scarselli, F., Gori, M., Tsoi, A.C., Hagenbuchner, M., Monfardini, G.: The graph neural network model. IEEE Trans. Neural Netw. **20**(1), 61–80 (2008)
27. Tang, Y., Agrawal, S., Faenza, Y.: Reinforcement learning for integer programming: learning to cut. In: ICML. PMLR (2020)
28. Wesselmann, F., Stuhl, U.: Implementing cutting plane management and selection techniques. Technical report. University of Paderborn (2012)
29. Ying, Z., You, J., Morris, C., Ren, X., Hamilton, W., Leskovec, J.: Hierarchical graph representation learning with differentiable pooling. In: NeurIPS, vol. 31 (2018)
30. Zarpellon, G., Jo, J., Lodi, A., Bengio, Y.: Parameterizing branch-and-bound search trees to learn branching policies. In: AAAI, vol. 35, pp. 3931–3939 (2021)

Learn, Compare, Search: One Sawmill's Search for the Best Cutting Patterns Across and/or Trees

Marc-André Ménard[1,2,3(✉)] iD, Michael Morin[1,2,4] iD, Mohammed Khachan[5] iD, Jonathan Gaudreault[1,2,3] iD, and Claude-Guy Quimper[1,2,3] iD

[1] FORAC Research Consortium, Université Laval, Québec, QC, Canada
`marc-andre.menard.2@ulaval.ca`
[2] CRISI Research Consortium for Industry 4.0 Systems Engineering, Université Laval, Québec, QC, Canada
[3] Department of Computer Science and Software Engineering, Université Laval, Québec, QC, Canada
[4] Department of Operations and Decision Systems, Université Laval, Québec, QC, Canada
[5] FPInnovations, Québec, QC, Canada

Abstract. A sawmilling process scans a wood log and must establish a series of cutting and rotating operations to perform in order to obtain the set of lumbers having the most value. The search space can be expressed as an and/or tree. Providing an optimal solution, however, may take too much time. The complete search for all possibilities can take several minutes per log and there is no guarantee that a high-value cut for a log will be encountered early in the process. Furthermore, sawmills usually have several hundred logs to process and the available computing time is limited. We propose to learn the best branching decisions from previous wood logs and define a metric to compare two wood logs in order to branch first on the options that worked well for similar logs. This approach (Learn, Compare, Search, or LCS) can be injected into the search process, whether we use a basic Depth-First Search (DFS) or the state-of-the-art Monte Carlo Tree Search (MCTS). Experiments were carried on by modifying an industrial wood cutting simulator. When computation time is limited to five seconds, LCS reduced the lost value by 47.42% when using DFS and by 17.86% when using MCTS.

Keywords: Monte Carlo search · And/or trees · Tree search algorithms · Sawmilling · Learning

1 Introduction

In North America, softwood lumber is a standardized commodity. Different dimensions and grades are possible and the hardware in the mill, thanks to embedded software, optimize cutting decisions for each log in order to maximize

M. Sellmann and K. Tierney (Eds.): LION 2023, LNCS 14286, pp. 552–566, 2023.
https://doi.org/10.1007/978-3-031-44505-7_37

Fig. 1. A basket of products obtained from the log at the exit of the sawmill represented in a virtual log.

profit. Each lumber type has a specific value on the market, and the equipment aims to maximize the total basket value given a log. Processing a log leads to a basket of these standardized lumber products (and byproducts such as sawdust). Figure 1 shows an example of cut for a given log leading to a specific basket of lumber products. Two different combinations of cutting decisions (e.g., differences in trimming, edging and/or sawing) can lead to two baskets of different values.

The optimizer of sawmills' equipment can be used offline to measure the impacts related to changes in the configuration. This can also be done using sawing simulators such as Optitek [11], RAYSAW [30], and SAWSIM [14]. Providing an optimal cutting solution may take too much time, whether it is for a real-time cutting decision purpose, or to get a suitable forecast for decision-making.

The cutting decision optimization problem has been addressed in the literature in several ways (optimization, simulation, prediction by neural networks or other ML approaches) [16]. In this paper, we are concerned with a specific case where the problem is solved by an algorithm enumerating all possible cuts.

Obtaining the best cuts for a given log implies that all log cutting possibilities must be tested according to the possible cutting choices of each machine. This complete search for all possibilities can take several minutes per log. In practice we need to use a time limit for the search and therefore the sawmill lose value as there is no guarantee that a good cut for a log will be encountered early in the process. In this paper, we address this challenge.

To get the best possible solution according to the computation time limit, we need to test first the most promising cut choices. Although the past is no guarantee of the future in many cases, when sawing similar logs at the same sawmill, it might as well be. Based on this observation, we suggest an informed search algorithm which learns from the previous similar logs to guide the search process for the actual log. The assumption being that for two similar logs the cutting decisions leading to the best value will be the same or at least similar enough, it makes sense to guide the search with our knowledge of the best decisions for these logs.

The rest of the paper is divided as follows. First, we describe the problem more thoroughly. Second, we present the preliminary concepts. Third, we present our method for learning from the optimal cutting decision of already cut logs. Fourth, we present our results and finally we conclude.

2   Problem Description

In many sawing optimization systems, such as Optitek, logs are represented by a surface scan: a point cloud of the log's surface structured as a sequence of circular sections (see Fig. 1 where the points of each section have been interpolated). A sawmill is defined as a set of machines. Each machine can make different transformations on the log. A machine that cuts the log into several parts creates different cutting sub-problems. Each part of the resulting log can go through the same or a different sequence of machines. At the end of the machine sequence, the sum of the values of each part of the initial log gives the value of the log. The value of a product is assessed through a value matrix according to the size and grade of the product. It is therefore not possible to know the exact value of the log before different cutting alternatives are explored.

The type of decisions varies from one machine to another. First, there are different sets of cutting patterns that can be applied by some machines. Second, for some machines, it is possible to rotate or to do a translation of the log before it passes through the machine. Each rotation or translation can greatly affect the basket of products obtained from the log.

The set of all decision sequence for a given log can be represented as a search tree. The root of the tree corresponds to the log at the entrance of the sawmill where no transformation has been done yet. Each level of the tree corresponds to a cutting decision taken on a part of the log by a machine. The decision can be a rotation, a translation or the choice of a cutting pattern. Each node of a level corresponds to a value for the decision at this level. A leaf corresponds to a solution, i.e., a basket of products for which a complete sequence of decisions has been made. To make sure we find the optimal solution, we have to go through the whole search tree (an exhaustive search is needed to prove the optimality).

The search tree can be represented by an and/or tree. When a decision cutting a log in more than one piece is made, we end up with several sub-problems. Each sub-problem (log part) then goes through different machines and cutting decisions. The sub-problems correspond to the "and" of an and/or tree. The subtree corresponding to each sub-problem must be searched to obtain a solution. Cutting decisions correspond to the "or" of an and/or tree. For an "or" node, we have a decision to make, whereas for an "and" node we have no choice but to solve the sub-problem, i.e., explore the subtree.

The width of the tree depends on the values to be tested for each cutting decision. Its maximum depth is bounded by the machine sequence and the number of unique types of cutting decisions to be tested. In our experiments and in practice, there are generally more cutting decisions to test than there are machines. The search tree is therefore wide and shallow.

Figure 2 shows a small search tree example. The log at the top of the tree enters the sawmill. Each circular node represents a decision that is possible to choose. For example, from the incoming log (root), we have the choice between two decisions. Each decision can then create sub-problems (rectangular nodes) by cutting the log until it reaches a leaf of the search tree. The sum of the leaves

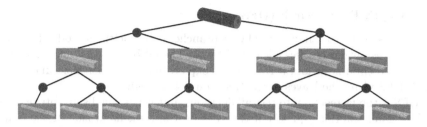

Fig. 2. Example of a search tree. The log at the top of the tree enters the sawmill. Decisions are made (circular nodes, "or") that can create new sub-problems (rectangular nodes, "and") by cutting the log.

corresponding to the best decisions for each sub-problem forms a product basket and the best product basket is returned.

Figure 3 shows an example of a (virtual) sawmill in the Optitek software. The rectangles are machines where cutting decisions can be optimized. The flow is generally from left to right (from the *Feed* to the *Sort* and *Chip* nodes), although there might be loops such as it is the case for the *Slasher* in our example. Each machine—except *Feed, Sort* and *Chip*—has inputs (left-hand side) and outputs (right-hand side). Products are routed from one machine to the other depending on their characteristics.

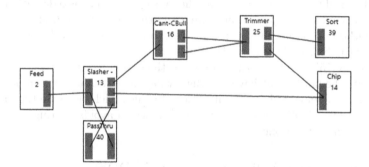

Fig. 3. A sawmill configuration. The rectangles are machines where cutting decisions can be optimized. The branches are the possible routes between a machine's output and another machine's input.

3 Preliminary Concepts

In this section, we present tree search algorithms from the literature. We will use them as a baseline of comparison for our approach we call Learn, Compare, Search, or LCS. Each algorithm without LCS will be compared to a version where we injected LCS into the search process. We also review and/or trees and related works on learning to guide a search tree traversal and adaptive search.

3.1   Depth-First Search (DFS)

DFS starts at the root node and always branches on the leftmost node of the possible choices until it reaches a leaf in the tree. Then it backtracks until there is an unexplored child node and it resumes its exploration on the leftmost unexplored child. DFS visits the leaves of the tree from left to right.

DFS is the fastest way to test all decisions, but it has the disadvantage of not finding good solutions quickly if they are on the right-hand side of the search tree. For example, if we have a rotational optimization on the first machine and we want to test all degrees between 0 and 180°, DFS will test the degrees in ascending order $(0, 1, 2, 3, \ldots)$. If the quality of a solution increases with rotation, DFS will find the optimal solution only at the end of the search and most of the time will be spent finding bad solutions.

3.2   Limited Discrepancy Search (LDS)

Harvey and Ginsberg [15] introduced the concept of discrepancy. A discrepancy is when the search heuristic is not followed and another node is visited instead. For example, in the DFS search algorithm, the heuristic is to branch left whenever possible. By branching right when we should have branched left in a binary tree, we deviate from the left-first heuristic of DFS, this is counted as one *deviation* (or discrepancy). If a node contains several children, there are two ways to count the deviations. First, the number of deviations can be the number of nodes skipped from left to right. For example, taking the first node counts as 0 deviation, taking the second counts as 1 deviation, taking the third counts as 2, and so on. The other method, and the one we use in this paper, is to count 1 deviation only whenever the first node is not taken.

The LDS [15] search algorithm explores the search tree iteratively. Each iteration visits the nodes that have less than k deviations. Starting from $k = 0$, the value of k is incremented at each iteration. In this article, we take the improved version of LDS (ILDS) presented by Korf [21] which avoids visiting a leaf of the search tree more than once.

3.3   Depth-Bounded Discrepancy Search (DDS)

DDS [31] is a search algorithm also based on deviations. At each iteration, DDS follows the DFS search algorithm until it reaches a node on level k where k is the iteration number. It then visits all the nodes on the next level, except the first node on the left, and continues the exploration by visiting only the nodes on the left for the remaining levels of the search tree. The idea of DDS is that the search heuristic is more likely to make a wrong choice at the top of the search tree than at the bottom.

3.4   Monte Carlo Tree Search (MCTS)

MCTS is a tree search algorithm using a compromise between exploration and exploitation [6]. MCTS works by iteration. Each iteration contains four phases:

selection, expansion, simulation and backpropagation. There are several ways to implement MCTS. Our implementation is inspired by Antuori et al. [2].

Selection. The selection phase starts at the root of the tree and ends when we reach a node that has not yet been visited in a previous iteration. When we are at a node, the next node to visit is chosen according to the formula (1) where $A(\sigma)$ represents a set of actions that can be done from the current node σ. Each action is represented by a branch in the search tree. Given node σ, $\sigma|a$ represents the child node reached by taking the action a. $\tilde{V}(\sigma|a)$ is the expected value of the node if the action a is chosen and corresponds to the exploitation term. $U(\sigma|a)$ corresponds to the exploration term. Finally, c is a parameter to balance the exploitation and the exploration (i.e., $\tilde{V}(\sigma|a)$ and $U(\sigma|a)$).

$$\arg\max_{a \in A(\sigma)} \tilde{V}(\sigma|a) + c \cdot U(\sigma|a) \tag{1}$$

There are different ways to compute an expected value $\tilde{V}(\sigma|a)$ for a node $\sigma|a$. For example, it is possible to take the average of the solutions found so far for the node. Instead, we use the best value found so far for this node as expected value. Keeping the best value is better in our case, because many of the solutions have a null value, i.e., the cutting decisions lead to a null value for one of the output products. The values of the different $\tilde{V}(\sigma|a)$ must be normalized to be compared with the exploration term $U(\sigma|a)$. We normalize this value between $[-1, 1]$ using the formula (2) where $N(\sigma)$ is the number of visits to the node σ, $V^+ = \max\{V(\sigma|a)|a \in A(\sigma), N(\sigma|a) > 0\}$ and $V^- = \min\{V(\sigma|a)|a \in A(\sigma), N(\sigma|a) > 0\}$ [2] which corresponds to the maximum and minimum value for the children nodes of the parent node.

$$\tilde{V}(\sigma|a) = \begin{cases} 2\frac{V^+ - V(\sigma|a)}{V^+ - V^-} - 1 & \text{if } N(\sigma|a) > 0 \\ 0 & \text{otherwise} \end{cases} \tag{2}$$

The exploration term $(U(\sigma))$ is computed with the formula (3) where $Pr(\sigma)$ corresponds to the priority probability biases, $p(\sigma)$ corresponds to the parent node of the current node σ, and $N(\sigma)$ corresponds to the number of visits to the current node.

$$U(\sigma) = Pr(\sigma)\frac{\sqrt{N(p(\sigma))}}{N(\sigma) + 1} \tag{3}$$

Expansion. The expansion phase creates a child node $\sigma|a$ for each action $a \in A(\sigma)$. For each child node $\sigma|a$, we initialize the number of visits $N(\sigma|a)$ to 0 and the expected objective value $V(\sigma|a)$ to 0. We also initialize the prior probability biases $Pr(\sigma|a)$ to a value if available and if not, we initialize $Pr(\sigma|a)$ according to a uniform distribution $\frac{1}{|A(\sigma)|}$.

Simulation. The simulation phase, also called rollout, aims at finding a possible expected value for the node. The simulation phase must then visit at least one leaf of the search tree from the current node. The simulation can be done by making a random choice of nodes to visit at each level of the tree. It is also possible to make a weighted random choice with the different probabilities of each node $Pr(\sigma)$ if these probabilities are available.

Backpropagation. The backpropagation phase will update the nodes visited during the selection phase. The number of visits $N(\sigma)$ is incremented by 1 and the expected value $V(\sigma)$ is updated if a better solution is found during the simulation phase.

3.5   Searching AND/OR Trees

The representation of the problem as an and/or tree allows having a search tree with less depth. The search algorithms can be adapted to the and/or tree.

For DFS, there is no change. For the LDS search algorithm, the algorithm must be modified. Larrosa et al. [22] presents the limited discrepancy and/or search (LDSAO) algorithm. This algorithm uses the LDS search algorithm, but for and/or trees. The difference between LDS and LDSAO is in the handling of the "and" nodes. For the "and" nodes, we have to find and solve these subproblems to get a solution. These nodes do not cause any discrepancy. For the "or" nodes, we do not cause any discrepancy if we follow the search heuristic. For the same number of discrepancies, Larrosa et al. [22] have shown that each iteration of LDSAO includes the search space of LDS on the original tree and more.

The same logic for going from LDS to LDSAO can be applied for DDS where we do not cause discrepancy when going through an "and" node, but only when we do not follow the search heuristic to go through an "or" node.

For the MCTS algorithm, in selection mode, we visit all the "and" nodes (subproblem) for a given level, but we visit only one "or" node (decisions). It is also the same principle in simulation mode, we visit all the "and" nodes (subproblem) for a given level, but we select only one of the decisions ("or" node).

3.6   Learning for Search Tree Traversal and Adaptive Search

Learning to guide a search tree traversal for search efficiency purposes is not a new concept. There are several adaptive search algorithms for variable choice heuristics and value choice heuristics such as *dom/wdeg* [5], solution counting based search [33], and activity-based search [25]. In this family of algorithms, we also find Impact-Based Search [27], Adaptive Discrepancy Search (ADS) [12], and the RBLS algorithm [3]. There is also Solution-guided multipoint constructive search (SGMPCS) [4] that guides the search from multiple solutions found during the search of the current instance. Our approach has some similarities to SGMPCS, but it starts with solutions found offline. MCTS [6], we described in Sect. 3.4, is also an adaptive algorithm. At the difference of the LCS approach, these algorithms, in their original form, tend to adapt at runtime. They do not use prior knowledge of the problem.

The alternative is to guide the search with known solutions. Loth et al. [24] presented Bandit-Based Search for CP (BASCOP), an adaptation of MCTS using Reinforcement Learning (RL) to know where to branch next. They use a reward function based on the failure depth. This approach is interesting to keep the learning from one problem instance to another. In our problem, however, the search tree is determined by the order of the machines and the cutting decisions. There are no failures.

Other approaches are specific to Constraint Programming (CP). Chu and Stuckey [8], for instance, have presented a value ordering approach to tree search. When branching, a score is assigned to each value in the domain of the variable using domain-based features and machine learning.

There is also a whole literature on learning in the context of a branch and bound for Mixed-Integer Linear Programming which leads to search trees where branching splits the problem into subproblems instead of assigning a value to a variable (as it is usually the case in CP branch and bounds). In their literature review on learning to improve variable and value selection, Lodi and Zarpellon [23] define two categories of approaches: (1) incorporating learning in more traditional heuristics [10,13,18], and (2) using machine learning [1,19,20].

Finally, LCS is closer to informed search methods with a priori information than to adaptive search methods that learn at runtime. The idea of informed search is to add domain-specific knowledge to help the search algorithm [9,29]. Recent examples of informed search includes the work of Silver et al. [28], Ontanón [26], and Yang et al. [32]. Silver et al. [28] introduced AlphaGo, a system using MCTS along with a neural network trained by supervised learning on human expert movements, and improved through self-play RL. Ontanón [26] presents Informed MCTS for real-time strategy (RTS) games. The approach consists of using several Bayesian models to estimate the probability distribution of each action played by an expert or search heuristics. It then initializes the prior probabilities of the possible actions in the MCTS search. Yang et al. [32] present the Guiding MCTS which uses one or more deterministic scripts based on the human knowledge of the game to know which node to visit first for a RTS game. Our approach is similar to these approaches since we rely on past decisions to initialize the prior probabilities of actions.

4   Learning from Past Decisions in LCS

To find the best solution faster, we propose to first learn from past instances/logs. This allows us to know which decisions are more promising and should be visited first. In the context of sawing, a decision is considered more promising than another if it is more often found in the optimal cutting decisions of previously cut logs. The assumption is that for similar logs, the same cutting decisions will be made to find the best log value.

It is possible to get information on the decisions made for previous logs, because the structure of the search tree is similar from one log to another. Indeed, the order of the machines and the cutting decisions remain the same. However, the best choices may differ according to the characteristics of the logs. Even in the case where two logs (therefore their search trees) differ substantially, a knowledge transfer might still be possible between the two.

4.1 Adaptation of Search Algorithms to Learning

Each search algorithm presented in Sect. 3 can make use of past optimal cutting decisions of already cut logs. For DFS, LDS and DDS, we ordered the "or" child nodes representing a decision of an "and" node in descending order of the number of times that decision is made to obtain a solution of a previously cut log. We name this value for a node $Q(\sigma)$ where σ represent the node. The first node is therefore the decision that is most often in a solution of the previously cut logs compared to the other possible decisions, i.e., $\max_{a \in A}(Q(\sigma|a))$. For the algorithms using discrepancy, no discrepancy is caused when we take the node that is most often in a solution of the previously cut logs.

For the MCTS search algorithm, the prior probability biases $Pr(\sigma|a)$ are initialized according to Eq. (4) representing the number of times a decision is chosen to obtain a solution from a previously cut log versus the number of times the parent decision is.

$$Pr(\sigma|a) = \frac{Q(\sigma|a)}{Q(\sigma)} \tag{4}$$

It may not be ideal to rely on the decisions made for all log sawed. Two logs that are very different may have different optimal cutting decisions. Therefore, it is best to rely only on the X decisions that led to the best cutting decisions for the logs most similar to the simulated log where X is a hyper-parameter. The following section presents the method for finding similar logs.

4.2 Finding Similar Logs

We assume each log is represented by a 3D point cloud. The point cloud is separated into sections and each section represents a part of the log. This is a standard representation for the logs in the industry.

Finding logs similar to the current log could be done by comparing the point clouds. For instance, by first using the *iterative closest point* (ICP) algorithm to align two point clouds, we could extract the alignment error and use it as a dissimilarity measure [7]. This, however, would be computationally intensive. The ICP algorithm is polynomial in the number of points [17], but each new log would need to be compared to every known log and log point clouds can have more than 30 thousand points. Our logs, for instance, have an average of 19326 points and the maximum number of points is 32893. To keep runtimes low, we compare the logs using a set of features that are easily computed. In fact, the features we use can be computed offline before the search. Table 1 shows the features used to represent a log.

Table 1. Features of a log, calculated from a surface scan

Features	Description
Length	Length of the log
Small-end diameter	Diameter of the log at the top end
Large-end diameter	Diameter of the log at the stump end
Diameter at 25%	Log diameter at 25% of length from the stump end
Diameter at 50%	Log diameter at 50% of length from the stump end
Diameter at 75%	Log diameter at 75% of length from the stump end
Max sweep	Maximum value of the curvature of the log
Location max sweep	Percentage of length at which curvature is greatest
Thickness	Thickness of the log calculated from the point cloud
Width	Log width calculated from the point cloud
Volume	Volume of the log
Volume homogeneity	Average volume difference from section to section
Taper	Difference between the diameters of the two ends

To calculate the similarity between two logs, we first apply a MinMax normalization (Eq. (5)) before comparing features.

$$b^c = \frac{b^c - \min(b^c)}{\max(b^c) - \min(b^c)} \quad \forall c \in C \tag{5}$$

This ensures two features that do not have the same order of magnitude have the same weight when computing the similarity, for example the length and diameter of a log. Then, we calculate the distance of the logs by taking the sum of their squared differences using Eq. (6) where C is the set of features, b_1^c is the value of the feature c of the first log and b_2^c is the value of the feature c of the second log.

$$S(b_1, b_2) = \sum_{c \in C} (b_1^c - b_2^c)^2 \tag{6}$$

5 Experiments

To demonstrate the potential of LCS on our log sawing problem, we implemented it—along with the DFS, LDS, DDS, and MCTS tree search algorithms—in a state-of-the-art commercial sawing simulator (Optitek). We have eight different sawmill configurations which allows testing the approach on eight different industrial contexts. For each sawmill we proceed as follows. We virtually cut one hundred logs, each time exploring the search tree completely thus finding the max/optimal value.

Then, we randomly separate the hundred logs into two datasets. Fifty logs will be used as a training dataset to find the best hyper-parameter values for

562 M.-A. Ménard et al.

each algorithm. The other fifty logs constitute the dataset used to compare the algorithms.

For these two datasets, another separation must be made. Thirty logs among the fifty are used to know the paths that most often led to the best solution for the methods that learn on the already simulated logs. The other twenty logs among the fifty will be used to compare the results obtained by each of the search algorithms. As the random separation impacts the method to learn on the already simulated logs, five replications of this separation are made to get an average of the results.

The hyper-parameters for MCTS are: the $c \in \{1, 2\}$ coefficient, used to balance exploitation and exploration; the number of simulations performed during the simulation phase, selected in $\{1, 2, 3, 4\}$; and the number of nodes visited for each level during the simulation, selected in $\{1, 2, 3\}$. If at a level, the number of child nodes is less than the number of child nodes to visit, we visit all of them.

When processing a new log, we tested with 10, 20, and 30 as the number of similar logs we will use to guide the search.

6 Results

Figure 4 presents the results obtained on the 8 sawmill configurations according to the search algorithm. Figure 5 presents the averaged for the 8 sawmill configurations. For each sawmill configuration and search algorithm, we plotted the percentage of optimality attained on average on the 20 test logs (y-axis) against the solving time in seconds (x-axis). Each curve represents an average over five replications. At a given point in time, a search algorithm at 100% optimality has found all the optimal solutions for the 20 logs.

MCTS is better than DFS, LDS and DDS except for the sawmill configurations #7 and #8 where DFS is slightly better. By comparing each search algorithm with or without learning (with or without LCS), we see that the results are always better except for DFS for the sawmill configuration #1 and LDS for the sawmill configuration #3. However, there is little difference between MCTS and MCTS with learning. The fact that MCTS learns and adapt during the search, balancing exploitation and exploration, should reduce the gains obtained with learning on decisions made on previous logs.

For each search algorithm, Table 2 shows the percentage of improvement (reduction of the lost value, in percentage) provided by LCS after n seconds (with $n \in \{1, 2, \ldots, 5\}$) according to the search algorithm. After 1 s, the improvement is 23.58% for MCTS, the best search algorithm, whereas it is 23.2% for DFS, the simplest search algorithm. After 5 s, the improvement is 17.86% for MCTS and 47.42% for DFS. In our experiments, completing the search took 1 h and 42 min per log on average. For industrial decisions, the "ideal" time limit depends on the hardware configuration of the sawmill and the available time to wait for a solution. With a time limit of one second, a thousand logs will take about 16.67 min whereas a time limit of 5 s would lead to a waiting time of 83.3 min.

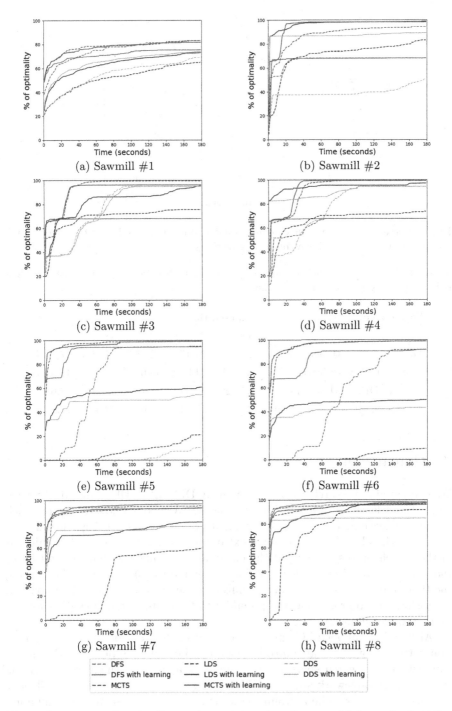

(a) Sawmill #1

(b) Sawmill #2

(c) Sawmill #3

(d) Sawmill #4

(e) Sawmill #5

(f) Sawmill #6

(g) Sawmill #7

(h) Sawmill #8

Fig. 4. Percentage of optimality against solving time (in seconds) for each algorithm and sawmill configurations; average of 5 replications on 20 logs.

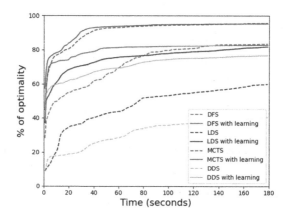

Fig. 5. Percentage of optimality against solving time (in seconds) for each tree search algorithm; averaged on all sawmill configurations.

Table 2. Reduction of the lost value when using LCS depending on the search algorithm according to the time in seconds.

Search algorithms	Elapsed time				
	1 s	2 s	3 s	4 s	5 s
DFS vs DFS with learning	23.2%	29.47%	42.45%	47.58%	47.42%
LDS vs LDS with learning	31.08%	44.55%	46.05%	46.62%	47.94%
DDS vs DDS with learning	35.22%	47.62%	47.79%	48.56%	48.63%
MCTS vs MCTS with learning	23.58%	23.73%	24.03%	22.44%	17.86%

7 Conclusion

We compared four search algorithms, namely DFS, LDS, DDS, and MCTS with and without learning (a total of eight variants). We showed how to improve each of them by learning from the best cutting decisions on previous logs, introducing a framework called LCS. In the context of our application, LCS, when injected in the search process, improved the cut quality obtained within the first 5 s of the search—a requirement for enabling better industrial decisions. Considering the thousands logs to process in sawmills and the need to forecast the impact of a decision, our approach has the potential to translate to tangible gains for the forest-product industry.

Although we applied LCS in the particular context of wood log cutting decision optimization, the framework is general and could be applied to other problems where the search space is represented as a tree. Learning from previous runs using a framework such as LCS, as long as the instances share sufficient similarities, appear to be a promising avenue for further research on combining learning and optimization.

References

1. Alvarez, A.M., Louveaux, Q., Wehenkel, L.: A machine learning-based approximation of strong branching. INFORMS J. Comput. **29**(1), 185–195 (2017)
2. Antuori, V., Hébrard, E., Huguet, M.J., Essodaigui, S., Nguyen, A.: Combining Monte Carlo tree search and depth first search methods for a car manufacturing workshop scheduling problem. In: International Conference on Principles and Practice of Constraint Programming (CP) (2021)
3. Bachiri, I., Gaudreault, J., Quimper, C.G., Chaib-draa, B.: RLBS: an adaptive backtracking strategy based on reinforcement learning for combinatorial optimization. In: 2015 IEEE 27th International Conference on Tools with Artificial Intelligence (ICTAI), pp. 936–942. IEEE (2015)
4. Beck, J.C.: Solution-guided multi-point constructive search for job shop scheduling. J. Artif. Intell. Res. **29**, 49–77 (2007)
5. Boussemart, F., Hemery, F., Lecoutre, C., Sais, L.: Boosting systematic search by weighting constraints. In: Proceedings of the 16th European Conference on Artificial Intelligence (ECAI), vol. 16 (2004)
6. Browne, C.B., et al.: A survey of Monte Carlo tree search methods. IEEE Trans. Comput. Intell. AI Games **4**(1), 1–43 (2012)
7. Chabanet, S., Thomas, P., El Haouzi, H.B., Morin, M., Gaudreault, J.: A kNN approach based on ICP metrics for 3D scans matching: an application to the sawing process. In: 17th IFAC Symposium on Information Control Problems in Manufacturing (INCOM) (2021)
8. Chu, G., Stuckey, P.J.: Learning value heuristics for constraint programming. In: Michel, L. (ed.) CPAIOR 2015. LNCS, vol. 9075, pp. 108–123. Springer, Cham (2015). https://doi.org/10.1007/978-3-319-18008-3_8
9. Drake, P., Uurtamo, S.: Move ordering vs heavy playouts: where should heuristics be applied in Monte Carlo go. In: Proceedings of the 3rd North American Game-On Conference, pp. 171–175. Citeseer (2007)
10. Fischetti, M., Monaci, M.: Backdoor branching. In: Günlük, O., Woeginger, G.J. (eds.) IPCO 2011. LNCS, vol. 6655, pp. 183–191. Springer, Heidelberg (2011). https://doi.org/10.1007/978-3-642-20807-2_15
11. FPInnovations: Optitek 10. In: User's Manual (2014)
12. Gaudreault, J., Pesant, G., Frayret, J.M., DAmours, S.: Supply chain coordination using an adaptive distributed search strategy. IEEE Trans. Syst. Man Cybern. Part C (Appl. Rev.) **42**(6), 1424–1438 (2012)
13. Glankwamdee, W., Linderoth, J.: Lookahead branching for mixed integer programming. In: Twelfth INFORMS Computing Society Meeting, pp. 130–150 (2006)
14. HALCO: Halco Software Systems Ltd. (2016). https://www.halcosoftware.com
15. Harvey, W.D., Ginsberg, M.L.: Limited discrepancy search. In: Proceedings of the Fourteenth International Joint Conference on Artificial Intelligence (IJCAI), vol. 1, pp. 607–615 (1995)
16. Hosseini, S.M., Peer, A.: Wood products manufacturing optimization: a survey. IEEE Access **10**, 121653–121683 (2022)
17. Jost, T., Hügli, H.: Fast ICP algorithms for shape registration. In: Van Gool, L. (ed.) DAGM 2002. LNCS, vol. 2449, pp. 91–99. Springer, Heidelberg (2002). https://doi.org/10.1007/3-540-45783-6_12
18. Karzan, F.K., Nemhauser, G.L., Savelsbergh, M.W.: Information-based branching schemes for binary linear mixed integer problems. Math. Program. Comput. **1**(4), 249–293 (2009)

19. Khalil, E., Le Bodic, P., Song, L., Nemhauser, G., Dilkina, B.: Learning to branch in mixed integer programming. In: Proceedings of the AAAI Conference on Artificial Intelligence, vol. 30 (2016)

20. Khalil, E.B.: Machine learning for integer programming. In: Proceedings of the Twenty-fifth International Joint Conference on Artificial Intelligence (IJCAI), pp. 4004–4005 (2016)

21. Korf, R.E.: Improved limited discrepancy search. In: Proceedings of the Thirteenth National Conference on Artificial Intelligence (AAAI), vol. 1, pp. 286–291 (1996)

22. Larrosa Bondia, F.J., Rollón Rico, E., Dechter, R.: Limited discrepancy and/or search and its application to optimization tasks in graphical models. In: Proceedings of the Twenty-Fifth International Joint Conference on Artificial Intelligence (IJCAI), pp. 617–623. AAAI Press (Association for the Advancement of Artificial Intelligence) (2016)

23. Lodi, A., Zarpellon, G.: On learning and branching: a survey. TOP 25(2), 207–236 (2017)

24. Loth, M., Sebag, M., Hamadi, Y., Schoenauer, M.: Bandit-based search for constraint programming. In: Schulte, C. (ed.) CP 2013. LNCS, vol. 8124, pp. 464–480. Springer, Heidelberg (2013). https://doi.org/10.1007/978-3-642-40627-0_36

25. Michel, L., Van Hentenryck, P.: Activity-based search for black-box constraint programming solvers. In: Beldiceanu, N., Jussien, N., Pinson, É. (eds.) CPAIOR 2012. LNCS, vol. 7298, pp. 228–243. Springer, Heidelberg (2012). https://doi.org/10.1007/978-3-642-29828-8_15

26. Ontanón, S.: Informed Monte Carlo tree search for real-time strategy games. In: 2016 IEEE Conference on Computational Intelligence and Games (CIG), pp. 1–8. IEEE (2016)

27. Refalo, P.: Impact-based search strategies for constraint programming. In: Wallace, M. (ed.) CP 2004. LNCS, vol. 3258, pp. 557–571. Springer, Heidelberg (2004). https://doi.org/10.1007/978-3-540-30201-8_41

28. Silver, D., et al.: Mastering the game of go with deep neural networks and tree search. Nature 529(7587), 484–489 (2016)

29. Świechowski, M., Godlewski, K., Sawicki, B., Mańdziuk, J.: Monte Carlo tree search: a review of recent modifications and applications. Artif. Intell. Rev. 1–66 (2022)

30. Thomas, R.E.: RAYSAW: a log sawing simulator for 3D laser-scanned hardwood logs. In: Proceedings of the 18th Central Hardwood Forest Conference, vol. 117, pp. 325–334 (2012)

31. Walsh, T.: Depth-bounded discrepancy search. In: Proceedings of the Fifteenth International Joint Conference on Artificial Intelligence (IJCAI), vol. 1, pp. 1388–1393 (1997)

32. Yang, Z., Ontanón, S.: Guiding Monte Carlo tree search by scripts in real-time strategy games. In: Proceedings of the AAAI Conference on Artificial Intelligence and Interactive Digital Entertainment (AAAI), vol. 15, pp. 100–106 (2019)

33. Zanarini, A., Pesant, G.: Solution counting algorithms for constraint-centered search heuristics. Constraints 14(3), 392–413 (2009)

Dynamic Police Patrol Scheduling with Multi-Agent Reinforcement Learning

Songhan Wong[ID], Waldy Joe[ID], and Hoong Chuin Lau[(✉)][ID]

School of Computing and Information Systems, Singapore Management University,
Singapore, Singapore
songhanwong.2020@mitb.smu.edu.sg, waldy.joe.2018@phdcs.smu.edu.sg,
hclau@smu.edu.sg

Abstract. Effective police patrol scheduling is essential in projecting police presence and ensuring readiness in responding to unexpected events in urban environments. However, scheduling patrols can be a challenging task as it requires balancing between two conflicting objectives namely projecting presence (*proactive* patrol) and incident response (*reactive* patrol). This task is made even more challenging with the fact that patrol schedules do not remain static as occurrences of dynamic incidents can disrupt the existing schedules. In this paper, we propose a solution to this problem using Multi-Agent Reinforcement Learning (MARL) to address the Dynamic Bi-objective Police Patrol Dispatching and Rescheduling Problem (DPRP). Our solution utilizes an Asynchronous Proximal Policy Optimization-based (APPO) actor-critic method that learns a policy to determine a set of prescribed dispatch rules to dynamically reschedule existing patrol plans. The proposed solution not only reduces computational time required for training, but also improves the solution quality in comparison to an existing RL-based approach that relies on heuristic solver.

Keywords: Reinforcement Learning · Multi-Agent · Dynamic Dispatch and Rescheduling · Proximal Policy Optimization · Police Patrolling

1 Introduction

Effective scheduling of police patrols is essential to project police presence and ensure readiness to respond to unexpected events in urban environments. Law enforcement agencies have the challenging task of balancing two conflicting objectives of projecting presence (*proactive* patrol) and incident response (*reactive* patrol). When an unexpected incident occurs, complex and effective response decisions must be made quickly while minimizing the disruption to existing patrol schedule. Such a decision is complex because each decision contains multiple components, namely which agent needs to be dispatched to respond to the incident and secondly which existing schedules are disrupted and/or require some

M. Sellmann and K. Tierney (Eds.): LION 2023, LNCS 14286, pp. 567–582, 2023.
https://doi.org/10.1007/978-3-031-44505-7_38

re-planning. In a real-world environment, the scale of the problem needs to consider multiple patrol areas and teams. Multiple patrol teams need to operate in cooperative manner to maximize the effectiveness of police patrolling. Hence, it is challenging to develop an efficient real-time strategy for reallocating resources when an incident occurs.

In this paper, we present a solution based on Multi-Agent Reinforcement Learning (MARL) that enables rescheduling of patrol timetable whenever dynamic events occur. The problem addressed in this paper is based on the Dynamic Bi-Objective Police Patrol Dispatching and Rescheduling Problem (DPRP) introduced in [9]. This problem is a variant of Dynamic Vehicle Routing Problem (DVRP) with the element of university time-tabling scheduling incorporated and in the context of cooperative multi-agent environment.

The key contribution of the paper is the successful application of Asynchronous Proximal Policy Optimization (APPO) policy-gradient method with dispatch-rules based actions for solving a dynamic patrol scheduling problem based on a real-world police patrolling environment. Our solution method emphasizes on the use of patrol dispatch rules to significantly reduce the computational time in making such a complex decision. In addition, RL is used to learn the policy in choosing the dispatch rule rather than relying on some fixed heuristic rules. We experimentally demonstrate that our proposed solution method is able to reduce training time by a factor of 40 while improving the quality of the solution by around 10% against the benchmark approach [9].

2   Background

The police patrol routing problem can generally be seen as an extension of the stochastic DVRP. In addition to route optimization, this routing problem also needs to consider scheduling aspect i.e. when and how long an agent remains in a particular node within a given route. A patrol unit patrolling in existing allocated area can be dispatched to an emergency call, and a redistribution of patrol resources is necessary to ensure optimal deployment. Existing works in the literature such as [2,5] mostly addresses the offline planning aspect of patrolling problem where the planned schedule is assumed to be static. These solutions include genetic algorithm, routing policies, and local search based on the network Voronoi diagram [7,16]. However, significant operational challenges lie mainly in the dynamic planning aspect, since there may be disruption from an unforeseen event that requires dispatch and re planing of the existing patrol schedules.

2.1   Scheduling Problem with Reinforcement Learning

The use of RL to solve dynamic scheduling problem has gained traction in the community [10,11,17] in recent years. Single-agent reinforcement learning (RL) algorithm suffers from a curse of dimensionality since the state-action space grows exponentially as the number of agents increases, particularly in the context of large-scale patrolling environment in modern metropolis. Cooperative multi-agent reinforcement learning (MARL) algorithms are necessary for solving such problem. Based on current works on MARL [1,8,12], MARL approach training schemes can generally be classified into centralized setting, decentralized setting with networked agents, and fully decentralized with independent learners setting. Common challenges of MARL include non-unique learning goals among agents, non-stationary environment, scalability, and partial observability [18].

In a closely related problem of the Job-Shop Scheduling Problem (JSSP), the use of Proximal Policy Optimization (PPO) [14] with Graph Neural Network algorithm was able to learn the priority dispatch rule for JSSP from first principles without intermediate handcrafted rules [17]. While such generalized learning approach is novel, it is important to note that the action-space of JSSP is reduced as the solution space narrows over time due to precedence constraint. This, however is not true for DPRP, where the same graph node can be visited multiple times and the action-space does not reduce over time. Another proposed method [11] based on actor-critic Deep Deterministic Policy Gradient (DDPG) algorithm in multi-agent environment only considers a set of simple dispatching heuristics rules in the agent action space. The constrained action-state space proved to be efficient for learning and sufficiently effective for solving complex scheduling problem. However, such a framework requires in-depth prior domain knowledge, and retraining of the model is needed when input parameters vary.

For our problem DPRP, [9] demonstrated successful application of a deep RL-based method with a rescheduling heuristic based on input and constraints from the real-world environment. The proposed method combines the value function approximation through Temporal-Difference (TD) learning with experience replay and an ejection chain heuristic solver. The solution is able to compute dispatch and rescheduling decisions instantaneously as required in real-world environment. There were also other works that addressed a similar variant of such problem. Most of these approaches adopted a two-stage decomposition that learns the dispatching and scheduling policies in separate stages (see [3,4]).

3   Problem Description

The problem being addressed in this paper closely represents the scenario of police patrolling in a modern and large city in which police needs to respond to incidents of various types within very short time frames (in less than 10 min within receiving the call for response). At the start of the day, police officers are assigned to different patrol areas under their jurisdiction. A centralized authority is tasked to plan the resource to ensure sufficient patrol presence for the entire city. In addition to patrol duties, the plan must also adapt to incidents arising in real time, to ensure that police officers are able to respond to incidents as soon as possible while not compromising the level of patrols within each jurisdiction.

Figure 1 shows an example of multiple patrol officers dispatched to different patrol areas based on the initial schedule given. We assume that all patrol agents have homogeneous capability.

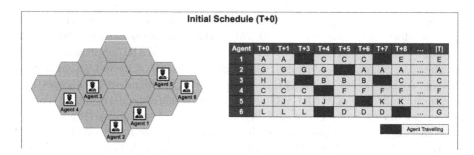

Fig. 1. Schematic diagram that shows multi-agent patrol environment and the initial timetable schedule at $T = 0$.

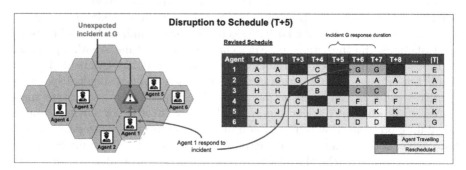

Fig. 2. An incident happened at $T + 5$ and patrol agent 1 is deployed to respond to the incident. Following the disruption, a rescheduling is made to the time table of the relevant patrol agents (patrol agent 1 and 3) at $T \geq 5$.

Figure 2 shows an example of a situation where an incident occurs at location G during in a given day or shift. In this case, Agent 1 is deployed to respond to the incident. The original patrol location of C is changed to G at $T = 5$ and this change may result in the need to reschedule the plans of other agents. For simplicity, we assume that only one incident can occur one at a time. In addition, we assume that the condition of partial observability does not exist with the presence of a central authority akin to a police headquarter.

4   Model Formulation

The objective of the problem is for every agent to make rescheduling and dispatching decision at every decision epoch in order to maximize both global patrol presence and response rate to dynamic incidents.

We model our problem as a fully cooperative multi-agent Markov Decision Process - $(\mathbf{S}, \mathbf{A}, \mathbf{T}, \mathbf{R})$ where $\mathbf{S}$ is a state for the timetable, $\mathbf{A}$ is a set of actions taken by the agents, $\mathbf{T}$ is the transition probability vector between state for different state-action pairs, and $\mathbf{R}$ is the immediate reward transitioning from state s to s' given action a.

The objective of the problem is for every agent to make rescheduling and dispatching decision at every decision epoch in order to maximize both global patrol presence and response rate to dynamic incidents (Table 1).

Table 1. Set of notations used in this paper.

Notation	Description
I	Set of patrol agents, $I \in \{1, 2, 3, \cdots, |I|\}$
J	Set of patrol areas, $J \in \{1, 2, 3, \cdots, |J|\}$
T	Set of time periods in a shift, $T \in \{1, 2, 3, \cdots, |T|\}$
k	Decision epoch
t_k	Time period in a shift where decision epoch k occurs, $t_k \in T$
$a(k)$	Set of dispatch actions taken by all agents at decision epoch k
$x_i(k)$	Dispatch action taken by agent i at decision epoch k
$\delta_i(k)$	A schedule of patrol agent i at decision epoch k
$\delta(k)$	A joint schedule of all patrol agents at decision epoch k
$\delta_{-i}(k)$	A joint schedule of all patrol agents except for agent i at decision epoch k
$\delta_i^a(k)$	A schedule of patrol agent i after executing action a at decision epoch k
$\delta^a(k)$	A joint schedule of all patrol agent after executing action a at decision epoch k
τ_{target}	A response time target
τ_{max}	A maximum buffer time for response time for incident
τ_k	Actual response time to incident at epoch k
$D_h(\delta', \delta)$	Hamming distance between schedules δ' and δ
$d(j, j')$	Travel time from patrol area j to another patrol area j'
Q_j	Minimum patrol time for patrol area j
σ_j	Patrol presence for area j in terms of ratio of the effective patrol time over Q_j
ω_k	State representation of dynamic incident that occurs at decision epoch k
$N(k)$	State representation of patrol agents availability at decision epoch k
$\Omega_{i,k}$	Patrol status of agent (patrolling or travelling) i at decision epoch k
$D_{i,k}$	Patrol or travel destination of agent i at decision epoch k
$M_{i,k}$	Travel arrival time of agent i at decision epoch k

4.1 State

The state of the MDP, S_k is represented as the following tuple: $\langle t_k, \delta(k), \omega(k), N(k) \rangle$. t_k is the time period in a shift where decision epoch k occurs. $\delta(k)$ is the joint schedule of all patrol agents, $\omega(k)$ is the dynamic incident, and $N(k)$ is the patrol agents' availability at decision epoch k.

Joint Schedule. The joint schedule, $\delta(k)$ has a dimension of $|T| \times |I| \times |J|$, which represents the time tables for all patrol agents.

Incident. A dynamic incident, $w(k)$ occurs at decision epoch k and is described as the following tuple: $\left\langle \omega_k^j, \omega_k^t, \omega_k^s \right\rangle$ where $\omega_k^j \in J$ refers to the location of the incident, $w_k^t \in T$ refers to the time period when the incident occurs, and w_k^s refers to the number of time periods required to resolve the incident.

Agents' Availability. The agents' availability, $N(k)$ represents the availability of patrol agents in the decision epoch k. It comprises the individual agent's availability, $N_i(k)$ for every agent i.

4.2   Action

$a_i(k)$ is a dispatch action taken by an agent i at decision epoch k. The set of action space is a set of dispatch rules described in Table 2. The action taken by the agent determines the state of the agent's individual time table at the next time step. The selection of action is made with an ϵ-greedy strategy.

Table 2. List of dispatch-rule-based actions for patrol agents.

Dispatch Rule	Description
a1. Respond to incident	Travel to the location of incident if there is an occurrence of incident. Otherwise, this action is not allowed
a2. Continue	Continue to patrol the same area or continue traveling to the destination patrol location if the agent was in the midst of traveling to another patrol area. This heuristics tries to minimizes deviation from initial schedule as the initial schedule was optimal for patrol presence in a situation where no unforeseen incident occurs
a3. Patrol an unmanned area	Travel to an unmanned patrol area that yields the best patrol presence utility. If multiple patrol areas have same the utility, randomly select one. This heuristics is a greedy approach that aims to maximize patrol presence utility of the schedule, with the assumption that unforeseen incidents may occur in unmanned patrol areas
a4. Patrol an existing manned area by other agents	Patrol an existing area currently being patrolled by other agents. Choose the area that has the best patrol presence utility. If multiple patrol areas have the same utility, randomly select one. This heuristics also aims to maximize patrol presence utility, but allowing the agent to take over an existing patrol area of another agent if the original agent needed to be dispatched elsewhere
a5. Nearest and least disruptive	Patrol the next nearest location such that it results in least deviation from the initial schedule. If multiple patrol areas have same the travel distance, randomly select one. This heuristics tries select a patrol area in such a way that it minimizes travel time and deviation from the initial schedule

4.3 Transition

A decision epoch k occurs at every time step. We move to the next decision epoch $k + 1$ after all agents have completed their actions, transiting from the pre-decision state S_k to the post-decision state S_{k+1}. In our formulation, the transition between state is deterministic, and we let transition probability $\mathbf{T} = 1$ for every state-action pair. It is deterministic as in if an agent has chosen an action, there is no possibility that the agent deviates from the chosen action.

4.4 Constraints

We subject our final schedule $\delta(T)$ at the end of the last decision epoch to the following soft constraint:

$$D_h(\delta(T), \delta(0)) \leq D_{h,\max} \tag{1}$$

where $D_h(\delta(T), \delta(0))$ is the Hamming distance of the final schedule with respect to the initial schedule $\delta(0)$. $D_{h,\max}$ is the maximum Hamming distance allowed. The constraint helps minimize disruption to our existing schedule caused by rescheduling.

4.5 Patrol Presence

Before discussing the reward function, we define patrol presence as the number of time periods each patrol area is being patrolled. Every patrol area j must be patrolled for a minimum of Q_j time periods in a given shift. A schedule with good patrol presence seeks to maximize the time spent patrolling while minimizing the travel time of patrol agents when moving to different patrol areas. The patrol presence utility function $f_p(\delta)$ is defined as the following

$$f_p(\delta) = \frac{\sum_{j \in J} U_p(j)}{|T| \times |I|} \tag{2}$$

where $U_p(j)$

$$U_p(j) = \min(\sigma_j, 1) + 1_j \times e^{-\beta(\sigma_j - 1)}$$
$$1_j = \begin{cases} 1, \sigma_j > 1 \\ 0, \sigma_j \leq 1 \end{cases} \tag{3}$$

where β is coefficient for patrol presence utility and patrol presence σ_j is defined as

$$\sigma_j = \frac{\sum_{t=1}^{|T|} p_{j,t}}{Q_j}$$
$$p_{j,t} = \begin{cases} 1, \text{patrol is present at area } j \text{ at time step } t \\ 0, \text{otherwise} \end{cases} \tag{4}$$

This utility function measures the utility of each patrol in each patrol area with respect to the minimum patrol requirement of that area, and additional patrol time comes with a diminishing return of utility beyond the minimum requirement.

4.6 Reward Function

The reward function $R(S_k, a_i(k))$ takes into consideration of three factors; patrol presence, incident response, and deviation from existing schedule. Note that the reward at $t < T$ only considers incident response, while the final reward when at the end of episodes at $t = T$ includes additional factors of patrol presence reward and schedule deviation penalty. $f_r(a_i(k))$ quantifies the success of an incident response when action $a_i(k)$ is taken by agent i. Similar to the patrol presence utility, any incident that is responded later than the target time will incur a reduced utility. The reward function is defined as

$$R(S_k, a_i(k)) = \begin{cases} f_r(a_i(k)) + f_p(\delta(t_k)) - p_r(\delta(t_k)), & t_k = T \\ f_r(a_i(k)), & t_k < T \end{cases}$$

$$f_r(a_i(k)) = \begin{cases} U_r = \exp^{-\alpha \times \max(0, \tau_k - \tau_{target})}, & \tau_k > 0 \\ 0, & \tau_k = 0 \text{ (Incident not responded)} \end{cases}$$
(5)

where α is the coefficient for response utility of a late response, τ_k is the response time taken at decision epoch k and τ_{target} is the target response time.

The penalty function for the deviation of the schedule $p_r(\delta(t_k))$ based on the Hamming distance D_h is defined as the following step functions:

$$p_r(\delta(t_k)) = C_1 \cdot D_h(\delta(t_k), \delta(0)) + C_2 \cdot 1_H(\delta(t_k), \delta(0))$$

$$1_H(\delta(t_k), \delta(0)) = \begin{cases} 0, D_h(\delta(t_k), \delta(0)) \leq D_{h,\max} \\ 1, D_h(\delta(t_k), \delta(0)) > D_{h,\max} \end{cases}$$
(6)

where C_1, C_2 are the weights of the Hamming distance penalty coefficients.

5   Solution Approach

The RL algorithm selected for our solution approach is an asynchronous variant of Proximal Policy Optimization (APPO) [15] algorithm based on IMPALA [6] actor-critic architecture. IMPALA is an off-policy actor-critic algorithm that decouples acting and learning, which allows multiple actors to generate experience in parallel, creating more trajectories over time. The off-policy RL algorithm uses the trajectories created by policy μ (behavior policy) to learn the value function of target policy π.

At the start of each trajectory, the actor updates its own policy μ in response to the latest policy from the learner, π, and uses it for n steps in the environment. After n steps, the actor sends the sequence of states, actions, and rewards along with the policy distributions to the learner through a queue. Batches of experiences collected from multiple actors are used to update the learner's policy. Such design allows the actors to be distributed across different machines. However, there is a delay in updating the policies between actors and the learner, as the learner policy π may have undergone several updates compared to the actor's policy μ at the time of the update. Therefore, an off-policy correction method called V-trace is used here.

V-Trace Target. For a trajectory state (x_t, a_t, x_{t+1}, r_t), the n–steps V-trace target for $V(x_s)$ at time s is given as,

$$v_s \overset{\text{def}}{=} V(x_s) + \sum_{t=s}^{s+n-1} \gamma^{t-s} \left(\prod_{i=s}^{t-1} c_i \right) \rho_t(r_t + \gamma V(x_{t+1}) - V(x_t)) \tag{7}$$

where $\rho_t = \min\left(\bar{\rho}, \frac{\pi(a_t|x_t)}{\mu(a_t|x_t)}\right)$ and $c_i = \min\left(\bar{c}, \frac{\pi(a_t|x_t)}{\mu(a_t|x_t)}\right)$ are truncated importance sampling weights. $\bar{\rho}$ and $\bar{c}$ are truncation constants with $\bar{\rho} \geq \bar{c}$. In the case of on-policy learning, then $c_i = 1$ and $\rho_t = 1$, and (7) becomes

$$v_s = \sum_{t=1}^{s+n-1} \gamma^{t-s} r_t + \gamma^n V(x_{s+n}) \tag{8}$$

which is the on-policy n–step Bellman target. By varying $\bar{\rho}$, we change the target of the value function to which we converge. When $\bar{\rho} = \infty$ (untruncated), the value function of the v trace will converge to the target policy V_π; When $\hat{\rho} \to 0$ (untruncated), the value function converges to behavior policy V_μ. Any value of $\hat{\rho} < \infty$ indicates the value function of the policy somewhere between μ and π. The truncation constant $\bar{c}$ changes the speed of convergence.

V-Trace Actor-Critic Algorithm. For every iteration update, $V_\theta(s)$ is the value function of state s parametrized by θ, which is updated by

$$\Delta\theta = (v_s - V_\theta(s))\nabla_\theta V_\theta(s) \tag{9}$$

and the policy parameters w is updated through policy gradient

$$\Delta w = \rho_s \nabla_w \log \pi_w(a_s|s)(r_s + \gamma v_{s+1} - V_\theta(s)) \tag{10}$$

Proximal Policy Optimization (PPO). The use of Proximal Policy Optimization (PPO) improve the training stability of our policy by avoiding excessive large policy update. PPO uses the following clipped surrogate objective function for policy update:

$$L^{CLIP}(\theta) = \hat{\mathbb{E}}_t[\min(r_t(\theta)\hat{A}_t, \text{clip}(r_t(\theta), 1-\epsilon, 1+\epsilon)\hat{A}_t)] \tag{11}$$

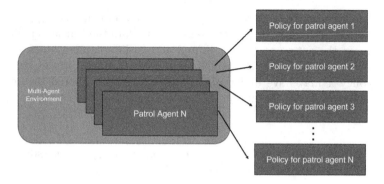

Fig. 3. Multi-agent setup with individual policy for the patrol agents.

Algorithm 1. Dispatch-rule based scheduling RL algorithm

---

for each episode do
    Set initial schedule $\delta \rightarrow \delta_0$ from a set of initial schedules, generate a scenario
    with a set of incident scenarios $\{\omega_k, \omega_{k+1}, \cdots\}$
    Set entries of $\delta_i(t \geq k) \rightarrow \varnothing$ $\forall i \in I$
    for $t = t_k$ to T do
        for each patrol agent i (in random order) do
            takes a feasible patrol action with ϵ-greedy strategy that decides the state
    of $\delta_i(t)$
        end for
    end for
    Revised schedule is completed $\delta \rightarrow \delta'$
end for

---

where θ is the policy parameter, $\mathbb{E}_t$ is the empirical expectation, r_t is the ratio of the probabilities under the new and old policies, $\hat{A}_t$ is the estimated advantage at time t, ϵ is a hyper parameter, usually 0.1 or 0.2.

As our environment is a multi-agent environment, we assign one policy network to each agent as shown in Fig. 3. Algorithm 1 describes the procedure for rescheduling according to the solution approach. We implemented our solution method using Ray RLlib library [13].

6   Experimental Setup

6.1   Environment

The patrol environment comprises hexagonal grids of size $2.2\,\mathrm{km} \times 2.2\,\mathrm{km}$ derived from local police patrol sectors, each grid representing a patrol area. We have chosen a large patrol setup with $|I| = 11, |J| = 51$ for our environment. This patrol setup represents an environment with a relatively low agent-to-area ratio. The duration of a day shift is $12\,\mathrm{h}$ and is divided into 72 discrete 10-min time units. The maximum Hamming distance for the revised schedule $D_{h,\max}$ is set at 0.4.

6.2   Model Parameters

The input state vectors are flattened into a one-dimensional array as a concatenated input to a fully-connected neural network encoder. The encoder neural network has a size of 256×256, with $tanh$ activation functions. The training batch size is 500. Learning rate $\alpha = 0.0001$, $\epsilon_{initial} = 0.6$, $\epsilon_{final} = 0.01$. We set the discount factor $\gamma = 1.0$ since the reward function for patrol presence and deviation from initial schedule is only evaluated at the end of the episode when $\tau_k = T$ when the revised schedule is complete. Agents must therefore take into consideration this final reward without discount when evaluating action choices in an earlier decision epoch. We set the maximum number of training episodes to be 5000 episodes.

6.3   Training and Test

During training, there are 100 samples of initial joint schedules for initialization. Each sample consists of an initial joint schedule for all patrol agents for the entire day. The initial schedules are obtained via a mixed linear integer program prior to training. We run a total of 5000 training episodes. During each training episode, an initial schedule is randomly sampled from the pool, and a set of incidents is generated based on Poisson distribution with λ set as 2 i.e. the rate of occurrences of incident is 2 per hour. A training episode ends when the time step reaches the end of the shift. The training results generally begin to converge after 3000 episodes. After training is completed, we evaluated the performance of our solution approach based on 30 samples of initial joint schedules separate from the training set.

6.4   Evaluation Metrics

Our evaluation considers the following metrics: **patrol presence score (%)**, **incidence response score (%)** and **deviation from original schedule (Hamming distance)**. A good solution should have both high patrol presence and incidence response scores (where 100% means all incidents are responded within the stipulated response time), and a low deviation from original schedule (where 0 means that the existing schedule remains unchanged). We benchmark the quality of our solution approach against two approaches:

- **Myopic rescheduling heuristic** - Baseline algorithm with ejection chain rescheduling heuristic for comparison with VFA-H and our approach APPO-RL;
- **Value Function Approximation heuristic (VFA-H)** [9] - An RL-based rescheduling heuristic with ejection chain based on a learnt value function.

7   Experimental Results

7.1   Solution Quality

The evaluation scores summarized in Table 3 show that our solution approach APPO-RL outperformed the VFA-H method. The patrol presence and incidence response scores of our method are higher than that of VFA-H by $+10.6\%$ ($+12.6\%$ vs $+2\%$) and $+6.6\%$ (-9.9% vs -16.5%) respectively. However, since schedule deviation is set as a soft constraint in our proposed solution, we see that the maximum Hamming distance $D_{h,\max}$ of 0.403 obtained by our approach exceeded the threshold of 0.4, implying that in some cases the maximum Hamming distance imposed has been violated, while this constraint is not violated in the VFA-H's method. On average, however, it is encouraging to see that the Hamming distance ($D_{h,\mathrm{mean}}$) is less than in VFA-H.

The biggest advantage of our solution approach is the improved computational efficiency in solving the DPRP problem. Compared to VFA-H, the training time required is about 40x less. This is due to the reduced search space as we limit the number of action states to a selected few dispatch rules.

Table 3. Evaluation metric scores and mean training time.

Metric	VFA-H	APPO RL
Δ in mean incidence response score over Myopic	$+2\%$	$+12.6\%$
Δ in mean patrol presence score over Myopic	-16.5%	-9.9%
$D_{h,\max}$	0.399	0.403
$D_{h,\mathrm{mean}}$	0.387	0.342
Mean training time per episode (s)	436	10.9

Figure 4 presents the performance of the three approaches with respect to the number of training episodes.

7.2   Constraint Sensitivity Analysis

We conducted a constraint sensitivity analysis to evaluate the trade-off between solution quality and constraint satisfaction by varying the value of C_2, the coefficient of a step function penalty term linked to the Hamming distance constraint threshold of $D_{h,\max} = 0.4$. As shown in Fig. 5, as we decreased the soft-constraint penalty coefficient C_2 on Hamming distance, the response score generally improved while the patrol presence score remained largely at similar levels. This is expected as agents have fewer constraints to consider when responding to incidents.

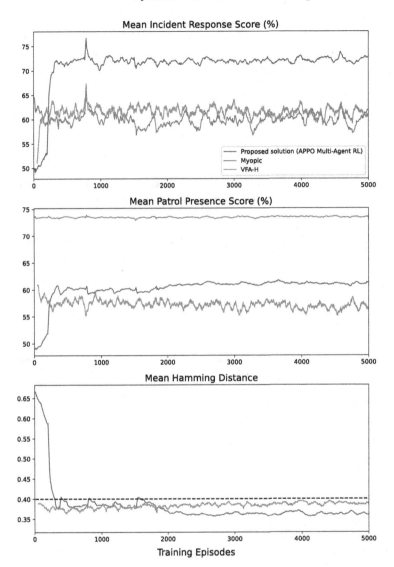

Fig. 4. Comparison of our solution approach (blue) with VFA-H (orange) and myopic (grey) baseline methods on various evaluation metrics. (Color figure online)

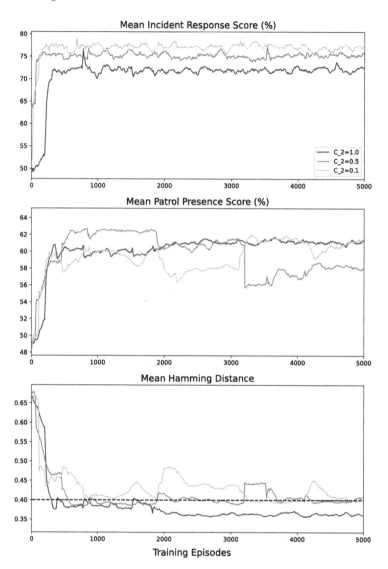

Fig. 5. Results of constraint sensitivity analysis based on our solution approach (darker colour indicates higher weightage of constraint penalty coefficient C_2). (Color figure online)

8   Discussion and Future Work

We discuss two major challenges of our work.

Generality. With a limited and handcrafted set of dispatch rules as actions, we are able to reduce the search space considerably even as the number of agents increases. Nonetheless, our approach requires some experimentation or prior domain knowledge in order to select a set of optimal dispatch rules suitable for the problem. This makes it challenging when we need to apply to a slightly different problem set, as the set of prescribed dispatch rules may have to be modified. It is also unclear if the learned policy is applicable as we vary the size of the environment. We may therefore have to train different policies as we change the size of patrol area and number of agents.

Constraint Satisfaction. The biggest drawback of our solution approach is that constraint satisfaction is not always guaranteed. In the DPRP problem, this constraint was set mainly to minimize disruption to patrol agents. Although such a violation of constraint is acceptable to some extent in our problem set, one can argue that this may not be applicable to other situations.

To conclude, we have demonstrated a successful application of multi-agent reinforcement learning technique in solving dynamic scheduling problem in the context of police patrolling. Our proposed method is able to improve solution quality while reducing training time considerably. We are in discussion with a local law enforcement agency to develop a prototype tool for real-world experimentation.

From the research standpoint, it would be worthwhile to apply this in a multi-agent environment where we have non-homogeneous patrol agents that need to collaborate with one another in order to respond to incidents. It would also be interesting to research further into the aspect of constrained RL to ensure that hard constraint satisfaction is met.

References

1. Canese, L., et al.: Multi-agent reinforcement learning: a review of challenges and applications. Appl. Sci. **11**(11), 4948 (2021)
2. Chase, J., Phong, T., Long, K., Le, T., Lau, H.C.: Grand-vision: an intelligent system for optimized deployment scheduling of law enforcement agents. In: Proceedings of the International Conference on Automated Planning and Scheduling, vol. 31, pp. 459–467 (2021)
3. Chen, X., Tian, Y.: Learning to perform local rewriting for combinatorial optimization. In: 33rd Conference on Neural Information Processing Systems (2019)
4. Chen, Y., et al.: Can sophisticated dispatching strategy acquired by reinforcement learning? - a case study in dynamic courier dispatching system. In: AAMAS 2019: Proceedings of the 18th International Conference on Autonomous Agents and MultiAgent Systems (2019)
5. Dewinter, M., Vandeviver, C., Vander Beken, T., Witlox, F.: Analysing the police patrol routing problem: a review. ISPRS Int. J. Geo Inf. **9**(3), 157 (2020)
6. Espeholt, L., et al.: IMPALA: scalable distributed deep-RL with importance weighted actor-learner architectures. In: Proceedings of the 35th International Conference on Machine Learning, pp. 1407–1416 (2018)
7. Ghiani, G., Guerriero, F., Laporte, G., Musmanno, R.: Real-time vehicle routing: solution concepts, algorithms and parallel computing strategies. Eur. J. Oper. Res. **151**(1), 1–11 (2003)

8. Gronauer, S., Diepold, K.: Multi-agent deep reinforcement learning: a survey. Artif. Intell. Rev. **55**(2), 895–943 (2022)

9. Joe, W., Lau, H.C., Pan, J.: Reinforcement learning approach to solve dynamic bi-objective police patrol dispatching and rescheduling problem. In: Proceedings of the International Conference on Automated Planning and Scheduling, vol. 32, pp. 453–461 (2022)

10. Li, W., Ni, S.: Train timetabling with the general learning environment and multi-agent deep reinforcement learning. Transp. Res. Part B: Methodol. **157**, 230–251 (2022)

11. Liu, C.L., Chang, C.C., Tseng, C.J.: Actor-critic deep reinforcement learning for solving job shop scheduling problems. IEEE Access **8**, 71752–71762 (2020)

12. OroojlooyJadid, A., Hajinezhad, D.: A review of cooperative multi-agent deep reinforcement learning. arXiv preprint arXiv:1908.03963 (2019)

13. Ray: Ray RLlib. https://www.ray.io/rllib

14. Schulman, J., Wolski, F., Dhariwal, P., Radford, A., Klimov, O.: Proximal policy optimization algorithms. arXiv preprint arXiv:1707.06347 (2017)

15. Silver, D., Lever, G., Heess, N., Degris, T., Wierstra, D., Riedmiller, M.: Deterministic policy gradient algorithms. In: Proceedings of the International Conference on Machine Learning, pp. 387–395. PMLR (2014)

16. Watanabe, T., Takamiya, M.: Police patrol routing on network Voronoi diagram. In: Proceedings of the 8th International Conference on Ubiquitous Information Management and Communication, pp. 1–8 (2014)

17. Zhang, C., Song, W., Cao, Z., Zhang, J., Tan, P.S., Chi, X.: Learning to dispatch for job shop scheduling via deep reinforcement learning. In: Advances in Neural Information Processing Systems, vol. 33, pp. 1621–1632 (2020)

18. Zhang, K., Yang, Z., Başar, T.: Multi-agent reinforcement learning: a selective overview of theories and algorithms. In: Vamvoudakis, K.G., Wan, Y., Lewis, F.L., Cansever, D. (eds.) Handbook of Reinforcement Learning and Control. SSDC, vol. 325, pp. 321–384. Springer, Cham (2021). https://doi.org/10.1007/978-3-030-60990-0_12

Analysis of Heuristics for Vector Scheduling and Vector Bin Packing

Lars Nagel[1]([✉]), Nikolay Popov[2], Tim Süß[3], and Ze Wang[1]

[1] Loughborough University, Loughborough, UK
l.nagel@lboro.ac.uk
[2] iC-Haus GmbH, Bodenheim, Germany
[3] Fulda University of Applied Science, Fulda, Germany

Abstract. Fundamental problems in operational research are vector scheduling and vector bin packing where a set of vectors or items must be packed into a fixed set of bins or a minimum number of bins such that, in each bin, the sum of the vectors does not exceed the bin's vector capacity. They have many applications such as scheduling virtual machines in compute clouds where the virtual and physical machines can be regarded as items and bins, respectively. As vector scheduling and vector bin packing are NP-hard, no efficient exact algorithms are known.

In this paper we introduce new heuristics and provide the first extensive evaluation of heuristics and algorithms for vector scheduling and bin packing including several heuristics from the literature. The new heuristics are a local search algorithm, a game-theoretic approach and a best-fit heuristic. Our experiments show a general trade-off between running time and packing quality. The new local search algorithm outperforms almost all other heuristics while maintaining a reasonable running time.

Keywords: vector scheduling · vector bin packing · heuristics · local search

1 Introduction

Many scheduling problems in computer science can be modelled as vector scheduling or vector bin packing problems. Examples are the scheduling of virtual machines or batch jobs on computer clusters where the resource requirements of a job as well as the resources available on machine can be represented as a vector. Vector scheduling and vector bin packing are defined as the problem of packing a given set of d-dimensional item vectors into d-dimensional bins. In vector scheduling the number of bins is predefined, in vector bin packing the task is to minimize the number of bins required. More formally:

Definition 1 (Vector Scheduling). *Let J be a set of items where each item $j \in J$ is a d-dimensional vector $(j_1, ..., j_d)$ with $j_i \in [0, 1]$ for $i \in [d]$. Further*

M. Sellmann and K. Tierney (Eds.): LION 2023, LNCS 14286, pp. 583–598, 2023.
https://doi.org/10.1007/978-3-031-44505-7_39

let m be the number of d-dimensional bins, each of capacity $(1, ..., 1)$. A valid packing of items is a partition of J into sets $B_1, ..., B_m$ such that

$$\forall k \in [m], \forall i \in [d], \quad \sum_{j \in B_k} j_i \leq 1. \tag{1}$$

The goal is to find such a valid packing.

Definition 2 (Vector Bin Packing). Let J be a set of items where each item $j \in J$ is a d-dimensional vector $(j_1, ..., j_d)$ with $j_i \in [0, 1]$ for $i \in [d]$. Further assume that bins have a d-dimensional capacity $(1, ..., 1)$. The goal is to find a valid packing while minimising the number of bins m.

Both problems are known to be NP-complete, and every exact algorithm known requires exponential time in the worst case. For this reason, researchers devised approximation algorithms [1,5] and heuristics, e.g. [3,12,17]. Some of these heuristics are in practical use. For example, *Sandpiper* [22], a resource management system for virtual machines, uses the *FFDProd* heuristic of Panigrahy et al. [17] for the migration of virtual machines from overloaded hosts.

In this paper we present three new heuristics for vector scheduling and compare them to a wide range of algorithms and heuristics from the literature. We analyze the scheduling quality of these algorithms as well as their running times in simulations using randomly generated data and benchmarks from the literature. The experiments show that our new *Local Search* algorithms outperform almost all other heuristics while maintaining a reasonable running time. The other two types of heuristics, called *Hybrid* and *Stable Pairing*, do not pack the jobs as successfully, but have shorter running times and can compete with similarly fast heuristics from the literature.

The paper is structured as follows: Existing algorithms are summarized in Sect. 2. The new algorithms are introduced in Sect. 3. They are evaluated in Sect. 4. Section 5 concludes the paper.

2 Algorithms and Complexity

In this section we survey the literature on algorithms and heuristics for vector scheduling and vector bin packing. We omit results for one-dimensional problems and for other related problems like the packing of orthotopes.

2.1 Theoretical Results and Exact Algorithms

Vector scheduling is well-known to be NP-hard for a constant number of dimensions d [7]. Vector bin packing is APX-hard as Woeginger [21] showed that already for $d = 2$ the existence of an asymptotic polynomial-time approximation scheme (PTAS) would imply $P = NP$. Sandeep [19] showed there is no asymptotic $o(\log d)$-approximation for vector bin packing if d is large.

Chekuri and Khanna [5] showed approximation bounds for vector bin packing and vector scheduling and introduced a PTAS for vector scheduling for constant *d*. It rounds the coordinates of the vectors and fits them into bins of size $1 + \epsilon$ (instead of 1). Bansal et al. [1] improved the scheme. But although its running time of $O(2^{(1/\epsilon)^{O(d \log \log d)}} + nd)$ is shown to be almost optimal, the algorithm is not feasible for practical purposes. Recently, Kulik et al. [13] introduced an asymptotic $(\frac{4}{3} + \epsilon)$-approximation for the 2-dimensional vector bin packing.

Although vector bin packing is NP-hard, there are exact algorithms that solve small instances in reasonable time. Heßler et al. [9] designed a branch-and-price algorithm. Brandão and Pedroso [2] used an arc-flow formulation to solve vector bin packing with integer vectors. In our evaluation in Sect. 4, we use their algorithm to compute the optimum for small instances.

2.2 First-Fit and Best-Fit Heuristics

Kou and Markowsky [12] introduced first-fit decreasing (FFD) and best-fit decreasing (BFD) heuristics which are suitable for vector bin packing and vector scheduling. The algorithms sort the vectors in decreasing order and assign them to the bins using first-fit or best-fit. In order to be comparable, each vector must be assigned a size. In the paper, the authors use the infinity norm (L_∞), the sum of all vector components (L_1) and lexicographical ordering.

Panigrahy et al. [17] changed the item-centric approach of Kou et al. to a bin-centric one. Their heuristics only open a new bin if no more item fits into the current one. Otherwise it assigns the largest fitting item to the current bin. If a norm (L_1, L_2, L_∞) is used, the largest item is the one that minimizes the distance between remaining capacity and item vector, possibly weighted by a vector w. They also apply the dot product where the largest item is the one whose vector maximises the dot product with the remaining capacity of the bin. The authors suggest two weight vectors for item-centric and bin-centric heuristics. The first one consists of the averages a_i in every dimension. The second one applies the exponential function to the average, i.e. $b_i = e^{\epsilon \cdot a_i}$, where ϵ is a suitable constant.

Some heuristics by Kou et al. and Panigrahy et al. are included in the evaluation in Sect. 4.

2.3 Genetic Algorithms

Wilcox et al. [20] designed the *Reordering Grouping Genetic Algorithm* (RGGA) to solve vector bin packing. In this genetic algorithm, the packing results, i.e. the individuals of the population, are represented as a sequence. The crossover operator is implemented using an exon shuffling approach. Three operators are equally used in the mutation phase. The first two methods involve swapping two items or moving one item to another position in the sequence. The third operator removes a number of bins and reassigns the removed items into the sequence.

Sadiq and Shahid [18] apply Simulated Evolution (SE) to vector bin packing, a genetic algorithm. It uses a first-fit heuristic for the initial solution before performing three steps in a loop: *Evaluation* assigns a goodness value to every allocated item, i.e. a probability for being moved into a reallocation pool by

Selection. The last step, *Allocation,* reallocates the items in the pool using again a first-fit scheme. The evaluation shows that SE outperforms FFD.

2.4   Local Search and Simulated Annealing

Mannai and Boulehmi [14] presented a *tabu search* algorithm for two-dimensional vector bin packing. As the description is not detailed enough, we did not implement it. Their experiments compare the algorithm only to an optimal algorithm which uses the CPLEX Optimizer [10].

Masson et al. [15] propose a *multi-start iterated local search* scheme for vector bin packing. It uses local search in combination with a shaking procedure of random reallocations.

Buljubašić and Vasquez [3] proposed a hybrid scheme named *consistent neighborhood search* (CNS) which combines a first-fit heuristic, tabu search and a technique for computing an optimal packing for a limited set of items. Results in the paper suggest that CNS outperforms the algorithm by Masson et al. [15] which is why we only evaluate CNS in this paper.

Simulated Annealing was often applied to geometric packing problems, but there seems to be only one paper applying it to vector bin packing [16]. It is shown to outperform first-fit without sorting, but since it did not perform well in our tests, we left it out of the evaluation.

3   New Algorithms

In this section, we propose three new vector scheduling strategies. The first algorithm is based on local search. The second algorithm is a combination of bin-centric and item-centric strategies. The last algorithm applies the stable roommate algorithm and the stable marriage algorithm from game theory.

3.1   Local Search

We propose two local-search algorithms for vector scheduling called *Local Search* (LS) and *Local Search with Randomization* (LSR). The strategy of LS is similar to the bin-centric approach in that it fills one bin at a time trying to minimize the bin's remaining capacity until no further improvement can be found. Before describing the logic of the algorithm, we describe the two main procedures of the algorithm: *Random Assignment* and *Swap.*

Random Assignment randomly packs items into the current bin. The procedure runs in a loop and in each iteration, it randomly selects an item from the pool of remaining items and tries to assign it to the bin without exceeding the bin's capacity. The number of failed assignments is counted and the procedure finishes when this count reaches the number of unpacked items. *Swap* is designed with the goal to reduce the bin's capacity further by swapping unpacked items with items in the current assignment. It takes a randomly chosen item from the remaining pool and tests it against every item in the bin. If there is at least one permissible swap that would reduce the remaining capacity (based on the L_2 norm), the best swap is performed.

LS starts by opening a new bin and placing the unpacked item with the biggest L_2 weight into it. Then *Random Assignment* is applied to produce an initial assignment. After the initial assignment is generated, the algorithm runs *Swap* n times in a loop to improve the current assignment, where n is the number of unpacked items. If a swap happens during a *Swap* run, the algorithm will call *Random Assignment* again to try if due to the change the assignment can be improved. After that the loop is continued. When it is finished, the algorithm will start a new loop if at least one successful swap was made in the current loop. Otherwise it will close the current bin and open the next one.

The *Swap* routine of LS only swaps items if the remaining capacity is reduced by the swap. For this reason, it can be trapped in a local optimum. To avoid this, we devised a second scheme, LSR, which also swaps items if the remaining capacity is not reduced. In such a case, the algorithm chooses the swap with the smallest capacity increase and performs it with a probability proportional to the increase. More precisely, the probability is the quotient of the L_2 norm of the previous remaining capacity and the changed remaining capacity. While these swaps do not pack the bin tighter, they create opportunities for the algorithm to escape from a local optimum. Since swaps may be performed in every loop of *Swap*, we require an additional termination condition and set the maximum number of loops to ten. During the run, the algorithm always memorises the assignment with the smallest remaining capacity which is used in the end.

3.2   Hybrid Heuristic

The second new heuristic is named *Hybrid Heuristic* and can be seen as a hybrid of bin-centric and item-centric heuristics. In the bin-centric scheme proposed by Panigrahy et al. [17], exactly one bin is open in every step, and the largest item the bin can fit is placed into the bin. If it cannot fit any of the remaining items, the bin is closed, and the next empty bin is opened. The Hybrid Heuristic differs from this scheme by opening more bins at the same time to generate more options for items and potentially find a better assignment.

The algorithm sorts the items and opens one empty bin in the beginning. Then it starts assigning the items. Whenever there is an unassigned item that is too large for all opened bins, a new bin is opened, and the item is placed in it. If there is no such item, then each opened bin determines its best-fitting item according to a cost function. For this we use the same cost functions as Panigrahy et al., i.e. we either minimize the L_1, L_2 or L_∞ norm of the remainder (i.e., remaining capacity of the bin minus item size), or we maximize the dot product of remaining capacity and item size. After that it remains to choose the bin that actually gets an item. The aim here is to choose an item that fits well, but also to favour bins, or at least not to neglect bins, that have not received many items yet. To achieve this, the heuristic compares all proposals to pick one which satisfies the following rules. For the norms, the heuristic evaluates the bins by the percentage by which the remaining capacity is reduced and picks the one with largest percentage. For the dot product, two cases are distinguished: (1) If two bins prefer the same item, the bin with the smaller dot product is chosen because it is a better fit. (2) If two bins prefer different items, the machine with the larger dot product is preferred to fill up lower-loaded bins more quickly. The algorithm applies rule (1) first, before applying rule (2) on all the remaining bins. If, after the assignment, the bin is full in the sense that it cannot fit any unpacked item, the bin will be closed.

3.3   Game-Theoretic Approach

Our game-theoretic approach is named *Stable Pairing Vector Scheduling* (SPVS) and combines algorithms for the stable roommate and the stable marriage problem [6,11] to place items where they feel most "comfortable".

The *stable roommate problem* deals with a set of $2n$ people looking for roommates. Every person has a preference list of all other people. A solution to the problem is a stable matching of n roommate pairs such that no two persons exist who prefer each other over their current partner. The *stable marriage problem* gets as input two sets of n people each of them having a preference list for all people from the other set. The task is again to find a stable matching such that no two persons would prefer to be paired with each other rather than the persons they are paired with. Gale and Shapley [6] proved that the stable marriage problem always has a solution. However, there is no guarantee for a stable matching in the stable roommate problem. Irving [11] designed an algorithm that finds a stable matching or proves that no such matching exists.

The new SPVS heuristic sorts all items in descending order based on the L_1 norm, creates m pairs from the first $2m$ items using Irving's stable roommate algorithm and places the pairs into the m bins. If Irving's algorithm does not produce a solution, the first $2m$ items are assigned using the first-fit heuristic from [12]. The remaining items are assigned in batches of size m using Gale and Shapley's stable marriage algorithm [6]. The algorithm matches the m items of the batch with the m bin loads. If bins cannot take any more items, they are removed from the set and correspondingly fewer items are selected in the next batch. If there are not enough items left to run the stable marriage algorithm, the remaining items are distributed using again the first-fit heuristic.

Two cost functions are needed to compute the preference lists for the bins and items involved in stable roommate and stable marriage. For stable roommate, the cost function is the cosine of the angle θ between the two item vectors. A smaller $\cos(\theta)$ value is preferred because then the items are more complementary to each other and leave more space for additional items. For stable marriage, the cost function is the cosine of the angle ϕ between the bin's remaining capacity and the item vector. A higher value of $\cos(\phi)$ is preferred because it indicates a higher similarity between the vectors, again filling the bin more evenly and leaving potentially more space for future items.

4   Evaluation

This section is dedicated to the evaluation of the algorithms and heuristics for vector scheduling and vector bin packing. It first describes the general settings of the experiments before the simulation results are presented and analyzed.

4.1   Simulator and Test Environment

Our simulator implemented in C++ is a micro-kernel capable of loading different data generators and algorithms which must be specified in a configuration file. Other configurable parameters include the number of bins, their capacities, the number of items and seeds for generating random input. The simulator is available https://bitbucket.org/vsvbp/vssimulator/src/master/.

Most algorithms that we implemented do not require libraries beyond the C++ standard. However, some algorithms solve integer / mixed-integer linear programming problems, for which they use the *IBM ILOG CPLEX Optimizer* version 12.10 [10] or the *Gurobi Optimizer* version 9.1.2 [8].

For the experiments we used random data and a benchmark [4,9]. For each random experiment, ten sets of data were generated to obtain more precise results. For each algorithm, we assured that the items were exactly the same by using the same random number seed.

The tests were performed on a single core of an Intel Core i7-10700 CPU with a frequency of 2.90 GHz. The PC provided 32 GiB RAM in total, ran on Linux Mint 20.2 Uma with a 5.4.0-86 kernel, and the compiler was GCC 10.2.0.

4.2   Data Sets

The random data sets were used for the vector scheduling experiments. They can be grouped into six classes that define how the vectors of the instances are generated. Five of them, C1-C5, are classes originally proposed by Caprara and Toth [4] and also used in [17]. They pick vector coordinates from a class-specific interval $[a; b]$ uniformly at random. The class C6 was created for the evaluation of exact and approximation algorithms. It is used for generating small items that are, however, large enough to reasonably limit the exponential running times.

C1: $[0.1; 0.4]$ **C2:** $[0.001; 1]$ **C3:** $[0.2; 0.8]$
C4: $[0.05; 0.2]$ **C5:** $[0.025; 0.1]$ **C6:** $[0.1; 0.25]$

We used the benchmark of Heßler et al. [4,9] in the vector bin packing experiments. The benchmark consists of 400 instances which can be evenly grouped into ten classes, where each class can be further divided into four subclasses of ten instances. Each instance contains many items of the same type, i.e. identical vectors, and the classes only differ by the item types used. In the subclasses of the first nine classes, the number of item types is 25, 50, 100 and 200, respectively; in the tenth class, it is 24, 51, 99 and 201. When the benchmark was generated, the number of items of each type was uniformly chosen from the set $\{1, ..., 100\}$. The item count of the instances ranges from 1193 to 10229.

4.3 Metric

The main metric used in the vector scheduling evaluation counts how often a certain number of items can be packed into a given number of bins. For every item set size, 10 sets are generated if not otherwise stated. If at least one of these sets can be placed completely, we increment the set size and repeat the experiment. This is also done if at least one experiment for the two previous set sizes was successful to make sure that we do not terminate prematurely. In the experiments with 15, 100 and 1,000 bins the increment is 1, 2 and 20, respectively. The time limit is set to one minute for one run and thus, algorithms taking longer than a minute are left out. This is because we did not observe significant changes in the results by allowing a longer runtime.

For a finer analysis of the simple heuristics we also count successful runs and display them in a table. The exact procedure is described in Sect. 4.5.

In the evaluation of vector bin packing algorithms, the algorithms are compared by the formula introduced in Sect. 4.7 which is based on the number of bins used for packing all the items.

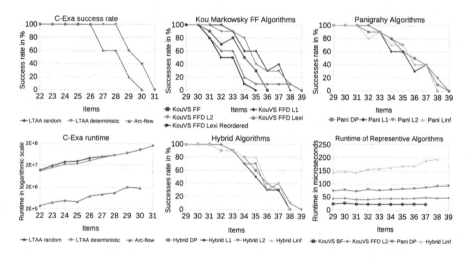

Fig. 1. Results of exact and approximation algorithms for C-Exa and results of heuristics for C1.

4.4 Exact and Approximation Algorithms

We start our evaluation with exact and approximation algorithms. The exact algorithm is Arc-Flow, the approximation algorithms are the random and the deterministic version of the Linear Time Approximation Algorithm (LTAA). Due to their long runtimes we only use 5 bins and slowly increase the number of items which are randomly picked from class C6. For the LTAAs, we choose $\epsilon = 0.1$, which means that the bins have size 1.1 in every dimension instead of 1.

The first plot of Fig. 1 shows that Arc-Flow is capable of scheduling up to 29 items. At first glance it is surprising that LTAA packed one item more than Arc-Flow, but this effect is explained by the larger capacity. The more interesting observations are made in the lower left plot of Fig. 1 which shows that all three algorithms are not suitable for real-time scheduling because of their rapidly increasing running times. Somewhat surprisingly, the approximation algorithms take more time than the exact algorithm. The reason is that the LTAAs only benefit from rounding when the number of items is much larger than the number of item types, which is not the case in these small examples. Without that benefit, rounding is a useless overhead. With an increasing number of items, however, the approximation algorithms would eventually outperform the exact algorithm.

4.5 First-Fit and Best-Fit Heuristics

Next we look at first-fit and best-fit heuristics which are described in Sect. 2.2 and 3.2 and include the new hybrid heuristic. The second plot in the first row of Fig. 1 shows the result for the item-centric FFD heuristics (Kou) using 10 bins and 3D items from class C1. (We do not show the BFD results because they closely resemble the FFD results.) We used the original size functions and added two new ones, L_2 and Lexi Reordered. While the performances differ for the other sorting methods, three methods always performed badly, the one that does not sort the items at all (FF) and, even worse, the ones that sort them in lexicographical order $(Lexi$ and $Lexi$ $Reordered)$, i.e. by the first component, then, if the first components are equal, by the second component, and so on. $Lexi$-$Reordered$ differs from $Lexi$ in that it changes the order of the components before sorting so that the ith component is the ith largest one on average. The better strategies sort the vectors by the L_1, L_2 or L_∞ $(Linf$ in the plots) norm.

Table 1. Comparison of item-centric and bin-centric heuristics. Each number is the percentage of successful runs among all runs in which at least one strategy was successful and one failed.

Items	Met.	10 Bins						50 Bins					
		3D			4D			3D			4D		
		Kou	Pani	Hyb	Kou	Pani	Hyb	Kou	Pani	Hyb	Kou	Pani	Hyb
C1	Linf	31.22	55.36	60.13	30.64	48.87	51.49	14.37	74.46	77.37	09.70	71.25	72.55
	L1	42.62	42.57	44.06	44.72	44.72	45.62	18.12	18.12	22.07	22.16	22.25	24.16
	L2	43.67	70.61	**70.90**	45.26	66.88	68.10	25.07	90.70	**91.92**	22.34	90.22	**91.95**
	DP	-	69.07	67.53	-	**70.31**	67.73	-	88.26	89.30	-	86.32	87.36
C3	Linf	79.27	35.98	54.88	**82.22**	33.33	42.22	64.24	31.13	45.03	61.47	36.45	48.29
	L1	79.27	79.27	79.27	75.56	75.56	75.56	71.52	71.52	71.52	77.40	77.40	77.56
	L2	**85.37**	72.56	71.34	78.89	66.67	71.11	**83.44**	58.28	54.97	**82.99**	64.80	70.73
	DP	-	78.05	79.88	-	74.44	72.22	-	72.85	73.50	-	78.07	79.23
C5	Linf	23.77	88.11	89.86	18.49	87.23	87.32	09.72	91.01	90.93	7.43	91.31	91.65
	L1	23.14	23.14	23.14	22.67	22.68	22.89	13.46	13.45	12.73	10.17	10.10	10.17
	L2	23.93	**93.50**	**93.50**	21.76	93.49	**93.69**	12.46	98.95	**99.06**	9.51	98.57	**98.80**
	DP	-	89.70	90.97	-	89.65	89.84	-	94.10	94.19	-	95.18	95.45

The reason for the poor performance of lexicographical ordering seems to be that it basically sorts all the items by the first component, especially when the vectors are randomly generated, while the other components are more or less in random order. So, the first items assigned will fill the first component of the first few bins while leaving the other components unfilled, and later there are often not enough items with small first components to fill these gaps.

The third plot in the first row and the second plot in the second row of Fig. 1 show results for the bin-centric approaches by Panigrahy et al. (Pani) and the new hybrid heuristics. They also indicate that the ordering of the items has a high impact on the quality of the scheduling. For both schemes, L_2 and *dot product* (DP) are usually better than L_1 and L_∞.

The last plot of Fig. 1 shows the running times. Since the heuristics of each group show basically the same behaviour, we only display one representative. The fastest ones are the item-centric heuristics without sorting, followed by the ones with sorting. The bin-centric and hybrid strategies are slower, but all algorithms perform well below 0.1 milliseconds in the experiments with 10 bins and items from C1. However, for an increasing number of bins, the hybrid algorithms run considerably longer than the other types which can be seen in Fig. 6.

In Table 1 we take a closer look at a selection which includes almost all heuristics except for the ones mentioned above that generally perform badly. All heuristics are run on the same set of 3-dimensional (or 4-dimensional) items randomly selected from the item classes C1, C3 and C5, respectively, and schedule them on 10 (or 50) bins. Each experiment is repeated 100 times. The numbers in the table give the percentage of successful allocations among all inputs for which at least one heuristic failed and at least one heuristic succeeded.

The table shows that the results for the hybrid heuristic are close to the ones of the bin-centric heuristic (Pani), usually outperforming them by a small margin. Most of them perform significantly better than the item-centric heuristics (Kou) for the item classes C1 and C5 as well as C4 which is not shown in the table. Suprisingly, the item-centric heuristics outperform the other ones for class C3 as well as C2 which is not shown. Since C1, C4 and C5 generate only or mostly small items and C2 and C3 generate mostly large items, it seems that the bin-centric and hybrid approaches are more suitable for the former while the item-centric approaches are more suitable for the latter.

However, the quality of the strategies often depends heavily on the size function. It can be observed that the dot product performs well in general. (Note that some table cells are left empty because it cannot be used with the item-centric heuristic.) For all types of heuristics, the L_2 norm appears more reliable than L_1 and L_∞. The L_∞ typically performs poorly in Kou's algorithm for all classes because it cannot effectively capture the size of items with multiple dimensions. In Pani and Hybrid, the performance of L_∞ increases with the size of items, while the performance of L_1 decreases with the size of items.

4.6   Comparison of All Heuristics

In this section we include the more complex heuristics which usually have longer running times. These are the new heuristics LS, LSR and SPVS as well as heuristics from the literature, CNS and SE. From the previous section we only consider the best-performing heuristics: Kou L_2 and Hybrid L_2.

Experiments with 15 Bins. In the first experiments shown in Fig. 2 only 15 bins were used. Every experiment was repeated ten times for each of the item classes C1, C3 and C5. We left out C2 and C4 due to lack of space and because the results are similar to C3 and C5, respectively. The low number of bins allowed to also run the exact algorithm Arc-Flow for comparison (except for class C5). Arc-Flow is clearly better than the heuristics when the vector components of the items are small (C1), but the heuristics are almost optimal when the items are larger (C3). The line of Arc-Flow is identical with the one of CNS and covered by it in the plot. CNS shows a strong performance in all classes, it is only worse than ArcFlow for dataset C1. Apart from CNS, the best heuristics for C1 and C5 are the new local search algorithms LS and LSR. They outperform the hybrid heuristic and SE, and they are significantly better than SPVS and Kou L_2. The local search algorithms perform worse for C3 because large vectors are harder to be swapped in the Swap procedure. The genetic algorithm SE outperforms Kou L_2 and is better than hybrid for dataset C1 but worse for C5.

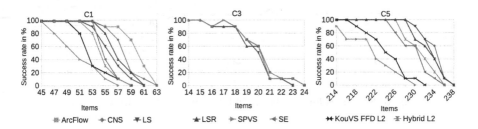

Fig. 2. Successfully scheduled 3d items for 15 bins.

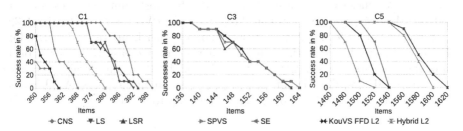

Fig. 3. Successfully scheduled 3d items for 100 bins.

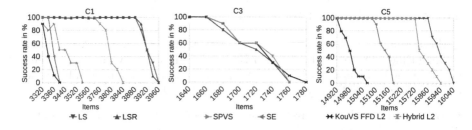

Fig. 4. Successfully scheduled 3d items for 1000 bins.

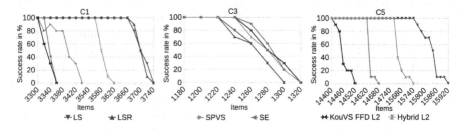

Fig. 5. Successfully scheduled 6d items for 1000 bins.

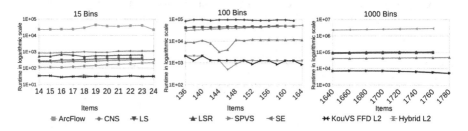

Fig. 6. Runtimes for 15, 100, 1000 bins with 3d items from class C3.

The runtimes of the algorithms are shown in the first plot of Fig. 6 for item class C3. We chose C3 because of the relatively short running times; for C1 and C5, they diverge more strongly. Kou L_2 and SPVS require only about 32 microseconds and are significantly faster than all other heuristics. The price for the improved packings of the hybrids is a runtime increase by a factor of 4 to 6, the price for LS is a factor between 9 and 12. The considerable runtime overhead of SE and CNS is not justified for these small instances. Arc-Flow is several orders of magnitude slower and requires up to 45 ms.

Experiments with 100 Bins. When the algorithms were run on 100 bins, we had to leave out Arc-Flow and, for C5, CNS because of their long running times. Figure 3 shows the success rates for C1, C3 and C5. The results are similar to the ones for 15 bins. All heuristics perform equally well when the items are larger (C3), and the performance can be differentiated in the other cases (C1 and C5). The two local search algorithms are only slightly worse than CNS, followed by the hybrid algorithm which outperforms the remaining algorithms for C1 and C5 and is nearly as good as LS and LSR for C5. The fastest algorithms, SPVS and Kou L_2, achieve the worst packing quality.

The runtimes of the algorithms are shown in the second plot of Fig. 6 for class C3. The relative differences in the runtimes are similar for the other classes. The plot shows the tradeoff between packing quality and running time: the slower the algorithm, the better the quality. However, there are exceptions. Although LSR runs longer than LS, its additional swaps to avoid local optima do not seem to improve the quality. And while the running times of Hybrid are better at the start, they grow faster and eventually exceed the ones of LS.

Table 2. Improvements by LS on the vector bin packing benchmark

Instance	BKLB	BKUB	LS
C4_100_8	642	645	**644**
C4_100_9	642	645	**644**
C4_200_1	1293	1297	**1296**
C4_200_4	1247	1253	**1250**
C4_200_9	1271	1275	**1274**
C5_100_2	325	327	**325***
C5_100_6	314	315	**314***
C5_100_8	321	323	**322**
C5_100_9	316	317	**316***
C5_100_10	327	328	**327***
C5_200_1	652	656	**653**
C5_200_2	624	629	**624***
C5_200_3	630	635	**631**
C5_200_4	630	631	**630***
C5_200_5	631	640	**631***
C5_200_6	626	630	**626***
C5_200_7	630	636	**631**
C5_200_8	635	640	**635***
C5_200_10	632	635	**632***

Experiments with 1,000 Bins. The measurements for 1,000 bins partially confirm the observations of the previous tests, but there are also some differences. We had to leave out LSR and SPVS for C5 because of their long running times. Figure 4 shows that the best success rates for C1 and C5 are achieved by LS. The hybrid is again in the middle, while the remaining algorithms form a third group. In this one SPVS tends to be slightly better than Kou L_2 and SE for C1. The running times for C3 are shown in the third plot of Fig. 6. For 1,000 bins, Hybrid takes much longer than the local search algorithms and SE. The fastest algorithms are again SPVS and Kou L_2.

Impact of Dimensions. In order to see the influence of the number d of dimensions, we ran the experiments also for $d = 6$ instead of $d = 3$. Since the influence is rather insignificant, we show only the plots for 1,000 bins in Fig. 5. Unsurprisingly, 1000 bins did not fit as many $6d$ items as $3d$ items. The local search algorithms are slightly worse for class C3, but still much better than the other algorithms for C1 and C5.

4.7   Vector Bin Packing Results

In this section, we evaluate the performance of non-exact algorithms using the vector bin packing benchmark produced by Heßler et al. [4,9]. SPVS is not included because it is only suitable for vector scheduling. To get the best performance of norm-based algorithms, we apply L_1, L_2 and DP for Hybrid, Pani and Kou and use the best result in each case. Because of the long running time of CNS, we had to limit the running time to one hour; if the algorithm has not terminated by then, the intermediate result is taken. The results shown in Table 3 only include the subclasses with the lowest and highest number of item types due to lack of space.

Table 3. Performance comparison on the vector bin packing benchmark

Class	Avg Size	CNS	LS	Hybrid	Pani	SE	Kou
C1_25	1270	**0.0092**	0.0303	0.0571	0.0564	0.0792	0.1183
C1_200	10203.5	0.0114	**0.0110**	0.0483	0.0492	0.0776	0.1103
C2_25	1270	0	0.0059	0	0.0003	0.0086	0
C2_200	10229	0.3394	0.0054	0.0035	0.0044	0.0083	**0.0013**
C3_25	1270	0	0.0031	0	0.0003	0.0079	0
C3_200	10229	0.1980	**0.0023**	0.0030	0.0042	0.0079	0.0026
C4_25	1270	**0.0008**	0.0071	0.0437	0.0444	0.0673	0.1233
C4_200	10203.5	0.0008	**0.0005**	0.0178	0.0178	0.0670	0.1396
C5_25	1270	**0.0014**	**0.0014**	0.0261	0.0261	0.0614	0.1237
C5_200	10151.1	0.1837	**-0.0053**	0.0036	0.0036	0.0777	0.1426
C6_25	1270	**0.0178**	0.0451	0.0437	0.0431	0.0647	0.0574
C6_200	10167.5	0.3960	**0.0331**	0.0436	0.0434	0.1000	0.0718
C7_25	1270	**0.0116**	0.0171	0.0372	0.0345	0.0323	0.0345
C7_200	9881.1	0.2864	**0.0155**	0.0249	0.0242	0.0292	0.0297
C8_25	1270	0	0	0.0115	0.0446	0.0375	0
C8_200	9503	0.2228	0	0	0.0010	0.0031	0
C9_25	1270	**0.0116**	0.0388	0.0745	0.0760	0.0817	0.1180
C9_200	10203.5	0.0307	**0.0112**	0.0440	0.0449	0.0836	0.1180
C10_24	1193	**0.0171**	0.0490	0.0741	0.0739	0.1192	0.1123
C10_201	9504	0.0291	**0.0270**	0.0579	0.0582	0.0664	0.0833

In Table 3, the first column identifies a subclass of 10 instances in the form of $class\_subclass$. The second column shows the average number of items per instance. The remaining columns display the results for the algorithms, more precisely the ratio between the number of bins used m_i and the best-known upper bound $BKUB_i$, averaged over all instances i; i.e. $\sum_{i=1}^{10} \frac{m_i - BKUB_i}{10 \cdot BKUB_i}$. Negative numbers imply that the upper bound was improved. In each row, the best result is highlighted in bold font. For all subclasses with 24 or 25 item types, CNS achieved the best results. The performance of CNS drops considerably for subclasses with more item types. CNS would generate better results if it ran more hours. For all subclasses with 200 or 201 item types (except for C3), LS outperforms all other algorithms. It also achieved the best results for C5 and C8 in general and was only worse than the best algorithm in all other cases. In C4 and C5, LS improved the best-known upper bound for 19 instances; and 10 of the results match the best-known lower bound (BKLB). For details see Table 2. Kou and Markowsky's algorithms and the hybrid algorithms performed very well for C2, C3, and C8.

Overall, the benchmark results resemble the results of the vector scheduling experiments. They suggest that LS and Hybrid are also competitive for the vector bin packing problem.

5 Conclusion

In this work we have presented three new algorithms for vector scheduling and vector bin packing and evaluated them together with a range of algorithms from the literature. The algorithms developed and chosen are of different types and include local search, genetic, stable-pairing, first-fit, best-fit, approximation and exact algorithms. The exact algorithms and the approximation schemes considered are as expected not suitable for practical purposes because of their long running times. Among the heuristics, the new local search algorithms outperformed all other algorithms except for CNS in terms of success rates for vector scheduling. They also improved results for 19 instances of a vector bin packing benchmark. While they are an order of magnitude slower than the fastest heuristics, they are much faster than CNS and have acceptable running times for many use cases. Our hybrid schemes combine ideas of bin-centric and item-centric heuristics from the literature and have slightly better success rates than those. The last new approach uses ideas from game theory to pack vectors into bins and is among the fastest algorithms in most cases. It tends to be better than equally fast first-fit schemes for larger instances.

References

1. Bansal, N., Oosterwijk, T., Vredeveld, T., van der Zwaan, R.: Approximating vector scheduling: almost matching upper and lower bounds. Algorithmica **76**, 1077–1096 (2016)
2. Brandão, F., Pedroso, J.P.: Bin packing and related problems: general arc-flow formulation with graph compression. Comput. Oper. Res. **69**, 56–67 (2016)
3. Buljubašić, M., Vasquez, M.: Consistent neighborhood search for one-dimensional bin packing and two-dimensional vector packing. Comput. Oper. Res. **76**, 12–21 (2016)
4. Caprara, A., Toth, P.: Lower bounds and algorithms for the 2-dimensional vector packing problem. Discret. Appl. Math. **111**(3), 231–262 (2001)

5. Chekuri, C., Khanna, S.: On multidimensional packing problems. SIAM J. Comput. **33**(4), 837–851 (2004)
6. Gale, D., Shapley, L.S.: College admissions and the stability of marriage. Am. Math. Mon. **69**(1), 9–15 (1962)
7. Garey, M.R., Johnson, D.S.: Computers and Intractability: A Guide to the Theory of NP-Completeness. Freeman, W. H (1979)
8. Gurobi Optimization Inc.: Gurobi Optimizer (2022)
9. Heßler, K., Gschwind, T., Irnich, S.: Stabilized branch-and-price algorithms for vector packing problems. Eur. J. Oper. Res. **271**(2), 401–419 (2018)
10. IBM: IBM ILOG CPLEX Optimizer (2022)
11. Irving, R.W.: An efficient algorithm for the stable roommates problem. J. Algorithms **6**(4), 577–595 (1985)
12. Kou, L.T., Markowsky, G.: Multidimensional bin packing algorithms. IBM J. Res. Dev. **21**(5), 443–448 (1977)
13. Kulik, A., Mnich, M., Shachnai, H.: An Asymptotic $(4/3+\epsilon)$-approximation for the 2-dimensional Vector Bin Packing Problem. preprint arXiv:2205.12828 (2022)
14. Mannai, F., Boulehmi, M.: A guided tabu search for the vector bin packing problem. In: The 2018 International Conference of the African Federation of Operational Research Societies (AFROS 2018) (2018)
15. Masson, R., et al.: An iterated local search heuristic for multi-capacity bin packing and machine reassignment problems. Expert Syst. Appl. **40**(13), 5266–5275 (2013)
16. Pandit, D., Chattopadhyay, S., Chattopadhyay, M., Chaki, N.: Resource allocation in cloud using simulated annealing. In: 2014 Applications and Innovations in Mobile Computing (AIMoC), pp. 21–27. IEEE (2014)
17. Panigrahy, R., Talwar, K., Uyeda, L., Wieder, U.: Heuristics for vector bin packing. research. microsoft. com (2011)
18. Sadiq, M.S., Shahid, K.S.: Optimal multi-dimensional vector bin packing using simulated evolution. J. Supercomput. **73**(12), 5516–5538 (2017)
19. Sandeep, S.: Almost optimal inapproximability of multidimensional packing problems. In: 2021 IEEE 62nd Annual Symposium on Foundations of Computer Science (FOCS), pp. 245–256. IEEE (2022)
20. Wilcox, D., McNabb, A., Seppi, K.: Solving virtual machine packing with a reordering grouping genetic algorithm. In: 2011 IEEE Congress of Evolutionary Computation (CEC), pp. 362–369. IEEE (2011)
21. Woeginger, G.J.: There is no asymptotic PTAS for two-dimensional vector packing. Inf. Process. Lett. **64**(6), 293–297 (1997)
22. Wood, T., Shenoy, P.J., Venkataramani, A., Yousif, M.S.: Black-box and gray-box strategies for virtual machine migration. In: 4th Symposium on Networked Systems Design and Implementation (NSDI). USENIX (2007)

Unleashing the Potential of Restart by Detecting the Search Stagnation

Yoichiro Iida[1]([✉]), Tomohiro Sonobe[2], and Mary Inaba[1]

[1] Graduate School of Information Science and Technology, The University of Tokyo,
Tokyo, Japan
yoichiro-iida@g.ecc.u-tokyo.ac.jp, mary@is.s.u-tokyo.ac.jp
[2] National Institute of Informatics, Tokyo, Japan
tomohiro_sonobe@nii.ac.jp

Abstract. SAT solvers are widely used to solve industrial problems owing to their exceptional performance. One critical aspect of SAT solvers is the implementation of restarts, which aims to enhance performance by diversifying the search. However, it is uncertain whether restarts effectively lead to search diversification. We propose to adapt search similarity index (SSI), a metric designed to quantify the similarity between search processes, to evaluate the impact of restarts. Our experimental findings, which employ SSI, reveal how the impact of restarts varies with respect to the number of restarts, instance categories, and employed restart strategies. In light of these observations, we present a new restart strategy called Break-out Stagnation Restart (BroSt Restart), inspired by a financial market trading technique. This approach identifies stagnant search processes and diversifies the search by shuffling the decision order to leave the stagnant search. The evaluation results demonstrate that BroSt Restart improves the performance of a sequential SAT solver, solving 19 more instances (+3%) than state-of-the-art solvers.

Keywords: SAT problem · SAT solver · restart · search similarity

1 Introduction

SAT solvers, particularly conflict-driven clause learning (CDCL) [3,15,17] solvers, are widely used in industry and academia to solve SAT problems. The *restart* is a critical technique in CDCL SAT solvers. A search of CDCL-solver consists of *decision, propagation,* and *backtrack* steps. *Restart* is the extreme *backtrack* because it deletes the entire decision history and resumes the search with a different decision assignment. One of the biggest objectives of *restart* is to diversify search activity, i.e., to attempt different search assignments. While deleting the search history sounds unreasonable, the effectiveness of restarts has been practically proved in multiple researches. Many restart strategies have been proposed, such as Luby [19], EMA [5], and MLR [13]. The primary focus of restart research has been on optimizing when to trigger a restart because its

© The Author(s), under exclusive license to Springer Nature Switzerland AG 2023
M. Sellmann and K. Tierney (Eds.): LION 2023, LNCS 14286, pp. 599–613, 2023.
https://doi.org/10.1007/978-3-031-44505-7_40

operation is quite simple — it cancels all variable assignments. To effectively coordinate the timing of restarts, high-frequency evaluation of the solver states is required during search. Therefore, states of search are assessed via lightweight processing, generally using heuristic methods based on simple features such as the number of conflicts or learnt clauses. We believe that the effect of search diversification by restart can be enhanced by a more precise analysis of search situations than simple features, thereby improving solver performance. Indeed, some studies [9,16] have attempted this in the context of portfolio-type parallel solvers. Portfolio solvers require search diversification for efficient paralleliza-tion. Specifically, search similarity index (SSI) [12] demonstrated the perfor-mance improvement of a parallel solver through the avoidance of the redundant parallel searches by checking search similarity.

We posit that the SSI has the potential to (1) quantitatively analyze the impact of restarts of sequential search and (2) enhance the effectiveness of restart as well as improve the solver performance. The analyses aim to provide insights into: the impact of diversification by single and accumulated restarts; the vari-ations in impact for different SAT instance categories and their satisfiability; and the differences caused by the chosen restart strategy. Next, considering the utilization of SSI to improve the solver in-process, computational expensiveness of SSI is the challenge. We resolve this problem by evaluating the overall trend of search not at a single restart but across multiple restarts. Our proposal con-tains a new way to detect the stagnation of search during some restarts, i.e., the situation of non-diversified search, using SSI. We call our proposal restart strat-egy as the Break-out Stagnation Restart ("BroSt Restart" or simply "BroSt"), employing a classical technical analysis method from financial market trading [18] to identify the trendless situation. Then, when we detect the stagnation, we initiate to change the order of the decision plan drastically to diversify the search. As the authors of ManySAT [10] state, "Contrary to the common belief, restarts are not used to eliminate heavy-tailed phenomena since after restarting SAT solvers dive in the part of the search space that they just left". In practice, a normal restart does not guarantee to diversify. This is why we assume that our Brost Restart strategy has potential to improve the performance of SAT solver.

In the remainder of this paper, Sect. 2 introduces techniques related to SAT solvers; Sect. 3 presents our analysis using SSI of the impact of restarts and differ-ences caused by instance type and restart strategy; Sect. 4 presents BroSt Restart and experimentally evaluates it; Sect. 5 discusses related work; and Sect. 6 sum-marizes the paper and discusses future research.

2 Preliminaries

2.1 SAT Problem and SAT Solver

The SAT problem determines whether a given propositional logic formula can be satisfied by a Boolean value assignment of variables. The formula is typically given in conjunctive normal form, where variables are combined into clauses with disjunctions, and the clauses are combined with conjunctions. If there is at least one valid assignment that satisfies all clauses, the formula is considered satisfiable (SAT), otherwise, it is unsatisfiable (UNSAT). The SAT problem is a famous NP-complete problem and considered difficult to solve in practical time when the formula has many variables. SAT solvers are applications to solve SAT problems and have been improved over decades through annual SAT competitions [22], etc. State-of-the-art solvers can often solve problems with more than one million variables and clauses. Owing to this efficiency, they are used to solve real-world problems such as mathematical problems [11] and binary neural network verification [6] encoded as SAT problems. The study and further improvement of SAT solvers is hence highly important.

2.2 Techniques of CDCL SAT Solvers

Most applicational SAT solvers are based on the Davis–Putnam–Logemann–Loveland (DPLL) algorithm [7]. DPLL searches for a SAT assignment or proof of UNSAT by repeating three steps: *decision*, *propagation*, and *backtrack*. In a *decision*, a variable without any assigned Boolean value is selected, and a true or false Boolean value is assigned to the variable as an assumption. Solvers employ heuristic methods to complete this step. During *propagation*, Boolean values are assigned to some variables whose value is decidable as the logical consequence of previous decisions and propagations. The solver repeats these steps until it finds a logical inconsistency (called a *conflict*) or a result concerning the satisfiability. When the solver finds a *conflict*, it proceeds to the *backtrack* step. *Backtrack* cancels the previous incorrect decision and subsequent propagations and returns to the *decision* step to attempt another assumption.

The conflict-driven clause learning (CDCL) SAT solver is based on the DPLL algorithm with some critical additions such as the *learnt clause* [3,15] and *restart* [8,14]. CDCL solver analyzes the root cause of the conflict and derives a new clause (called the *learnt clause*) to avoid the same conflict in subsequent searches. *Restart* is the extreme type of *backtrack*. It cancels all previous decisions and propagations and initiates a search again. The solver maintains efficiency by retaining information from previous searches, such as the learnt clauses and the effective polarities of Boolean value assignments.

2.3 Restart Strategies

Although the restart has been practically proven to substantially improve solver performance, its underlying theoretical explanation remains a subject of ongoing research. Initially, restart was introduced to infuse randomness into the search [8] to smooth runtime fluctuations due to variable assumptions such as decision order and Boolean value polarity. Restarts are also believed to foster learning efficiency. Longer decision trees often generate longer learnt clauses with more literals. As restarts erase the decision tree, they help retrieve shorter clauses which are believed to prune the search space more efficiently than longer ones and enhance the solver's performance. Moreover, restarts help obtain a wider variety of learnt clauses through different assignments. Therefore, the objective of executing a restart can be paraphrased as "search diversification".

Various heuristic restart methods have been proposed. The most basic statistical method is the uniform method, a constant interval policy [17]. It initiates restarts when the solver reaches a certain number of intervals, with conflicts typically serving as intervals. MiniSAT 1.14 [19] adopts a policy that geometrically increases the intervals as $c * a^i$, where c is the initial interval, a denotes the increment value, and i denotes the number of executed restarts. Its successor, MiniSAT 2.0, employs the Luby restart, which is based on the randomized heuristic Luby algorithm [14]. This algorithm generates the following sequence of periodic values: 1, 1, 2, 1, 1, 2, 4, 1, 1, 2, 1, 1, 2, 4, 8, ... The value of this sequence multiplied by the base constant value (e.g., 32 conflicts) yields the intervals. EMA [5] is a state-of-the-art restart strategy. Its interval is determined based on the concept of exponential moving average, as $EMA(n, \alpha) = \alpha \sum_{i=0}(1 - \alpha)^i \times t_{n-i}$, where n denotes the number of current conflict, α is a smoothing parameter, and t_i donates the LBD score of learnt clause in the i-th conflict.

2.4   Search Similarity Index

This section briefly introduces SSI [12], originally proposed for comparing searches of parallel nodes. We adapt the idea to sequential solver to quantify the impact of restart. The basic idea of SSI is to compare two search points and quantify how different they are. Therefore, it requires the definition of the search, but generally search contains several kinds of states. We assumed and simplified that a search can be represented by the value-assignment plan in the *decision* because same value-assignment results in the same decision and thus same search. We named the set of information of the value-assignment plan the current search direction (CSD) because the information determines the direction of the future search. CSD_i denotes the plan at the ith step while searching (e.g., the ith restart or ith conflict). Two similar CSDs indicate that the searches are similar and vice versa. $SSI_{i,j}$ is calculated by comparing CSD_i and CSD_j for arbitrary i and j. The value assignment in the *decision* can be decomposed into *polarity* and *priority* of each variable. —*polarity* is the prospective of Boolean value assignment of variables, and *priority* is the order of variables used in the coming decisions. Therefore, CSD_i consists of $polarity_i(v)$ and $priority_i(v)$ s.t. $v \in V$ (all variables). Moreover, $polarity_i(v)$ is defined as $\{True, False\}$, and $priority_i(v)$ is defined as $(order\ of\ variable\ v)/|V_i|$, where $|V_i|$ is the set of valid variables at the ith step in the search.

The SSI is defined as the weighted sum of variable similarities. The variable similarity of v between i and j, $similarity_{i,j}(v)$, is defined as the similarity of $polarity(v)$ and $priority(v)$ between i and j. The similarity of $polarity_{i,j}(v)$ is 1 if $polarity_i(v) = polarity_j(v)$; otherwise, it is 0 because we assumed that a different Boolean assignment results in a totally different search. The similarity of $priority_{i,j}(v)$ is $1 - |priority_i(v) - priority_j(v)|$, indicating the difference of decision order. *Importance* is a weight factor for $similarity_{i,j}(v)$ because we assumed that some variables are important to determine the search while some others can be decided easily by others. Here, we provide higher *importance* scores to higher *priority* variables, defining $importance_{i,j}(v)$ as $2^{-priority_i \times c} + 2^{-priority_j \times c}$, where the value of constant c is set to 0.1. To use SSI as a metric, its value is normalized to a value between 0 and 1 by dividing it by the sum of *importance*. 0 represents zero similarity, whereas 1 represents identical searches. In conclusion, $SSI_{i,j}$ is defined as (1).

In this paper, we define $SSI_{|k|}$ as the average value of $SSI_{a,a+k}$, where a denotes all intra-search restarts unless specifically mentioned. As $SSI_{|k|}$ is the similarity between two searches k restarts apart, it represents the average impact of diversification of k restarts.

$$SSI_{i,j} := normalize \sum_{v} \{similarity_{i,j}(v) \times importance_{i,j}(v)\} \tag{1}$$

3   Restart Analysis

We conducted experiments to analyze the impact of search diversification by restart using SSI. We used CaDiCaL [1] as the base solver. The solver finished its search when it found a solution or 3600 s of search time elapsed (excluding SSI computation time for the analysis in this section). As the benchmark, 400 industrial instances from the main track in SAT competition 2021 [22] were used.

3.1   Impact of Restart on Search Similarity

Impact of a Single Restart. First, we experimentally confirmed that restart diversifies the search. It is generally assumed that the search changes as it progresses. This change can be captured as the change in search similarity as the value of SSI. Assume an SSI value $SSI_{conf_i,conf_j}$ for two CSDs at two conflicts $conf_i$ and $conf_j$ that include j - i conflicts between them. $SSI_{i,j}$ denote $SSI_{conf_i,conf_j}$ in this experiment. If restart diversifies the search, there should be a difference in the SSI value of $SSI_{i,j}$ and $SSI_{i',j'}$, where $SSI_{i,j}$ includes a restart between conflicts i and j and $SSI_{i',j'}$ does not. We denote these SSI values as $SSI_{i,j}^r$ and $SSI_{i',j'}^{nr}$, where r means that the SSI contains a restart and nr means that it does not standing for "no restart". We set the interval between i and j to 100 conflicts (i.e., $j - i = 100$). Figure 1 illustrates the SSI^r and SSI^{nr}. We adopted uniform restart policy and fixed the restart interval to 128 conflicts to ensure at most one restart during the 100 conflicts. Consider c, the index of conflicts that a restart is triggered. We stored every CSD at every conflict and every c. We calculated two SSIs for any c, where $(i,j) = (c-1, c+99)$ and $(i',j') = (c - 101, c - 1)$. For readability, we use SSI^{nr} and SSI^r for the average values of the $SSI_{i,j}^r$ and $SSI_{i',j'}^{nr}$ of all c, respectively. A comparison of SSI^r and SSI^{nr} quantifies the average impact of restarts over the same interval (100 conflicts). Assuming restart diversifies the search, SSI^r should be less than SSI^{nr}.

Figure 2 presents the results. 67 instances containing fewer than 1000 restarts were excluded. This is because we assumed that the value of SSI fluctuates; thus,

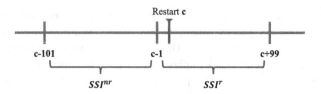

Fig. 1. Illustration of SSI^r and SSI^{nr} around a restart at conflict index c

only 1000 restarts were too small to find meaningful insight statistically. The average SSI values of an instance are plotted against instances sorted by SSI^r. The result indicates that SSI^r was less than SSI^{nr} in most instances, implying that restart diversifies the search. The average SSI^r and SSI^{nr} values of all instances are 0.78 and 0.80, respectively, and only 12 instances had an SSI^r higher than the SSI^{nr} (3.6% of 337 instances).

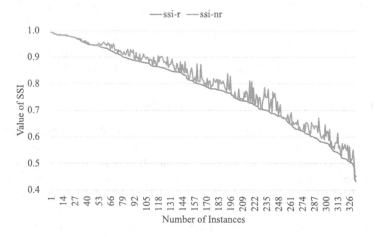

Fig. 2. Values of SSI^r and SSI^{nr} across instances

Impact of Accumulated Restarts. Next, we calculated $SSI_{|1|}$, $SSI_{|10|}$, and $SSI_{|100|}$ as defined in Sect. 2.4 and compared them to analyze the impact of accumulated restarts. Recall that $SSI_{|x|}$ is the average of $SSI_{i,j}$ of an arbitrary i, where $j - i = x$, and hence this comparison quantifies the degree of diversification obtained by accumulated x restarts (i.e., by $x = 1, 10, 100$ restarts).

Figure 3 presents the results, where the average SSI values of an instance are plotted against instances sorted by $SSI_{|1|}$. 35 instances are excluded from the result since they have no restart observed in their search. Our observations are as follows. First, as restarts accumulate, the search changes further, as expected. Second, the impact of restart varies substantially by instance. For example, instance E02F17.cnf yielded 0.95, 0.83, and 0.77 for $SSI_{|1|}$, $SSI_{|10|}$, and

$SSI_{|100|}$, respectively, whereas size_5_5_5_i223_r12.cnf yielded 0.81, 0.51, and 0.41. This supports Oh [20], who stated that the effectiveness of restart differs for each problem. Lastly, the impact of accumulated restarts decays (i.e., the gap between $SSI_{|10|}$ and $SSI_{|100|}$ is generally less than that of $SSI_{|1|}$ and $SSI_{|10|}$).

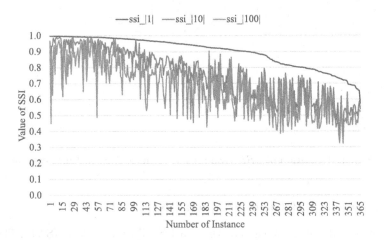

Fig. 3. Values of $SSI_{|1|}$, $SSI_{|10|}$, $SSI_{|100|}$ across instances. These three SSI values for an instance are plotted on the same vertical axis, and instances are sorted based on the value of $SSI_{|1|}$

3.2 Differences Among SAT Instance Categories

We further investigated the difference in the impact of restart and its decay. We categorized instances based on their families and observed how $SSI_{|x|}$ changes, where $x \in \mathbb{N}$ and $x \leq 1000$. In this experiment, i was fixed at $i = \{1000n + 1 | n \in \mathbb{Z}_{\geq 0}\}$, and j was an integer ranging from $i + 1$ to $i + 1000$. Note that i and j represent the i-th and j-th restart in a search. We picked i only at the $1000n + 1$-th restart to reduce computation. We depict the dots of $x = \{1, 2, .., 9, 10, 20, .., 90, 100, 200, .., 900, 1000\}$ and use a log-scale x-axis in Fig. 4.

Figure 4 presents the results. Note that we selected categories that include at least 10 instances and plotted the average $SSI_{|x|}$ in each category. The SSI is plotted against x, revealing that SSI decreases as the number of restarts increases for all categories. However, the degree and timing of the decrease depends on the category. For example, category "vlsat2" drops gradually: 0.97, 0.94, 0.90, and 0.87 at $x = 1, 10, 100, 1000$, respectively. Category "edit_distance" first drops sharply, then decreases more slowly from $x = 100$, 0.91, 0.62, 0.52, and 0.50 at $x = 1, 10, 100, 1000$. The decrease in category "at-least-two" accelerates as x increases, 0.98, 0.92, 0.87, and 0.79 at $x = 1, 10, 100, 1000$. This confirms the necessity of dynamic restart strategies.

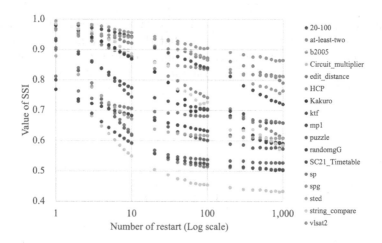

Fig. 4. Transition of $SSI_{|x|}$ values according to x by instance category

3.3 Differences Between SAT and UNSAT

We categorized instances based on their satisfiability (SAT or UNSAT). The scatter plot in Fig. 5 shows the variance in the $SSI_{|100|}$ of each instance. The vertical axis is $SSI_{|100|}$, and the points are colored according to their satisfiability. The horizontal axis is search runtime, and dashed lines indicate the linear prediction of each category. Instances that the solver could not solve were excluded.

The instances solved in less than 1000 s runtime— relatively easy instances— have widely dispersed SSI values (0.4–.9) both for SAT and UNSAT. However, as runtime increases— more difficult instances— the SSI values of the SAT instances remain dispersed, but higher SSI values are observed for UNSAT instances. This trend is reflected in the prediction lines. This result implies that a high-similarity search — intensive search — is essential to solving UNSAT instances, whereas an extensive search is required for SAT. This corroborates with the result of [20] and explains why gradually reducing the frequency of restarts as search progresses improves performance; this allows the solver to focus on the current search space intensively for a longer period as runtime increases. Furthermore, it suggests new restart strategies: restarts to ensure more extensive search to solve SAT instances and intensive search for UNSAT ones.

3.4 Differences in Restart Strategies

Lastly, we compared three well-known restart strategies, *uniform*, *Luby*, and *EMA*. *Uniform* invokes a restart at constant intervals; in this experiment, we used 128 conflicts. *Luby* uses a sequence generated by the Luby algorithm as intervals, with an initial interval of 32 conflicts. *EMA* is a state-of-the-art strategy used in CaDiCaL. We used the default settings of EMA [5].

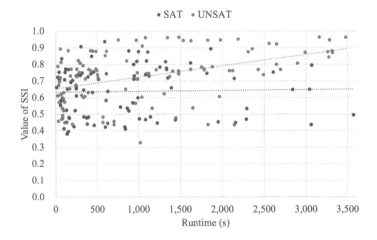

Fig. 5. Variance in $SSI_{|100|}$ of each instance according to satisfiability

Figure 6 shows the variance in $SSI_{|x|}$ of instances as a boxplot, where $x = \{1, 10, 100, 1000\}$, for the three restart strategies. The mean $SSI_{|1|}$ of EMA is the largest, and its variance is the smallest. However, its mean and variance for $SSI_{|1000|}$ are largely equivalent to the other results. Importantly, this implies that EMA changes the search in 1000 restarts faster than the others. Because EMA performed more frequent restarts than the others (only nine conflicts on average were observed between consecutive EMA restarts), EMA's 1000 restarts required fewer conflicts than the other methods, generally reducing runtime. This explains at least one reason for the effectiveness of EMA.

4   Proposal and Evaluation – BroSt Restart

We propose a new restart strategy, BroSt Restart strategy, in this section and experimentally evaluate the effectiveness of it.

4.1   Observations

Figure 7 depicts the typical change in SSI values of randomly sampled instances. The light-blue line indicates the actual value of $SSI_{i,j}$, where i is fixed at the 10,001st restart because the initial search can fluctuate substantially. The values of j are arbitrary integers from 10,002 to 20,001, sequentially. The red dashed line indicates the moving average of 100 SSI values. This preliminary experiment was conducted under the same conditions as the previous ones.

These results yield several observations. First, search similarities gradually decrease as search progresses overall. Here, progress is represented by the number of restarts. Second, after a sufficient number of restarts, this trend becomes flat or narrowly changes. Third, but most importantly, the trend is not a steady decline

608   Y. Iida et al.

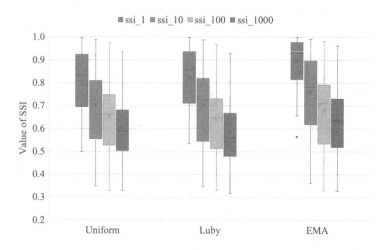

Fig. 6. Variance in the $SSI_{|x|}$ of instances by strategy

but can plateau in places. We posit that these plateaus indicate that the search remains stagnant, and avoiding them can reinforce the objective of restarts — diversifying the search — and ultimately improve solver performance. Note that these plateaus and declines rely on the instance. Whereas Fig. 7b indicates a relatively gradual decrease over 2000 restarts, Fig. 7c is overall flat.

4.2   BroSt Restart

To break out the stagnation in search observed in Sect. 4.1, we propose break-out stagnation restart, BroSt, which consists of two parts: detecting search stagnation and breaking it. For detection, we are inspired by a classical method of technical analysis from financial market trading to detect stagnant situation in a chart. Stagnation occurs when there are no clear upward or downward trends in the market, called trendless in [18]. The method uses the historical maximum (resistance line or peak) and minimum values (support line or troughs) to identify if the market is stagnant (i.e., if neither value is updated for a while, the market is assumed to be stagnant). The previous observation revealed both up-down and flat trends in the SSI values that are similar to trends in market charts. Therefore, we assume that a comparison of the current SSI value with the historical maximum and minimum values can detect the trendless situation in the SSI value, that is, "search stagnation". To break out a stagnant situation, we use the decision order shuffling function that is included in the original CaDiCaL. The shuffling function swaps the order of variables in the decision queue. Changing the decision order directly affects the performance of solver, and we show that shuffling using the default option reduced performance drastically in the following experiment. Thus, properly triggering the shuffling function is essential and is achieved by the detection method described above.

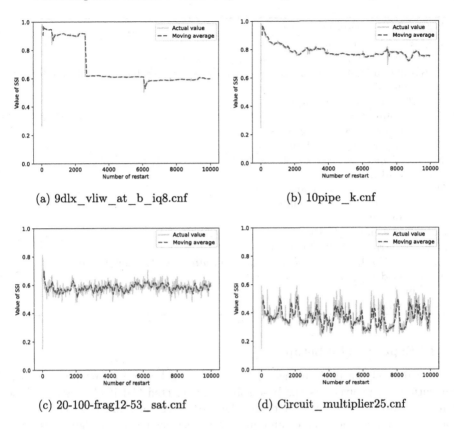

(a) 9dlx_vliw_at_b_iq8.cnf (b) 10pipe_k.cnf

(c) 20-100-frag12-53_sat.cnf (d) Circuit_multiplier25.cnf

Fig. 7. SSI values and their moving average over 10,000 restarts

Algorithm 1 presents the pseudocode of BroSt Restart. During the search, the SSI of the last saved CSD and current CSD is calculated every 10 restarts. We set 10 restarts as interval because of the calculation cost for SSI. The SSI between every restart costs too much, and we assumed that it deteriorates the performance of the solver. However, every 100 restarts are too long as interval. The setting of this frequency has much room to improve. The CSD is saved at the first restart or just after the previous shuffling. The maximum and minimum SSI values, max_{ssi} and min_{ssi}, are stored and updated during the search. If the calculated SSI value remains between max_{ssi} and min_{ssi} for a certain duration, the solver triggers the shuffling function. In this research, we used c restarts as the duration, where c is set to 300. We selectively decided to use 300 through preliminary experiments to determine c by testing 100, 300, 1000 to check the frequency of Brost detection. After shuffling, the saved CSD is updated to the current CSD, and the counter, max_{ssi}, and min_{ssi} are reset. Because the calculation of SSI is computationally expensive and proportional to the number of variables, we did not execute BroSt for instances with more than 5 million variables.

Algorithm 1. BroSt Restart

Require: limit of counter c
1: initialize $CSD_{saved} \leftarrow CSD_1$, $max_{ssi} \leftarrow 0$, $min_{ssi} \leftarrow 1$, counter $\leftarrow 0$
2: **while** search at $restart_i$ s.t. $i \geq 1$ and number of variables ≤ 5 millions **do**
3:     counter++
4:     **if** i (mod 10) $= 0$ **then**
5:         obtain CSD_i
6:         calculate $SSI_{i,saved}$ as ssi
7:         **if** $ssi > max_{ssi}$ **then**
8:             $max_{ssi} \leftarrow ssi$, counter $\leftarrow 0$
9:         **end if**
10:        **if** $ssi < min_{ssi}$ **then**
11:            $min_{ssi} \leftarrow ssi$, counter $\leftarrow 0$
12:        **end if**
13:        **if** counter $> c$ **then**
14:            trigger shuffle function
15:            $CSD_{saved} \leftarrow CSD_i$, $max_{ssi} \leftarrow 0$, $min_{ssi} \leftarrow 1$, counter $\leftarrow 0$
16:        **end if**
17:    **end if**
18: **end while**

4.3 Experimental Setup

We evaluated the performance of a solver using BroSt Restart by comparing it with a base solver. As the base solver, we selected CaDiCaL, a state-of-the-art sequential SAT solver that won the 2019 SAT race and has been used as a base solver in the SAT community since 2020 owing to its high source-code readability, structural simplicity, and prominent performance. Experiments were conducted on a computer with an AMD Threadripper Pro 3995WX processor (64 core) and 512GB (128GB 4 slots, DDR4-3200MHz) RAM. A total of 1200 instances from the main tracks of SAT competitions 2020–2022 (400 instances per year) were used for the benchmark. The necessary functions for SSI and BroSt Restart were added on top of the base solver. For other configurations, the default CaDiCaL configuration was used. The performance was evaluated in terms of the number of instances solved within a time limit (3600 s on the CPU clock) and the PAR-2 score, which is defined as the total time required to solve all instances with a penalty (additional 3600 s) for each unsolved instance.

4.4 Evaluation

Table 1 summarizes the results of the experiments. We ran three solvers: Base (CaDiCaL 1.4.1 without any options), Shuffle (CaDiCaL 1.4.1 with the"shuffle" option *-shuffle=true*), and BroSt (CaDiCaL 1.4.1 plus BroSt Restart). The number of instances identified as SAT or UNSAT for each year's benchmark is reported along with the PAR-2 score.

The results for all benchmarks indicate that the BroSt Restart is superior to the base solver. Greater improvement is observed for the SAT instances, which is

Table 1. Number of instances solved and PAR-2 scores for each solver

Solver	sc20		sc21		sc22		Total	PAR-2
	SAT	UNSAT	SAT	UNSAT	SAT	UNSAT		
Base	110	94	111	**124**	128	112	679	4,325,684
Shuffle	96	91	104	119	121	108	639	4,625,936
BroSt	**116**	**96**	**117**	**124**	**130**	**115**	**698**	**4,198,437**

consistent with the analysis in Sect. 3.3. A SAT instance can have many solutions distributed throughout the search space, and the proposed method diversifies the search by breaking out the stagnation. This helps the solver to encounter at least one solution with a higher probability than the base solver. By contrast, UNSAT proof is achieved by the resolving of an empty clause, and it is believed to require good learning. Search diversification using our method does not necessarily contribute to this. The deteriorated performance of the Shuffle solver verified that the improvement of BroSt comes from its algorithm, not just the shuffle function.

5 Related Work

In addition to the three methods analyzed above, many dynamic restart strategies have been proposed. Ryvchin and Strichman [21] used conflict frequency within a specific period as the criterion to increase conflict encounters. Biere [4] focused on the number of Boolean polarity switches, initiating restarts when it reaches a threshold. Sinz and Iser [23] used many features such as conflict information, decision levels, and backtrack levels to control to trigger restarts. Audemard and Simon [2] secured intensive searches by stopping restart triggers when the search depth exceeds a criterion. Liang et al., [13] integrated machine-learning techniques into dynamic restarts. While most restart research focuses on controlling the timing restart triggers, counter implication restart (CIR) [24] attempts to directly secure sufficient diversification after a restart. To do this, it analyzes the implication graph and directly changes value assignments.

6 Conclusion

In this study, we used SSI to quantify the level of diversification achieved through restarts. Our analysis revealed that the effect of restarts varies depending on the instance, its family, and its satisfiability (SAT or UNSAT). Moreover, different types of search activity are required to solve SAT and UNSAT instances, with SAT instances generally requiring more extensive search, whereas UNSAT instances require more intensive search. Finally, existing restart methods exhibit varying degrees of impact on search diversification. Building on these insights, we confirmed that SSI is applicable to restart, and we hence proposed the BroSt Restart strategy to maximize the effect of diversification. Experiments demonstrated that BroSt Restart improves the performance of sequential SAT solvers, solving 20 more instances (+3%) than state-of-the-art solvers. It implies that the effect of diversification obtained by existing restart strategies can be insufficient. There remains scope for further improvements in our proposed method. We assume that the search shuffling function and configurations such as judgment constant c could be refined by dynamically tuned parameters. Furthermore, while our proposal strategy was more effective on SAT instances, our insights could lead to a new restart strategy for UNSAT instances that considers their higher need for intensive search.

References

1. Armin, B., Fazekas, K., Fleury, M.: CaDiCaL. https://github.com/arminbiere/cadical. Accessed 26 Jan 2023
2. Audemard, G., Simon, L.: Refining restarts strategies for SAT and UNSAT. In: Milano, M. (ed.) CP 2012. LNCS, pp. 118–126. Springer, Heidelberg (2012). https://doi.org/10.1007/978-3-642-33558-7_11
3. Bayardo, R.J., Schrag, R.C.: Using CSP look-back techniques to solve real-world sat instances. In: Proceedings of the Fourteenth National Conference on Artificial Intelligence and Ninth Conference on Innovative Applications of Artificial Intelligence, pp. 203–208. AAAI 1997/IAAI 1997, AAAI Press (1997)
4. Biere, A.: Adaptive restart strategies for conflict driven SAT solvers. In: Kleine Büning, H., Zhao, X. (eds.) SAT 2008. LNCS, vol. 4996, pp. 28–33. Springer, Heidelberg (2008). https://doi.org/10.1007/978-3-540-79719-7_4
5. Biere, A., Frohlich, A.: Evaluating CDCL restart schemes. In: Proceedings of Pragmatics of SAT 2015 and 2018. EPiC Series in Computing, vol. 59, pp. 1–17. EasyChair (2019). https://doi.org/10.29007/89dw
6. Bright, C., Kotsireas, I., Heinle, A., Ganesh, V.: Complex golay pairs up to length 28: a search via computer algebra and programmatic SAT. J. Symb. Comput. **102**, 153–172 (2021). https://doi.org/10.1016/j.jsc.2019.10.013
7. Davis, M., Logemann, G., Loveland, D.: A machine program for theorem-proving. Commun. ACM **5**(7), 394–397 (1962). https://doi.org/10.1145/368273.368557
8. Gomes, C.P., Selman, B., Kautz, H.: Boosting combinatorial search through randomization. In: 15th National Conference on Artificial Intelligence and 10th Conference on Innovative Applications of Artificial Intelligence, pp. 431–437. AAAI (1998)
9. Guo, L., Lagniez, J.M.: Dynamic polarity adjustment in a parallel SAT solver. In: 2011 IEEE 23rd International Conference on Tools with Artificial Intelligence, pp. 67–73 (2011). https://doi.org/10.1109/ICTAI.2011.19
10. Hamadi, Y., Jabbour, S., Sais, L.: ManySAT: a parallel SAT solver. J. Satisfiability, Boolean Model. Comput. **6**(4), 245–262 (2009). https://doi.org/10.3233/sat190070
11. Heule, M.J.H., Kullmann, O., Marek, V.W.: Solving and verifying the Boolean Pythagorean triples problem via cube-and-conquer. In: Creignou, N., Le Berre, D. (eds.) SAT 2016. LNCS, vol. 9710, pp. 228–245. Springer, Cham (2016). https://doi.org/10.1007/978-3-319-40970-2_15
12. Iida, Y., Sonobe, T., Inaba, M.: Diversification of parallel search of portfolio SAT solver by search similarity index. In: Khanna, S., Cao, J., Bai, Q., Xu, G. (eds.) PRICAI 2022. Lecture Notes in Computer Science, vol. 13629, pp. 61–74. Springer, Cham (2022). https://doi.org/10.1007/978-3-031-20862-1_5

13. Liang, J.H., Oh, C., Mathew, M., Thomas, C., Li, C., Ganesh, V.: Machine learning-based restart policy for CDCL SAT solvers. In: Beyersdorff, O., Wintersteiger, C.M. (eds.) SAT 2018. LNCS, vol. 10929, pp. 94–110. Springer, Cham (2018). https://doi.org/10.1007/978-3-319-94144-8_6

14. Luby, M., Sinclair, A., Zuckerman, D.: Optimal speedup of las Vegas algorithms. Inf. Process. Lett. **47**(4), 173–180 (1993). https://doi.org/10.1016/0020-0190(93)90029-9

15. Marques-Silva, J., Sakallah, K.: Grasp: a search algorithm for propositional satisfiability. IEEE Trans. Comput. **48**(5), 506–521 (1999). https://doi.org/10.1109/12.769433

16. Moon, S., Inaba, M.: Dynamic strategy to diversify search using a history map in parallel solving. In: Festa, P., Sellmann, M., Vanschoren, J. (eds.) LION 2016. LNCS, vol. 10079, pp. 260–266. Springer, Cham (2016). https://doi.org/10.1007/978-3-319-50349-3_21

17. Moskewicz, M., Madigan, C., Zhao, Y., Zhang, L., Malik, S.: Chaff: engineering an efficient SAT solver. In: Proceedings of the 38th Design Automation Conference (IEEE Cat. No.01CH37232), pp. 530–535 (2001). https://doi.org/10.1145/378239.379017

18. Murphy, J.J.: Technical Analysis of the Financial Markets: A Comprehensive Guide to Trading Methods and Applications. Prentice Hall Press, Hoboken (1999)

19. Niklas Eén, N.S.: MiniSat. https://minisat.se/MiniSat.html. Accessed 26 Jan 2023

20. Oh, C.: Between SAT and UNSAT: the fundamental difference in CDCL SAT. In: Heule, M., Weaver, S. (eds.) SAT 2015. LNCS, vol. 9340, pp. 307–323. Springer, Cham (2015). https://doi.org/10.1007/978-3-319-24318-4_23

21. Ryvchin, V., Strichman, O.: Local restarts. In: Kleine Büning, H., Zhao, X. (eds.) SAT 2008. LNCS, vol. 4996, pp. 271–276. Springer, Heidelberg (2008). https://doi.org/10.1007/978-3-540-79719-7_25

22. SAT-competition: https://www.satcompetition.org/. Accessed 26 Jan 2023

23. Sinz, C., Iser, M.: Problem-sensitive restart heuristics for the DPLL procedure. In: Kullmann, O. (ed.) SAT 2009. LNCS, vol. 5584, pp. 356–362. Springer, Heidelberg (2009). https://doi.org/10.1007/978-3-642-02777-2_33

24. Sonobe, T., Inaba, M.: Counter implication restart for parallel SAT solvers. In: Hamadi, Y., Schoenauer, M. (eds.) LION 2012. LNCS, pp. 485–490. Springer, Heidelberg (2012). https://doi.org/10.1007/978-3-642-34413-8_49

Author Index

© The Editor(s) (if applicable) and The Author(s), under exclusive license
to Springer Nature Switzerland AG 2023
M. Sellmann and K. Tierney (Eds.): LION 2023, LNCS 14286, pp. 615–616, 2023.
https://doi.org/10.1007/978-3-031-44505-7

Printed in the United States
by Baker & Taylor Publisher Services